I0895136

Carbon Quantum Dots With Tailored Heterogeneous Atoms

Surendran Pandiyan
https://orcid.org/0000-0001-6022-0855
Arignar Anna Government Arts College, India

Karthik Kannan
https://orcid.org/0000-0001-5438-2460
National Chung Cheng University, Taiwan

IGI Global
Scientific Publishing
Publishing Tomorrow's Research Today

Vice President of Editorial Melissa Wagner
Director of Acquisitions Mikaela Felty
Director of Book Development Jocelynn Hessler
Production Manager Mike Brehm
Cover Design Jose Rosado

Published in the United States of America by
IGI Global Scientific Publishing
701 East Chocolate Avenue
Hershey, PA, 17033, USA
Tel: 717-533-8845 | Fax: 717-533-7115
Website: https://www.igi-global.com E-mail: cust@igi-global.com

Library of Congress Cataloging-in-Publication Data

LCCN: 2025041407 (CIP Data Pending)
ISBN13: 9798337342474
Isbn13Softcover: 9798337342481
EISBN13: 9798337342498

British Cataloguing in Publication Data
A Cataloguing in Publication record for this book is available from the British Library.

Table of Contents

Detailed Table of Contents

Carbon quantum dots (CQDs) have gained significant attention as an emerging class of carbon-based nanomaterials due to their unique microstructure, tunable physicochemical properties, facile synthesis, and excellent biocompatibility. Their strong photoluminescence, high water solubility, low cytotoxicity, and stability make them promising for applications like LEDs, photovoltaics, sensors, and bioimaging etc. This chapter explores the fundamentals of CQDs, including the influence of core and surface states on luminescence, the role of surface passivation, and the effects of heteroatom doping on optoelectronic properties. The chapter also looks into the evolutionary milestones in CQDs research, such as different synthesis breakthroughs and related findings. Key features like, phosphorescence, upconversion luminescence etc. are examined critically, while current challenges and future prospects are highlighted to explore further CQDs as next-generation multifunctional nanomaterials.

Carbon Quantum Dots (CQDs), less than 10 nm carbon nanomaterials, exhibit high surface area, conductivity, tunable photoluminescence, low toxicity, cost-effective synthesis, and strong environmental compatibility, making them ideal for bioimaging, drug delivery, photovoltaics, photocatalysis, sensors, supercapacitors, and LEDs. CQD synthesis follows two routes: top-down and bottom-up. Top-down decomposes bulk carbon via arc discharge, laser ablation, or electrochemical exfoliation, offering size/surface control but demanding high energy and advanced tools. Bottom-up assembles CQDs from molecular precursors through carbonization or pyrolysis (hydrothermal, microwave, thermal, template-assisted), enabling flexible precursor use and doping for structural/functional tuning. Emerging methods plasma treatment, biomass-derived green synthesis, and chemical vapor deposition enhance scalability, sustainability, and tailored properties. This chapter reviews recent advances, emphasizing principles, merits, and challenges driving next-generation CQD technologies.

Precise modifications in the CD structures can be induced to enhance their properties. The Luminescent emission characteristics, and all other properties rely on CDs morphology and functional groups attached to it. The functional groups can be modified according to the applications or which property has to be modified a need to concentrate on the concentrations and the proportions of materials choose for selection. The non-radiative path ways reductions and targeted functionalities can be introduced to have a smart, reliable, multi-functional CD synthesis. The advancement of aggregation from nano size to macro structures and ACQ phenomenon are advantages of controlled aggregation. Hence this engineered material finds applications in almost all sectors, alone and in a matrix to enhance the property of CDs.

Characterisation of carbon dots involves the process of determining their physical, chemical and optical properties using different spectroscopic, microscopic and surface analytical approaches. Characterization of carbon dots (CDs) is crucial for understanding their properties, behavior, and potential applications. Different spectroscopic techniques used in the study of carbon dots are UV-Visible Spectroscopy,

Photoluminescence (PL) Spectroscopy, Fourier Transform Infrared Spectroscopy (FTIR), Raman Spectroscopy, X-ray Photoelectron Spectroscopy (XPS), and Nuclear Magnetic Resonance (NMR) Spectroscopy. Microscopic techniques used for characterisation of carbon dots include Transmission Electron Microscopy (TEM), Scanning Electron Microscopy (SEM), and Atomic Force Microscopy (AFM). Moreover, different kinds of surface analytical techniques are also used for the study of carbon dots.

Numerous studies have previously reported the applications of carbon-based materials, in diagnostic imaging and therapies as standalone treatments. This chapter specifically explores the capable role of carbon nanostructures, to synergistically integrate both the platforms, for a combined treatment. The chapter classifies the five structural classes of carbon, and their respective physiochemical characteristics such as size, morphology, surface and optical properties. It further explores the top-down and bottom-up fabrication methods, like arc discharge, CVD, hydrothermal and detonation methods. Several studies are reported to support successful theragnostic applications such as photothermal, photodynamic and chemodynamic treatments, synergistically enabling MRI, fluorescence and photoacoustic imaging. Mechanistic roles such as EPR effect, ROS-generation, are elaborately discussed. Limitations around lack of clinical studies, long-time exposure and cost-effective functionalization are challenges to be critically evaluated to accelerate translational research.

Green synthesis of quantum dots (QDs) uses environmentally benign, low-toxicity, and energy efficient routes plant extracts, agricultural residues and harmless solvents to prepare nanomaterials designed for future biomedical applications. Carbon- and graphene-based QDs provide excellent fluorescence, photostability, and surface functionalities with varied versatility, enabling the synthesis of multifunctional drug delivery vehicles. Their inherent biocompatibility and low cytotoxicity render them suitable for deep-tissue bioimaging, targeted therapy of cancer and gene delivery, with real-time monitoring of therapy. Renewable biomass valorization to produce QDs advocates the concepts of circular economy, reducing both environmental impact and industrial dependency on toxic precursors. Green-synthesized QDs are a viable approach for the provision of high-performance and eco-friendly drug delivery systems for human health and environmental stewardship.

Zero-dimensional (0D) carbon quantum dots (CQDs) have attracted significant attention due to their nontoxicity, tunable photoluminescence, chemical stability, biocompatibility, low cost, and ease of surface functionalization. These properties enable applications in biomedicine, drug delivery, sensing, optoelectronics, photocatalysis, and energy storage. In supercapacitors, batteries, and hybrid systems, CQDs contribute to enhance charge storage by improving ion transport, increasing active sites, and boosting conductivity. In hybrid systems, CQDs help achieve both high energy and power densities, with minimal environmental impact. This chapter explores on the role of CQDs-based composites in electrochemical systems, highlighting their mechanisms, contributions to energy storage devices, and the challenges and future prospects for real-world application.

This chapter reviews heteroatom-engineered carbon quantum dots (CQDs) for energy storage, bridging the gap between high-energy batteries and high-power supercapacitors. We examine how N, S, P, and B doping modifies electronic

Sreejitha Raj, Department of Mechanical Engineering, Amrita Vishwa Vidyapeetham, Amritapuri, India
Akhila Raman, Department of Mechanical Engineering, Amrita Vishwa Vidyapeetham, Amritapuri, India
Gopika Venu, Department of Mechanical Engineering, Amrita Vishwa Vidyapeetham, Amritapuri, India
Lakshmi Krishnaa Suresh, Department of Mechanical Engineering, Amrita Vishwa Vidyapeetham, Amritapuri, India
Saritha Appukuttan, Department of Mechanical Engineering, Amrita Vishwa Vidyapeetham, Amritapuri, India
A. Santhy, Department of Mechanical Engineering, Amrita Vishwa Vidyapeetham, Amritapuri, India

Preface

Carbon quantum dots (CQDs) have rapidly evolved from an academic curiosity into one of the most versatile and impactful classes of carbon-based nanomaterials. Their quantum-confined structures, rich surface chemistry, tunable optoelectronic behavior, and exceptional biocompatibility have positioned them at the forefront of next-generation technologies across healthcare, energy, environment, and sensing. In recent years, a major thrust in CQD research has shifted toward heteroatom engineering, sustainable synthesis, advanced characterization, and device-level integration—areas that collectively define the future of CQD-enabled technologies.

Carbon Quantum Dots With Tailored Heterogeneous Atoms brings together leading researchers from diverse scientific backgrounds to offer a comprehensive, multidisciplinary understanding of CQDs and their functional potential. This edited volume spans fundamental theory, experimental breakthroughs, scalable synthesis strategies, and real-world applications, making it a valuable resource for students, researchers, technologists, and industry professionals working in nanoscience and nanotechnology.

The book is structured to follow the logical progression of CQD research from fundamental principles and synthetic methodologies to property modification, characterization, and a broad spectrum of application domains. Each chapter offers deep insights supported by contemporary research, while also identifying open challenges and promising directions for the next decade of CQD innovations.

CHAPTER OVERVIEWS

Chapter 1: Advanced Insights into Carbon Quantum Dots – Fundamental Principles, Heteroatom Effects & Milestones

Jayanta Hazarika and Koushik Saikia establish the scientific foundation of CQDs, explaining their microstructure, optical behavior, and the influence of surface and core states on photoluminescence. The chapter critically explores heteroatom doping, surface passivation, phosphorescence, upconversion luminescence, and the milestones that have shaped CQD research. It provides a lucid entry point into the field while highlighting emerging theoretical and experimental challenges.

Chapter 2: Advanced Synthesis Approaches for Carbon Quantum Dots – Exploring Top-Down, Bottom-Up, and Novel Cutting-Edge Techniques

Subin Stephen, Matbiangthew Shadap, and Sakunthala Ayyasamy present a detailed survey of the synthetic landscape of CQDs. From top-down methods like arc discharge and electrochemical exfoliation to bottom-up carbonization strategies and modern plasma-based or biomass-derived routes, the authors compare mechanisms, advantages, sustainability considerations, and opportunities for scalable manufacturing.

Chapter 3: Enhancing Carbon Quantum Dots Properties – Surface Passivation, Functionalization, Doping, and Aggregation

Rajni K S and Amala S address structural engineering strategies that tailor CQD properties for specific applications. Through controlled passivation, functionalization, doping, and aggregation, the chapter explores how emission behavior, non-radiative pathways, morphological tuning, and ACQ (aggregation-caused quenching) phenomena can be manipulated to design multifunctional CQDs with enhanced reliability and performance.

Chapter 4: Spectroscopic, Microscopic, and Surface Analytical Characterization Techniques for Carbon Dots

Julin Joseph, Cicy Varghese, and Bincy John provide an essential reference on characterization methodologies used to probe CQD composition, structure, and functionality. Techniques spanning UV–Vis, PL, Raman, FTIR, XPS, NMR, TEM, SEM, AFM, and surface analytical tools are explained in the context of understanding CQD behavior, enabling readers to connect material properties with synthesis outcomes.

Chapter 5: Carbon Quantum Nanostructures with Theranostic Applications in Cancer

Tanmay Urs and colleagues examine the translational potential of carbon nanostructures, including CQDs, in cancer theranostics. The chapter classifies carbon structures, discusses fabrication techniques, and highlights combined diagnostic and therapeutic modalities such as photothermal, photodynamic, and chemodynamic therapy. Mechanisms including ROS generation and EPR effects are clearly outlined, alongside the current barriers limiting clinical deployment.

Chapter 6: Green Synthesis of Quantum Dots for Sustainable Drug Delivery

Gayathri T and collaborators explore eco-friendly, low-toxicity routes for synthesizing QDs using plant extracts, biomass waste, and benign solvents. The chapter emphasizes the environmental advantages, circular-economy compatibility, and biomedical relevance of green-synthesized carbon-based QDs in imaging, targeted drug delivery, gene therapy, and real-time therapeutic monitoring.

Chapter 7: Carbon Quantum Dots in Energy Storage – Supercapacitors, Batteries, and Hybrid Systems

Sathish Kumar Ponnaiah, Balakumar Vellaichamy, and Sungwon Lee investigate the incorporation of CQDs into energy storage devices. Their analysis demonstrates how CQDs improve ion transport, active surface area, conductivity, and overall device performance in supercapacitors and various battery systems. The chapter also outlines challenges in durability, integration, and real-world applicability.

Chapter 8: Heteroatom-Engineered Carbon Quantum Dots for High-Performance Energy Storage – From Quantum-Level Design to Scalable Device Architectures

Matbiangthew Shadap, Subin Stephen, and Sakunthala Ayyasamy focus specifically on heteroatom-modified CQDs for high-energy storage systems. The chapter explains how N, S, P, and B doping alters charge distribution, redox-active sites, and conductivity. It also explores hybrid electrode architectures, scalability, techno-economic assessments, and emerging AI-driven material discovery for future commercial energy technologies.

Chapter 9: Carbon Quantum Dots with Tailored Heterogeneous Atoms for Electrochemical Sensors

Sreejitha Raj and co-authors conclude the volume with an in-depth discussion on heteroatom-engineered CQDs for advanced electrochemical sensing platforms. The chapter links synthesis, electronic structure modulation, surface chemistry tuning, and redox behavior to enhanced sensitivity, selectivity, and stability. Applications span biosensing, environmental monitoring, and chemical detection, with forward-looking perspectives on next-generation sensor architectures.

Together, these chapters provide a unified, interdisciplinary perspective on the science and engineering of carbon quantum dots—bridging fundamental principles with cutting-edge applications. This volume aims to inspire continued innovation in the design and deployment of CQDs, empowering researchers to harness their extraordinary capabilities for societal and technological advancement.

Acknowledgment

The editors extend their deepest gratitude to everyone who contributed to the success of this project, with special appreciation to the authors and reviewers who took part in the review process. This book would not have been possible without their invaluable support and commitment.

First and foremost, we sincerely thank all the authors for their meaningful contributions. We truly value the time, effort, and expertise they invested in developing their chapters and enriching the overall quality of this book.

In addition, the editors wish to recognize the vital role of the reviewers in refining the clarity, coherence, and scholarly quality of the chapters. We are especially thankful to those contributors who graciously served both as authors and reviewers, and we deeply appreciate their dedication in fulfilling these dual responsibilities.

Editor Surendran Pandiyan is thankful to the administration of Arignar Anna Government Arts College, India, for continuous support. Last but not least, the editor extends their sincere gratitude to their parents and to the universe for its generous blessings.

Editor Karthik Kannan wishes to express his sincere gratitude to Professor Yong-Song Chen, Department Chair and Associate Dean of the College of Engineering, Department of Mechanical Engineering, National Chung Cheng University, Taiwan, for his continuous encouragement and support. He also extends heartfelt thanks to the Advanced Institute of Manufacturing with High-Tech Innovations (AIM-HI), Department of Mechanical Engineering, National Chung Cheng University, Taiwan, for their unwavering support and collaboration throughout this project.

Editor Dr. Karthik Kannan also extends heartfelt thanks to the Department of Physics, Center for Energy and Environment, Karpagam Academy of Higher Education, Coimbatore, Tamil Nadu, India, for their unwavering support and collaboration throughout this project.

Chapter 1
Advanced Insights Into Carbon Quantum Dots:
Fundamental Principles, Heteroatom Effects, and Milestones

Jayanta Hazarika
Department of Physics, Pandu College, Guwahati, India

Koushik Saikia
https://orcid.org/0000-0001-5223-7614
Department of Physics, The Assam Royal Global University, Guwahati, India

ABSTRACT

Carbon quantum dots (CQDs) have gained significant attention as an emerging class of carbon-based nanomaterials due to their unique microstructure, tunable physicochemical properties, facile synthesis, and excellent biocompatibility. Their strong photoluminescence, high water solubility, low cytotoxicity, and stability make them promising for applications like LEDs, photovoltaics, sensors, and bioimaging etc. This chapter explores the fundamentals of CQDs, including the influence of core and surface states on luminescence, the role of surface passivation, and the effects of heteroatom doping on optoelectronic properties. The chapter also looks into the evolutionary milestones in CQDs research, such as different synthesis breakthroughs and related findings. Key features like, phosphorescence, upconversion luminescence etc. are examined critically, while current challenges and future prospects are highlighted to explore further CQDs as next-generation multifunctional nanomaterials.

DOI: 10.4018/979-8-3373-4247-4.ch001

1. INTRODUCTION

Quantum dots are zero-dimensional semiconductor nanostructures, typically ranging from 1 to 10 nm in size, that exhibit unique optical and electronic properties arising from the strong quantum confinement effect (Mimona et al, 2025; Quesada-González & Merkoçi, 2025). The Nobel Prize in Chemistry 2023 was awarded for the discovery and synthesis of quantum dots, highlighting the significance of this class of nanomaterials in the present time. However, despite their high demand in fundamental research and diverse applications, research on semiconductor quantum dots has gradually slowed down in recent years. Niveria et al. (2023) and Dager et al. (2019) discussed in their research that the most of the high-performing semiconductor quantum dots (QDs) require toxic heavy metals such as cadmium (Cd), lead (Pb), mercury (Hg), and selenium (Se) as their core constituents. In response, and in line with the pursuit of sustainable materials development, researchers have turned their attention to a new class of quantum dots, carbon quantum dots (CQDs). As the name suggest, CQDs contain carbon as building block and are classified as a new class of carbon nanomaterial after graphene, carbon nanotube and fullerene. CQDs have successfully overcome many of the limitations of inorganic quantum dots by offering a non-toxic, biocompatible, and environmentally friendly alternative, while retaining many of the desirable optical and electronic properties. Notably, CQDs exhibit excellent optical features, including strong photoluminescence, longer luminescence lifetimes, and high photostability (Meng et al., 2024). These characteristics with some additional properties as shown in the Figure 1 make CQDs ideal candidate for the replacement of semiconductor quantum dots and find applications in the fields like, light-emitting diodes (LEDs), photovoltaic devices, sensitive chemical and biological sensing, and high-resolution bioimaging etc. (Kumar et al., 2022). Beyond these applications, CQDs have also shown remarkable promise in drug delivery, antimicrobial treatments, photocatalysis, and energy storage systems such as supercapacitors and lithium-ion batteries. Their low toxicity, environmental friendliness, and multifunctional nature position them as a convincing nanomaterial for next-generation technologies across biomedical, environmental, and energy sectors.

Figure 1. Different optical properties of CQDs

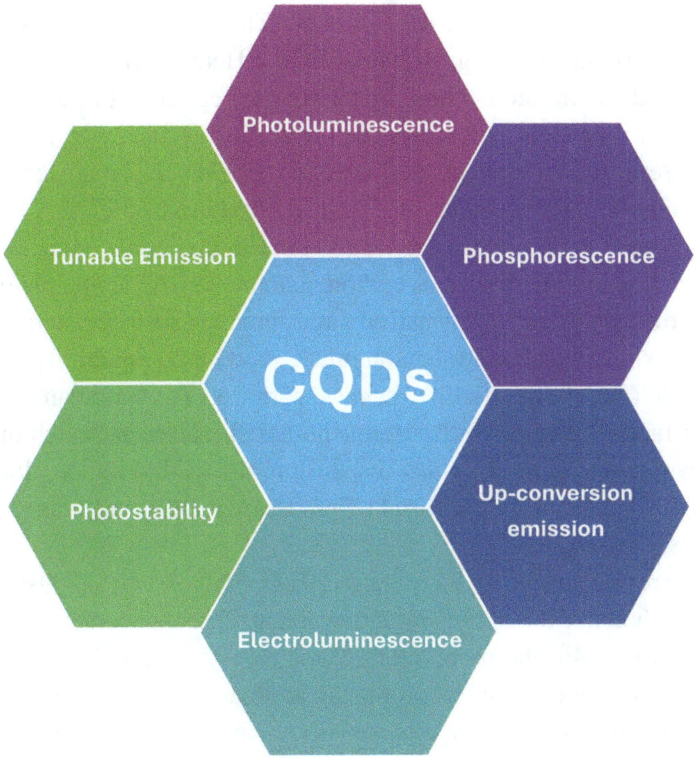

Structurally, CQDs are of spherical and cuboidal shape and consist of a carbonaceous core and a disordered shell. CQDs inherently possess prolong colloidal stability due to their ultrasmall size and availability of abundance of functional groups on their surfaces. Ding et al. (2020) discussed in their work that the optoelectronic properties such as absorbance and photoluminescence (PL) of CQDs are influenced by the core as well surface electronic states. The intrinsic electronic states of CQDs depend on the size of the core as well on the sp^2 to sp^3 boned carbon ratio, while the surface electronic states are influenced by the edge effect and nature of functional groups present on the surface (Mondal & Pramanik, 2023). The excitation energy-dependent photoluminescence (PL) emission observed in CQDs highlights the critical role of surface electronic states in governing their optical behavior (Mehta et al., 2023). Their surface composition tunable functionalities, acting either as electron donors or acceptors, also make them attractive electrochemiluminescence (ECL) nano-emitters (Redondo-Fernandez et al., 2023). Beyond their size and surface tunable optoelectronic characteristics, the properties of CQDs can be further effectively tailored through other strategies. These include- (i) heteroatom doping with nonmetal elements such as nitrogen (N), phosphorus (P), boron (B), sulfur (S), and fluorine (F) etc., or with metals such as iron (Fe), zinc (Zn), gadolinium (Gd), and nickel (Ni) etc.; and (iii) controlled aggregation of CQDs. Heteroatom doping can tune PL emission as well as emission energy by passivation effect of the surface states or introducing new core electronic states. Lan et al. (2017) discussed in their studies that phosphorescence (persistent luminescence) and upconversion luminescence are two prominent phenomena arising from heteroatom doping in CQDs. These properties of doped CQDs make them superior nanomaterials compared to other carbon-based nanomaterials such as carbon nanotubes (CNTs) and nanodiamonds (NDs). Moreover, due their ultra-small size and abundant surface functional groups, doped CQDs exhibit excellent water solubility, good biocompatibility, and low cytotoxicity, which are often difficult to achieve with CNTs or nanodiamonds without extensive surface treatment. Unlike graphene-based materials, which often suffer from aggregation and limited optical response in the visible region, doped CQDs maintain high photostability and bright fluorescence even under harsh chemical environments.

The authors of Guo et al. (2019), Ru et al. (2022), Arshada and Sk (2019) discussed in their research about the aggregation-dependent PL enhancement and emission wavelength shifting. Furthermore, hierarchical assembly and chiral structure formation of CQDs have emerged as promising approaches for developing advanced functional materials. Collectively, these modification strategies significantly expand the potential of CQDs in tunable optoelectronic devices and bioimaging applications.

The growing popularity of CQDs has been further driven by their facile, low-cost, and scalable synthesis methods. One of the most compelling advantages of CQDs lies in their ability to be synthesized from a wide variety of abundant and inexpensive carbon sources, including natural precursors such as fruits, plant extracts, and carbohydrates, as well as industrial and agricultural waste materials (Çeşme et al., 2022; Pandiyan et al., 2020, 2021). These green synthesis routes not only reduce production costs but also align with the principles of sustainability and environmental responsibility, making CQDs highly attractive in the context of current global efforts toward green nanotechnology. Furthermore, such eco-friendly synthesis approaches often impart inherent biocompatibility and water solubility to the CQDs, eliminating the need for complex post-synthesis functionalization.

Despite over two decades of intensive research on CQDs, many fundamental questions still remain unresolved. In particular, the precise origin of their fluorescence along with its dependence on surface defects, functional groups, and the interplay between core and surface states has yet to be fully clarified. Achieving consistent and reproducible fluorescence also remains challenging, largely due to the strong sensitivity of CQDs to the synthesis method, choice of carbon source, and reaction conditions during synthesis. Addressing of these issues requires systematic investigations to establish convincing structure–property correlations, which are essential for understanding the optoelectronic and physicochemical properties of CQDs. This chapter provides a detailed exploration of these aspects by reviewing significant research breakthroughs in the field from its inception. It also offers a comprehensive overview of the evolution of CQD research, their properties, and underlying principles, with particular emphasis on recent advancements in heteroatom doping.

2. HISTORICAL BACKGROUND AND EVOLUTION OF CARBON QUANTUM DOTS RESEARCH

Xu et al. (2004) first accidentally identified carbon quantum dots while segregating single-wall carbon nanotubes (SWCNT) using gel electrophoresis. Thereafter, Sun et al. (2006) synthesized carbon dots of different sizes via a laser ablation technique and formally coined the name "carbon dots". This work reported that the strong fluorescence emission is obtained from carbon dots by surface passivation. This work further demonstrated the good water dispersibility, chemical stability, and low cytotoxicity of CQDs. Since then, the field has seen rapid growth, with intensive research focused on various synthesis strategies, and investigations into their optical, electronic, catalytic properties and theranostic agents in biomedical applications. In all the synthesis processes, CQDs have been prepared from basically three class of precursor materials; (i) carbon materials such as graphite, carbon nanotubes (both SWCNT and MWCNT), graphene oxide (Facure et al., 2022), candle soot (Pankaj et al., 2018), single-walled or multi-walled carbon fibers and fullerene etc., (ii) carbon rich molecules such as glucose (Cho et al., 2024), citrate (Ren et al., 2021), glycerol (Uriarte et al., 2019), various amines etc. and (iii) biodegradable materials such as, various food waste, coffee seeds (Ramírez Sanabria et al., 2025), banana peel (Atchudan et al., 2021), Fennel seeds (Dager et al., 2019), orange juice (Sahu et al. 2012), cow milk (Kumar et al. 2022), silk, honey (Surendran et al., 2020), watermelon/grape peels (Puchaicela et al., 2025) have been reported. Top-down approaches (i) Chemical ablation, (ii) Electrochemical carbonization, (iii) Laser synthesis, (iv) Arc discharge have been mostly used for the synthesis of CQDs from mentioned carbon materials. On the other hand, the bottom-up approaches like, (i) Microwave irradiation method, (ii) Hydrothermal/solvothermal method, (iii) Pyrolysis method have been mostly used to synthesis CQDs using either carbon rich molecules and biodegradable precursors as mentioned above.

In the early stages of CQDs research, most studies relied on top-down synthesis techniques, such as Chemical ablation and Laser ablation etc. In chemical ablation, large carbon structures like graphite, candle soot, carbon black, or other organic carbon sources are broken down using strong acids or oxidants. Common reagents included HNO_3, H_2O_2/AcOH, H_2SO_4, and $HNO_3/HClO_4$ for chemical ablation. For instance, Liu et al. (2007) used candle soot as a precursor for synthesis of CQDs. They applied oxidative acid treatment using HNO_3 or H_2O_2/AcOH to produce CQDs. Peng and Sejdic (2009) synthesized CQDs from carbohydrates. In their method, the sugars were dehydrated using concentrated H_2SO_4 to form a carbon-rich material and then oxidized with HNO_3 and passivated with amine-functionalized molecules such as 4,7,10-trioxa-1,13-tridecanediamine. Surface passivation was used in this method to enhance quantum yield (QY) of the as synthesized CQDs. Laser ablation was another widely used top-down method. In this approach, a laser beam irradiates a carbon-rich target (solid or liquid), breaking it down into nanoscale carbon particles. Hu et al. (2009) used a Nd: YAG pulsed laser to irradiate graphite powders dispersed in various solvents, namely, Diamine hydrate, Diethanolamine, and PEG200N to produce CQDs. Another top-down technique, electrochemical carbonization was also used for synthesis of CQDs. This method involves the controlled electrochemical cutting of carbon materials like graphite, CNTs, or carbon fibers. This is carried out in an electrochemical cell with three electrodes: a carbon-based working electrode, a Pt counter electrode, and a reference electrode, in an electrolyte solution. Zhou et al. (2007) reported the synthesis of CQDs using MWCNTs on carbon paper as the working electrode and TBAP as the electrolyte., Zheng et al. (2009) used graphite as the working electrode and Pt mesh as the counter electrode in a phosphate buffer solution (PBS) at neutral pH. The arc discharge method, which had already been widely used to produce MWCNTs, was also adapted for CQD synthesis. Bottini et al. (2006) synthesized CQDs from pristine and SWCNTs using arc discharge. The resulting CQDs showed bright photoluminescence in the violet-blue and blue-green spectral regions.

Alongside these top-down methods, researchers also began developing bottom-up strategies such as Hydrothermal, Microwave-assisted, Thermal decomposition, and Pyrolysis techniques. Zhu et al. (2009) first synthesized CQDs using bottom-up synthesis, where they used a microwave irradiation method where an aqueous solution of saccharides (like glucose or fructose) and PEG200 was heated at 500 W for 2–10 minutes. Wang et al. (2012) synthesized fluorescent, water-soluble CQDs from eggshell membranes using the same microwave approach. Roshni et al. (2017) produced nitrogen-doped CQDs from sesame seeds, using microwave-assisted synthesis, showing strong fluorescence and excellent water solubility. Liu et al. (2009) demonstrated a pyrolysis-based method using resol (as the carbon precursor) supported on silica spheres. This route produced CQDs through thermal decomposition. Pan et al. (2010) introduced a one-step pyrolysis method using EDTA salts, yielding highly fluorescent CQDs at low temperatures. The same year, Zhang et al. (2010) reported a hydrothermal synthesis of CQDs using L-ascorbic acid in ethanol. This one-pot method did not require acid treatment or additional surface passivation, making it a cleaner and simpler approach. Though other techniques like, pyrolysis, microwave assisted synthesis etc. are used by many research groups, hydrothermal synthesis is one of the most widely used methods for producing high-quality pristine as well as heteroatom doped CQDs. Recent research efforts have focused on making this process more efficient by lowering reaction temperatures, minimizing the use of chemical precursors, and still achieving superior-quality CQDs. In addition, high quality CQDs synthesis through green synthesis approaches have become key priorities in contemporary CQDs research. A critical analysis reveals that the use of bottom-up synthesis methods has grown exponentially, whereas the application of top-down approaches remains comparatively minimal (as shown in Figure 2). Alongside the advancements in synthesis approaches, significant progress has also been made in understanding the microstructure as well as the physicochemical and optoelectronic properties of CQDs. Over the years, several breakthrough findings have emerged, such as the discovery of phosphorescence and up-conversion emission in CQDs, excitation wavelength dependent emission, tunable emission through size, surface passivation and heteroatom doping, and their applications in bioimaging, sensing, catalysis, and energy conversion. Collectively, these advancements have greatly accelerated the growth of CQDs research and expanded their potential applications. As a whole, the evolution of CQDs research can be broadly categorized into four phases-(i) Development of top-down synthesis methods and comparison of their optoelectronic properties with the semiconductor quantum dots, (ii) The exploration of bottom-up synthesis approaches to achieve better control over CQDs microstructure and optical behavior, (iii) advancement of green synthesis strategies for sustainable and eco-friendly production and (iv) tuning of CQDs optoelectronic and chemical properties through heteroatom doping and also through their controlled clustering. It is observed that the majority of CQDs research in last phases are on heteroatom doping. Heteroatom doping has emerged as a promising strategy, offering significant potential to tailor existing properties and induce entirely new functionalities in CQDs.

Figure 2. Number of publications/reports on bottom-up and top-down synthesis approaches from 2004 to 2024 (after each two years).

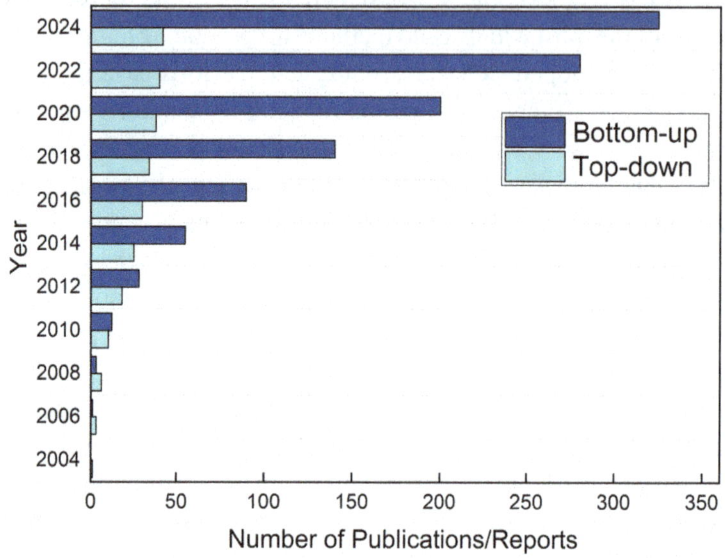

3. MICROSTRUCTURE AND MORPHOLOGY OF CQDs

CQDs generally possess a core–shell structure in which the carbon core is surrounded by a "self-passivated" shell composed of diverse functional groups (Tang et al., 2012). Detailed investigations have shown that the core may be either graphitic crystalline with sp_2 hybridized carbons or amorphous/mixed depending on the fraction of sp^3-hybridized carbon present. Most studies have reported predominantly graphitic crystalline sp^2 cores, typically 2-3 nm in size with a characteristic lattice spacing of about 0.2 nm (Hola et al., 2014 and Sciortino et al., 2016). The exact nature of the core is strongly influenced by the synthesis strategy, precursor chemistry, and reaction conditions such as temperature, duration of reaction, and pH. In general, higher reaction temperatures favor sufficient graphitization and the formation of sp^2 carbon domains, whereas lower temperatures yield more amorphous cores. The choice of synthesis approach, whether, top-down or bottom-up and types of carbon sources also plays a decisive role. Top-down methods (e.g., laser ablation, electrochemical oxidation, acid treatment) usually preserve graphitic crystalline domains inherited from bulk carbon materials, while bottom-up methods (e.g., pyrolysis, hydrothermal carbonization of molecular precursors) often generate more amorphous structures enriched with functional groups (Modi et al., 2023). Thus, synthesis directly dictates the balance between sp^2/sp^3 domains, crystallinity, and surface chemistry.

Xia et al. (2019) discussed in their work that CQDs represent one of the four major classes of carbon dots, the others being Graphene quantum dots (GQDs), Carbon nanodots (CNDs), and Carbonized polymer dots (CPDs). GQDs are nanoscale graphene fragments with lateral dimensions much larger than their thickness. They typically consist of single or few-layered graphene sheets with well-defined lattice structures and edge-functional groups, which give rise to pronounced quantum confinement and edge effects. CQDs, in contrast, are generally quasi-spherical nanoparticles that exhibit distinct crystalline lattices in their carbon cores, along with abundant surface chemical groups (as shown in Figure 3). CNDs are small, highly carbonized nanoparticles that may carry surface functional groups but usually lack both clear lattice structures and polymeric features. CPDs, on the other hand, possess a carbonized core surrounded by aggregated or cross-linked functional groups or polymer chains. Since the focus of this chapter is specifically on CQDs, the other types of carbon dots are not covered in this chapter.

The understanding of CQD cores is largely derived from various advanced characterization techniques, such as High-Resolution Transmission Electron Microscope (HRTEM), X-ray diffraction technique, Raman Spectroscopy etc. HRTEM provides evidence of lattice fringes corresponding to sp_2 graphitic planes. X-ray diffraction (XRD) patterns often show two broad peaks at ~25° and ~ 44°, attributed to the (002) and (100) reflections of low-graphitic carbon (Hou et al., 2015). As per the work of Ferrari (2007), Raman spectroscopy distinguishes structural order and disorder through the D (~1360 cm^{-1}) and G (~1560 cm^{-1}) bands, and the intensity ratio (I_D / I_G) serves as an indicator of the relative proportions of sp^3 and sp^2 hybridized carbon. In addition to these, X-ray photoelectron spectroscopy (XPS) and Fourier-Transform Infrared Spectroscopy (FTIR) are used to investigate the surface chemistry of CQDs. Surrounding the carbon core, the shell of CQDs is typically amorphous, rich in functional groups such as oxygenated species (–OH, –COOH, –C=O), amino-based groups, or polymer chains (Gayen et al., 2019). The shell not only enhances dispersibility but also defines the surface states, thereby governing many of the optical and electronic properties of CQDs. The shell can also act as either an electron donor or acceptor, and its modification through surface functionalization, passivation agents, or solvent environment leads to substantial variation in photoluminescence (PL) behavior. In particular, the nature of surface moieties directly tunes the surface-related electronic states, thereby modulating emission wavelength, quantum yield, and charge transfer efficiency. The core-shell structure results in improved photoluminescence (PL), quantum yield, higher emission stability, and better control over emission wavelength through modulation of shell composition and thickness. In optical applications, the shell can induce excitation-dependent emission and prevent fluorescence quenching by isolating emissive centers from environmental effects. From a catalytic standpoint, the core–shell configuration promotes efficient charge separation and directional charge transfer between the core and shell regions, enhancing photocatalytic and electrocatalytic activity. Moreover, the presence of a doped or functionalized carbon shell increases the density of active sites and facilitates adsorption of reactant molecules, leading to superior catalytic efficiency.

Figure 3. A general structure of a doped CQDs. Green and blue balls represent dopants while the brown balls represent carbon atoms.

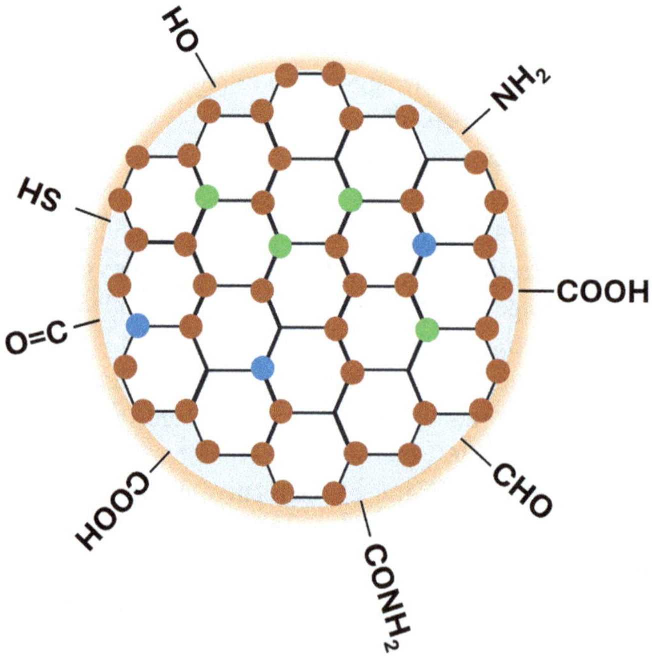

4. OPTICAL PROPERTIES OF CQDs

The unique optical properties of CQDs arise fundamentally from their ultrasmall size and the resulting quantum confinement effects, combined with surface states introduced by surface defects and abundant functional groups. Quantum confinement occurs when the dimension of a material becomes comparable to or smaller than the exciton Bohr radius, such that the motion of charge carriers (electrons and holes) is restricted in all three spatial dimensions. In bulk carbonaceous material, such as graphite the electronic states are continuous, but when the particle size is reduced to the nanoscale (typically below 10 nm for CQDs), the energy bands discretize into quantized levels, leading to a widening of the effective bandgap and alter the optoelectronic properties such as absorbance and photoluminescence. The UV-visible absorbance spectrum of CQDs typically shows a strong peak between 200–260 nm attributed to the π–π* transition of the aromatic C=C bonds within the sp^2-hybridized domains, and a broader shoulder peaks at ~ 350 nm corresponding to the n–π* transitions of C=O and C=N bonds (N- and O-containing structures present as functionalities) at the edge of carbon structures (Keenan et al, 2019). In certain cases, additional absorbance bands can appear in the visible region (400–600 nm), which are generally attributed to surface-state transitions and the formation of mid-gap energy levels induced by functional groups or dopants.

The photoluminescence (PL) of CQDs is also attributed to two major contributions; the intrinsic bandgap emission from the sp^2-hybridized carbon core governed by quantum confinement, and extrinsic surface-state-related emission. Typically, when CQDs are excited in the n– π* absorption band around 350 nm, they exhibit a strong blue emission centered at ~ 450 nm (Mmelesi et al., 2024). In contrast, excitation in the far-UV region (~250 nm), corresponding to the π–π* absorption transition, generally produces very weak or negligible photoluminescence. Therefore, unlike conventional semiconductor-based quantum dots, CQDs exhibit pronounced surface effects that often dominate their emission spectra. As a result, the photoluminescence (PL) of CQDs is highly dependent on the excitation wavelength. The emission intensity and hence quantum yield of CQDs can be improved by surface passivation with thin insulating or protective layer composed of materials such as oligomers (e.g., polyethylene glycol), thionyl chloride, thiols, or spiropyrans (Yadav et al., 2023). These protective coatings not only prevent the adhesion of impurities but also enhance stability and suppresses the dissipation of photoinduced carriers from surface sites, thereby improving their quantum yields.

Introducing functional groups such as carbonyl, carboxyl, hydroxyl, and amines onto the surface of CQDs significantly enhances their absorption and emission in the UV–visible region by inducing surface doping. CQDs enriched with oxygen-containing groups are typically p-doped, owing to the higher electronegativity of oxygen compared to carbon, whereas nitrogen-containing surface groups can readily induce n-doping in CQDs. In contrast to bare CQDs, functionalized CQDs display excellent photo-reversibility, high stability, good biocompatibility, and low toxicity. Surface passivation or functionalization helps stabilize defects, thereby promoting more efficient radiative recombination of surface-confined electrons and holes.

Beyond quantum confinement and surface-state effects, heteroatom doping has emerged as a highly effective strategy for tailoring optical properties of CQDs. By incorporating metal and non-metal heteroatoms such as, Nitrogen (N), Sulfur (S), Phosphorous (P), Boron (B), Zinc (Zn), Copper (Cu), Manganese (Mn), Iron (Fe), Gadolinium (Gd) etc. into either the core or the surface, the electronic structure can be modulated through bandgap engineering, introduction of mid-gap states, or modification of charge distribution. Quantum confinement, surface functionalization, and heteroatom doping thus provide a triad of mechanisms that collectively govern and expand the functional landscape of CQDs.

Phosphorescence or long-lived photoluminescence is a fascinating optical property of CQDs, which is a synergistic effect of doping and surface functionalization. The mechanism of phosphorescence in carbon quantum dots (CQDs) can be explained on the basis of singlet–triplet transitions and intersystem crossing (ISC) (Zuo et al., 2023). Upon photoexcitation, electrons are promoted from the ground singlet state (S_0) to an excited singlet state (S_1), followed by rapid relaxation to the lowest vibrational level of S_1. Owing to the presence of surface defects, heteroatom doping, or strong spin–orbit coupling, efficient ISC occurs, transferring excited electrons from the singlet state (S_1) to the triplet state (T_1). The triplet state is long-lived because the return to the ground state ($T_1 \rightarrow S_0$) is spin-forbidden, resulting in delayed radiative emission known as phosphorescence. Embedding doped CQDs into a supporting matrix such as silica and PVA has also been shown to promote room-temperature phosphorescence (Cheng et al., 2023). In this configuration, the covalent bonds and hydrogen-bonding network formed between amorphous silica and CQDs effectively shield the triplet excitons from environmental quenchers while suppressing non-radiative transitions.

In addition to heteroatom doping and matrix embedding strategies that enable room-temperature phosphorescence in CQDs, another intriguing optical phenomenon observed in these nanomaterials is upconversion luminescence. Upconversion CQDs are capable of converting low-energy excitation, typically in the near-infrared (NIR) region, into higher-energy visible emission. This anti-Stokes luminescence is generally attributed to multi-photon absorption facilitated by surface molecular states due to the presence of various fluorescent molecular fragments on CQDs surface. The efficiency of this process is also strongly influenced by heteroatom doping which enhance nonlinear optical responses.

5. HETEROATOMS DOPING AND THEIR TYPES

Doping has long been recognized as an effective strategy for modulating optical and electrical properties of semiconductor nanomaterials. Cánchig et al. (2025) and

Verma et al. (2024) showed in their research that doping plays a pivotal role in tailoring their optical and electronic behavior CQDs also by introducing heteroatoms into the carbon framework. Nitrogen (N), phosphorus (P), boron (B), and sulfur (S) are the closest neighbors to carbon (C) in the periodic table, which have been applied widely for doping CQDs, converting them to practical tools for a broader range of applications. By introducing these impurities into CQDs, their electronic structures can be adjusted generating n-type or p-type carriers. Doping can alter the highest occupied molecular orbital (HOMO) and lowest unoccupied molecular orbital (LUMO) levels. More specifically, heteroatom doping can substantially alter the electronic structure by introducing new energy levels, modifying the π–π^* and n–π^* transitions, and creating additional radiative recombination pathways in CQDs. A typical representation of various electronic absorption and emission pathways of N-doped CQDs is shown in Figure 4. The research of Sharma et al. (2025) have showed that heteroatom doping alters the optoelectronic properties of CQDs not only by changing their intrinsic electronic states related to the core but also by changing their surface electronic states. Heteroatom doping results in improved fluorescence and causes considerable red-shifting of the emission peak towards red by introducing intermediate radiative states (Cánchig et al., 2025). The development of red-emitting carbon quantum dots (CQDs) through the introduction of suitable dopants is currently a hot topic in CQD research, as red-emissive quantum dots are highly valuable for bioimaging applications. In contrast to this, Kamaruzaman et al. (2025) demonstrated blue-shifting of the emission peak after heteroatom doping. The shifting, whether red or blue, depend entirely on the type of dopants and on their bonding configuration. In addition, heteroatom doping has been reported to induce new physicochemical properties in CQDs, such as phosphorescence and up-conversion luminescence, which hold potential for use as bioimaging probes (Patra et al., 2024). Furthermore, the work of Tavan et al. (2025) showed that doping with elements like Gd^{3+} and Mn^{2+} results in efficient MRI–fluorescence dual-mode bioimaging probes). Wang et al. (2012) explored Nitrogen doping into the CQDs and its effects on the optical properties for the first time. Their study observed high photo luminescence with a quantum yield of 22% and showed potential applications of N doped CQDs in multi-photon single molecule imaging. Since then, many research groups have worked on heteroatom doping to improve the functionality of CQDs through to control or improve their optical, chemical, and electrical properties.

Figure 4. Energy band structure and possible absorption and emission processes for CQDs

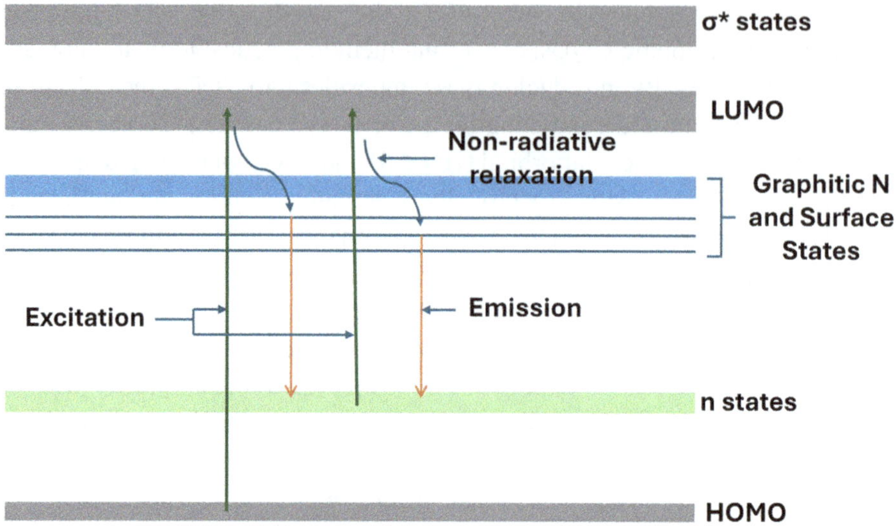

Doped CQDs can be classified according to the number of heteroatoms incorporated into their structure. When a single element is introduced, they are referred to as single-doped CQDs, whereas the incorporation of two or more elements leads to co-doped CQDs. Similarly, depending on the synthesis route, their preparation can be broadly categorized into one-step and multistep methods. Single doped CQDs include both non-metals and metals, such as nitrogen (N), Sulfur (S), Phosphorus (P), Boron (B), Germanium (Ge), Zinc (Zn), Copper (Cu), Gadolinium (Gd), Silicon (Si), Manganese (Mn), Cobalt (Co), Europium (Eu), and Gallium (Ga) as dopants. Non-metals can effectively change the electronic structure of CQDs because of the similarity of their size with that of the carbon atom (C) and they are comparatively easy to synthesis than metal/transition metal doping. But metal ion doping is even interesting as metal ions have more electrons and unoccupied orbitals compared to non-metals, leading to enhanced optoelectronic and catalytic properties. Table 1 shows some of the recent works on single doped CQDs. In addition to the single doped CQDs, researchers were also simultaneously exploring the possibilities of co-doping to CQDs. Synergistic effect of both non-metal and metal ions can be harnessed by co-doping strategy. Commonly reported metal-nonmetal co-dopants include N–Mg, N–Cu, N–Zn, N–Tb, N–Gd, N–I, N–Fe, and N–Co. Co-doping with pair of non-metal ions like N–S, N–P, N–B are also widely reported. These doping combinations provide versatile platforms for fine-tuning the band structure, surface states, and functional performance of CQDs for diverse applications (Wang et al., 2012). Depending on the type of dopant the optical properties of the CQDs changes differentially. For example, CQDs doped with highly electronegative atoms (e.g., N and S) usually cause blue-shifted PL emission; while, those doped with atoms of low electronegativity (e.g., P, B, and Se) result in a redshift (Kou et al., 2020). Some of the recent works on co-doping along with type of synthesis method, used precursors and quantum yields have been shown in the table 2. It is to be also mentioned that the type of dopant precursor influences the dopant content and bonding configuration in heteroatom doped CQDs, which in turn impacts the particle size, optical properties, and electronic structure. Some of the widely explored single and co-doped CQD systems have been discussed below-

5.1 Single Doped CQDs

(a) **N doped CQDs:** The primary reason for nitrogen (N) doping in carbon quantum dots (CQDs) lies in the close electronic resemblance between nitrogen and carbon. Among various dopants, nitrogen is the most widely employed to enhance the optical, chemical, and electrical properties of CQDs. Incorporation of nitrogen atoms injects electrons into the CQD framework, thereby altering the internal electronic environment and significantly improving fluorescence properties. Nitrogen doping also increases surface-state defects and enhances electronic characteristics and chemical reactivity by introducing numerous reactive groups (–OH, –COOH, –N=C–, –CN, $-NH_2$). These functional groups, in turn, contribute to a higher fluorescence quantum yield.

Nitrogen in N-doped CQDs typically exists in three primary forms: (i) nitrogen atoms integrated within the carbon lattice, (ii) nitrogen atoms incorporated into surface functional groups such as amines and amides, and (iii) nitrogen atoms embedded in heteroaromatic rings, which include pyrrolic, pyridinic, and graphitic nitrogen configurations. Pyridinic and pyrrolic nitrogen bond with two carbon atoms (C–N=C) in hexagonal and pentagonal arrangements, respectively (Nasrin et al., 2020). In the graphitic configuration, one nitrogen atom bonds with three carbon atoms in an sp^2 arrangement inside the CQDs. Additionally, nitrogen oxides and aminic nitrogen species can also form in N-doped CQDs. Because of these diverse chemical configurations, nitrogen doping imparts a broad range of physicochemical properties and sometimes leads to unexpected features. For instance, graphitic nitrogen exhibits an electron-doping effect that modifies the electronic energy levels of the system and induces a pronounced red-shift (Liu et al., 2017).

(b) **Sulfur doping CQDs**: In recent years, sulfur-doped carbon quantum dots (S-CQDs) have attracted increasing attention, as sulfur atoms can introduce energy or emissive trap states for photoexcited electron capture, thereby modifying the electronic structure of CQDs. Moreover, S-CQDs retain most of the advantages of pristine CQDs while avoiding self-quenching due to their large ensemble Stokes shift. However, compared with nitrogen-doped CQDs, the synthesis of S-CQDs is more challenging because C–S bonds are relatively difficult to form. The larger atomic radius of sulfur compared with carbon results in a C–S bond length approximately 25% longer than that of a C–C bond. Furthermore, since sulfur and carbon possess similar electronegativities, the C–S bond provides limited charge transfer (Sarkar et al., 2016).

Mechanistic studies have revealed that the orbital mismatch between the outermost S and C atoms leads to an uneven distribution of spin density, thereby imparting surface reactivity and catalytic properties to S-CQDs. Reports have also shown that sulfur doping can increase the optical band gap and induce excitation-dependent emission. Although the number of studies on S-CQDs is relatively limited compared to other heteroatom-doped CQDs, they have emerged as promising candidates for catalytic applications owing to their unique electronic properties and high surface chemical activity.

(c) P-doped CQDs: Phosphorus functions as an *n*-type donor in carbon quantum dots (CQDs), thereby enhancing their photoelectric properties. In addition, P-doping modifies the electronic structure of CQDs and introduces numerous active sites (Gao et al., 2012), which contribute to improved fluorescence efficiency and stability. With increasing phosphorus concentration, the number of sp^2-hybridized carbon atoms in P-doped CQDs also increases. Phosphorus doping further facilitates the separation of sp^2 carbon clusters, leading to a coexistence of defect sites and isolated sp^2 domains within the CQD matrix. As a result, the bandgap of P-CQDs can be tuned effectively from the UV to the visible region, enabling an emission shift from blue to green.

(d) *B-doped CQDs:* Unlike nitrogen, phosphorus, and sulfur, which act as *n*-type dopants and increase the electron concentration of CQDs, boron (a group III element) reduces the electron density. The B–C bond is about 0.5% longer than the C–C bond, and the electron deficiency of boron introduces defects in the energy states of CQDs. These defect states give rise to surface-related emissions, which can induce a certain degree of chemical disturbance. Consequently, B-doping modifies the optical properties of CQDs by generating a large number of active sites (Zhou et al., 2014). Furthermore, the creation of new electronic states through B-doping enhances the nonlinear optical response of B-CQDs compared to undoped CQDs. Notably, the absorption and emission spectra of B-CQDs under UV excitation resemble those of inorganic quantum dots.

(e) **Florine doped CQDs:** Fluorine is an important element in the medical field due to its ease of incorporation into organic compounds, which enhances the selectivity between drugs and their receptors. Owing to the large electronegativity difference between carbon and fluorine, F-doping can significantly modify the chemical, structural, and electronic properties of CQDs, thereby enhancing their fluorescence, magnetic behavior, and biological functionality. Consequently, F-CQDs hold great promise for medical applications.

5.2 Co-Doped CQDs

(a) N and S co-doped CQDs (NS -CQDs): Both sulfur (S) and nitrogen (N) contain lone pair electrons, which can cooperatively donate electrons, promote radiative recombination, and enhance the fluorescence quantum yield of CQDs (Zhou et al., 2017). Nitrogen-doped CQDs (N-CQDs) generally exhibit higher fluorescence intensity than undoped CQDs, whereas sulfur-doped CQDs (S-CQDs) display lower fluorescence intensity. This behavior is evident from fluorescence spectra and the proposed luminescence mechanisms of CQDs, N-CQDs, and S-CQDs. Undoped CQDs typically possess an O-state, which gives rise to broad UV and visible absorption bands as well as excitation-dependent emissions, corresponding to the distribution of various energy levels. Nitrogen doping introduces a new surface state (the N-state), where electrons are captured and undergo radiative transitions, thereby increasing the fluorescence yield. In contrast, sulfur doping eliminates the O-state, rendering the original surface state negligible and resulting in a lower fluorescence intensity compared to undoped CQDs. Nevertheless, S-doping can induce a more pronounced red-shift of emission. Thus, the synergistic effect of N and S co-doping not only enhances the fluorescence quantum yield of CQDs but also enables the generation of tunable emissions across different colors (Dong et al., 2013).

(b) N and B co-doped CQDs (NB-CQDs): Because boron (B) and nitrogen (N) have atomic radii comparable to that of carbon, they can be homogeneously incorporated into the CQD framework. Research of Lu et al. (2015) have shown that both N and B not only integrate effectively into the carbon skeleton but also contribute to the formation of surface functional groups such as –COOH, –CONH, –OH, and boric acid. Boron facilitates charge transfer in the excited state, and consequently, the fluorescence quantum yield of NB-CQDs increases with rising B content. The simultaneous incorporation of B (lower electronegativity) and N (higher electronegativity) generates a unique electronic structure through a cooperative coupling effect between the heteroatoms, thereby enhancing the catalytic activity of the CQDs. As a result, BN-CQDs are considered promising alternatives to metal-based catalysts.

Guo et al. (2017) reported a hydrothermal synthesis route for BN-CQDs that exhibited yellow-green emission. Increasing the reaction temperature led to a red-shift in the fluorescence emission while reducing fluorescence intensity. To date, most NB-CQDs have been synthesized via hydrothermal methods, although combustion treatment using aminophenylboronic acid (APBA) has also been employed to prepare NB-CQDs.

(c) Gd and Yb doped CQDs: Gadolinium (Gd) and Ytterbium (Yb) co-doped CQDs have attracted significant attention due to their unique multifunctional properties combining the merits of rare-earth elements with carbon nanostructures (Bouzas-Ramos et al., 2019). Gd, with its high magnetic moment, imparts strong paramagnetic behavior, making these CQDs useful as contrast agents in magnetic resonance imaging (MRI). On the other hand, Yb, a well-known near-infrared (NIR) emitter, introduces excellent optical properties, especially in the NIR-II region, which is highly favorable for deep-tissue bioimaging and photonic applications. The synergistic effect of Gd and Yb co-doping not only enhances the photoluminescence quantum yield of CQDs but also provides dual-modal functionality, enabling simultaneous optical and magnetic imaging.

5.3 Three Heteroatom-doped CQDs

Though dual-atom co-doped CQDs, such as N–S, N–B, N–P, and N–Se systems, have been extensively investigated, multi-atom doping remains relatively unexplored due to the complexity of synthesis processes and stringent reaction conditions. Compared with dual-doped CQDs, three-atom co-doped CQDs possess a greater diversity of surface functional groups, which generate multiple excited states. Excited electrons in these systems can undergo inter-state conversion and radiative recombination, resulting in fluorescence emissions across different wavelengths.

Nitrogen plays a crucial role in electronic modification because of its atomic size similarity with carbon. N-doping introduces new surface states that facilitate high-yield radiative recombination. Sulfur, when co-doped, further enhances the effect of nitrogen through cooperative interactions. The addition of a third dopant not only reduces the CQD size but also significantly enhances fluorescence quantum yield and overall performance. For instance, fluorine doping within the carbon framework has been shown to improve oxygen reduction reaction (ORR) catalysis.

These unique features of multi-doped CQDs offer distinct advantages for applications in sensing and bioimaging. N–S–P, N–S–F, N–S–B and N–O–P are some combinations of three heteroatom co-doping which have been studied. Table 3 shows some of the recent works on triply doped CQDs.

Thus, heteroatom doping in CQDs, whether through single-, dual-, or multi-atom incorporation, effectively tunes their electronic structure, surface chemistry, and fluorescence behavior, thereby broadening their functional versatility. Collectively, these advances in doping strategies have positioned CQDs as highly versatile nanomaterials with broad potential in bioimaging, sensing, catalysis, and optoelectronic applications.

5.4 Surface and Core Doping

Heteroatom doping in CQDs can be also categorized as surface doping and core doping, each influencing the structural and optoelectronic properties differently. In surface doping, heteroatoms such as N, S, P, B etc. are introduced onto the surface or edge sites of CQDs, typically through functional groups or chemical modification, thereby modulating surface states, enhancing solubility, and tuning photoluminescence through surface-related electronic transitions. In contrast, core doping involves the incorporation of heteroatoms directly into the sp^2 carbon lattice during synthesis, which alters the intrinsic electronic structure by modifying the π-conjugated system, introducing new energy levels, and improving charge transfer dynamics. While surface doping predominantly affects interfacial interactions and emission stability, core doping results in more profound changes in the band structure and optical absorption. The extent to which heteroatoms are incorporated into the surface or the core of carbon quantum dots is strongly governed by both the synthetic route and the precursor chemistry. In bottom-up methods (such as hydrothermal, solvothermal, or microwave-assisted synthesis), the type of organic precursor dictates whether the heteroatoms become part of the carbon framework or remain as surface functionalities. For example, using nitrogen-rich precursors like urea, ethylenediamine, or citric acid–urea mixtures during high-temperature carbonization often leads to core doping, as heteroatoms integrate into the conjugated sp^2 lattice. In contrast, if doping agents are introduced through post-synthesis methods or at lower synthesis temperatures, they tend to attach as surface dopants through functional groups ($-NH_2$, –OH, –COOH, $-SO_3H$, etc.).

On the other hand, in top-down approaches (such as chemical oxidation of graphite, arc discharge, or laser ablation), the starting carbon source is relatively rigid, so heteroatoms introduced during chemical treatments (e.g., nitric acid oxidation with ammonia) are more likely to localize on surface defects or edge sites, resulting in surface doping rather than core incorporation. Thus, precursors with heteroatom-rich frameworks (like amino acids, proteins, or heteroaromatic compounds) under harsh conditions promote core doping, while post-treatment functionalization or heteroatom-containing surface modifiers favor surface doping.

Table 1. Methods of synthesis and performance parameters for CQDs doped with single element.

Precursor	Synthesis methods	Quantum yield (%)	Particle size(nm)	Excitation /Emission	Dopant	Ref.
Orthophenylene diamine	Hydrothermal synthesis	28	1-2	420/477	N	(Dai et al., 2017)
Maltitol and H_2SO_4	Hydrothermal synthesis	-	2.75	-	S	(Hu et al., 2013)
Thioacetic acid	Hydrothermal synthesis	-	3.8-7.4	-	S	(Bourlinos et al., 2015)
Triethyl phosphonoacetate and phosphoric acid	Hydrothermal synthesis	-	3.5	-	P	(Li et al., 2017)
Ethylenediamine and boric acid	Hydrothermal synthesis	50	1-6	340/430	B	(Shen et al. 2015)
Selenocystine	Hydrothermal synthesis	7.6	1–5	398/490	Se	(Van Tam et al. 2017)
Fluorinated graphene oxide	Hydrothermal synthesis	-	5.6	330/445	F	(Li et al. 2017)
Degreasing cotton and hydrofluoric acid	Hydrothermal synthesis	-	3.25	340/428	F	(Sun et al. 2015)
Iodixanol and glycine	Hydrothermal synthesis	-	2.7	370/475	I	(Zuo et al., 2017)
Acetylacetone and $NH_3{\cdot}H_2$	Microwave synthesis	51.61	< 4.31	380/420	N	(He et al., 2017)
Citric acid, boric acid and urea	Microwave synthesis	10-15	2-6	350/450	B	(He et al., 2017)
Sodium phytate	Electrochemical synthesis	-	2-4	-	P	(Li et al., 2017)
Phosphorous tribromide and hydroquinone	Solvent-thermal method	25	5-15	-	P	(Jana et al. 2017)
Silicon tetrachloride and hydroquinone	Solvent-thermal method	19.2	7±2	-	Si	(Qian et al., 2014)

Table 2. Methods of synthesis and performance parameters for CQDs doped with N\S

N\S precursor	C precursor	Method	QY (%)	Particle Size	Excitation /Emission	Ref.
Methionine	Acrylic acid	Hydrothermal synthesis	10.55	2.3 nm	420/ 485	(Lu et al., 2015)
Cysteine	Citric acid	Pyrolysis	67	-	346/415	(Chen et al., 2017)
L-cysteine	Glycerol	Hydrothermal synthesis	15.2	4.7 nm	350/440	(Zou et al., 2017)

continued on following page

Table 2. Continued

N\S precursor	C precursor	Method	QY (%)	Particle Size	Excitation /Emission	Ref.
Sulfuric acid	M-phenylenediamine	Hydrothermal synthesis	43	4.3 nm		(Zhou et al., 2018)
Ammonium persulfate /urea	Caffeine	Hydrothermal synthesis	69	13 nm	370/426	(Dang et al., 2018)
L-cysteine	Citric acid	Hydrothermal synthesis	57.2	-	355/450	(Zhao et al., 2019)
Methionine and ethylenediamine		Hydrothermal synthesis	-	5 nm	320/390	(Liu et al., 2018)

Table 3. Methods of synthesis and performance parameters for triply doped CQDs

Precursors	Method	QY (%)	PL color/ Size	Surface doping	Excitation /Emission	Ref.
Cucumber juice	Hydrothermal	-	< 10 nm	N/S/P	-	(Wang et al., 2014)
Anthracite coal Concentrated sulphuric acid and nitric acid	Wet chemical plus dialysis	-	1-7 nm	N/S/P	-	(Xu et al., 2018)
A customized ionic liquid Multiwalled	Microwave treatment	70	2 nm	N/S/F	-	(Kundu et al., 2015)
Thiourea and 3-amino phenyl boronic acid citric acid	Hydrothermal	28	-	N/S/B	-	(Das et al., 2017)
3-aminobenzeneboronic acid and 2,5-diaminobenzen	Hydrothermal	11.6	2.84 nm	N/S/B	520/605	(Huang et al., 2018)
Egg powder and phosphorus acid	Hydrothermal	11.1	3.1 ± 0.2 nm	N/S/B	360/443	(Fu et al., 2018)

6. CARBON QUANTUM DOTS AS NEXT-GENERATION MATERIALS IN EMERGING FIELDS

Advancement in CQDs research demonstrate their growing demand in diverse fields, particularly energy storage and conversion, optoelectronics, sensing, and catalysis. In energy-related systems, CQDs have been widely utilized as electrode materials, interfacial modifiers, and conductive additives in supercapacitors and batteries, where their unique characteristics such as tunable bandgap, excellent electron transfer capability, and abundant surface functionalities significantly enhance energy density, stability, and charge transport (Rasal et al., 2021). In sensing, surface-engineered CQDs provide selective and sensitive detection of a wide range of analytes including metal ions, biomolecules, and environmental pollutants, owing to their strong photoluminescence, high aqueous solubility, and functional group versatility (Feng and Qian, 2018). Meanwhile, in catalysis, CQDs function as efficient electron mediators, co-catalysts, or active catalytic sites in photocatalysis and electrocatalysis, improving reaction kinetics and stability under mild conditions (Phang and Tan, 2019). Collectively, these advances highlight CQDs as versatile, sustainable nanomaterials with immense potential possibilities for next-generation technologies. The following sections discuss their applications across various sectors.

(a) Energy Storage

CQDs have already proven to be a novel class of fluorescent nanomaterials, distinguished by their unique optical, electronic, and surface characteristics. They have received widespread attention in energy storage devices such as supercapacitors, batteries (Gurushankar et al., 2025; Kannan et al., 2024) and conversion such as hydrogen production and carbon dioxide conversion applications (Varadharajan et al., 2024; Kannan et al., 2023; Kannan et al., 2020), where they exhibit tunable redox activity, rich surface functional groups, small sizes, uniform dispersion, unique structure and composition, low toxicity, and low production cost. Their integration improves energy storage efficiency, cycling stability, and charge transport. Various studies on the applications of CQDs in supercapacitors and batteries as energy storage devices, as well as in solar cells for energy conversion, are discussed below.

Supercapacitor: Supercapacitors are valued for their high-power density, fast charge–discharge rate, long cycle life, and excellent durability. However, their relatively low energy density limits their practical applications, motivating strategies to improve energy storage capacity (Kannan et al., 2024). CQDs, with their graphene-like sp^2 and diamond-like sp^3 hybridized structures and diverse surface functionalities, offer unique electrical properties suitable for supercapacitor electrodes (Xu et al., 2022). For instance, Abid Inayat et al. reported CQD-based electrodes with a specific capacitance of 302.0 F g^{-1} at 0.5 A g^{-1}, energy density of 41.9 Wh kg^{-1}, and power density of 250 W kg^{-1}, while maintaining nearly 100% efficiency over thousands of cycles (Inayat et al., 2023). CQDs also can form effective nanohybrids with metal oxides/sulfides which can significantly boost the performance of supercapacitors. For example, Quan et al. (2020) fabricated CQD@CuS nanohybrids with a specific capacitance of 920.5 F g^{-1}, power density of 397.75 W kg^{-1}, and 92.8% capacitance retention after 10,000 cycles. Similarly, Xu et al. (2017) developed CQDs/MnO_2 nanoflowers with 345 F g^{-1} capacitance at 0.5 A g^{-1} and 90.3% retention after 10,000 cycles. In another report, CQDs/NiO nanorods achieved 1858 F g^{-1} at 1 A g^{-1} and retained 93% capacity over 1000 cycles (Xu et al., 2016).

Furthermore, integrating CQDs with conducting polymers exhibits great potential as electrode materials for supercapacitors, offering improved mechanical flexibility compared to CQD–metal oxide/sulfide nanohybrids. Wang et al. synthesized N-doped CQDs/polyaniline (N-CQD/PAni), which demonstrated superior capacitance and cycling stability (Wang et al., 2017). Jian et al. (2017) fabricated CQDs/PPy nanocomposites for flexible, solid-state supercapacitors, achieving 308 F g^{-1} at 0.2 mA cm^{-2} and 85.7% retention after 2000 cycles Similarly, Zhao and Xie (2017) prepared CQDs–PAni hybrid nanowires with 738.3 F g^{-1} at 1 A g^{-1} and 78% retention after 1000 cycles.

Batteries: CQDs also play a crucial role in battery technologies by promoting ion storage and transfer, owing to their high specific surface area, abundant functional groups, high chemical activity. The use of CQDs in various types of batteries and their resulting improvements in performance characteristics are discussed separately as follows.

Lithium/Sodium-ion batteries (LIBs/SIBs): Javed et al. (2019) developed CQD-based anodes with capacities of 864.9 mAh g^{-1} for lithium-ion batteries (LIBs) and 323.9 mAh g^{-1} for sodium-ion batteries (SIBs) at 0.5 C, showing 91.6% and 72.4% capacity retention after 500 cycles, respectively.

Potassium-ion batteries: Wang et al. (2022) synthesized N-CQD exhibited specific capacity of 280.5 mAh g^{-1} at 0.1 A g^{-1} with excellent rate capability (136 mAh g^{-1} at 5 A g^{-1}).

Al–air batteries: Guo et al. (2025) demonstrated that varying N-CQDs concentrations improved anode utilization of Al-air batteries from 41.4% to 85.0%, increasing energy density from 1540.2 to 3367 Wh kg^{-1}.

(b) Solar Cells

CQDs also have emerged as promising components in solar cells, thanks to their tunable bandgap, high chemical stability, eco-friendly synthesis, dispersibility in solvents, and trap-healing ability (Guan et al., 2023). Unlike toxic semiconductor quantum dots (Cd, Pb-based), CQDs offer a safer, cost-effective alternative They can function as sensitizers, carrier transport layers, photon absorbers, and energy-band modulators, thereby enhancing device performance (Kumar et al., 2023). For instance, Mirtchev et al. (2012) reported the functionalized CQDs as sensitizers in TiO_2-based DSSCs, improving PCE from 0.03% to 0.13%. In another study, Nguyen et al. (2021) incorporated N-CQDs into PEDOT:PSS hole-transport layers, enhancing charge transport and energy alignment, achieving PCE of 8.57%, compared to 8.17% (O-CQDs) and 7.26% (pristine PEDOT:PSS).

(c) Sensing

Environmental pollution has emerged as one of the most critical global issues, with contaminants affecting the earth's surface, water, and atmosphere. Even at extremely low concentrations, hazardous pollutants such as heavy metal ions, viruses, and other toxic species can cause irreversible harm to human health. Therefore, development of sensitive, real-time, and rapid detection systems is of urgent importance to minimize unintentional exposure. The use of metal-free sensing materials is especially attractive since they avoid introducing secondary contamination into the tested samples. In this context, carbon quantum dots (CQDs) represent an ideal alternative to traditional semiconductor QDs. Unlike Cd or Pb-based QDs, CQDs are non-toxic, environmentally friendly, and highly water-soluble, making them excellent candidates for sensing applications. Their unique optical features including strong photoluminescence, tunable emission, and abundant surface functional groups enable high sensitivity and selectivity toward a wide range of analytes. Remarkably, CQD-based probes have achieved limits of detection (LOD) in the nanomolar, picomolar, and even femtomolar ranges, making them highly suitable for precision analytical and biosensing applications (Chen et al., 2018). Recent studies have increasingly emphasized the promising role of CQDs in diverse sensing applications as addressed below.

Fe^{3+} detection: Kalaiyarasan et al. (2020) developed phosphorus-doped CQDs (P-CQDs) as fluorometric probes with a quantum yield (QY) of 16.1% (at 310/440 nm), achieving rapid and economic Fe^{3+} detection in blood, urine, and industrial wastewater.

Hg^{2+} detection: Hou et al. (2015) prepared functionalized CQDs with a QY of 11.9% as a label-free probe, achieving selective Hg^{2+} sensing with an exceptionally low LOD of 3.3 nM. In another study, Dhandapani et al. (2015) also developed highly luminescent CQDs (QY = 77.2%) as a fluorescence probe for Hg^{2+} sensing, with an LOD of 6.2 nM.

Cr^{6+} detection: Ming et al. (2019) synthesized nitrogen-doped CQDs (N-CQDs) with a QY of 47.5% for ultra-sensitive Cr^{6+} detection in water. The system achieved rapid detection (< 1 min) with excellent reproducibility and stability. Similarly, John et al. (2019) fabricated N-doped CQDs with a QY of 18% for Cr^{6+} sensing, achieving an LOD of 300 nM.

(d) Sustainable Catalysts

Catalysts play a vital role in dye removal by accelerating the degradation of dye molecules into less harmful or colorless products without being consumed during the process. CQDs have gained considerable attention as catalysts due to their economic and environmental sustainability, nontoxicity, water dispersibility arising from abundant oxygen-containing groups (carbonyl, carboxyl, hydroxyl), and excellent biocompatibility. Doping CQDs with multiple heteroatoms further improves their catalytic activity by accelerating electron transfer via interior interactions. For example, Preethi et al. (2022) synthesized blue-fluorescent CQDs that served as efficient photocatalysts for Rhodamine B, achieving 99.11% degradation under sunlight and 83.04% under ultrasonication. Abd Rani et al. (2021) reported the synthesis of N-doped CQDs that degraded methylene blue (97%) and malachite green (98%) under sunlight within 180 and 120 minutes, respectively. Furthermore, Sabet and Mahdavi (2019) synthesized high-yield, highly photoluminescent N-doped CQDs, which effectively removed heavy metal ions (37% Cd^{2+} and 75% Pb^{2+}) from drinking water. Thus, the above studies highlight that CQDs, particularly when doped with heteroatoms, serve as sustainable and highly efficient catalysts for organic dye degradation and heavy metal ion removal, offering great potential for environmental remediation.

7. CHALLENGES AND FUTURE PROSPECTS OF CARBON QUANTUM DOTS

Despite extensive research on CQDs due to their unique photoluminescence, biocompatibility, tunable surface chemistry, and eco-friendly nature, as well as their potential applications in photocatalysis, bioimaging, drug delivery, energy storage and conversion, sensing, and optoelectronics, the field remains in its early stages. Several key challenges along with their future prospects for the advancement of CQDs suitable for commercial potential applications are addressed as below:

(a) Standardization and Controlled Synthesis: Although CQDs can be synthesized from a wide variety of natural and synthetic precursors, reproducibility and large-scale production remain key challenges (Sead et al., 2025 and Hu et al., 2025). Future research should focus on developing standardized synthesis protocols, where control over particle size distribution, surface functionalities, and heteroatom doping can be reliably achieved. Advanced in-situ monitoring techniques and automated large-scale production could enable CQDs to transition from laboratory-scale curiosities to industrially viable materials.

(b) Bandgap Engineering and Multicolor Emission: Despite significant progress in CQDs research, the precise origin of photoluminescence in CQDs is not yet fully understood. Obtaining tunable emission across the entire visible–near-infrared spectrum will be critical for future optoelectronic and biomedical applications. Tailored heteroatom doping, surface passivation, and core–shell designs are expected to unlock new pathways for bandgap engineering, enabling CQDs to replace traditional semiconductor quantum dots in displays, lighting, and quantum communication technologies due to their low-cost and eco-friendly nature (Shariq et al., 2025).

(c) Energy Storage and Conversion: With remarkable progress in their optical and electronic tunability, CQDs have emerged as promising candidates for energy storage and conversion. Their tunable redox properties, rich surface functionalities, and exceptional electron-transfer efficiency make them highly suitable for sustainable energy technologies, including supercapacitors and batteries (Siwal & Bishnoi, 2025). In solar cells and photocatalytic systems, CQDs can act as light absorbers, charge mediators, or co-catalysts, thereby improving overall efficiency and operational stability. Furthermore, integrating CQDs with other functional nanomaterials, such as two-dimensional materials, metal oxides or sulfides, and conducting polymers, is expected to further accelerate their utilization in commercial energy devices.

(d) Environmental Remediation and Green Technology: Transitioning from energy-related applications, CQDs also hold great promise in environmental protection and green technology. Future research must focus on the eco-friendly synthesis of CQDs from biomass, agricultural waste, and other renewable sources, aligning with the use of green and clean materials for sustainable environment. In environmental monitoring and remediation, CQDs can act as sensitive probes for pollutant detection or as catalysts for the degradation of hazardous compounds (Mankoti et al., 2024). Their scalability and low-cost production further strengthen their prospects for large-scale deployment in environmental applications.

(e) Biomedical Applications: Extending beyond environmental applications, CQDs also exhibit remarkable prospects in the biomedical domain owing to their excellent biocompatibility and low toxicity. They are suitable candidates for next-generation bioimaging, biosensing, and targeted drug delivery systems. Future research is expected to integrate CQDs into theranostic systems, where a single nanoplatform can perform simultaneous diagnosis and therapy (e.g., imaging-guided drug delivery, photothermal or photodynamic therapy) (Alsuraifi et al., 2025). Further investigation into long-term toxicity, in-vivo stability, and clearance mechanisms are important for their real time applications in into clinical diagnostics (Li et al. 2025).

In summary, the future advancement of CQDs research should focus on overcoming the current challenges through standardized, eco-friendly, scalable and sustainable synthesis approaches, precise bandgap engineering, integration with advanced materials to unlock their full commercial and multidisciplinary potential in energy, environmental, and biomedical applications.

8. CONCLUSION

CQDs have already proven to be a unique class of carbon-based nanomaterials that can be produced through simple and versatile fabrication methods. They offer useful photophysical features of traditional semiconductor quantum dots, along with superior advantages such as low toxicity, environmental compatibility, and versatile surface chemistry. Their ultrasmall size imparts quantum confinement effects, while abundant functional groups and heteroatom doping provide further tunability of optical, electronic, and catalytic properties. As discussed in the chapter, heteroatom doping, using elements such as N, S, P, B, and F, offers a powerful strategy to precisely modulate the electronic structure, surface states, and fluorescence behavior of CQDs. While the single doping influence charge density, band structure, and emission wavelength, co-doping or multi-doping introduces synergistic effects that enhance quantum yield, tune emission colors, improve catalytic and sensing performance.

As highlighted in this chapter, CQDs have shown immense promise across a wide spectrum of applications. In energy storage and conversion, they improve charge transport, redox activity, and interfacial stability in supercapacitors, batteries, and solar cells. In sensing, their strong photoluminescence, aqueous dispersibility, and surface functionalization enable rapid, selective, and ultrasensitive detection of hazardous pollutants, biomolecules, and metal ions. In catalysis, CQDs function as active sites, electron mediators, or co-catalysts, accelerating degradation and remediation processes under mild and sustainable conditions. Their ability to integrate with metal oxides, sulfides, conducting polymers, and other nanomaterials further enhances their performance, offering hybrid platforms for next-generation technologies.

REFERENCES

Abd Rani, U., Ng, L. Y., Ng, C. Y., Mahmoudi, E., Ng, Y. S., & Mohammad, A. W. (2021). Sustainable production of nitrogen-doped carbon quantum dots for photocatalytic degradation of methylene blue and malachite green. *Journal of Water Process Engineering*, *40*, 101816. DOI: 10.1016/j.jwpe.2020.101816

Alsuraifi, A., Mouzan, M. M., Algzaare, A., Aqeel, Z., Al-Essa, U. I. R., Mohammed, N. A., & Ayad, A. (2025). Quantum Dots: Catalysts for a New Era of Precision Medicine and Biomedical Innovation. *Journal of Fluorescence*, •••, 1–64. DOI: 10.1007/s10895-025-04452-2 PMID: 41051600

Arshada, F., & Sk, M. P. (2019). Aggregation-induced red shift in N, S-doped chiral carbon dot emissions for moisture sensing. *New Journal of Chemistry*, *43*(33), 13240–13248. DOI: 10.1039/C9NJ03009C

Atchudan, R., Immanuel Edison, T. N. J., Shanmugam, M., Perumal, S., Somanathan, T., & Lee, Y. R. (2021). Sustainable synthesis of carbon quantum dots from banana peel waste using hydrothermal process for in vivo bioimaging. *Physica E, Low-Dimensional Systems and Nanostructures*, *126*, 114417. DOI: 10.1016/j.physe.2020.114417

Bao, Z., Lu, L., Wang, J., Zhu, H., Dang, X., Chen, Y., Yin, X., Song, Y., & Wen, X. (2025). Application of quantum dots in cancer diagnosis and treatment: Advances and perspectives. *Nano Research*, *18*(1), 10–26599. DOI: 10.1007/s10895-025-04598-z PMID: 41139377

Bottini, M., Balasubramanian, C., Dawson, M. I., Bergamaschi, A., Bellucci, S., & Mustelin, T. (2006). Isolation and characterization of fluorescent nanoparticles from pristine and oxidized electric arc-produced single-walled carbon nanotubes. *The Journal of Physical Chemistry B*, *110*(2), 831–836. DOI: 10.1021/jp055503b PMID: 16471611

Bourlinos, A. B., Trivizas, G., Karakassides, M. A., Baikousi, M., Kouloumpis, A., Gournis, D., Bakandritsos, A., Hola, K., Kozak, O., Zboril, R., Papagiannouli, I., Aloukos, P., & Couris, S. (2015). Green and simple route toward boron doped carbon dots with significantly enhanced non-linear optical properties. *Carbon*, *83*, 173–179. DOI: 10.1016/j.carbon.2014.11.032

Bouzas-Ramos, D., Canga, J. C., Mayo, J. C., Sainz, R. M., Encinar, J. R., & Costa-Fernandez, J. M. (2019). Carbon quantum dots co-doped with nitrogen and lanthanides for multimodal imaging. *Advanced Functional Materials*, *29*(38), 1903884. DOI: 10.1002/adfm.201903884

Cánchig, M. B., López, F., Morales-Navarro, Z., Debut, A., Vizuete, K., Terencio, T., Caetano, M., & Saucedo-Vázquez, J. P. (2025). Enhanced selectivity of carbon quantum dots for metal ion detection through surface modification by heteroatom doping: A study on optical properties and theoretical approach. *Carbon Trends*, *18*, 100445. DOI: 10.1016/j.cartre.2024.100445

Çeşme, M., Eskalen, H., & Başkaya, S. K. (2022). Fluorescent carbon dots from vegetable and fruit wastes and their applications. In Ray, R. C. (Ed.), *Fruits and vegetable wastes*. Springer., DOI: 10.1007/978-981-16-9527-8_15

Chen, C., Zhao, D., Hu, T., Sun, J., & Yang, X. (2017). Highly fluorescent nitrogen and sulfur co-doped graphene quantum dots for an inner filter effect-based cyanide sensor. *Sensors and Actuators. B, Chemical*, *241*, 779–788. DOI: 10.1016/j.snb.2016.11.010

Chen, X., Gong, F., Cao, Z., Zou, W., & Gu, T. (2018). Highly cysteine-selective fluorescent nanoprobes based on ultrabright and directly synthesized carbon quantum dots. *Analytical and Bioanalytical Chemistry*, *410*(12), 2961–2970. DOI: 10.1007/s00216-018-0980-3 PMID: 29523940

Cheng, H., Chen, S., Li, M., Lu, Y., Chen, H., Fang, X., Qiu, H., Wang, W., Jiang, C., & Zheng, Y. (2023). Ultra-stable dual-color phosphorescence Carbon-Dot@Silica material for advanced anti-counterfeiting. *Dyes and Pigments*, *208*, 110827. DOI: 10.1016/j.dyepig.2022.110827

Cho, S., Kim, H., Song, D., Jung, J., Park, S., Jo, H., Seo, S., Han, C., Park, S., Kwon, W., & Han, H. (2024). Insights into glucose-derived carbon dot synthesis via Maillard reaction: From reaction mechanism to biomedical applications. *Scientific Reports*, *14*(1), 31325. DOI: 10.1038/s41598-024-82767-z PMID: 39733004

Dager, A., Uchida, T., Maekawa, T., & Kumar, D. S. (2019). Synthesis and characterization of mono-disperse carbon quantum dots from fennel seeds: Photoluminescence analysis using machine learning. *Scientific Reports*, *9*(1), 14004. DOI: 10.1038/s41598-019-50397-5 PMID: 31570739

Dai, B., Wu, C., Lu, Y., Deng, D., & Xu, S. (2017). Synthesis and formation mechanism of s-doped carbon dots from low-molecule-weight organics. *Journal of Luminescence*, *183*, 1–6. DOI: 10.1016/j.jlumin.2017.04.054

Dang, D. K., Sundaram, C., Ngo, Y. L. T., Chung, J. S., Kim, E. J., & Hur, S. H. (2018). One-pot solid-state synthesis of highly fluorescent N and S co-doped carbon dots and their use as a fluorescent probe for Ag^+ detection in aqueous solution. *Sensors and Actuators. B, Chemical*, *255*, 3284–3291. DOI: 10.1016/j.snb.2017.09.155

Das, R. K., & Mohapatra, S. (2017). Highly luminescent, heteroatom-doped carbon quantum dots for ultrasensitive sensing of glucosamine and targeted imaging of liver cancer cells. *Journal of Materials Chemistry. B*, *5*(11), 2190–2197. DOI: 10.1039/C6TB03141B PMID: 32263692

Dhandapani, E., Maadeswaran, P., Raj, R. M., Raj, V., Kandiah, K., & Duraisamy, N. (2023). A potential forecast of carbon quantum dots (CQDs) as an ultrasensitive and selective fluorescence probe for Hg (II) ions sensing. *Materials Science and Engineering B*, *287*, 116098. DOI: 10.1016/j.mseb.2022.116098

Ding, H., Li, X.-H., Chen, X.-B., Wei, J.-S., Li, X.-B., & Xiong, H.-M. (2020). Surface states of carbon dots and their influences on luminescence. *Journal of Applied Physics*, *127*(23), 231101. DOI: 10.1063/1.5143819

Dong, Y., Pang, H., Yang, H., Guo, C., Shao, J., Chi, Y., Li, C., & Yu, T. (2013). Carbon-based dots co-doped with nitrogen and sulfur for high quantum yield and excitation-independent emission. *Angewandte Chemie International Edition*, *52*(30), 7800–7804. DOI: 10.1002/anie.201301114 PMID: 23761198

Facure, M. H. M., Schneider, R., Mercante, L. A., & Correa, D. S. (2022). Rational hydrothermal synthesis of graphene quantum dots with optimized luminescent properties for sensing applications. *Materials Today. Chemistry*, *23*, 100755. DOI: 10.1016/j.mtchem.2021.100755

Feng, H., & Qian, Z. (2018). Functional carbon quantum dots: A versatile platform for chemosensing and biosensing. *Chemical Record (New York, N.Y.)*, *18*(5), 491–505. DOI: 10.1002/tcr.201700055 PMID: 29171708

Ferrari, C. (2007). Raman spectroscopy of graphene and graphite: Disorder, electron–phonon coupling, doping and nonadiabatic effects. *Solid State Communications*, *143*(1-2), 47–57. DOI: 10.1016/j.ssc.2007.03.052

Fu, X., Sheng, L., Yu, Y., Ma, M., Cai, Z., & Huang, X. (2018). Rapid and universal detection of ovalbumin based on N, O, P-co-doped carbon dots-fluorescence resonance energy transfer technology. *Sensors and Actuators. B, Chemical*, *269*, 278–287. DOI: 10.1016/j.snb.2018.04.134

Gao, H., Liu, Z., Song, L., Guo, W., Gao, W., Ci, L., Rao, A., Quan, W., Vajtai, R., & Ajayan, P. M. (2012). Synthesis of S-doped graphene by liquid precursor. *Nanotechnology*, *23*(27), 275605. DOI: 10.1088/0957-4484/23/27/275605 PMID: 22710561

Gayen, B., Palchoudhury, S., & Chowdhury, J. (2019). Carbon dots: A mystic star in the world of nanoscience. *Journal of Nanomaterials, 3451307*, 1–19. Advance online publication. DOI: 10.1155/2019/3451307

Guan, X., Li, Z., Geng, X., Lei, Z., Karakoti, A., Wu, T., Kumar, P., Yi, J., & Vinu, A. (2023). Emerging trends of carbon-based quantum dots: Nanoarchitectonics and applications. *Small, 19*(17), 2207181. DOI: 10.1002/smll.202207181 PMID: 36693792

Guo, L., Huang, Y., Su, A., Shang, T., Qu, M., Shi, W., Qiang, Y., & Marzouki, R. (2025). Boosting the performance of alkaline Al-air batteries by interfacial reconstruction via N-doped carbon quantum dots. *Journal of Energy Storage, 127*, 117136. DOI: 10.1016/j.est.2025.117136

Guo, R., Li, T., & Shi, S. (2019). Aggregation-induced emission enhancement of carbon quantum dots and applications in light emitting devices. *Journal of Materials Chemistry. C, Materials for Optical and Electronic Devices, 7*(17), 5148–5154. DOI: 10.1039/C9TC01138B

Guo, Y., Chen, Y., Cao, F., Wang, L., Wang, Z., & Leng, Y. (2017). Hydrothermal synthesis of nitrogen and boron doped carbon quantum dots with yellow-green emission for sensing Cr (VI), anti-counterfeiting and cell imaging. *RSC Advances, 7*(76), 48386–48393. DOI: 10.1039/C7RA09785A

Gurushankar, K., Karuppaiya, C., Tari, V., & Kannan, K. (2025). Exploring MXenes and MBenes in Metal Sulfur and Selenium Battery Technologies. In *Novel Energy Storage and Conversion Technologies for Two-Dimensional MXenes and MBenes* (pp. 105–120). IGI Global Scientific Publishing. DOI: 10.4018/979-8-3693-9546-2.ch005

He, G., Shu, M., Yang, Z., Ma, Y., Huang, D., Xu, S., Wang, Y., Hu, N., Zhang, Y., & Xu, L. (2017). Microwave formation and photoluminescence mechanisms of multi-states nitrogen doped carbon dots. *Applied Surface Science, 422*, 257–265. DOI: 10.1016/j.apsusc.2017.05.036

He, G., Shu, M., Yang, Z., Ma, Y., Huang, D., Xu, S., Wang, Y., Hu, N., Zhang, Y., & Xu, L. (2017). Microwave formation and photoluminescence mechanisms of multi-states nitrogen doped carbon dots. *Applied Surface Science, 422*, 257–265. DOI: 10.1016/j.apsusc.2017.05.036

Hola, K., Bourlinos, A. B., Kozak, O., Berka, K., Siskova, K. M., Havrdova, M., Tucek, J., Safarova, K., Otyepka, M., Giannelis, E. P., & Zboril, R. (2014). Photoluminescence effects of graphitic core size and surface functional groups in carbon dots: COO^- induced red-shift emission. *Carbon, 70*, 279–286. DOI: 10.1016/j.carbon.2014.01.008

Hou, H., Banks, C. E., Jing, M., Zhang, Y., & Ji, X. (2015). Carbon quantum dots and their derivative 3D porous carbon frameworks for sodium-ion batteries with ultralong cycle life. *Advanced Materials*, *27*(47), 7861–7866. DOI: 10.1002/adma.201503816 PMID: 26506218

Hou, Y., Lu, Q., Deng, J., Li, H., & Zhang, Y. (2015). One-pot electrochemical synthesis of functionalized fluorescent carbon dots and their selective sensing for mercury ion. *Analytica Chimica Acta*, *866*, 69–74. DOI: 10.1016/j.aca.2015.01.039 PMID: 25732694

Hu, C., Channa, A. I., Xia, L., Li, X., Li, Z., Wang, Z. M., & Tong, X. (2025). Colloidal InAs quantum dots: Synthesis, properties, and optoelectronic devices. *Advanced Functional Materials*, *35*(32), 2500280. DOI: 10.1002/adfm.202500280

Hu, S., Tian, R., Dong, Y., Yang, J., Liu, J., & Chang, Q. (2013). Modulation and effects of surface groups on photoluminescence and photocatalytic activity of carbon dots. *Nanoscale*, *5*(22), 11665–11671. DOI: 10.1039/c3nr03893a PMID: 24101143

Hu, S. L., Niu, K. Y., Sun, J., Yang, J., Zhao, N. Q., & Du, X. W. (2009). One-step synthesis of fluorescent carbon nanoparticles by laser irradiation. *Journal of Materials Chemistry*, *19*(4), 484–488. DOI: 10.1039/B812943F

Huang, S., Yang, E., Yao, J., Liu, Y., & Xiao, Q. (2018). Red emission nitrogen, boron, sulfur co-doped carbon dots for "on-off-on" fluorescent mode detection of Ag^+ ions and l-cysteine in complex biological fluids and living cells. *Analytica Chimica Acta*, *1035*, 192–202. DOI: 10.1016/j.aca.2018.06.051 PMID: 30224139

Inayat, A., Albalawi, K., Rehman, A. U., Saad, A. Y., Saleh, E. A. M., Alamri, M. A., El-Zahhar, A. A., Haider, A., & Abbas, S. M. (2023). Tunable synthesis of carbon quantum dots from the biomass of spent tea leaves as supercapacitor electrode. *Materials Today. Communications*, *34*, 105479. DOI: 10.1016/j.mtcomm.2023.105479

Jana, J., Ganguly, M., Chandrakumar, K. R. S., Rao, G. M., & Pal, T. (2017). Boron Precursor-Dependent Evolution of Differently Emitting Carbon Dots. *Langmuir*, *33*(3), 573–584. DOI: 10.1021/acs.langmuir.6b04100 PMID: 28024393

Javed, M., Saqib, A. N. S., Ali, B., Faizan, M., Anang, D. A., Iqbal, Z., & Abbas, S. M. (2019). Carbon quantum dots from glucose oxidation as a highly competent anode material for lithium and sodium-ion batteries. *Electrochimica Acta*, *297*, 250–257. DOI: 10.1016/j.electacta.2018.11.167

Jian, X., Yang, H. M., Li, J. G., Zhang, E. H., Cao, L. L., & Liang, Z. H. (2017). Flexible all-solid-state high-performance supercapacitor based on electrochemically synthesized carbon quantum dots/polypyrrole composite electrode. *Electrochimica Acta*, *228*, 483–493. DOI: 10.1016/j.electacta.2017.01.082

John, B. K., John, N., Mathew, S., Korah, B. K., Punnoose, M. S., & Mathew, B. (2022). Fluorescent carbon quantum dots as a novel solution and paper strip-based dual sensor for the selective detection of Cr (VI) ions. *Diamond and Related Materials*, *126*, 109138. DOI: 10.1016/j.diamond.2022.109138

Kalaiyarasan, G., Joseph, J., & Kumar, P. (2020). Phosphorus-doped carbon quantum dots as fluorometric probes for iron detection. *ACS Omega*, *5*(35), 22278–22288. DOI: 10.1021/acsomega.0c02627 PMID: 32923785

Kamaruzaman, N. A. N., Fadzil, N. I., Alli, Y. A., Tan, T., Abdullah, J., & Abdul Rashid, S. (2025). Optimized synthesis and characterization of red-shifted nitrogen-doped carbon quantum dots for highly sensitive detection of Fe^{3+} ions in aqueous media. *Sensors and Actuators. A, Physical*, *394*, 116895. DOI: 10.1016/j.sna.2025.116895

Kannan, K., Chanda, D., Gautam, J., Behera, A., Meshesha, M. M., Jang, S. G., & Yang, B. (2023). Hydrothermally synthesized mixed metal oxide nanocomposites for electrochemical water splitting and photocatalytic hydrogen production. *International Journal of Hydrogen Energy*, *48*(93), 36412–36426. DOI: 10.1016/j.ijhydene.2023.06.042

Kannan, K., Chanda, D., Gautam, J., Meshesha, M. M., Jang, S. G., & Yang, B. L. (2023). Facial synthesis of pp heterojunction composites: Evaluation of their electrochemical properties with photovoltaics-electrolyzer water splitting using two-electrode system. *International Journal of Hydrogen Energy*, *48*(37), 13814–13826. DOI: 10.1016/j.ijhydene.2022.12.235

Kannan, K., Chinnaiah, K., Gurushankar, K., Krishnamoorthi, R., Chen, Y. S., Murphin Kumar, P. S., & Li, Y. Y. (2024). Investigation of the electrochemical behavior of CuO-NiO-Co3O4 nanocomposites for enhanced supercapacitor applications. *Materials (Basel)*, *17*(16), 3976. DOI: 10.3390/ma17163976 PMID: 39203154

Kannan, K., Gautam, J., Chanda, D., Meshesha, M. M., Jang, S. G., & Yang, B. L. (2023). Two-dimensional MAX supported copper oxide/nickel Oxide/MAX as an efficient and novel photocatalyst for hydrogen evolution. *International Journal of Hydrogen Energy*, *48*(20), 7273–7283. DOI: 10.1016/j.ijhydene.2022.10.131

Kannan, K., Sliem, M. H., Abdullah, A. M., Sadasivuni, K. K., & Kumar, B. (2020). Fabrication of ZnO-Fe-MXene based nanocomposites for efficient CO2 reduction. *Catalysts*, *10*(5), 549. DOI: 10.3390/catal10050549

Kannan, K., & Tari, V. (Eds.). (2024). Novel energy storage and conversion technologies for two-dimensional MXenes and MBenes. ISBN: 9798369395462, eISBN: 9798369395486DOI: 10.4018/979-8-3693-9546-2

Keenan, J. M., Zhou, Y., & Leblanc, R. M. (2019). Recent development of carbon quantum dots regarding their optical properties, photoluminescence mechanism, and core structure. *Nanoscale*, *11*(11), 4634–4652. DOI: 10.1039/C8NR10059D PMID: 30834912

Kou, X., Jiang, S., Park, S.-J., & Meng, L.-Y. (2020). A review: Recent advances in preparations and applications of heteroatom-doped carbon quantum dots. *Dalton Transactions (Cambridge, England)*, *49*(21), 6915–6938. DOI: 10.1039/D0DT01004A PMID: 32400806

Kumar, A., Kumar, I., & Gathania, A. K. (2022). Synthesis, characterization and potential sensing application of carbon dots synthesized via the hydrothermal treatment of cow milk. *Scientific Reports*, *12*(1), 22495. DOI: 10.1038/s41598-022-26906-4 PMID: 36577768

Kumar, P., Bhatt, G., Kaur, R., Dua, S., & Kapoor, A. (2020). Synthesis and modulation of the optical properties of carbon quantum dots using microwave radiation. *Fullerenes, Nanotubes, and Carbon Nanostructures*, *28*(9), 724–731. DOI: 10.1080/1536383X.2020.1752679

Kumar, P., Dua, S., Kaur, R., Kumar, M., & Bhatt, G. (2022). A review on advancements in carbon quantum dots and their application in photovoltaics. *RSC Advances*, *12*(8), 4714–4732. DOI: 10.1039/D1RA08452F PMID: 35425490

Kumar, P., Mahapatra, A., Kumar, S., & Pradhan, B. (2023). Photovoltaic application of carbon quantum dots. In *Carbon Quantum Dots for Sustainable Energy and Optoelectronics* (pp. 135–158). Woodhead Publishing., DOI: 10.1016/B978-0-323-90895-5.00017-5

Kundu, S., Yadav, R. M., Narayanan, T. N., Shelke, M. V., Vajtai, R., Ajayan, P. M., & Pillai, V. K. (2015). Synthesis of N, F and S co-doped graphene quantum dots. *Nanoscale*, *7*(27), 11515–11519. DOI: 10.1039/C5NR02427G PMID: 26087457

Lan, M., Zhao, S., Zhang, Z., Yan, L., Guo, L., Niu, G., Zhang, J., Zhao, J., Zhang, H., Wang, P., Zhu, G., Lee, C.-S., & Zhang, W. (2017). Two-photon-excited near-infrared emissive carbon dots as multifunctional agents for fluorescence imaging and photothermal therapy. *Nano Research*, *10*(9), 3113–3123. DOI: 10.1007/s12274-017-1528-0

Li, F., Li, T., Sun, C., Xia, J., Jiao, Y., & Xu, H. (2017). Facile synthesis of carbon dots with superior sensing ability. *Applied Nanoscience*, *8*(5), 1189–1196. DOI: 10.1007/s13204-018-0755-3

Li, Y., Li, S., Wang, Y., Wang, J., Liu, H., Liu, X., Wang, L., Liu, X., Xue, W., & Ma, N. (2017). Electrochemical synthesis of phosphorus-doped graphene quantum dots for free radical scavenging. *Physical Chemistry Chemical Physics*, *19*(22), 11631–11638. DOI: 10.1039/C6CP06377B PMID: 28430285

Li, Y., Lin, H., Luo, C., Wang, Y., Jiang, C., Qi, R., Huang, R., Travass-Sejdic, J., & Peng, H. (2017). Aggregation induced red shift emission of phosphorus doped carbon dots. *RSC Advances*, *7*(51), 32225–32228. DOI: 10.1039/C7RA04781A

Liu, G., Li, S., Cheng, M., Zhao, L., Zhang, B., Gao, Y., Yan, X., Liu, F., & Lu, G. (2018). Facile synthesis of nitrogen and sulfur co-doped carbon dots for multiple sensing capacities: alkaline fluorescence enhancement effect, temperature sensing, and selective detection of Fe^{3+} ions. *New Journal of Chemistry, 42*, 13147–13156. DOIhttps://DOI: 10.1039/C8NJ02086H

Liu, H., Ye, T., & Mao, C. (2007). Fluorescent carbon nanoparticles derived from candle soot. *Angewandte Chemie International Edition*, *46*(34), 6473–6475. DOI: 10.1002/anie.200701271 PMID: 17645271

Liu, J., Li, J., Xu, L., Qiao, Y., & Chen, J. (2017). Facile Synthesis of N, B-Doped Carbon Dots and Their Application for Multisensor and Cellular Imaging. *Industrial & Engineering Chemistry Research*, *56*(14), 3905–3912. DOI: 10.1021/acs.iecr.6b04752

Liu, R., Wu, D., Liu, S., Koynov, K., Knoll, W., & Li, Q. (2009). An aqueous route to multicolor photoluminescent carbon dots using silica spheres as carriers. *Angewandte Chemie International Edition*, *48*(25), 4598–4601. DOI: 10.1002/anie.200900652 PMID: 19388019

Lu, W., Gong, X., Nan, M., Liu, Y., Shuang, S., & Dong, C. (2015). Low temperature synthesis of phosphorous and nitrogen co-doped yellow fluorescent carbon dots for sensing and bioimaging. *Analytica Chimica Acta*, *898*, 116–127. DOI: 10.1016/j.aca.2015.09.050 PMID: 26526917

Lu, W., Gong, X., Nan, M., Liu, Y., Shuang, S., & Dong, C. (2015). Comparative study for N and S doped carbon dots: Synthesis, characterization and applications for Fe^{3+} probe and cellular imaging. *Analytica Chimica Acta, 898*, 116–127. DOI: 10.1016/j.aca.2015.09.050 PMID: 26526917

Luo, B., Yang, H., Zhou, B., Ahmed, S. M., Zhang, Y., Liu, H., Liu, X., He, Y., & Xia, S. (2020). Facile synthesis of Luffa sponge activated carbon fiber-based carbon quantum dots with green fluorescence and their application in Cr (VI) determination. *ACS Omega, 5*(10), 5540–5547. DOI: 10.1021/acsomega.0c00195 PMID: 32201847

Mankoti, M., Meena, S. S., & Mohanty, A. (2024). Exploring the potential of eco-friendly carbon dots in monitoring and remediation of environmental pollutants. *Environmental Science and Pollution Research International, 31*(31), 43492–43523. DOI: 10.1007/s11356-024-33448-x PMID: 38713351

Mehta, J., Dilbaghi, N., Singhal, N. K., Marrazza, G., Kaushik, A., & Kumar, S. (2023). Electrochemiluminescent quantum dots as emerging next generation sensing platforms. *Chemical Engineering Journal, 477*, 146958. DOI: 10.1016/j.cej.2023.146958

Meng, L., Xu, Q., Zhang, J., & Wang, X. (2024). Colloidal quantum dot materials for next-generation near-infrared optoelectronics. *Chemical Communications, 60*(9), 1072–1087. DOI: 10.1039/D3CC04315K PMID: 38174780

Mimona, M. A., Rimon, M. I. H., Zohura, F. T., Sony, J. M., Rim, S. I., Arup, M. M. R., & Mobarak, M. H. (2025). Quantum dot nanomaterials: Empowering advances in optoelectronic devices. *Chemical Engineering Journal Advances, 21*, 100704. DOI: 10.1016/j.ceja.2025.100704

Ming, F., Hou, J., Hou, C., Yang, M., Wang, X., Li, J., Huo, D., & He, Q. (2019). One-step synthesized fluorescent nitrogen doped carbon dots from thymidine for Cr (VI) detection in water. *Spectrochimica Acta. Part A: Molecular and Biomolecular Spectroscopy, 222*, 117165. DOI: 10.1016/j.saa.2019.117165 PMID: 31185440

Mirtchev, P., Henderson, E. J., Soheilnia, N., Yip, C. M., & Ozin, G. A. (2012). Solution phase synthesis of carbon quantum dots as sensitizers for nanocrystalline TiO 2 solar cells. *Journal of Materials Chemistry, 22*(4), 1265–1269. DOI: 10.1039/C1JM14112K

Mmelesi, O. K., Mguni, L. L., Li, F., Nkosi, B., & Liu, X. (2024). Recent development in fluorescent carbon quantum dots-based photocatalysts for water and energy applications. *Materials Science in Semiconductor Processing, 181*, 108661. DOI: 10.1016/j.mssp.2024.108661

Modi, P. D., Mehta, V. N., Prajapati, V. S., Patel, S., & Rohit, J. V. (2023). Bottom-up approaches for the preparation of carbon dots. In Kailasa, S. K., & Hussain, C. M. (Eds.), *Carbon dots in analytical chemistry* (pp. 15–29). Elsevier., DOI: 10.1016/B978-0-323-98350-1.00022-0

Mondal, M., & Pramanik, S. (2023). A mechanism for excitation-dependent emission from carbon nanodots. *Materials Letters: X, 18,* 100195. DOI: 10.1016/j.mlblux.2023.100195

Nasrin, A., Hassan, M., Mann, G., & Gomes, V. G. (2020). Conjugated ternary doped carbon dots from vitamin B derivative: Multispectral nanoprobes for targeted melanoma bioimaging and photosensitization. *Journal of Luminescence*, *217*, 116811. DOI: 10.1016/j.jlumin.2019.116811

Nguyen, D. N., Roh, S. H., Kim, D. H., Lee, J. Y., Wang, D. H., & Kim, J. K. (2021). Molecular manipulation of PEDOT: PSS for efficient hole transport by incorporation of N-doped carbon quantum dots. *Dyes and Pigments*, *194*, 109610. DOI: 10.1016/j.dyepig.2021.109610

Niveria, K., Singh, P., Yadav, M., & Verma, A. K. (2023). Quantum Dot (QD)-Induced Toxicity and Biocompatibility. In Korotcenkov, G. (Ed.), *Handbook of II-VI Semiconductor-Based Sensors and Radiation Detectors*. Springer., DOI: 10.1007/978-3-031-19531-0_8

Pan, D., Zhang, J., Li, Z., Wu, C., Yan, X., & Wu, M. (2010). Observation of pH, solvent, spin and excitation dependent blue photoluminescence from carbon nanoparticles. *Chemical Communications*, *46*(21), 3681–3683. DOI: 10.1039/c000114g PMID: 20396809

Pandiyan, S., Arumugam, L., Srirengan, S. P., Pitchan, R., Sevugan, P., Kannan, K., Pitchan, G., Hegde, T. A., & Gandhirajan, V. (2020). Biocompatible carbon quantum dots derived from sugarcane industrial wastes for effective nonlinear optical behavior and antimicrobial activity applications. *ACS Omega*, *5*(47), 30363–30372. DOI: 10.1021/acsomega.0c03290 PMID: 33283084

Pankaj, A., Tewari, K., Singh, S., & Singh, S. P. (2018). Waste candle soot derived nitrogen doped carbon dots based fluorescent sensor probe: An efficient and inexpensive route to determine Hg (II) and Fe (III) from water. *Journal of Environmental Chemical Engineering*, *6*(4), 5561–5569. DOI: 10.1016/j.jece.2018.08.059

Patra, R., Panda, P. K., Hung, C.-Y., Dash, P., Yang, P.-C., & Hsieh, C.-T. (2024). Facile synthesis of yellow emissive heteroatom doped carbon quantum dots fluorescent probe with superior quantum yield for fast-track ionic detection in aqueous media. *Materials Research Bulletin*, *180*, 113040. DOI: 10.1016/j.materresbull.2024.113040

Peng, H., & Sejdic, J. T. (2009). Simple aqueous solution route to luminescent carbogenic dots from carbohydrates. *Chemistry of Materials, 21*(23), 5563–5565. DOI: 10.1021/cm901593y

Phang, S. J., & Tan, L. L. (2019). Recent advances in carbon quantum dot (CQD)-based two dimensional materials for photocatalytic applications. *Catalysis Science & Technology, 9*(21), 5882–5905. DOI: 10.1039/C9CY01452G

Preethi, M., Viswanathan, C., & Ponpandian, N. (2022). A metal-free, dual catalyst for the removal of Rhodamine B using novel carbon quantum dots from muskmelon peel under sunlight and ultrasonication: A green way to clean the environment. *Journal of Photochemistry and Photobiology A Chemistry, 426*, 113765. DOI: 10.1016/j.jphotochem.2021.113765

Puchaicela, M., Lara, D., Cevallos, V. J., Garzón, A., Gardener, J., Solorzano, G., Spencer, L., & Chimborazo, J. (2025). Synthesis and physical characterization of carbon quantum dots from watermelon seed towards a biological application. *Carbon Trends, 20*, 100493. DOI: 10.1016/j.cartre.2025.100493

Qian, Z., Shan, X., Chai, L., Ma, J., Chen, J., & Feng, H. (2014). Si-Doped Carbon Quantum Dots: A Facile and General Preparation Strategy, Bioimaging Application, and Multifunctional Sensor. *ACS Applied Materials & Interfaces, 6*(8), 6797–6805. DOI: 10.1021/am500403n PMID: 24707855

Quan, Y., Wang, G., Lu, L., Wang, Z., Xu, H., Liu, S., & Wang, D. (2020). High-performance pseudo-capacitor energy storage device based on a hollow-structured copper sulfide nanoflower and carbon quantum dot nanocomposite. *Electrochimica Acta, 353*, 136606. DOI: 10.1016/j.electacta.2020.136606

Quesada-González, D., & Merkoçi, A. (2025). Quantum dots for biosensing: Classification and applications. *Biosensors & Bioelectronics, 273*, 117180. DOI: 10.1016/j.bios.2025.117180 PMID: 39864309

Ramírez Sanabria, A. E., Narváez, S. U., Gutiérrez Portilla, J. V., & Torres Rodríguez, G. A. (2025). Sonochemical synthesis of CQDs from coffee husk: Insights in aggregation mechanism, optimization and sustainability analysis. *Heliyon, 11*(1), e41000. DOI: 10.1016/j.heliyon.2024.e41000 PMID: 39807508

Rasal, A. S., Yadav, S., Yadav, A., Kashale, A. A., Manjunatha, S. T., Altaee, A., & Chang, J. Y. (2021). Carbon quantum dots for energy applications: A review. *ACS Applied Nano Materials, 4*(7), 6515–6541. DOI: 10.1021/acsanm.1c01372

Redondo-Fernandez, G., Cigales Canga, J., Soldado, A., Ruiz Encinar, J., & Costa-Fernandez, J. M. (2023). Functionalized heteroatom-doped carbon dots for biomedical applications: A review. *Analytica Chimica Acta*, *1284*, 341874. DOI: 10.1016/j.aca.2023.341874 PMID: 37996151

Ren, J., Malfatti, L., & Innocenzi, P. (2021). Citric acid derived carbon dots, the challenge of understanding the synthesis-structure relationship. *Journal of Carbon Research*, *7*(2), 2. Advance online publication. DOI: 10.3390/c7010002

Rong, M. C., Zhang, K. X., Wang, Y. R., & Chen, X. (2017). The synthesis of B, N-carbon dots by a combustion method and the application of fluorescence detection for Cu^{2+}. *Chinese Chemical Letters*, *28*(7), 1119–1124. DOI: 10.1016/j.cclet.2016.12.009

Roshni, V., & Divya, O. (2017). One-step microwave-assisted green synthesis of luminescent N-doped carbon dots from sesame seeds for selective sensing of Fe (III). *Current Science*, *112*(2), 385–390. DOI: 10.18520/cs/v112/i02/385-390

Ru, Y., Waterhouse, G. I. N., & Lu, S. (2022). Aggregation in carbon dots. *Aggregate*, *3*(6), e296. DOI: 10.1002/agt2.296

Sabet, M., & Mahdavi, K. (2019). Green synthesis of high photoluminescence nitrogen-doped carbon quantum dots from grass via a simple hydrothermal method for removing organic and inorganic water pollutions. *Applied Surface Science*, *463*, 283–291. DOI: 10.1016/j.apsusc.2018.08.223

Sahu, S., Behera, B., Maiti, T. K., & Mohapatra, S. (2012). Simple one-step synthesis of highly luminescent carbon dots from orange juice: Application as excellent bio-imaging agents. *Chemical Communications*, *48*(70), 8835–8837. DOI: 10.1039/c2cc33796g PMID: 22836910

Sarkar, S., Sudolská, M., Dubecký, M., Reckmeier, C. J., Rogach, A. L., Zbořil, R., & Otyepka, M. (2016). Graphitic nitrogen doping in carbon dots causes red-shifted absorption. *The Journal of Physical Chemistry. C, Nanomaterials and Interfaces*, *120*(2), 1303–1308. DOI: 10.1021/acs.jpcc.5b10186

Sciortino, E., Marino, B., Dam, P., Schall, M., Cannas, F., & Messina, F. (2016). Solvatochromism unravels the emission mechanism of carbon nanodots. *The Journal of Physical Chemistry Letters*, *7*(17), 3419–3423. DOI: 10.1021/acs.jpclett.6b01590 PMID: 27525451

Sead, F. F., Jadeja, Y., & Kumar, A., MM, R., Kundlas, M., Saini, S., Joshi K. K. & Noorizadeh, H. (. (2025). Carbon quantum dots for sustainable energy: Enhancing electrocatalytic reactions through structural innovation. *Nanoscale Advances*, *7*(13), 3961–3998. DOI: 10.1039/D5NA00205B PMID: 40444140

Shariq, M., AlGhamdi, G. S., Alotaibi, R., Almuqati, N. S., Toraba, M., Albarqan, M., Abdullah, S. B., Alshamrani, K. M., & Ali, S. K. (2025). Design, synthesis, and structural engineering of carbon quantum dots and multifunctional N, P, B, S-doped CQDs for photocatalytic water purification. *Journal of Water Process Engineering*, *77*, 108450. DOI: 10.1016/j.jwpe.2025.108450

Sharma, A. K., Kuamri, N., Chauhan, P., Thakur, S., Kumar, S., & Shandilya, M. (2025). Comprehensive Insights into Carbon Quantum Dots: Synthesis Strategies and Multidomain Applications. *Journal of Fluorescence*. Advance online publication. DOI: 10.1007/s10895-025-04432-6 PMID: 40616654

Shen, C., Wang, J., Cao, Y., & Lu, Y. (2015). Facile access to B-doped solid-state fluorescent carbon dots toward light emitting devices and cell imaging agents. *Journal of Materials Chemistry. C, Materials for Optical and Electronic Devices*, *3*(26), 6668–6675. DOI: 10.1039/C5TC01156F

Siwal, S. S., & Bishnoi, P. (2025). *Nanoscopic Wonders: Carbon Quantum Dots as Catalysts and Charge Carriers in Advanced Energy Storage Systems*. CleanMat., DOI: 10.1002/clem.70003

Sun, H., Ji, H., Ju, E., Guan, Y., Ren, J., & Qu, X. (2015). Synthesis of fluorinated and nonfluorinated graphene quantum dots through a new top-down strategy for long-time cellular imaging. *Chemistry (Weinheim an der Bergstrasse, Germany)*, *21*(9), 3791–3797. DOI: 10.1002/chem.201406345 PMID: 25614445

Sun, Y. P., Zhou, B., Lin, Y., Wang, W., Fernando, K. A. S., Pathak, P., Meziani, M. J., Harruff, B. A., Wang, X., Wang, H., Lu, P. G., Yang, H., Kose, M. E., Chen, B., Veca, L. M., & Xie, S. Y. (2006). Quantum-sized carbon dots for bright and colorful photoluminescence. *Journal of the American Chemical Society*, *128*(24), 7756–7757. DOI: 10.1021/ja062677d PMID: 16771487

Surendran, P., Lakshmanan, A., Sakthy Priya, S., Geetha, P., Rameshkumar, P., Kannan, K., Hegde, T. A., & Vinitha, G. (2021). Fluorescent carbon quantum dots from Ananas comosus waste peels: A promising material for NLO behaviour, antibacterial, and antioxidant activities. *Inorganic Chemistry Communications*, *124*, 108397. DOI: 10.1016/j.inoche.2020.108397

Surendran, P., Lakshmanan, A., Vinitha, G., Ramalingam, G., & Rameshkumar, P. (2020). Facile preparation of high fluorescent carbon quantum dots from orange waste peels for nonlinear optical applications. *Luminescence*, *35*(2), 196–202. DOI: 10.1002/bio.3713 PMID: 31591819

Surendran, P., Lakshmanan, A., Vinitha, G., Ramalingam, G., & Rameshkumar, P. (2020). Bioinspired fluorescence carbon quantum dots extracted from natural honey: Efficient material for photonic and antibacterial applications. *Nano-Structures & Nano-Objects*, *24*, 100589. DOI: 10.1016/j.nanoso.2020.100589

Tang, L., Ji, R., Cao, X., Lin, J., Jiang, H., Li, X., Teng, K. S., Luk, C. M., Zeng, S., Hao, J., & Lau, S. P. (2012). Deep ultraviolet photoluminescence of water-soluble self-passivated graphene quantum dots. *ACS Nano*, *6*(6), 5102–5110. DOI: 10.1021/nn300760g PMID: 22559247

Tavan, M., Yousefian, Z., Mirjalili, M. H., Bakhtiar, Z., Rahmandoust, M., & Mirjalili, M. H. (2025). Carbon quantum dots: Multifunctional fluorescent nanomaterials for sustainable advances in biomedicine and agriculture. *Industrial Crops and Products*, *231*, 121207. DOI: 10.1016/j.indcrop.2025.121207

Uriarte, D., Domini, C., & Garrido, M. (2019). New carbon dots based on glycerol and urea and its application in the determination of tetracycline in urine samples. *Talanta*, *201*, 143–148. DOI: 10.1016/j.talanta.2019.04.001 PMID: 31122404

Van Tam, T., Kang, S. G., Babu, K. F., Oh, E. S., Lee, S. G., & Choi, W. M. (2017). Synthesis of B-doped graphene quantum dots as a metal-free electrocatalyst for the oxygen reduction reaction. *Journal of Materials Chemistry. A, Materials for Energy and Sustainability*, *5*(21), 10537–10543. DOI: 10.1039/C7TA01485F

Varadarajan, S., Kavitha, A., Selvaraju, P., Muthu, S. E., Gurushankar, K., Shanmugan, S., & Kannan, K. (2024). Enhanced Photocatalytic Hydrogen Evolution by TiO2: A Synergistic Approach with Defect-Rich SnS2 and Ti3C2 MXene Cocatalysts. *Hydrogen*, *5*(4), 940–957. DOI: 10.3390/hydrogen5040050

Verma, C., Dubey, S., Alfantazi, A., & Rhee, K. Y. (2024). Heteroatoms-doped carbon dots: Fundamental, properties, coordination bonding and corrosion protection. *Journal of Industrial and Engineering Chemistry*, *133*, 90–111. DOI: 10.1016/j.jiec.2023.12.018

Wang, C., Sun, D., Zhuo, K., Zhang, H., & Wang, J. (2014). Simple and green synthesis of nitrogen-, sulfur-, and phosphorus-co-doped carbon dots with tunable luminescence properties and sensing application. *RSC Advances*, *4*(96), 54060–54065. DOI: 10.1039/C4RA10885J

Wang, C., Wu, X., Li, X., Wang, W., Wang, L., Gu, M., & Li, Q. (2012). Upconversion fluorescent carbon nanodots enriched with nitrogen for light harvesting. *Journal of Materials Chemistry*, *22*(31), 15522–15525. DOI: 10.1039/c2jm30935a

Wang, D., Wang, Q., Tan, M., Wang, S., Luo, S., Hou, P., Zhang, Y., Yan, S., & Liu, X. (2022). Biomass CQDs derivate carbon as high-performance anode for K-ion battery. *Journal of Alloys and Compounds*, *922*, 166260. DOI: 10.1016/j.jallcom.2022.166260

Wang, Q., Wang, H., Liu, D., Du, P., & Liu, P. (2017). Synthesis of flake-shaped nitrogen-doped carbon quantum dot/polyaniline (N-CQD/PANI) nanocomposites via rapid-mixing polymerization and their application as electrode materials in supercapacitors. *Synthetic Metals*, *231*, 120–126. DOI: 10.1016/j.synthmet.2017.06.018

Xia, C., Zhu, S., Feng, T., Yang, M., & Yang, B. (2019). Evolution and Synthesis of Carbon Dots: From Carbon Dots to Carbonized Polymer Dots. *Advancement of Science*, *6*(23), 1901316. DOI: 10.1002/advs.201901316 PMID: 31832313

Xu, J., Hou, K., Ju, Z., Ma, C., Wang, W., Wang, C., Cao, J., & Chen, Z. (2017). Synthesis and electrochemical properties of carbon dots/manganese dioxide (CQDs/MnO_2) nanoflowers for supercapacitor applications. *Journal of the Electrochemical Society*, *164*(2), A430–A437. DOI: 10.1149/2.1241702jes

Xu, J., Xue, Y., Cao, J., Wang, G., Li, Y., Wang, W., & Chen, Z. (2016). Carbon quantum dots/nickel oxide (CQDs/NiO) nanorods with high capacitance for supercapacitors. *RSC Advances*, *6*(7), 5541–5546. DOI: 10.1039/C5RA24192H

Xu, Q., Niu, Y., Li, J., Yang, Z., Gao, J., Ding, L., Ni, H., Zhu, P., Liu, Y., Tang, Y., Lv, Z. P., Peng, B., Hu, T. S., Zhou, H., & Xu, C. (2022). Recent progress of quantum dots for energy storage applications. *Carbon Neutrality*, *1*(1), 13. DOI: 10.1007/s43979-022-00002-y

Xu, X., Ray, R., Gu, Y., Ploehn, H. J., Gearheart, L., Raker, K., & Scrivens, W. A. (2004). Electrophoretic analysis and purification of fluorescent single–walled carbon nanotube fragments. *Journal of the American Chemical Society*, *126*(40), 12736–12737. DOI: 10.1021/ja040082h PMID: 15469243

Xu, Y., Wang, S., Hou, X., Sun, Z., Jiang, Y., Dong, Z., Tao, Q., Man, J., & Cao, Y. (2018). Coal-derived nitrogen, phosphorus and sulfur co-doped graphene quantum dots: A promising ion fluorescent probe. *Applied Surface Science*, *445*, 519–526. DOI: 10.1016/j.apsusc.2018.03.156

Yadav, P. K., Chandra, S., Kumar, V., Kumar, D., & Hasan, S. H. (2023). Carbon quantum dots: Synthesis, structure, properties, and catalytic applications for organic synthesis. *Catalysts*, *13*(2), 422. DOI: 10.3390/catal13020422

Zhang, B., Liu, C. Y., & Liu, Y. (2010). A novel one-step approach to synthesize fluorescent carbon nanoparticles. *European Journal of Inorganic Chemistry*, *4411–4414*(28), 4411–4414. Advance online publication. DOI: 10.1002/ejic.201000622

Zhao, Y., Zou, S., Huo, D., Hou, C., Yang, M., Li, J., & Bian, M. (2019). Simple and sensitive fluorescence sensor for methotrexate detection based on the inner filter effect of N, S co-doped carbon quantum dots. *Analytica Chimica Acta*, *1047*, 179–187. DOI: 10.1016/j.aca.2018.10.005 PMID: 30567648

Zhao, Z., & Xie, Y. (2017). Enhanced electrochemical performance of carbon quantum dots-polyaniline hybrid. *Journal of Power Sources*, *337*, 54–64. DOI: 10.1016/j.jpowsour.2016.10.110

Zheng, L., Chi, Y., Dong, Y., Lin, J., & Wang, B. (2009). Electrochemiluminescence of water-soluble carbon nanocrystals released electrochemically from graphite. *Journal of the American Chemical Society*, *131*(13), 4564–4565. DOI: 10.1021/ja809073f PMID: 19296587

Zhou, J., Booker, C., Li, R., Zhou, X., Sham, T. K., Sun, X., & Ding, Z. (2007). An electrochemical avenue to blue luminescent nanocrystals from multiwalled carbon nanotubes (MWCNTs). *Journal of the American Chemical Society*, *129*(4), 744–745. DOI: 10.1021/ja0669070 PMID: 17243794

Zhou, J., Shan, X., Ma, J., Gu, Y., Qian, Z., Chen, J., & Feng, H. (2014). Facile synthesis of P-doped carbon quantum dots with highly efficient photoluminescence. *RSC Advances*, *4*(11), 5465–5468. DOI: 10.1039/c3ra45294h

Zhou, J., Zhou, H., Tang, J., Deng, S., Yan, F., Li, W., & Qu, M. (2017). Carbon dots doped with heteroatoms for fluorescent bioimaging: A review. *Microchimica Acta*, *184*(2), 343–368. DOI: 10.1007/s00604-016-2043-9

Zhou, W., Zhuang, J., Li, W., Hu, C., Lei, B., & Liu, Y. (2017). Towards efficient dual-emissive carbon dots through sulfur and nitrogen co-doping. *Journal of Materials Chemistry. C, Materials for Optical and Electronic Devices*, *5*(32), 8014–8021. DOI: 10.1039/C7TC01819C

Zhu, H., Wang, X., Li, Y., Wang, Z., Yang, F., & Yang, X. (2009). Microwave synthesis of fluorescent carbon nanoparticles with electrochemiluminescence properties. *Chemical Communications*, *5118–5120*(34), 5118. Advance online publication. DOI: 10.1039/b907612c PMID: 20448965

Zou, S., Hou, C., Fa, H., Zhang, L., Ma, Y., Dong, L., Li, D., Hou, D., & Yang, M. (2017). An efficient fluorescent probe for fluazinam using N, S co-doped carbon dots from l-cysteine. *Sensors and Actuators. B, Chemical, 239*, 1033–1041. DOI: 10.1016/j.snb.2016.07.169

Zuo, K., Liu, W., Liu, X., & Liu, X. (2023). Phosphorescence of carbon dot: The intrinsic mechanism and recent progress. *Carbon Trends, 12*, 100278. DOI: 10.1016/j.cartre.2023.100278

Zuo, W., Tang, L., Xiang, J., Ji, R., Luo, L., Rogee, L., & Lau, S. P. (2017). Nitrogen-doped hollow carbon nanoparticles as efficient counter electrodes in quantum dot sensitized solar cells. *Applied Physics Letters, 110*(21), 211901. DOI: 10.1016/j.solener.2024.112699

Chapter 2

Advanced Synthesis Approaches for Carbon Quantum Dots:

Exploring Top-Down, Bottom-Up, and Novel Cutting-Edge Techniques

Subin Stephen
https://orcid.org/0009-0003-6411-7814
Karunya Institute of Technology and Sciences, India

Matbiangthew Shadap
https://orcid.org/0000-0002-9159-6668
Karunya Institute of Technology and Sciences, India

Sakunthala Ayyasamy
https://orcid.org/0000-0002-9945-6272
Karunya Institute of Technology and Sciences, India

ABSTRACT

Carbon Quantum Dots (CQDs), less than 10 nm carbon nanomaterials, exhibit high surface area, conductivity, tunable photoluminescence, low toxicity, cost-effective synthesis, and strong environmental compatibility, making them ideal for bioimaging, drug delivery, photovoltaics, photocatalysis, sensors, supercapacitors, and LEDs. CQD synthesis follows two routes: top-down and bottom-up. Top-down decomposes bulk carbon via arc discharge, laser ablation, or electrochemical exfoliation, offering size/surface control but demanding high energy and advanced tools. Bottom-up assembles CQDs from molecular precursors through carbonization or

DOI: 10.4018/979-8-3373-4247-4.ch002

pyrolysis (hydrothermal, microwave, thermal, template-assisted), enabling flexible precursor use and doping for structural/functional tuning. Emerging methods plasma treatment, biomass-derived green synthesis, and chemical vapor deposition enhance scalability, sustainability, and tailored properties. This chapter reviews recent advances, emphasizing principles, merits, and challenges driving next-generation CQD technologies.

INTRODUCTION

Carbon-based materials including carbon nanotubes, carbon nanofibers, carbon nanosheets, and graphene have received a lot of interest in recent years because of their structural, physical, and chemical characteristics (Lim et al., 2015). Carbon quantum dots (CQDs) have emerged as a potential new class of carbon-based nanomaterials (Y. Wang & Hu, 2014). CQDs were found unintentionally by Xu et al. in 2004 while purifying single-walled carbon nanotubes (SWCNT). However, at the time, they were referred to as fluorescent carbon nanoparticles rather than this term. Since then, substantial research efforts have enhanced our understanding and exact control of CQD characteristics, allowing for systematic examination and expanding their potential across a wide range of applications. In 2006, Sun *et al.* synthesized non-toxic carbon nanoparticles of different sizes using a laser ablation technique, which are now recognized as carbon quantum dots (P. Kumar et al., 2022). CQD are small, highly fluorescent nanomaterials that generally measure between 2 and 8 nm in size. They typically have a core-shell structure, with the carbon core measuring around 2 - 3 nm. According to studies, its core might be graphitic, crystalline, or amorphous (Dua et al., 2023). Synthesis approaches for CQD are divided into two categories: top-down and bottom-up. The top-down strategy uses physical or chemical approaches to break down bulk carbon compounds into small-sized CQD. In contrast, the bottom-up technique depends on a sequence of chemical events that polymerize and carbonize tiny organic molecules to generate CQD **(Figure 1)**.

Figure 1. Schematic illustration of CQD synthesis processes

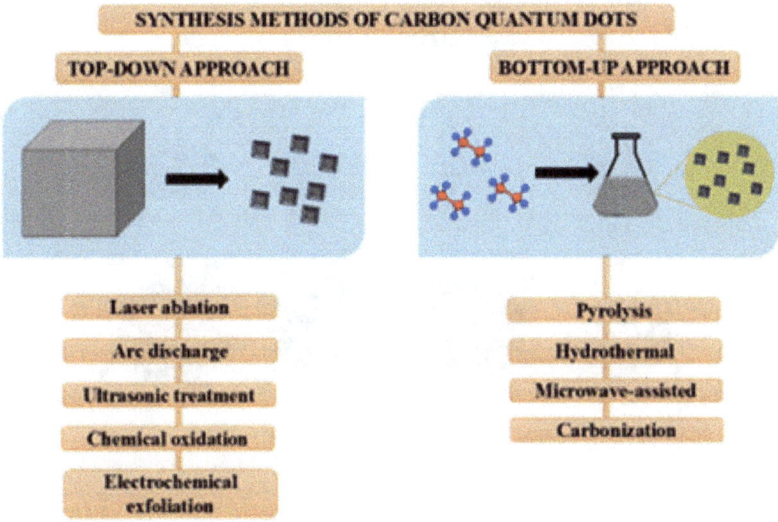

In addition to these standard techniques, innovative synthesis procedures have evolved to allow for the environmentally friendly, cost-effective, and scalable manufacture of CQD. These developing solutions focus on developing sustainable precursors and improved energy sources to enhance the structural and optical features of CQD, thereby broadening its potential for a wide variety of applications (Magesh et al., 2022). CQD is classified into three forms based on their structure: graphene quantum dots (GQD), carbon nanodots (CND), and polymer dots (PD) (**Figure 2**). GQDs are composed of graphene layers with different chemical groups attached to their edges, which are usually derived from the carbon-based reactants employed in production. CNDs have a spherical shape and can be crystalline, graphite-like or amorphous, like other carbon nanoparticles. In contrast, PD can be formed as cross-linked polymers or as polymer chains clustered around a spherical carbon core (Koutsogiannis et al., 2020).

Figure 2. Classification of carbon dots

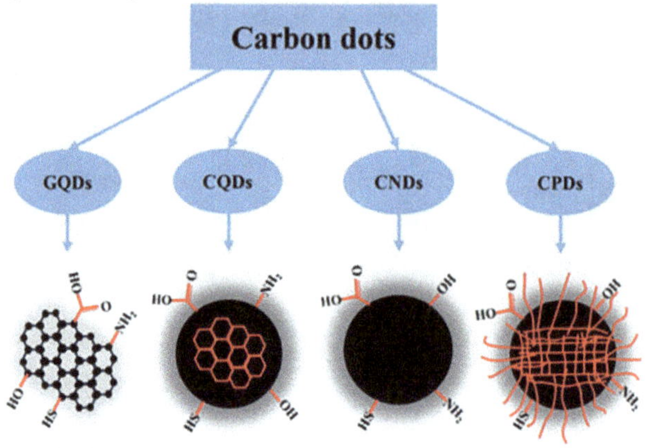

(G. Li et al., 2023) ©2024 The Authors.

In comparison, semiconductor quantum dots produce strong and controllable fluorescence emissions that are important for biosensing and bioimaging. However, their usage is limited by the high toxicity of the heavy metals used in their manufacture (Lim et al., 2015). Because of their superior characteristics, these benefits have prompted the development of CQD as prospective alternatives for semiconductor quantum dots (Surendran et al., 2021a). Typically, CQD is formed by functionalizing carbon surfaces with organic or polymeric molecules. Furthermore, CQD's abundance of surface functional groups (such as -OH and -NH_2) provides various active sites and anchoring points, making them excellent for constructing multi-component, high performance composites (L. Tian et al., 2021). CQD features and characteristics can vary significantly based on a number of parameters, including the synthesis method's time, pH, precursor selection, solvent employed, and precursor composition. CQD exhibits a range of advantageous optical, structural, chemical, and biocompatibility features (Mohammadpour et al., 2024; Kannan et al., 2023; Kannan et al., 2022; Kannan et al., 2021), including low toxicity, strong photostability, multiphoton excitation, and great dispersibility, making them appealing for diverse applications (Dua et al., 2023). There are three major obstacles to consider when preparing for CQD. First, carbonaceous aggregation can develop during carbonization; however, this problem can be addressed by employing procedures such as electrochemical synthesis and pyrolysis (Kannan et al., 2024). Second, regulating the size and establishing uniformity in CQD is critical for maintaining consistent characteristics and allowing valid fundamental research; this may be achieved by post-treatment procedures like as centrifugation and dialysis. Third, surface characteristics can be modified during synthesis or by post-synthesis treatment (Koutsogiannis et al., 2020).

This chapter offers a thorough assessment of advanced synthesis techniques for CQDs, including both traditional top-down and bottom-up approaches as well as developing novel methodologies **(Table 1)**. It demonstrates how new techniques provide fine control over CQD's structural and optical features while improving scalability and environmental sustainability. This paper provides a complete overview for researchers looking for a CQD with specialized features for specific applications in energy storage, sensing, and medicinal domains.

Table 1. Comparative analysis of top-down, bottom-up, and new approaches

Aspect	Top-Down Approach	Bottom-Up Approach	Novel Methods
Principle	Breaking down bulk carbon materials into nanoscale CQDs (H. Li et al., 2012a)	Building CQDs from molecular or atomic precursors (Baker & Baker, 2010)	Advanced or hybrid methods integrating precision, scalability, and sustainability (Das et al., 2024)
Methods	Laser ablation, arc discharge, electrochemical exfoliation, ultrasonic treatment (H. Li et al., 2012b; Y. P. Sun et al., 2006)	Hydrothermal/solvothermal, microwave-assisted synthesis, thermal decomposition, carbonization (Dong et al., 2013; Zhao et al., 2015)	Plasma-assisted synthesis, ultrasonic spray pyrolysis, microreactor synthesis, green synthesis using biomass (Chahal et al., 2021a; Duong et al., 2025; L. Yan et al., 2016)
Size Control	Limited, broad distribution (Baker & Baker, 2010)	Better control over size and morphology (Das et al., 2024)	Precise size and morphology control through advanced process parameters (Das et al., 2024)
Yield	Generally low to moderate (H. Li et al., 2012a)	Generally higher yield (Baker & Baker, 2010)	Moderate to high yield with process optimization (H. Li et al., 2012a)
Crystallinity	Can produce highly crystalline CQDs (H. Li et al., 2012a)	Amorphous or crystalline depending on method (Das et al., 2024)	Crystalline or defect-engineered CQDs achievable (H. Li et al., 2012a)
Scalability	Challenging for large-scale production (Das et al., 2024)	More suitable for scale-up (H. Li et al., 2012a)	Designed for scalability (Das et al., 2024)
Environmental Impact	Cleaner, fewer reagents (H. Li et al., 2012a)	May require solvents/acids (Baker & Baker, 2010)	Emphasis on green solvents, lower emissions and sustainability (Das et al., 2024)

TOP-DOWN APPROACH

This method includes converting bulk materials into much tiny, nano-sized particles. While top-down approaches are often regarded as simple and easy to use, they frequently failed to generate particles with extremely tiny dimensions or irregular morphologies. One of the primary disadvantages of this approach is the difficulty of accurately managing both particle size and shape, which can impact the homogeneity, surface characteristics, and overall performance of the produced nanomaterials (Abid et al., 2022).

Laser Ablation

The laser is used as an external energy source to ablate solid target materials in laser ablation. In this approach, a strong laser beam is focused on a precise location of a solid surface, causing the light-absorbing substance to evaporate. The word "ablation" refers to the removal of surface atoms and includes both single-photon and multi-photon excitation techniques (Kim et al., 2017). This approach produces high-purity nanoparticles because the purity of the resultant particles is mostly controlled by the purity of the target material and the medium, which helps to reduce contamination from the reactor (Semaltianos, 2010). Following the synthesis of CQD via laser ablation, the prepared nanomaterials typically undergo processing to facilitate subsequent characterization and device integration. For example, (Cui et al., 2020) synthesized CQD via laser ablation of low-cost carbon fabric utilizing an orthogonal pulsed dual-beam laser system, with the procedure carried out in dimethyl sulfoxide (DMSO) and ethyl acetate (EA), to generate DMSO-CQD and EA-CQD. The Pulsed laser ablation (PLA) was conducted using a Q-switched Nd: YAG laser operating at 10 Hz with a wavelength of 1064 nm, a pulse width of 6 ns and a pulse energy of 20 mJ for a duration of 10 min. In this work a single laser beam is divided into two laser beams for shortening laser ablation time an improving the yield in the preparation process of CQD. Notably, the CQD exhibit improved photoluminescence quantum yield of 35.4%, surpassing that of CQD synthesized using a single-beam pulsed laser ablation. The implementation of a dual beam laser system for the synthesis of CQD represents a promising advancement, offering enhanced ablation efficiency and improved photoluminescence quantum yield. (Cortes et al., 2024) synthesized CQD using the pulsed laser fragmentation in liquid (PLFL) technique. In this method, short laser pulses (ranging from10^{-3} –10^{-8} s) are directed at a solid bulk target submerged in a liquid medium such as acetone, deionized water and DMF. During laser irradiation, the majority of the energy is absorbed by the solid target, resulting in the formation of a plasma plume. The interaction between the plasma plume and the surrounding liquid generates a cavitation bubble, which undergoes rapid expansion and collapses ultimately leading to the release of nanoparticles into the liquid. Compared to PLA, the PLFL method enables the production of CQD with sizes down to the nanometer scale, uniformly dispersed in the liquid medium (**Figure 3**). The particle size reduction is primarily governed by two mechanisms: (1) **Photothermal vaporization**, which occurs when the temperature of the particles exceeds the boiling point of the bulk material, causing material ejection. (2) **Coulomb explosion**, wherein the ejection of electrons during ionization induces high internal charge repulsion within the nanoparticles, leading to spontaneous fragmentation into nanoscale particles. The PLFL method is an efficient way to produce CQD with tiny and homogeneous particle sizes. It uses high-energy laser pulses to fracture a solid target submerged in a liquid media, removing the need for dangerous chemical reagents. Despite its benefits, the PLFL approach has certain drawbacks, including the need for expensive laser equipment, significant energy consumption, and difficulties in scaling up the process for larger-scale production (X. Ren et al., 2019) prepared nitrogen doped, microporous carbon quantum dot (NM-CQD) with high quantum yield and dual-wavelength emission, employing a bio-mass-derived precursor by a PLA process **Figure 3**. Waste *Platanus* biomass was first carbonized at 600 °C, followed by activated at 900 °C with KOH under a nitrogen atmosphere to yield porous carbon powder. This powder was then dispersed in 5 mL of formamide, and the resulting mixture was subjected to laser ablation using a Nd:YAG pulsed laser with a wavelength of 1064 nm, repetition rate of 10 Hz, pulse energy of ~20 mJ, and pulse duration between 3-6 ns, sustained for approximately 30 minutes. The ablated solution was centrifuged at 10,000 rpm, resulting in a yellow solution containing NM-CQD. Notably, the NM-CQD displayed two distinct and strong photoluminescence emission bands in the indigo – blue wavelength region, attributable to excited electrons transitioning from the intrinsic π^* orbital to surface states created by saturation passivation. The quantum yield of these NM-CQD reached 32.4%, surpassing that of undoped carbon dots. This enhancement indicates that the incorporation of nitrogen-containing functional groups effectively improves quantum yield, demonstrating the benefits of nitrogen doping for optical performance. PLA employing a dual beam configuration can significantly reduce processing time, enhance material yield, and improve the photoluminescence quantum yield of CQD. In contrast pulsed laser fragmentation in liquid facilitates the synthesis of uniformly dispersed nanoscale CQD through mechanism such as photothermal vaporization and coulomb explosion; however, this method is constrained by the need for high energy expensive laser systems and presents challenges in scalability. Furthermore, nitrogen doping via PLA using biomass-derived porous carbon precursors has been shown to enhance optical performance, producing dual-wavelength emission and a high quantum yield. Despite these advantages, laser-based synthesis approaches remain limited by substantial equipment costs, elevated energy consumption and difficulties associated with large-scale production.

Figure 3. (a) Synthesis process of N-doped CQD derived from biomass, (b-g) Confocal microphotographs of live HeLa cells, L02 cells, and macrophage cells without NM-CQD incubation, (h) UV-Visible spectra of CQD, (i-j) PL dependent spectra of CQD, (k) PL profile comparison between CQD and oxidized CQD and (l) PL independent profile of CQD.

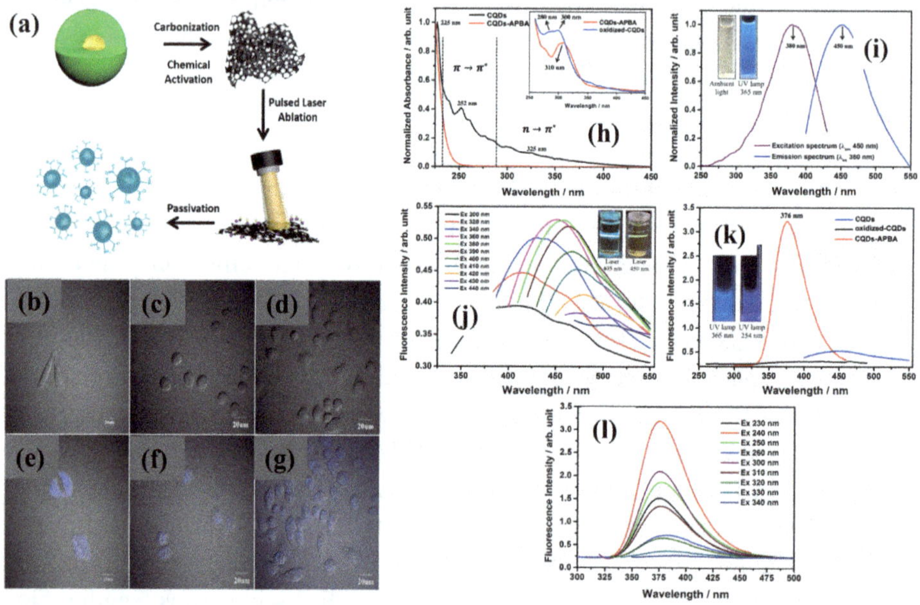

(a), (b-g) (X. Ren et al., 2019) ©2019 The Authors; (h), (i-j), k, (l) (Cortes et al., 2024) ©2024 The Authors.

Arc Discharge

Arc discharge is defined as a method that produces a continuous plasma discharge through electrical breakdown of gas in a nonconductive environment, typically using a graphite electrode in an inert medium. This process results in the sublimation of carbon and the synthesis of carbon nanotubes and fullerenes temperature exceeding 4000 °C (J. Yan et al., 2012). In arc discharge method, once the electric arc is initiated, a gaseous bubble forms around the arc within the liquid medium. The internal pressure of this bubble typically ranges between 0.5-1 atm. The gas-liquid interface of the bubble function as a temporary container wall. When water is used as the liquid medium, the bubble primarily consists of carbon monoxide, carbon vapour, hydrogen and water vapour (Hernandez-Tabares et al., 2021). Only a limited number of studies have reported the preparation of CQD using the arc discharge method. Notably, (Chao-Mujica et al., 2021) synthesized CQD using the submerged arc discharge in water (SADW) method. The experimental setup consisted of two pure graphite electrodes (cathode and anode) with a density of 1.69 g/cm^3, diameters of 6-12 mm, and lengths of 20 – 100 mm. The electrodes were placed in distilled water inside a rectangular stainless-steel chamber. During operation, the arc discharge produced a gas bubble around the arc that acted as a microreactor, isolating the reaction zone from the surrounding environment. An electric stabilization mechanism was used to maintain the arc stable, and the discharge was maintained until the anode was completely consumed, which normally took 40 - 60 minutes. To avoid overheating, the water temperature was carefully managed to stay below 40 °C. Following synthesis, the products were divided into three different fractions: floating material, suspended material, and precipitated material. The suspended fraction was primarily composed of CQDs, with a minor presence of graphene oxide sheets. The floating material was first removed from the water surface, after which the solution was left undisturbed for 24 hours to allow larger particles to settle. Finally, the supernatant containing the suspended fraction was carefully decanted from the precipitate.

The solvent-free nature of this process and its inherent product fractionation represent clear advantages, simplifying the recovery of CQDs without the need for chemical reagents. Nevertheless, the method presents several limitations, including low selectivity, broad particle size distribution, and the generation of by-products such as graphene oxide, which complicates purification. Additionally, scalability is constrained by continuous electrode consumption, high energy demand, and the need to strictly regulate the water temperature below 40 °C during operation. Compared to chemical or hydrothermal routes, SADW provides less control over particle size uniformity, surface functionalization, and doping, which limits the ability to optimize CQD properties for specific applications.

Ultrasonic Treatment

Ultrasonic treatment is a widely employed technique for synthesizing well-dispersed nanomaterials with fine particles size. The incorporation of ultrasonic treatment into the homogeneous precipitation process facilitates uniform cation distribution, thereby improving the structural and compositional homogeneity of the resulting material (Lu et al., 2010). In most reports, ultrasonic methods have been employed for nanoparticles synthesis, while some studies have demonstrated their use in reducing particle sized to approximately 10 μm. The underlying mechanism of particle refinement through ultrasound is primarily attributed to acoustic radiation forces and the cavitation effect induced by ultrasonic waves (G. Yang et al.,

2021). The effectiveness of ultrasonic treatment in promoting uniform nanoparticle synthesis has been widely recognized. In this study (Huang et al., 2018) synthesized poly (ethylene oxide) – functionalized CQD from cigarette ash using ultrasonic treatment. Initially, PEG-SH was prepared by reacting PEG with mercaptoacetic in the presence of DCC and DMAP in anhydrous toluene at 45 °C for 24 h under a nitrogen atmosphere. The resulting product was precipitated with diethyl ether, centrifuged (6000 rpm, 8 min) and washed to obtain pure PEG-SH. Subsequently, cigarette ash and PEG-SH were ultrasonically treated for 30 mL of DMF in 30 min. The supernatant was obtained via centrifugation, dialyzed with distilled water, and concentrated using rotary evaporation to yield the final product. The incorporation of PEG not only enhanced the water dispersibility of the CQD but also provides excellent biocompatibility. Comprehensive characterization confirmed that the materials exhibited uniform nanosized morphology, strong fluorescence stability, excellent aqueous dispersibility and favourable biocompatibility. (Hoang Huy & Bark, 2024) prepared one-step, alkali-assisted ultrasonic process using glucose as the precursor. The glucose solution was ultrasonicated (300 W, 40 kHz, 4 h) in the presence of NaOH, followed by ethanol addition and neutralization with HCL. Post-treatment with $MgSO_4$ facilitated salt and water removal, yielding purified CQD (**Figure 4**). This combination step is advantageous as it integrates carbonization, purification and stabilization within a relatively mild and low-cost process, underscoring its scalability and environmental compatibility. (Lu & Zhou, 2019) ultrasonic-assisted synthesis of dopamine-derived NCQDs offers an efficient and relatively mild strategy, utilizing dopamine hydrochloride as a dual carbon–nitrogen precursor to achieve intrinsic doping without the need for external additives. High-power ultrasonication (600 W, 20 kHz) generates cavitation-induced localized heating and pressure fluctuations, thereby accelerating nucleation and carbonization in DMF, while the use of an ice bath provides effective thermal regulation to prevent uncontrolled decomposition. Post-synthesis purification through centrifugation and dialysis successfully removes aggregates and low-molecular-weight impurities, resulting in a stable aqueous dispersion of NCQDs with high dispersibility. Nevertheless, the prolonged irradiation time (8 h) raises concerns regarding energy efficiency and large-scale applicability, while the reliance on toxic DMF as the reaction medium poses challenges for environmental sustainability and biomedical translation. Furthermore, dependency on cavitation-driven nucleation may contribute to variations in particle size distribution and surface functionalities, emphasizing the importance of optimizing reaction parameters and investigating greener solvent systems to improve reproducibility, scalability, and environmental compatibility. Overall, ultrasonic-assisted synthesis offers several benefits, including gentle reaction conditions, precise control over particle size, improved dispersibility, and the ability to easily functionalize. Dependence on cavitation-driven nucleation frequently causes unpredictability in size distribution and surface chemistry, while lengthy processing durations and excessive energy usage restrict scaling. In addition, the frequent use of toxic solvents such as DMF raises concerns regarding environmental sustainability and biomedical applicability. Therefore, although ultrasonic-assisted synthesis represents a promising strategy for producing high-quality CQDs, future research should prioritize the optimization of sonication parameters, the adoption of green or solvent-free, and the integration of ultrasonication with other energy-efficient approaches (Surendran et al., 2024).

Figure 4. (a) Schematic illustration of the UVC photodetector (UVCPD) structure, (b) cross-sectional SEM image of the optimized device, (c) photocurrent response at zero bias under varying light intensities, (d) spectral responsivity of the self-powered photodetector, (e) detectivity under different applied biases, (f) rise and decay time profiles, (g) time-dependent photocurrent of the device under UVC and UVA illumination (0.6 mW cm^{-2}), and (h) long-term stability over 50 on/off cycles (4 s illumination per cycle)

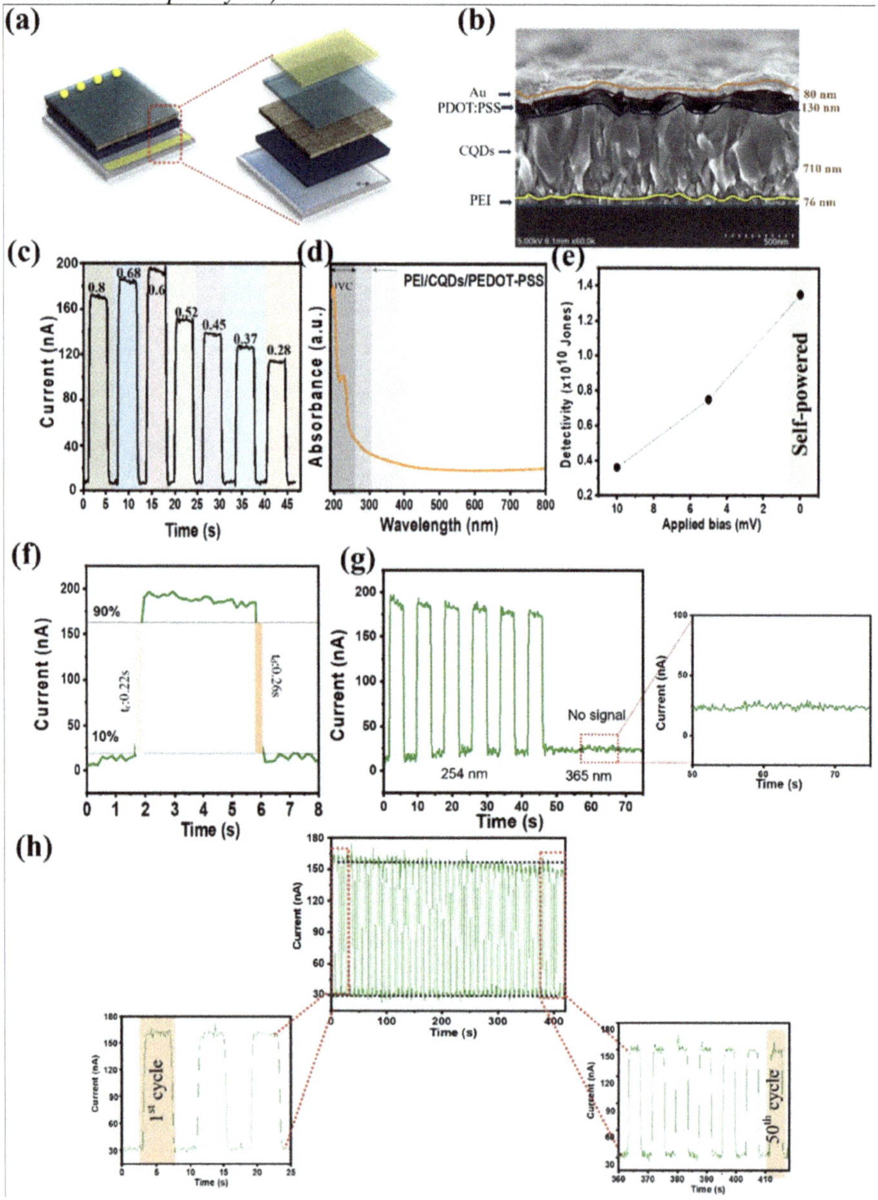

(Hoang Huy & Bark, 2024) ©2024 The Authors.

Chemical Oxidation

The chemical oxidation method utilizes bulk carbon materials, such as graphite and carbon fibers, as raw precursors and breaks them down into nanoscale CQDs by controlling oxidation conditions. This process typically involves harsh treatments with strong oxidizing agents, such as nitric acid, sulfuric acid, or their mixtures, which effectively convert both large and small molecular structures into CQDs. Owing to its simplicity, this method does not require specialized equipment and can accommodate a wide range of readily available carbon sources, making it practical and scalable. However, the dependent on effective oxidants increases synthesis costs and poses possible environmental problems (Kong et al., 2024). (Tan et al., 2019) report a systematic study on the preparation of multi-coloured carbon quantum dots via $HNO_3/HClO_4$ oxidation of graphitized activated carbon, emphasizing the influence of precursor microstructure on fluorescence properties. A key strength of this work lies in the materials selection strategy. By employing coconut shell–derived activated carbon subjected to high-temperature graphitization (1500-2500 °C). This precursor-dependent approach highlights the inheritance effect, wherein highly graphitized carbons yield CQDs with ordered sp_2 domains, reduced self-absorption quenching, and tunable emission (yellow to green). The study thereby advances understanding of structure–property correlations in CQDs, which is often overlooked in top-down oxidation methods. The use of a mixed $HNO_3/HClO_4$ oxidant under microwave heating significantly reduces reaction time and enables efficient etching of highly graphitized carbons, even those treated at 2500 °C. This underscores the effectiveness of strong oxidants for accessing high-yield CQDs (up to 11%) from carbon sources. Moreover, the incorporation of post-reduction using $NaBH_4$ demonstrates a materials modification pathway, where surface carbonyl and epoxy groups are converted to hydroxyls, thereby enhancing quantum yield (from 3.9% to 8.0%)

Importantly, the CQDs exhibit low cytotoxicity and strong cellular uptake, confirming their promise for bioimaging applications. The use on highly corrosive oxidants ($HNO_3/HClO_4$) raises safety, cost, and environmental sustainability concerns, which could hinder scale-up and green applicability (Surendran et al., 2021b). Although graphitization improves structural order and photoluminescence stability, it simultaneously reduces CQD yield, suggesting a better crystallinity and production efficiency. Additionally, while $NaBH_4$ reduction enhances quantum yield, its scalability and environmental impact are questionable compared to greener reduction routes.

Table 2. Performance of top-down synthesis methods for CQD.

Method	Particle Size (nm)	Quantum Yield (QY, %)	Reaction Time (min)	Yield (%)	Highlights	References
Laser ablation	~ 3 – 5	~ 35.4	10 – 30	~ 10 – 15	High crystallinity, costly setup, scalable issues	(Cui et al., 2020)
Pulsed Laser Fragmentation (PLFL)	2 – 6	~20 – 30	10 – 20	~ 12	Produces uniform CQD, high energy demand	(Cortes et al., 2024)
Arc discharge	2 – 10	5 – 10	40 – 60	~ 5 – 8	Solvent-free, but broad size distribution	(Chao-Mujica et al., 2021)
Ultrasonic treatment	3 – 8	~ 3.6	30 – 4h	~ 10 – 12	Good dispersibility, but energy-intensive	(Lu et al., 2010)
Chemical oxidation	3 – 6	~ 3.9 – 8	1 – 2h	~ 11	Strong oxidants	(Tan et al., 2019)

BOTTOM-UP APPROACH

The bottom-up approach is an additive method where smaller units serve as building blocks that combine to form nanostructures. Some techniques based on bottom-up approaches include the sol-gel method, chemical vapour deposition, plasma spraying, micro-emulsion. Bottom-up approach is also described as a constructive technique, since it involves the formation of an assembly using smaller units (Escudero et al., 2021).

Pyrolysis

Pyrolysis is a thermochemical conversion process that degrades agriculture residues into its constituent elements by heating in the absence of oxygen to produce end products such as biochar, bio-oil, and non-condensable gaseous mixture. Various operating parameters such as reactor configuration, heating rate, operating temperature, and agricultural residue particle size, significantly influence the composition of end products (K. P. Kumar & Remya, 2023). Pyrolysis can be classified into several categories, such as carbonization, conventional, fast, flash-liquid, flash-gas, ultra, vacuum, hydro pyrolysis, and methanopyrolysis, based on process environment, parameters and products obtained (Yadav & Jain, 2024). (Vibhute et al., 2022) prepared CQDs using a one-pot pyrolysis method. Anhydrous glucose powder was taken in a beaker, placed on a hot plate, and heated at 100 °C for 10 min.

The powder changed to a brown residue, which was cooled to room temperature. The residue was then stirred with distilled water for 10 min at 500 rpm. For CQD, 500 μL of poly-L-lysine was added to the CQD solution, followed by magnetic stirring at 27 °C and 400 rpm for 1 h. The resulting solution was stored at 4 °C for further use. The synthesized CQD possess a bright blue fluorescence and a quantum yield of 19.35% (**Table 3**). (M. Tian et al., 2018) synthesized N-CQD via pyrolysis using edetic acid and acrylamide. The mixture was heated to 300° C for 12 min in a heating mantle, producing a black solution. This solution was then subjected to ultrasonic treatment in acetone for 30 min. Finally, the ark brown solution was centrifuged and filtered. (P. Li et al., 2020) synthesized CQD from lysine and arginine via a pyrolytic process at 240° C, followed by purification step such as sonicated, centrifugation and dialysis, which produced amino-functionalized CQD. The approach is widespread for its simplicity, scalability, and environmental friendliness. However, the synthesis route lacks precise control over particle size distribution and consistency, which limited its use. The studied literature demonstrates a broad range of techniques for carbon quantum dot (CQD) synthesis, each with different benefits and inherent constraints. (Rao et al., 2019) developed a two-step pyrolytic route combined with optically active surface passivation, employing tartaric or citric acid as carbon precursors and D/L-penicillamine as chiral ligands. This approach effectively incorporated chirality into CQDs, expanding their applications in chiral catalysis and enantioselective sensing. Nonetheless, the dependence on pure ligands and precisely controlled pyrolytic conditions presents substantial scaling and cost problems. Alternatively, (Brachi, 2020) explored a catalyst-assisted thermal conversion of agro-industrial residues using γ-alumina spheres produces luminous CQDs at low temperatures (250-300 °C) while yielding biofuel by-products. This dual valorisation pathway underscores its sustainability potential, though dependence on fixed-bed catalytic systems constrains industrial upscaling. (R. Ren et al., 2019) reported a straightforward one-step biomass pyrolysis process, yielding luminescent CQDs with superior surfactant properties, particularly effective in Pickering emulsion stabilization. Despite its cost-effectiveness and suitability for bulk production, the optical performance of the CQDs was highly sensitive to pyrolysis temperature, and achieving precise size control remained problematic.

Biomass-derived pyrolysis has attracted particular attention for its alignment with green chemistry principles. (Guerrero-Gonzalez et al., 2023) converted *agave bagasse*, a by-product of the tequila industry, into CQDs that displayed strong UV–blue luminescence. When applied in greenhouse studies, these CQDs enhanced seed germination and plant growth by transforming harmful ultraviolet radiation into photosynthetically useful blue light. Such findings highlight the promise of pyrolytic CQDs in agriculture, although particle size variation often reduces reproducibility in optical behavior. Pyrolysis has also proven effective in valorising non-biomass waste streams. (Parra et al., 2020) demonstrated the conversion of waste tires into graphene-like CQDs with a high photoluminescence quantum yield of nearly 19%, suitable for applications in optoelectronics and fluorescence imaging. However, residual metallic impurities and heterogeneous morphologies necessitate careful post-synthetic purification. In the environmental sector has also benefited from pyrolytic CQDs. Ahlawat et al. reported the facile synthesis of CQDs that were successfully employed for the photocatalytic degradation of Rhodamine B, indicating the method ability to produce functional nanomaterials with catalytic activity. However, the surface chemistry of these CQDs is highly dependent on precursor type and processing conditions, making it difficult to ensure consistency across batches. Hybrid pyrolytic approaches further demonstrate the versatility of this method. (Shang et al., 2018) described the one-pot pyrolysis of citric acid to produce CQD/graphene oxide composites with superior lubricating properties, reducing friction and wear. These results point to the adaptability of pyrolytic synthesis in designing multifunctional nanostructures, though balancing the formation of CQDs with other carbon phases remains a critical challenge. Beyond undoped systems, metal-doped CQDs synthesized through pyrolysis have shown enhanced catalytic behavior. (L. Wang et al., 2025) reported that Co-, Ni-, and Mn-doped CQDs significantly reduced the activation energy of oil shale pyrolysis, thereby increasing oil yields. While such doping methods demonstrate how surface engineering may guide catalytic pathways, the use of transition metals raises toxicity concerns, limiting their applicability in medicinal domains despite obvious benefits in energy-related applications.

Overall, pyrolytic synthesis offers a scalable, low-cost, waste-to-resource conversion approach for CQDs, with proven applications in agriculture, catalysis, environmental remediation, lubrication, and energy recovery. However, constraints such as variable particle size, unpredictable surface chemistry, and the risk of contaminants remain significant barriers. Addressing these challenges through careful precursor design, efficient thermal processing, and post-synthetic functionalization will be critical for progressing pyrolytic CQDs in high-value applications such as biomedicine and sophisticated electronics.

Microwave-Assisted

Recent research underscores the versatility, efficiency, and eco-friendliness of microwave-assisted synthesis for CQD, highlighting its rapid, cost-effective, and scalable nature across a wide range of applications. This method enables precise control over particle size, surface chemistry, and photoluminescent properties by carefully choosing precursor materials and reaction conditions. For example, (Prekodravac et al., 2019) produced nitrogen-doped CQDs by exposing an aqueous glucose solution to microwave irradiation for just one minute at low power (~100-200

W) and ~100 °C, achieving highly active nanodots with remarkable pH-dependent photocatalytic performance for organic dye degradation. Under microwave conditions, nitrogen-rich CQDs are formed from a combination of citric acid and thiourea before being integrated with TiO_2 nanosheets to improve charge separation and considerably increase CO_2 photoreduction efficiency. In another green approach, (Başoğlu et al., 2020) used roasted chickpeas as a natural carbon source, applying 350 W microwave pyrolysis for 120 seconds to produce photostable, water-soluble CQDs suitable for Fe^{3+} ion sensing. (Gyulai et al., 2019) conducted a comparative study of 5 min microwave heating versus 16 h solvothermal refluxing, revealing that microwave synthesis delivers smaller, more uniform particles with tunable surface polarity and optimized heteroatom doping, thereby influencing optical properties and cellular uptake behaviours. (Ko et al., 2018) leveraged silk fibroin protein as a biopolymer precursor to create biocompatible, blue-emitting CQDs under short microwave irradiation, while Architha et al. employed a microwave-assisted reflux using Mexican mint extract to generate crystalline CQDs with bright blue emission and excellent Fe^{3+} sensing capability, alongside low cytotoxicity in bioimaging assays. Similarly, (Kamarol Zaman et al., 2021) prepared CQDs from empty fruit bunch (EFB) biochar through a simple microwave process, coupling experimental characterization with density functional theory (DFT) modeling to better understand their molecular structure, fluorescence behavior, and sensitivity to Cu^{2+} ions. (Pajewska-Szmyt et al., 2020) advanced the field by introducing sulfur and nitrogen co-doping via microwave irradiation of citric acid combined with thiourea or glutathione, resulting in CQDs with enhanced fluorescence for highly sensitive mercury detection in environmental samples. More recently, (Salman, 2023) synthesized Cu-N doped CQDs from plum juice and copper sulfate, achieving a high quantum yield for pharmaceutical sensing, while (Nazar et al., 2024) utilized coffee-ground-derived activated carbon exposed to 5-10 minutes of microwave heating to produce CQDs capable of selectively detecting Fe^{3+}, Pb^{2+}, and Cr^{3+} ions through fluorescence quenching (**Figure 5**). (Hagiwara et al., 2021) developed a low-pressure microwave solvothermal (MWS) method for the synthesis of monodispersed, yellow-emitting carbon quantum dots (CQDs) ranging from 2 to 9 nm in size. This approach achieved exceptional crystallinity and quantum yields of up to 61.1%, significantly outperforming the traditional autoclave solvothermal (CAS) method by reducing reaction time and enhancing luminescence efficiency through surface functionalization with polyethylene glycol (PEG-6000). In a similar effort to promote greener chemistry, (Ramezani et al., 2018) introduced a one-pot microwave-assisted synthesis using quince fruit powder as a natural carbon source. Their method, optimized through a Taguchi orthogonal array design that adjusted temperature, microwave power, and irradiation time, produced small-sized CQDs (~4.85 nm) with multicolour fluorescence. However, the relatively low quantum yield (~8.5%) indicated the need for further optimization, such as enhanced surface passivation or heteroatom doping.

Figure 5. (a) Schematic representation of CQD synthesis through microwave heating, (b) UV–vis absorption spectrum alongside the fluorescence intensity of CQDs, (c) fluorescence response of CQDs under different excitation wavelengths, (d) Tauc plot analysis for direct band gap estimation, and (e) normalized fluorescence intensity of CQDs at various excitation wavelengths

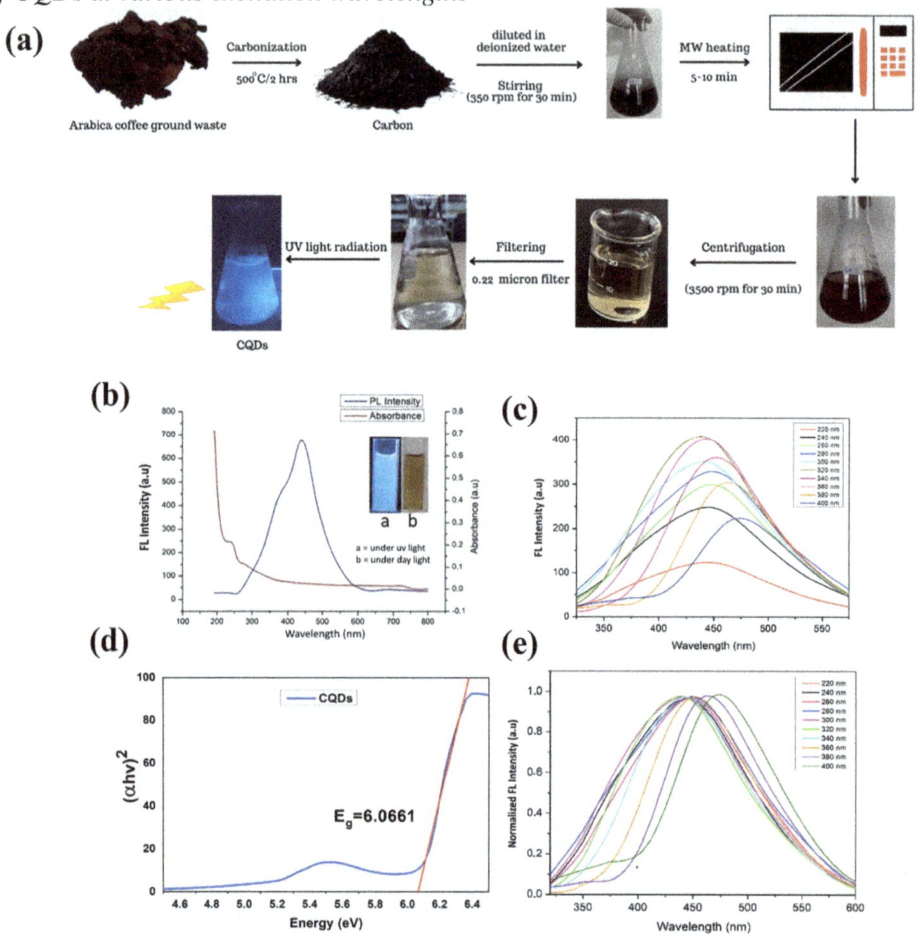

(Nazar et al., 2024) ©2024 The Authors.

Pushing the boundaries of speed and efficiency, (Dager et al., 2020) reported an innovative microwave plasma-enhanced decomposition (MPED) process that generated highly crystalline, monodispersed CQDs in just five minutes using fenugreek seeds as a sustainable carbon source. Unlike conventional pyrolysis or thermal decomposition, MPED utilizes high-energy electrons and hydrogen ions within a plasma field to drive rapid carbonization, achieving superior crystallinity, a narrow size distribution, and tunable dual-mode photoluminescence. This approach not only demonstrated remarkable efficiency but also showed strong potential for scalable industrial production. Several studies have also focused on heteroatom doping to further enhance CQD performance. (Zhou et al., 2019) synthesized nitrogen and sulfur co-doped CQDs using a rapid microwave-assisted oil-bath method with citric acid and glutathione as precursors. Their approach yielded CQDs with an exceptional quantum yield of 76.5%, high photobleaching resistance, and saline stability, making them ideal for advanced sensing and imaging applications. Likewise, (Salman et al., 2023) reported a green, one-pot microwave synthesis of nitrogen-doped CQDs (N@ CQDs) from *Eruca sativa* leaves, achieving a quantum yield of 41.39% in just four minutes at 700 W. These N@CQDs showed good sensitivity for modafinil detection using fluorescence in pharmaceutical formulations and human plasma, demonstrating CQDs' analytical flexibility while adhering to sustainable synthesis principles.

(Janus et al., 2019) expand the extent of functionalization. Microwave-assisted production of poly(L-lysine)-based CQD composites resulted in biocompatible, water-soluble nanomaterials with adjustable fluorescence, antioxidant activity, and the capacity to detect heavy metal ions. However, their study noted the challenge of residual byproducts affecting the accuracy of optical characterization, emphasizing the importance of purification. (Ali et al., 2021) took a different approach, pioneering in-situ microwave growth of nitrogen-doped CQDs on multiwalled carbon nanotubes (MWCNTs). This direct-growth technique eliminated complex multistep processing, reduced synthesis time and energy consumption, and produced highly conductive hybrids that enhanced electrocatalytic activity in dye-sensitized solar cells (DSSCs) by approximately 50% compared to unmodified MWCNTs. Similarly, (Magdy et al., 2023) demonstrated the rapid microwave-assisted pyrolysis of orange juice and urea to produce nitrogen-doped CQDs with a narrow particle size distribution (2-5 nm) and a quantum yield of 29.3%. These CQDs were successfully applied as fluorescent nano sensors for the detection of palbociclib and exhibited low cytotoxicity, supporting their use in cellular imaging and potential cancer diagnostic applications.

Collectively, these findings demonstrate that microwave-assisted synthesis not only provides superior control and reproducibility compared to conventional methods but also aligns with the principles of green chemistry, paving the way for broader applications in catalysis, sensing, and bioimaging. Future work, however, should focus on standardizing synthesis protocols and deepening the mechanism understanding of CQD luminescence to further enhance performance and scalability (Lakshmanan et al., 2024).

Carbonization

Carbonization is a thermal technique that increases the carbon content of a material by processing the precursor in an environment devoid of oxygen, effectively eliminating non-carbon elements from its structure. The lack of oxygen is essential to prevent unwanted carbon oxidation and to minimize the creation of by-products like water and other gaseous substances. Generally, carbonization results in a carbon material with a relatively smaller surface area, pore volume, and pore size compared to its activated versions, although these features still significantly influence the final product properties. The process outcome is largely determined by crucial operational factors, such as the carbonization temperature, the inert gas flow rate, the heating rate, and the duration of the treatment (Malini et al., 2023). (Chai et al., 2019) employed a straightforward one-step HTC process, treating finely ground sugarcane bagasse (200 mesh) with water under varying liquid–solid ratios (4:1-8:1), temperatures (140-180 °C), and residence times (0-4 h), followed by rapid cooling. The supernatant containing sugars and quantum dots was centrifuged and dialyzed, while the remaining solid residue was impregnated with phosphoric acid and carbonized at 500 °C under nitrogen to produce porous carbon. This approach achieved fermentable sugar yields of up to 41.85% and generated graphene quantum dots (~2.26 nm) exhibiting bright blue fluorescence, though the absence of statistical optimization or kinetic modelling limited opportunities for efficiency enhancement (**Table 3**). (Zhu et al., n.d.-a) refined this strategy by integrating a mild acidolysis step using o-amino benzenesulfonic acid prior to HTC, enabling selective lignin fractionation. Subsequent HTC at 200 °C for 12 h, followed by filtration, dialysis (1000 Da), and freeze-drying, yielded green CQDs (G-CQDs) with a 30.6% yield and blue CQDs (B-CQDs) with a 15.2% yield. These CQDs exhibited tunable fluorescence (green emission around 475 nm and blue around 400 nm), with their optical properties supported by quantum-chemical modelling; however, the introduction of acidolysis added complexity and raised potential environmental considerations. In contrast, (Gomes et al., 2019a) adopted a design-driven, statistically optimized approach, employing a 3^2 full factorial design and response surface methodology (RSM) to optimize precursor ratios of chitin/chitosan (1:1, 1:3, and 1:7) with graphite (0.06-0.1 g) in ethanol. The mixture underwent HTC at 200 °C for 6 h, followed by centrifugation to separate the CQDs. This method enabled precise control over band-gap tuning (3.16-3.38 eV) and achieved a quantum yield of approximately 17.1%, demonstrating strong reproducibility. However, the study provided limited mechanistic insights into the synthesis process and did not fully address the scalability of the method for industrial applications.

Hydrothermal

Hydrothermal synthesis is a crystal growth technique that relies on the solubility of minerals in hot, pressurized water. One of its key advantages over conventional methods is its ability to produce crystalline phases that are unstable at their melting points, making it highly suitable for growing large, high-quality crystals with precise control over their composition. In this technique, a sealed vessel is used to heat water or another solvent to temperatures much over its boiling point, resulting in significant autogenous pressure. When water is employed as the solvent, the technique is known as hydrothermal synthesis; for other solvents, it is known as solvothermal synthesis. Despite its efficiency, the approach has certain drawbacks. These include the expensive cost of autoclave equipment and the impossibility to directly see crystal formation throughout the development phase (Sonawane et al., 2018). (Baragau et al., 2020) explored the use of Continuous Hydrothermal Flow Synthesis (CHFS), a rapid and scalable bottom-up technique in which biomass precursors, such as glucose or citric acid, are rapidly carbonized in supercritical water under continuous flow conditions. This approach generates uniform, blue-emitting CQDs (~2-3 nm) with excitation-independent luminescence and excellent stability, making them ideal for Cr(VI) and Fe(II) sensing. CHFS is not only faster, with reactions conducted in milliseconds, but it is also environmentally harmless than typical batch processes. However, the requirement for high-pressure equipment and the limited ability to fine-tune surface functionalities without additional post-synthesis modifications remain key challenges. In contrast, (Q. Yang et al., 2018) and (Wu et al., 2025) utilized a simpler one-pot hydrothermal carbonization approach, another bottom-up synthesis method, employing low-cost biomass sources such as *Eichhornia crassipes* and distillers grains (**Figure 6**). This process leverages naturally occurring carbon and nitrogen content in biomass, resulting in an eco-friendly and economical synthesis route. The CQDs obtained are typically 1.7-2.8 nm in size, with good water solubility and biocompatibility. However, these materials exhibit lower quantum yields (QY, ~7%) and broader size distributions compared to those synthesized via CHFS. Moreover, the reaction temperature (110-220 °C) plays a critical role in defining the optical properties: higher temperatures enhance fluorescence intensity but simultaneously reduce the co-produced biochar yield. (Zhu et al., 2023) proposed a mild acidolysis-assisted hydrothermal carbonization strategy that combines top-down and bottom-up pathways to achieve efficient lignin valorisation. In this method, lignin is first partially hydrolysed into smaller fragments, which are then subjected to hydrothermal carbonization to generate two types of CQDs simultaneously: green-emitting CQDs with yields up to 30.6% and blue-emitting CQDs with yields of about 15.2%. This dual-pathway approach not only maximizes biomass conversion efficiency but also offers deeper mechanistic insights into CQD formation and emission behavior.

Figure 6. **Schematic illustration of the one-step hydrothermal process for the simultaneous synthesis of CDs and BC**

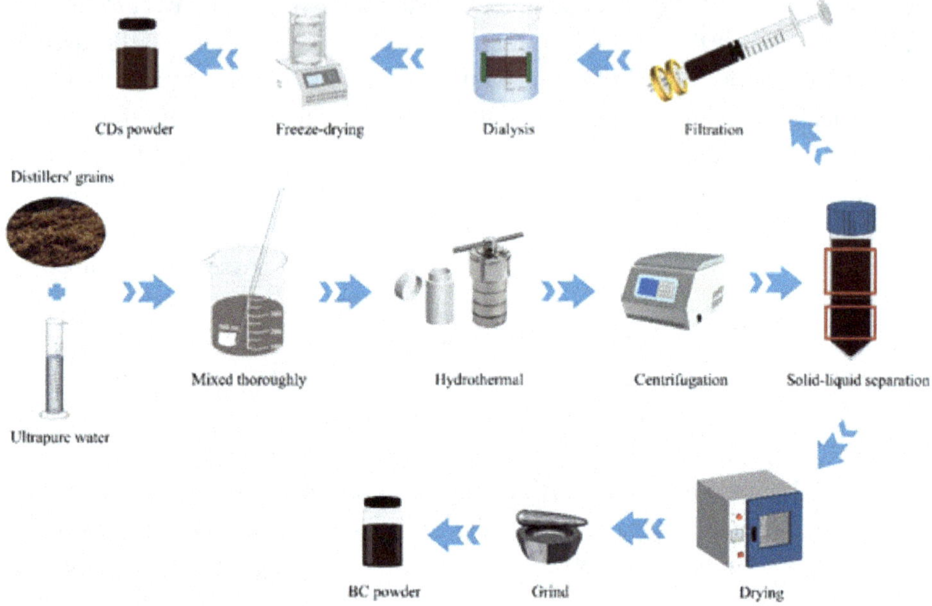

(Wu et al., 2025) ©2024 The Authors.

Nitrogen doping is another widely adopted strategy to enhance CQD performance. (Vercelli et al., 2021) synthesized N-doped CQDs through a hydrothermal reaction of citric acid and urea, revealing that specific nitrogen configurations such as graphitic, pyrrolic, and pyridinic nitrogen significantly influence the electrochemical and optical characteristics of the CQDs, enabling their tunability for optoelectronic and catalytic applications. Similarly, (Aghamali et al., 2018) utilized citric acid and diethylenetriamine (DETA) in a bottom-up hydrothermal synthesis to produce N-doped CQDs with exceptionally high quantum yields (~95%) and remarkable photostability, making them highly effective for photocatalytic dye degradation. Nevertheless, nitrogen doping introduces challenges, particularly in maintaining precise control over dopant type and distribution, which can hinder reproducibility and consistency in large-scale production. Other heteroatom doping techniques have also been reported. (Hong et al., 2019) developed fluorine-doped CQDs (F-CQDs) through a one-step hydrothermal route using perfluorooctanoic sulfonate (PFOS), achieving excitation-independent down- and up-conversion luminescence along with high sensitivity for Fe^{3+} detection. Although this method improves fluorescence performance, the environmental and toxicological issues linked to PFOS raise concerns about its potential for widespread and sustainable use (Gao et al., 2022) demonstrated an innovative application by integrating CQDs into Ni-based catalysts via an in-situ hydrothermal synthesis. The abundant surface functional groups of the CQDs facilitated uniform metal dispersion within the catalyst structure, significantly improving catalytic activity and stability in CO methanation processes. (Smagulova et al., 2019) synthesized carbon quantum dots (CQDs) from glucose using a conventional hydrothermal method, where glucose was dissolved in distilled water, sealed in a Teflon-lined autoclave, and heated at 180 °C for 6 hours. The CQDs produced displayed a mix of amorphous and crystalline structures, enriched with numerous surface functional groups, which resulted in their vivid green luminescence. While this method is both economical and simple, it offers limited control over the uniformity of particle size and the ability to tune emissions, thus limiting its potential for precise functional optimization. Building on this foundation, (Shen et al., 2021) employed hydrothermal carbonization using both glucose and citric acid as precursors, integrating the resulting CQDs with TiO_2 to enhance photocatalytic activity. In this synthesis, aqueous precursor solutions were heated in an autoclave at 200 degrees Celsius for 8 hours. Their findings revealed that the selection of precursor is critical in determining the crystallinity and charge transfer behavior of CQDs, with citric acid-derived CQDs exhibiting superior charge separation and photocatalytic efficiency when compared to the glucose-derived counterparts.

Similarly, (Sabet & Mahdavi, 2019) adopted a green hydrothermal synthesis to prepare nitrogen-doped CQDs (N-CQDs) from grass biomass. In this procedure, powdered grass was dispersed in deionized water and hydrothermally treated at 200 °C for 10 hours. The resultant N-CQDs demonstrated significant photoluminescence, great water dispersibility, and high efficacy in removing organic dyes and heavy metals like Pb^{2+} and Cd^{2+} from wastewater. However, establishing homogeneous nitrogen incorporation and maintaining regular surface passivation are substantial hurdles that hinder reproducibility. (Slewa, 2024) developed lemon- and onion-derived CQDs (LCQDs and OCQDs) through a one-step green hydrothermal synthesis for application in antifungal food packaging. Fresh lemon and onion juices were centrifuged to eliminate heavier particles, then 75 mL of each extract was placed in a 100 mL Teflon-lined stainless-steel autoclave and heated to 200 °C for 6 hours. After cooling, the samples were filtered, centrifuged, and freeze-dried, yielding CQDs with average particle sizes of 4.3 nm for LCQDs and 3.3 nm for OCQDs. The smaller OCQDs exhibited enhanced antifungal activity, significantly reducing fungal growth and slowing strawberry spoilage and weight loss during storage, highlighting their potential in food preservation technologies.

Table 3. Performance of bottom-up synthesis methods for CQD.

Method	Particle Size (nm)	Quantum Yield (QY, %)	Reaction Time (min)	Yield (%)	Highlights	References
Hydrothermal (citric acid + DETA)	2 – 5	~ 95	6 – 10	~ 15 – 30	Excellent QY, stable and biocompatible	(Aghamali et al., 2018)
Hydrothermal (grass biomass)	~ 4 – 5	~ 30.6	8 – 10h	~ 20	Green route, good for wastewater treatment	(Zhu et al., 2023)
Pyrolysis (glucose amino acids)	3 – 6	~ 19	10 – 20	~ 15	Simple, scalable, less control on size	(Vibhute et al., 2022)
Pyrolysis (biomass)	2 – 8	10 – 25	30 – 60	~ 20 – 25	Sustainable, low cost, variable size	(Parra et al., 2020)
Microwave assisted	2 – 6	~ 61.1	1 – 5	~ 25 – 30	Fast, eco-friendly, reproducible	(Hagiwara et al., 2021)

Overall, hydrothermal synthesis remains the leading, scalable, and environmentally friendly method for CQD production, offering versatility in precursor selection, doping strategies, and reaction parameters. This approach facilitates the production of CQDs with tunable optical and surface properties, but challenges persist in achieving precise particle size control, consistent heteroatom doping, and reproducibility in large-scale production. Future research should focus on optimizing reaction conditions, improving dopant precision, and enhancing emission tunability to expand the applicability of CQDs in industrial processes, catalysis, advanced sensing, and biomedical imaging.

NOVEL SYNTHESIS METHOD

High-performance CQD have become highly demanded and these demands pose a challenge to the current limitations of high production yield, size control, and tunability, among other possible conditions of the traditionally conducted synthesis approaches. Such novel strategies encompass green synthesis, in where renewable or waste-derived precursors are used to produce sustainable processes; chemical vapor deposition (CVD), where the process provides tight control over structure and abatement is uniform; and plasma-based synthesis, where rapid and solvent free synthesis is possible. Collectively, these approaches represent scalable and efficient pathways to CQDs with improved optical features, robustness, and functional capabilities that can expand their utility in sensing, catalysis, and biomedical applications.

Green Synthesis

Carbon quantum dots (CQDs) green synthesis is a green sustainable, and environmentally clean method based on natural biomass or renewable starting material synthesized under mild reaction conditions, without using harmful reagents. CQDs have also been synthesized using walnut oil via one step hydrothermal method by (Arkan et al., 2018). In this method, walnut oil was mixed with distilled water in a Teflon-lined autoclave and heated at 200 °C in 6 h, producing ~12 nm with a QY of 14.58%, which were applied to bioimaging and biosensing purposes because of the outstanding fluorescence stability (**Table 4**). Nevertheless, the small degree of control over particle size and functional surface features land them low in their transfer applicability to precision optoelectronics. Equally, (Arumugham et al., 2020) used the Hydrothermal carbonization method on a leaf extract of Catharanthus roseus at 180 °C and 8 hours, resulting in smaller CQDs (~5 nm) and high QY (~28.2 percent), exceptional aqueous solubility, and high biocompatibility. These characteristics generated the CQDs which proved to be exquisite in multi-ion sensing and bioimaging but the variation in chemical composition of the natural precursor embedded batch-to-batch incongruency in fluorescence. Building on these methods, (Chahal et al., 2021b) studied green synthesis methods in the literature such as the modes of hydrothermal, microwave-assisted, and ultrasonic-assisted. They stressed that, the renewable precursors like fruit extracts, amino acids, agricultural wastes can be used, which is consistent with the 12 Principles of Green Chemistry, ensuring cost-effective production, high sustainability, and non-toxicity, as well as scalability. However, there are still outstanding issues such as comparatively low QYs and difficulties with managing particle size and surface characteristics, as well as inconsistencies due to the complexity of biomass feedstocks. One such issue was alleviated by (Salman et al., 2025) who used microwave-assisted hydrothermal synthesis of ultra-small CQDs (~2.8 nm) with high QY (~37.1%) and good stability, based on Prunus armeniaca (apricot) juice. Remarkably, heating at 700 W only 5 minutes was sufficient.

CQD produced by (Mindivan & Göktaş, 2023) using rosehip fruit extracts under aqueous hydrothermal carbonization using a Teflon-lined autoclave at 180 °C 8 hours. This synthesis produced ultra-small CQDs (~5 nm) with a favorable QY (~23.8%) and great biocompatibility, making them suitable for biological applications. In a comparable manner, (Pandiyan et al., 2020) transformed sugarcane bagasse waste into fluorescent CQDs through a green hydrothermal process at 200 °C in 6 hours of incubation, producing nanomaterials with interesting nonlinear optical characteristics and considerable antimicrobial activity, which signify their multifunctionality. A different strategy was presented by (Shahba & Sabet, 2022) who developed a Ball-milling-aided green synthesis, which involved crushing pine fruit biomass, dispersing it in deionized water, and carbonizing it at 180-200 °C, produced nitrogen-doped CQDs with stronger fluorescence and high heavy metal adsorption capacity, making them useful as environmental cleaners. To continue green processing, (S. Sun et al., 2022) synthesized the nitrogen-doped CQDs using a water-based hydrothermal process and backing it up with Box-Behnken statistical optimization that enhanced the quantum yield (QY), ~12.2 percent) and produced high sensitivity in the detection of Fe^{3+} (detection limit 0.44 mM). green surface modifications, such as passivation with amino acids, enhanced photoluminescence and photostability, broadening applicability to sensing and biomedical fields. Overall, these green synthesis methods, hydrothermal, ball-milling, or solvent-based, show the effectiveness of a cost-effective scalable and biocompatible production route of CQDs.

Plasma-Based Synthesis

Plasma-based green synthesis of CQD has rapidly emerged as a clean, efficient, and highly tunable approach for producing high-quality nanomaterials under mild, solvent-free, and environmentally friendly conditions. (Weerasinghe et al., 2022) showed a room-temperature air plasma discharge technique, where carbon precursors are bombarded by an atmospheric plasma jet roughly 20 minutes. The simple filtrated and freeze-dried CQDs are characterized by nanoscale size (~3 nm) and narrow size distribution but tunable light at blues or blue-green. Although this approach is cost-effective and reagent-free, quantum yields (QY ~1.5%) are relatively low and optimization of plasma parameters is required to optimize optical efficacies.

Similarly, (Hoa et al., 2025) carried out the micro plasma-aided synthesis of biomass-derived CQDs by subjecting citrus extract in water to a low-voltage (~1.6 kV) micro plasma discharge to perform about 15 minutes at temperatures below 60 °C. This is a fast (<12 h) and scalable method that produced blue-green emitting CQDs that showed promising anti-bacterial activity and were highly sensitive to detecting Cd^{2+}. The quantum yields were, however, moderate and the surface functional control was poor. (R. Wang et al., 2024) also studied continuous-flow plasma jet synthesis, in which high-frequency plasma was used to treat carbon-containing precursor solutions, thus allowing the production of uniform CQDs in copious quantities with adjustable optical characteristics that could be applied in photocatalysis and energy devices. The limitations of the previous studies by incorporating low-pressure methane/argon plasma reactors, which facilitated controlled carbonization and controlled surface functionalization resulting in highly luminophore CQDs with the level of photon emission matching what would be needed by applications in optoelectronics. Nevertheless, the experimental setup remains complex and costly, limiting practical scalability.

Table 4. Performance of novel synthesis methods for CQD.

Method	Particle Size (nm)	Quantum Yield (QY, %)	Reaction Time (min)	Yield (%)	Highlights	References
Green synthesis	2 – 5	~ 86	6 – 10	~ 15 – 30	Excellent QY, stable and biocompatible	(V. B. Kumar et al., 2024)
Plasma-based	~ 4 – 5	~25 – 30	8 – 10h	~ 20	Green route, good for wastewater treatment	(Alaqel et al., 2023)
CVD	3 – 6	~ 17.1	10 – 20	~ 15	Simple, scalable, less control on size	(Gomes et al., 2019b)

COMPARATIVE ANALYSIS

Summary Comparison Analysis

The synthesis methods of Carbon Quantum Dots (CQDs) can be classified into top-down, bottom-up, and novel approaches, where each one of them has its advantages and also limitations:

Top-Down Approaches

Methods like laser ablation, arc discharge, ultrasonic exfoliation, and chemical oxidation are used to break down bulk carbon materials into nanoscale quantum dots. These techniques can create highly crystalline CQDs with adjustable surface properties. Nonetheless, they face challenges such as low yield, high energy consumption, limited control over particle size, and restricted scalability.

Bottom-Up Approaches

Traditional techniques such as hydrothermal/solvothermal processes, pyrolysis, carbonization, and microwave-assisted preparation begin with molecular or polymeric materials and yield CQDs under specific conversion conditions. These methods enable control over particle size, shape, and heteroatom doping to suit particular applications. Moreover, they are generally cost-effective and capable of high throughput, except for certain protocols that require toxic solvents or acidic environments, which can limit their environmental compatibility.

New and Emerging Methods

Recent strategies such as green synthesis, plasma-assisted synthesis, and chemical vapor deposition (CVD) focus on improving sustainability, surface functional flexibility, and efficiency. These approaches are especially useful in applications requiring customized functionalities or low-contamination conditions. Nonetheless, they still face challenges including moderate-to-low quantum yields, process complexity, and batch-to-batch variability, particularly when natural biomass feedstocks are employed.

Criteria for Selecting a Synthesis Method

Considerations of Costs

Cheaper routes involve hydrothermal synthesis, pyrolysis and green biomass-based routes that use cheap starting materials and modest instrumentation.

Laser ablation, plasma-based synthesis that involves use of specialized reactors, and CVD belong to the high-cost pathways category due to their operating costs.

Scalability

Methods that support large-scale production include continuous hydrothermal flow synthesis (CHFS), microwave-assisted carbonization, and bulk pyrolysis, all of which enable consistent quality at high volumes. By comparison, pulsed laser fragmentation in liquids (PLFL) and ultrasonic-assisted synthesis are limited by high energy demands and their inherent restriction to small-batch processing.

Environmental Impact

Green synthesis approaches are increasingly valued for sustainable nanomaterial production, particularly those employing biomass-derived feedstocks, solvent-free conditions, or energy-efficient methods (**Figure 7**). Plasma-assisted synthesis (requiring minimal reagents and producing low emissions) and microwave-assisted processes (characterized by shorter reaction times and reduced waste) are also promising. In contrast, acid-based oxidation using HNO_3 or H_2SO_4, as well as pyrolysis involving metal dopants, can generate hazardous by-products that necessitate stringent waste treatment protocols.

Figure 7. Comparative analysis of Top-down, Bottom-up and novel method

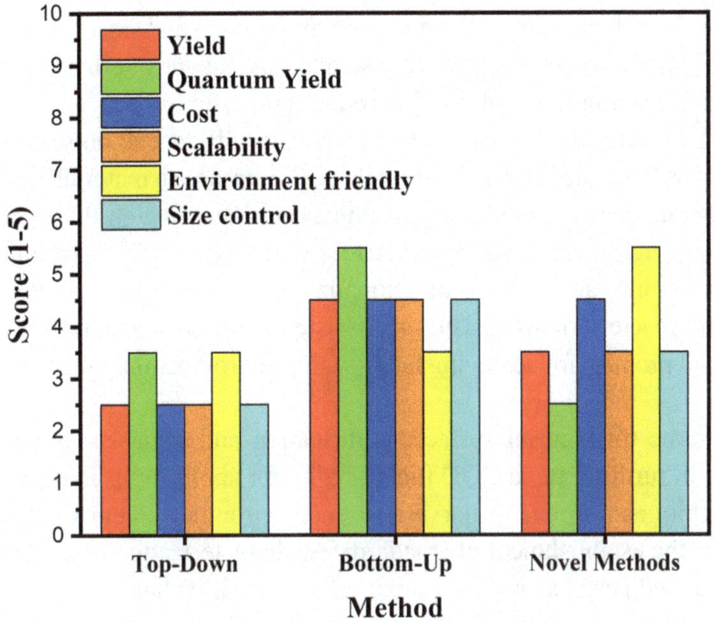

CHALLENGES AND FUTURE OUTLOOK

Despite the remarkable progress achieved in the synthesis and functionalization of CQD, several critical challenges remain unresolved, limiting their widespread application. A major issue is the lack of consistent control over size, crystallinity, and surface chemistry across different synthesis routes. This often results in batch-to-batch variability and unpredictable optical or electronic properties.

Green and plasma-assisted methods are appealing due to their sustainability and eco-friendly nature; however, they generally suffer from low quantum yields (QYs), limited photostability, and poor emission control, restricting their applicability in high-performance domains such as optoelectronics and targeted bioimaging. Furthermore, the fundamental mechanisms governing nucleation, growth, and surface-state evolution remain poorly understood, particularly in bottom-up and plasma-based processes, hindering the rational design of optimized synthesis strategies.

From an industrial standpoint, top-down methods (e.g., laser ablation) are constrained by low throughput and high energy demands, while plasma and chemical vapor deposition (CVD) techniques face challenges related to complex reactor systems and high operational costs. In addition, chemical oxidation methods commonly employed in oil spill cleanup and certain hydrothermal processes often rely on strong acids and toxic solvents, raising environmental and safety concerns. Similarly, biomass-based syntheses, while sustainable, frequently suffer from feedstock heterogeneity, leading to variability in product properties.

Looking forward, the future of CQD synthesis will depend on scalable, reproducible, and sustainable approaches that do not compromise material performance. Integrating computational modelling, machine learning, and high-throughput experimental optimization can enable predictive control over particle size distribution, heteroatom doping, and surface functionalization. Advances in continuous-flow hydrothermal reactors, optimized plasma-assisted synthesis, and microwave-assisted methods hold promise for achieving large-scale, environmentally responsible CQD production.

At the same time, novel surface modification and multi-heteroatom doping strategies will further extend CQD functionality for applications in photocatalysis, biosensing, bioimaging, energy harvesting, and environmental remediation. Equally important is the establishment of standardized characterization protocols, toxicity assessments, and performance benchmarks, which will enhance safety, reliability, and cross-laboratory comparability, thereby accelerating translation from laboratory-scale research to industrial and clinical applications.

Ultimately, realizing the full potential of CQDs as next-generation nanomaterials will require strong collaborations between academia, industry, and regulatory bodies to overcome current bottlenecks and foster responsible, large-scale deployment.

CONCLUSION

Carbon quantum dots represent a new generation of multifunctional nanomaterials with unique optical, electronic, and biocompatible properties, enabling rapid integration into diverse applications such as sensing, bioimaging, energy conversion, catalysis, and environmental remediation. This review highlights the progressive development of CQD synthesis, spanning top-down, bottom-up, and emerging methods, each with distinct strengths and limitations. Top-down approaches yield highly crystalline CQDs but are restricted by high energy demands and limited scalability. In contrast, bottom-up strategies particularly hydrothermal, microwave-assisted, and pyrolytic methods offer greater flexibility in controlling size, morphology, and doping while maintaining cost-effectiveness and scalability. Meanwhile, green and plasma-assisted methods are steering the field toward sustainable, tunable, and environmentally responsible synthesis.

Despite these advances, challenges remain in achieving uniform structural control, reproducibility, high quantum yields, and scalable eco-friendly processes. Future directions will likely rely on the integration of design, process optimization, and standardized protocols to bridge the gap between laboratory research and industrial manufacturing. Additionally, advances in green synthesis, surface functionalization, and heteroatom engineering are expected to further enhance CQD performance and broaden their application spectrum. As collaborations between academia, industry, and regulatory agencies strengthen, CQDs hold immense potential to serve as key enablers of next-generation nanotechnologies, offering innovative solutions for pressing global challenges in healthcare, sustainable energy, and environmental management.

REFERENCES

Abid, . (2022). Synthesis of nanomaterials using various top-down and bottom-up approaches, influencing factors, advantages, and disadvantages: A review. In *Advances in Colloid and Interface Science* (Vol. 300). Elsevier B.V., DOI: 10.1016/j.cis.2021.102597

Aghamali, , Khosravi, M., Hamishehkar, H., Modirshahla, N., & Behnajady, M. A. (2018). Synthesis and characterization of high efficient photoluminescent sunlight driven photocatalyst of N-Carbon Quantum Dots. *Journal of Luminescence*, *201*, 265–274. DOI: 10.1016/j.jlumin.2018.04.061

Alaqel, , Alqahtani, A. S., Alharbi, A., Althobaiti, Y. S., Bamaga, A. K., Algarni, M. A., Almrasy, A. A., & Almalki, A. H. (2023). Spectrofluorometric quantitative analysis of aripiprazole based on quenching of natural derived carbon quantum dots in spiked human plasma. *Scientific Reports*, *13*(1), 21048. Advance online publication. DOI: 10.1038/s41598-023-47392-2 PMID: 38030673

Ali, , Riaz, R., Anjum, A. S., Sun, K. C., Li, H., Ahn, S. J., Jeong, S. H., & Ko, M. J. (2021). Microwave-assisted ultrafast in-situ growth of N-doped carbon quantum dots on multiwalled carbon nanotubes as an efficient electrocatalyst for photovoltaics. *Journal of Colloid and Interface Science*, *586*, 349–361. DOI: 10.1016/j.jcis.2020.10.098 PMID: 33187666

Arkan, , Barati, A., Rahmanpanah, M., Hosseinzadeh, L., Moradi, S., & Hajialyani, M. (2018). Green synthesis of carbon dots derived from walnut oil and an investigation of their cytotoxic and apoptogenic activities toward cancer cells. *Advanced Pharmaceutical Bulletin*, *8*(1), 149–155. DOI: 10.15171/apb.2018.018 PMID: 29670850

Arumugham, , Alagumuthu, M., Amimodu, R. G., Munusamy, S., & Iyer, S. K. (2020). A sustainable synthesis of green carbon quantum dot (CQD) from Catharanthus roseus (white flowering plant) leaves and investigation of its dual fluorescence responsive behavior in multi-ion detection and biological applications. *Sustainable Materials and Technologies*, *23*, e00138. Advance online publication. DOI: 10.1016/j.susmat.2019.e00138

Baker, , & Baker, G. A. (2010). Luminescent Carbon Nanodots: Emergent Nanolights. *Angewandte Chemie International Edition*, *49*(38), 6726–6744. DOI: 10.1002/anie.200906623 PMID: 20687055

Baragau, , Power, N. P., Morgan, D. J., Heil, T., Lobo, R. A., Roberts, C. S., Titirici, M.-M., Dunn, S., & Kellici, S. (2020). Continuous hydrothermal flow synthesis of blue-luminescent, excitation-independent nitrogen-doped carbon quantum dots as nanosensors. *Journal of Materials Chemistry. A, Materials for Energy and Sustainability*, *8*(6), 3270–3279. DOI: 10.1039/C9TA11781D

Başoğlu, , Ocak, Ü., & Gümrükçüoğlu, A. (2020). Synthesis of Microwave-Assisted Fluorescence Carbon Quantum Dots Using Roasted–Chickpeas and its Applications for Sensitive and Selective Detection of Fe^{3+} Ions. *Journal of Fluorescence*, *30*(3), 515–526. DOI: 10.1007/s10895-019-02428-7 PMID: 32152829

Brachi., (2020). Synthesis of fluorescent carbon quantum dots (CQDs) through the mild thermal treatment of agro-industrial residues assisted by γ-alumina. *Biomass Conversion and Biorefinery*, *10*(4), 1301–1312. DOI: 10.1007/s13399-019-00503-4

Chahal., ... (2021b). Green synthesis of carbon dots and their applications. In *RSC Advances* (Vol. 11, Issue 41, pp. 25354–25363). Royal Society of Chemistry. DOI: 10.1039/D1RA04718C

Chahal, , Macairan, J.-R., Yousefi, N., Tufenkji, N., & Naccache, R. (2021a). Green synthesis of carbon dots and their applications. *RSC Advances*, *11*(41), 25354–25363. DOI: 10.1039/D1RA04718C PMID: 35478913

Chai, , He, H., Fan, H., Kang, X., & Song, X. (2019). A hydrothermal-carbonization process for simultaneously production of sugars, graphene quantum dots, and porous carbon from sugarcane bagasse. *Bioresource Technology*, *282*, 142–147. DOI: 10.1016/j.biortech.2019.02.126 PMID: 30852334

Chao-Mujica, F. J., Garcia-Hernández, L., Camacho-López, S., Camacho-López, M., Camacho-López, M. A., Reyes Contreras, D., Pérez-Rodríguez, A., Peña-Caravaca, J. P., Páez-Rodríguez, A., Darias-Gonzalez, J. G., Hernandez-Tabares, L., Arias de Fuentes, O., Prokhorov, E., Torres-Figueredo, N., Reguera, E., & Desdin-García, L. F. (2021). Carbon quantum dots by submerged arc discharge in water: Synthesis, characterization, and mechanism of formation. *Journal of Applied Physics*, *129*(16), 163301. Advance online publication. DOI: 10.1063/5.0040322

Cortes, , Falomir, E., Lancis, J., & Mínguez-Vega, G. (2024). Pulsed laser fragmentation synthesis of carbon quantum dots (CQDs) as fluorescent probes in non-enzymatic glucose detection. *Applied Surface Science*, *665*, 160326. Advance online publication. DOI: 10.1016/j.apsusc.2024.160326

Cui, L., Ren, X., Wang, J., & Sun, M. (2020). Synthesis of homogeneous carbon quantum dots by ultrafast dual-beam pulsed laser ablation for bioimaging. *Materials Today Nano, 12*, 100091. Advance online publication. DOI: 10.1016/j.mtnano.2020.100091

Dager, , Baliyan, A., Kurosu, S., Maekawa, T., & Tachibana, M. (2020). Ultrafast synthesis of carbon quantum dots from fenugreek seeds using microwave plasma enhanced decomposition: Application of C-QDs to grow fluorescent protein crystals. *Scientific Reports, 10*(1), 12333. Advance online publication. DOI: 10.1038/s41598-020-69264-9 PMID: 32704038

Das, , Mondal, S., & Ghosh, D. (2024). Carbon quantum dots in bioimaging and biomedicines. *Frontiers in Bioengineering and Biotechnology, 11*, 1333752. DOI: 10.3389/fbioe.2023.1333752 PMID: 38318419

Dong, , Pang, H., Yang, H. B., Guo, C., Shao, J., Chi, Y., Li, C. M., & Yu, T. (2013). Carbon-based dots co-doped with nitrogen and sulfur for high quantum yield and excitation-independent emission. *Angewandte Chemie International Edition, 52*(30), 7800–7804. DOI: 10.1002/anie.201301114 PMID: 23761198

Dua., ... (2023). Stability of carbon quantum dots: a critical review. In *RSC Advances* (Vol. 13, Issue 20, pp. 13845–13861). Royal Society of Chemistry. DOI: 10.1039/D2RA07180K

Duong, , Minh Hoa, N., Toan, L. D., Chuyen, P. T., Hien, N. T. T., Hong Minh, P., Thanh Binh, N., & Anh Thi, L. (2025). Plasma-assisted green fabrication of fluorescent carbon quantum dot–silver nanocomposites. *Journal of Dispersion Science and Technology, 46*(9), 1356–1364. Advance online publication. DOI: 10.1080/01932691.2024.2440422

Escudero., ... (2021). Molecular bottom-up approaches for the synthesis of inorganic and hybrid nanostructures. In *Inorganics* (Vol. 9, Issue 7). DOI: 10.3390/inorganics9070058

Gao, , Yin, Q., Meng, X., He, X., & Xin, Z. (2022). Excellent behaviors of highly dispersed Ni-based catalyst in CO methanation synthesized by in-situ hydrothermal method with carbon quantum dots assisted. *Fuel, 310*, 121813. Advance online publication. DOI: 10.1016/j.fuel.2021.121813

Gomes, M. F., Gomes, Y. F., Lopes-Moriyama, A., de Barros Neto, E. L., & de Souza, C. P. (2019a). Design of carbon quantum dots via hydrothermal carbonization synthesis from renewable precursors. *Biomass Conversion and Biorefinery, 9*(4), 689–694. DOI: 10.1007/s13399-019-00387-4

Gomes, M. F., Gomes, Y. F., Lopes-Moriyama, A., de Barros Neto, E. L., & de Souza, C. P. (2019b). Design of carbon quantum dots via hydrothermal carbonization synthesis from renewable precursors. *Biomass Conversion and Biorefinery*, *9*(4), 689–694. DOI: 10.1007/s13399-019-00387-4

Guerrero-Gonzalez, R., Vázquez-Dávila, F., Saucedo-Flores, E., Ruelas, R., Ceballos-Sánchez, O., & Pelayo, J. E. (2023). Green approach synthesis of carbon quantum dots from agave bagasse and their use to boost seed germination and plant growth. *SN Applied Sciences*, *5*(8), 204. Advance online publication. DOI: 10.1007/s42452-023-05428-2

Gyulai, , Ouanzi, F., Bertóti, I., Mohai, M., Kolonits, T., Horváti, K., & Bősze, S. (2019). Chemical structure and in vitro cellular uptake of luminescent carbon quantum dots prepared by solvothermal and microwave assisted techniques. *Journal of Colloid and Interface Science*, *549*, 150–161. DOI: 10.1016/j.jcis.2019.04.058 PMID: 31029843

Hagiwara, , Horikoshi, S., & Serpone, N. (2021). Luminescent monodispersed carbon quantum dots by a microwave solvothermal method toward bioimaging applications. *Journal of Photochemistry and Photobiology A Chemistry*, *415*, 113310. Advance online publication. DOI: 10.1016/j.jphotochem.2021.113310

Hernandez-Tabares, , Darias-Gonzalez, J. G., Chao-Mujica, F. J., Ledo-Pereda, L. M., Antuch, M., Carrillo-Barroso, E., Chong-Quero, J. E., Reguera, E., & Desdin-Garcia, L. F. (2021). Stabilization Methods in the Submerged Arc Discharge Synthesis of Carbon Nanostructures. *Journal of Nanomaterials*, *2021*, 1–12. Advance online publication. DOI: 10.1155/2021/6550809

Hoa, , Toan, L. D., Tran, N., Hung, L. X., & Thi, L. A. (2025). Microplasma-synthesized Citrus-derived carbon quantum dots: Antibacterial properties and nanoprobe sensitivity. *RSC Advances*, *15*(31), 25362–25371. DOI: 10.1039/D5RA02150B PMID: 40677949

Hong, , Deng, X., Liang, J., Li, J., Tao, Y., & Tan, K. (2019). One-step hydrothermal synthesis of down/up-conversion luminescence F-doped carbon quantum dots for label-free detection of Fe3+. *Microchemical Journal*, *151*, 104217. Advance online publication. DOI: 10.1016/j.microc.2019.104217

Huang, , Cui, Y., Liu, M., Chen, J., Wan, Q., Wen, Y., Deng, F., Zhou, N., Zhang, X., & Wei, Y. (2018). A one-step ultrasonic irradiation assisted strategy for the preparation of polymer-functionalized carbon quantum dots and their biological imaging. *Journal of Colloid and Interface Science*, *532*, 767–773. DOI: 10.1016/j.jcis.2018.07.099 PMID: 30130727

Huy, H.. (2024). A self-powered photodetector through facile processing using polyethyleneimine/carbon quantum dots for highly sensitive UVC detection. *RSC Advances*, *14*(18), 12360–12371. DOI: 10.1039/D3RA08538D PMID: 38633486

Janus, , Piątkowski, M., & Radwan-Pragłowska, J. (2019). Microwave-assisted synthesis and characterization of poly(L-lysine)-based polymer/carbon quantum dot nanomaterials for biomedical purposes. *Materials (Basel)*, *12*(23), 3825. Advance online publication. DOI: 10.3390/ma12233825 PMID: 31766363

Kannan et al., (2021). Gd3+ and Y3+ co-doped mixed metal oxide nanohybrids for photocatalytic and antibacterial applications. Nano express, 2(1), 010014.

Kannan, , Radhika, D., Gnanasangeetha, D., Krishna, L. S., & Gurushankar, K. (2021). Y3+ and Sm3+ co-doped mixed metal oxide nanocomposite: Structural, electrochemical, photocatalytic, and antibacterial properties. *Applied Surface Science Advances*, *4*, 100085. DOI: 10.1016/j.apsadv.2021.100085

Kannan, , Radhika, D., Gnanasangeetha, D., Lakkaboyana, S. K., Sadasivuni, K. K., Gurushankar, K., & Hanafiah, M. M. (2021). Photocatalytic and antimicrobial properties of microwave synthesized mixed metal oxide nanocomposite. *Inorganic Chemistry Communications*, *125*, 108429. DOI: 10.1016/j.inoche.2020.108429

Kannan, , Radhika, D., Kasai, R. D., Gnanasangeetha, D., Palani, G., Gurushankar, K., Koutavarapu, R., Lee, D.-Y., & Shim, J. (2022). Facile fabrication of novel ceria-based nanocomposite (CYO-CSO) via co-precipitation: Electrochemical, photocatalytic and antibacterial performances. *Journal of Molecular Structure*, *1256*, 132519. DOI: 10.1016/j.molstruc.2022.132519

Kannan, K., & Tari, V. (Eds.). (2024). Novel energy storage and conversion technologies for two-dimensional MXenes and MBenes. ISBN: 9798369395462, eISBN: 9798369395486DOI: 10.4018/979-8-3693-9546-2

Kim., ... (2017). Synthesis of nanoparticles by laser ablation: A review. In *KONA Powder and Particle Journal* (Vol. 2017, Issue 34, pp. 80–90). Hosokawa Powder Technology Foundation. DOI: 10.14356/kona.2017009

Ko, , Nafiujjaman, M., Cherukula, K., Lee, S. J., Hong, S.-J., Lim, H.-N., Park, C. H., Park, I.-K., Lee, Y.-K., & Kwon, I. K. (2018). Microwave-Assisted Synthesis of Biocompatible Silk Fibroin-Based Carbon Quantum Dots. *Particle & Particle Systems Characterization*, *35*(3), 1700300. Advance online publication. DOI: 10.1002/ppsc.201700300

Kong., ... (2024). Carbon Quantum Dots: Properties, Preparation, and Applications. In *Molecules* (Vol. 29, Issue 9). Multidisciplinary Digital Publishing Institute (MDPI). DOI: 10.3390/molecules29092002

Koutsogiannis., ... (2020). Advances in fluorescent carbon dots for biomedical applications. In *Advances in Physics: X* (Vol. 5, Issue 1). Taylor and Francis Ltd. DOI: 10.1080/23746149.2020.1758592

Kumar., ... (2022). A review on advancements in carbon quantum dots and their application in photovoltaics. In *RSC Advances* (Vol. 12, Issue 8, pp. 4714–4759). Royal Society of Chemistry. DOI: 10.1039/D1RA08452F

Kumar et al., (2023). Technoeconomic analysis of biofuel production from agricultural residues through pyrolysis. *Green Approach to Alternative Fuel for a Sustainable Future*, 463–469. DOI: 10.1016/B978-0-12-824318-3.00001-1

Kumar, , Mirsky, S. K., Shaked, N. T., & Gazit, E. (2024). High Quantum Yield Amino Acid Carbon Quantum Dots with Unparalleled Refractive Index. *ACS Nano*, *18*(3), 2421–2433. DOI: 10.1021/acsnano.3c10792 PMID: 38190624

Lakshmanan, A., Surendran, P., Sakthy Priya, S., Balakrishnan, K., Rameshkumar, P., Kannan, K., Mahalakshmi, K., Gayathri, V., & Vinitha, G. (2024). Investigations on structural, dielectric, and third-order nonlinear optical studies of $Ni_{0.5}Zn_{0.5}Fe_2O_4$ nanoparticles by microwave combustion method with different fuels (glycine, urea and citric acid). *Materials Letters*, *360*, 135973. DOI: 10.1016/j.matlet.2024.135973

Li., ... (2023). Physiological Functions of Carbon Dots and Their Applications in Agriculture: A Review. In *Nanomaterials* (Vol. 13, Issue 19). Multidisciplinary Digital Publishing Institute (MDPI). DOI: 10.3390/nano13192684

Li, , Han, F., Cao, W., Zhang, G., Li, J., Zhou, J., Gong, X., Turnbull, G., Shu, W., Xia, L., Fang, B., Xing, X., & Li, B. (2020). Carbon quantum dots derived from lysine and arginine simultaneously scavenge bacteria and promote tissue repair. *Applied Materials Today*, *19*, 100601. Advance online publication. DOI: 10.1016/j.apmt.2020.100601

Li, , Kang, Z., Liu, Y., & Lee, S.-T. (2012a). Carbon nanodots: Synthesis, properties and applications. *Journal of Materials Chemistry*, *22*(46), 24230–24253. DOI: 10.1039/c2jm34690g

Li, , Kang, Z., Liu, Y., & Lee, S.-T. (2012b). Carbon nanodots: Synthesis, properties and applications. *Journal of Materials Chemistry*, *22*(46), 24230–24253. DOI: 10.1039/c2jm34690g

Lim., ... (2015). Carbon quantum dots and their applications. In *Chemical Society Reviews* (Vol. 44, Issue 1, pp. 362–381). Royal Society of Chemistry. DOI: 10.1039/C4CS00269E

Lu, , & Zhou, L. (2019). One-step sonochemical synthesis of versatile nitrogen-doped carbon quantum dots for sensitive detection of Fe^{2+} ions and temperature in vitro. *Materials Science and Engineering C, 101*, 352–359. DOI: 10.1016/j.msec.2019.03.109 PMID: 31029328

Lv, , Luo, Z., Yang, H., Liu, B., Weng, W., & Liu, J. (2010). Effect of processing conditions on sonochemical synthesis of nanosized copper aluminate powders. *Ultrasonics Sonochemistry*, *17*(2), 344–351. DOI: 10.1016/j.ultsonch.2009.06.006 PMID: 19570706

Magdy, , Belal, F., & Elmansi, H. (2023). Rapid microwave-assisted synthesis of nitrogen-doped carbon quantum dots as fluorescent nanosensors for the spectrofluorimetric determination of palbociclib: Application for cellular imaging and selective probing in living cancer cells. *RSC Advances*, *13*(7), 4156–4167. DOI: 10.1039/D2RA05759J PMID: 36744285

Magesh, . (2022). Recent Advances on Synthesis and Potential Applications of Carbon Quantum Dots. In *Frontiers in Materials* (Vol. 9). Frontiers Media S.A., DOI: 10.3389/fmats.2022.906838

Malini et al., (2023). Activated carbon from biomass: Preparation, factors improving basicity and surface properties for enhanced CO_2 capture capacity - A review. *Journal of CO2 Utilization, 67*. DOI: 10.1016/j.jcou.2022.102318

Mindivan, F., & Göktaş, M. (2023). The green synthesis of carbon quantum dots (CQDs) and characterization of polycaprolactone (PCL/CQDs) films. *Colloids and Surfaces. A, Physicochemical and Engineering Aspects*, *677*, 132446. Advance online publication. DOI: 10.1016/j.colsurfa.2023.132446

Mohammadpour, , Kazemi, A., Baghapour, M. A., Samaei, M. R., Kannan, K., & Mousavi Khaneghah, A. (2024). Bioengineered FeZn/GA@ Cu nanocomposite utilizing spent coffee ground extract and gum arabic: Enhanced nitrate removal via (RSM) and machine learning optimization. *International Journal of Biological Macromolecules*, *277*, 134060. DOI: 10.1016/j.ijbiomac.2024.134060 PMID: 39097464

Nazar, , Hasan, M., Wirjosentono, B., Gani, B. A., & Nada, C. E. (2024). Microwave Synthesis of Carbon Quantum Dots from Arabica Coffee Ground for Fluorescence Detection of Fe^{3+}, Pb^{2+}, and Cr^{3+}. *ACS Omega*, *9*(18), 20571–20581. DOI: 10.1021/acsomega.4c02254 PMID: 38737012

Pajewska-Szmyt, , Buszewski, B., & Gadzała-Kopciuch, R. (2020). Sulphur and nitrogen doped carbon dots synthesis by microwave assisted method as quantitative analytical nano-tool for mercury ion sensing. *Materials Chemistry and Physics*, *242*, 122484. Advance online publication. DOI: 10.1016/j.matchemphys.2019.122484

Pandiyan, , Arumugam, L., Srirengan, S. P., Pitchan, R., Sevugan, P., Kannan, K., Pitchan, G., Hegde, T. A., & Gandhirajan, V. (2020). Biocompatible Carbon Quantum Dots Derived from Sugarcane Industrial Wastes for Effective Nonlinear Optical Behavior and Antimicrobial Activity Applications. *ACS Omega*, *5*(47), 30363–30372. DOI: 10.1021/acsomega.0c03290 PMID: 33283084

Parra, , Silva, K., Valezin, P., Martins, R., Gomes, R., Pereira, R., de Melo, F., Morandim-Giannetti, A., dos Santos, R., Panzarini, L. C., & Toyama, M. (2020). Preparation of fluorescent carbon-based dots from waste tire pyrolysis. *Journal of the Brazilian Chemical Society*, *31*(11), 2224–2231. DOI: 10.21577/0103-5053.20200072

Prekodravac, , Vasiljević, B., Marković, Z., Jovanović, D., Kleut, D., Špitalský, Z., Mičušik, M., Danko, M., Bajuk-Bogdanović, D., & Todorović-Marković, B. (2019). Green and facile microwave assisted synthesis of (metal-free) N-doped carbon quantum dots for catalytic applications. *Ceramics International*, *45*(14), 17006–17013. DOI: 10.1016/j.ceramint.2019.05.250

Ramezani, , Qorbanpour, M., & Rahbar, N. (2018). Green synthesis of carbon quantum dots using quince fruit (Cydonia oblonga) powder as carbon precursor: Application in cell imaging and As^{3+} determination. *Colloids and Surfaces. A, Physicochemical and Engineering Aspects*, *549*, 58–66. DOI: 10.1016/j.colsurfa.2018.04.006

Rao, , Yuan, M., Jiang, H., Li, L., & Liu, Z. (2019). A universal strategy to obtain chiroptical carbon quantum dots through the optically active surface passivation procedure. *New Journal of Chemistry*, *43*(35), 13735–13740. DOI: 10.1039/C9NJ03434J

Ren, , Zhang, F., Guo, B., Gao, N., & Zhang, X. (2019). Synthesis of N-doped Micropore carbon quantum dots with high quantum yield and dual-wavelength photoluminescence emission from biomass for cellular imaging. *Nanomaterials (Basel, Switzerland)*, *9*(4), 495. Advance online publication. DOI: 10.3390/nano9040495 PMID: 30939724

Ren, , Zhang, Z., Zhao, P., Shi, J., Han, K., Yang, Z., Gao, D., & Bi, F. (2019). Facile and one-step preparation carbon quantum dots from biomass residue and their applications as efficient surfactants. *Journal of Dispersion Science and Technology*, *40*(5), 627–633. DOI: 10.1080/01932691.2018.1475239

Sabet, , & Mahdavi, K. (2019). Green synthesis of high photoluminescence nitrogen-doped carbon quantum dots from grass via a simple hydrothermal method for removing organic and inorganic water pollutions. *Applied Surface Science, 463*, 283–291. DOI: 10.1016/j.apsusc.2018.08.223

Salman, , Batakoushy, H. A., Saraya, R. E., Hassan, A. I., Al-Harrasi, A., & Ibrahim, A. E. (2025). Comprehensive investigation of Prunus armeniaca for natural green synthesis of carbon quantum dots; Applications as fluorescent nano-probes for ramipril. *Talanta, 292*, 128014. Advance online publication. DOI: 10.1016/j.talanta.2025.128014 PMID: 40154049

Salman, , Hassan, A. I., Hassan, Y. F., Saraya, R. E., & Batakoushy, H. A. (2023). Rapid One-Pot Microwave Assisted Green Synthesis Nitrogen Doped Carbon Quantum Dots as Fluorescent Precursor for Estimation of Modafinil as Post-Covid Neurological Drug in Human Plasma with Greenness Assessments. *Journal of Fluorescence, 33*(3), 1101–1110. DOI: 10.1007/s10895-022-03128-5 PMID: 36576682

Salman (2023). A Novel Design Eco-friendly Microwave-assisted Cu–N@CQDs Sensor for the Quantification of Eravacycline via Spectrofluorimetric Method; Application to Greenness Assessments, Dosage Form and Biological Samples. *Journal of Fluorescence, 33*(5), 1887–1896. DOI: 10.1007/s10895-023-03190-7

Semaltianos (2010). Nanoparticles by laser ablation. *Critical Reviews in Solid State and Materials Sciences, 35*(2), 105–124. DOI: 10.1080/10408431003788233

Shahba et al., (2022). *Two-Step and Green Synthesis of Highly Fluorescent Carbon Quantum Dots and Carbon Nanofibers from Pine Fruit*. https://doi.org/DOI: 10.1007/s10895-020-02562-7/Published

Shang, , Cai, T., Zhang, Y., Liu, D., & Liu, S. (2018). Facile one pot pyrolysis synthesis of carbon quantum dots and graphene oxide nanomaterials: All carbon hybrids as eco-environmental lubricants for low friction and remarkable wear-resistance. *Tribology International, 118*, 373–380. DOI: 10.1016/j.triboint.2017.09.029

Shen, , Jia, P.-Y., Chen, D.-S., & Wang, L.-N. (2021). Hydrothermal synthesis of N-doped carbon quantum dots and their application in ion-detection and cell-imaging. *Spectrochimica Acta. Part A: Molecular and Biomolecular Spectroscopy, 248*, 119282. Advance online publication. DOI: 10.1016/j.saa.2020.119282 PMID: 33316652

Slewa (2024). Antifungal films for strawberry packaging using carbon quantum dots derived from lemon and onion juice via green hydrothermal method. *Food Bioscience, 61*. DOI: 10.1016/j.fbio.2024.104653

Smagulova, S., Egorova, M., & Tomskaya, A. (2019). Investigation of the properties of carbon quantum dots synthesized by the hydrothermal method. *IOP Conference Series. Materials Science and Engineering*, *693*(1), 012031. Advance online publication. DOI: 10.1088/1757-899X/693/1/012031

Sonawane et al., (2018). Nanocomposites and Its Applications. *Applications of Nanomaterials: Advances and Key Technologies*, 1–22. DOI: 10.1016/B978-0-08-101971-9.00001-6

Sun, , Guo, S., Qin, Q., Liao, Y., Li, M., & Du, F. (2022). Box–Behnken Design Optimizing Sugarcane Bagasse-Based Nitrogen-Doped Carbon Quantum Dots Preparation and Application in Ferric Ion Detection. *Chemosensors (Basel, Switzerland)*, *10*(11), 453. Advance online publication. DOI: 10.3390/chemosensors10110453

Sun, , Zhou, B., Lin, Y., Wang, W., Fernando, K. A. S., Pathak, P., Meziani, M. J., Harruff, B. A., Wang, X., Wang, H., Luo, P. G., Yang, H., Kose, M. E., Chen, B., Veca, L. M., & Xie, S.-Y. (2006). Quantum-sized carbon dots for bright and colorful photoluminescence. *Journal of the American Chemical Society*, *128*(24), 7756–7757. DOI: 10.1021/ja062677d PMID: 16771487

Surendran, P., Lakshmanan, A., Priya, S. S., Balakrishnan, K., Rameshkumar, P., Kannan, K., Mahalakshmi, K., Gayathri, V., & Vinitha, G. (2024). Synthesis of fluorescent carbon quantum dots from Manihot esculenta waste peels for nonlinear optical and biological applications. *Chemical Physics Impact*, *8*, 100515. DOI: 10.1016/j.chphi.2024.100515

Surendran, P., Lakshmanan, A., Priya, S. S., Geetha, P., Rameshkumar, P., Kannan, K., Hegde, T. A., & Vinitha, G. (2021). Fluorescent carbon quantum dots from Ananas comosus waste peels: A promising material for NLO behaviour, antibacterial, and antioxidant activities. *Inorganic Chemistry Communications*, *124*, 108397. DOI: 10.1016/j.inoche.2020.108397

Tan, , Zuo, S., Zhao, Y., & Shen, B. (2019). Preparation of multicolored carbon quantum dots using HNO3/HClO4 oxidation of graphitized carbon. *Journal of Materials Research*, *34*(20), 3428–3438. DOI: 10.1557/jmr.2019.261

Tian, . (2021). Carbon quantum dots for advanced electrocatalysis. In *Journal of Energy Chemistry* (Vol. 55, pp. 279–294). Elsevier B.V., DOI: 10.1016/j.jechem.2020.06.057

Tian, , Wang, Y., & Zhang, Y. (2018). Synthesis of Fluorescent Nitrogen-Doped Carbon Quantum Dots for Selective Detection of Picric Acid in Water Samples. *Journal of Nanoscience and Nanotechnology*, *18*(12), 8111–8117. DOI: 10.1166/jnn.2018.16342 PMID: 30189927

Vercelli, , Donnini, R., Ghezzi, F., Sansonetti, A., Giovanella, U., & La Ferla, B. (2021). Nitrogen-doped carbon quantum dots obtained hydrothermally from citric acid and urea: The role of the specific nitrogen centers in their electrochemical and optical responses. *Electrochimica Acta*, *387*, 138557. Advance online publication. DOI: 10.1016/j.electacta.2021.138557

Vibhute, , Nille, O., Kolekar, G., Rohiwal, S., Patil, S., Lee, S., & Tiwari, A. P. (2022). Fluorescent Carbon Quantum Dots Functionalized by Poly L-Lysine: Efficient Material for Antibacterial, Bioimaging and Antiangiogenesis Applications. *Journal of Fluorescence*, *32*(5), 1789–1800. DOI: 10.1007/s10895-022-02977-4 PMID: 35689742

Wang, , Guo, J., Zhou, Y., Zhang, X., Gao, C., Xiong, R., Song, H., Chen, X., & Luo, J. (2025). Enhancing oil shale pyrolysis using metal-doped carbon quantum dot catalysts: A comprehensive behavioral, kinetic, and mechanistic analysis. *Fuel*, *381*, 133464. Advance online publication. DOI: 10.1016/j.fuel.2024.133464

Wang, , & Hu, A. (2014). Carbon quantum dots: Synthesis, properties and applications. *Journal of Materials Chemistry. C, Materials for Optical and Electronic Devices*, *2*(34), 6921–6939. DOI: 10.1039/C4TC00988F

Wang, , Shen, J., Ma, Y., Qin, X., Qin, X., Yang, F., Ostrikov, K. K., Zhang, Q., He, J., & Zhong, X. (2024). Cancer-targeting carbon quantum dots synthesized by plasma electrochemical method for red-light-activated photodynamic therapy. *Plasma Processes and Polymers*, *21*(4), 2300174. Advance online publication. DOI: 10.1002/ppap.202300174

Weerasinghe, , Scott, J., Deshan, A. D. K., Chen, D., Singh, A., Sen, S., Sonar, P., Vasilev, K., Li, Q., & Ostrikov, K. K. (2022). Monochromatic Blue and Switchable Blue-Green Carbon Quantum Dots by Room-Temperature Air Plasma Processing. *Advanced Materials Technologies*, *7*(4), 2100586. Advance online publication. DOI: 10.1002/admt.202100586

Wu, , Mu, L., Luo, X., Zhang, Y., Huang, J., Peng, H., & Tao, J. (2025). Effect of temperature on synthesis of carbon quantum dots and biochar through one-step hydrothermal treatment of distillers' grains. *Industrial Crops and Products*, *227*, 120832. Advance online publication. DOI: 10.1016/j.indcrop.2025.120832

Yadav et al., (2024). Emerging ecofriendly approaches for biomass conversion and waste management. *Sustainable Management of Agro-Food Waste: Fundamental Aspects and Practical Applications*, 39–55. DOI: 10.1016/B978-0-443-23679-2.00004-5

Yan et al., (2012). Functionalized Carbon Nanotubes and Their Enhanced Polymers. *Polymer Science: A Comprehensive Reference: Volume 1-10, 1–10*, 439–478. DOI: 10.1016/B978-0-444-53349-4.00217-X

Yan, , Yang, Y., Ma, C.-Q., Liu, X., Wang, H., & Xu, B. (2016). Synthesis of carbon quantum dots by chemical vapor deposition approach for use in polymer solar cell as the electrode buffer layer. *Carbon, 109*, 598–607. DOI: 10.1016/j.carbon.2016.08.058

Yang, , Duan, J., Yang, W., Li, X., Mo, J., Yang, P., & Tang, Q. (2018). Nitrogen-doped carbon quantum dots from biomass via simple one-pot method and exploration of their application. *Applied Surface Science, 434*, 1079–1085. DOI: 10.1016/j.apsusc.2017.11.040

Yang, , Lin, W., Lai, H., Tong, J., Lei, J., Yuan, M., Zhang, Y., & Cui, C. (2021). Understanding the relationship between particle size and ultrasonic treatment during the synthesis of metal nanoparticles. *Ultrasonics Sonochemistry, 73*, 105497. Advance online publication. DOI: 10.1016/j.ultsonch.2021.105497 PMID: 33677187

Zaman, K.. (2021). Properties and molecular structure of carbon quantum dots derived from empty fruit bunch biochar using a facile microwave-assisted method for the detection of Cu2+ ions. *Optical Materials, 112*, 110801. Advance online publication. DOI: 10.1016/j.optmat.2021.110801

Zhao, , Chen, Z., Zhao, C., Gao, N., Ren, J., & Qu, X. (2015). Recent advances in bioapplications of C-dots. *Carbon, 85*, 309–327. DOI: 10.1016/j.carbon.2014.12.045

Zhou, , Qiao, M., Zhang, L., Sun, L., Zhang, Y., & Liu, W. (2019). Green and efficient synthesis of carbon quantum dots and their luminescent properties. *Journal of Luminescence, 206*, 158–163. DOI: 10.1016/j.jlumin.2018.10.057

Zhu et al., (2000). *Graphic abstract Mild acidolysis-assisted hydrothermal carbonization of lignin for simultaneous preparation of green and blue fluorescent carbon quantum dots.*

Chapter 3
Enhancing Carbon Quantum Dots Properties:
Surface Passivation, Functionalization, Doping, and Aggregation

K. S. Rajni
https://orcid.org/0000-0002-9046-8461
Department of Physics, Amrita School of Physical Sciences Coimbatore, Amrita Vishwa Vidyapeetham, India

S Amala Devi
Department of Physics, Amrita School of Physical Sciences Coimbatore, Amrita Vishwa Vidyapeetham, India

ABSTRACT

Precise modifications in the CD structures can be induced to enhance their properties. The Luminescent emission characteristics, and all other properties rely on CDs morphology and functional groups attached to it. The functional groups can be modified according to the applications or which property has to be modified a need to concentrate on the concentrations and the proportions of materials choose for selection. The non-radiative path ways reductions and targeted functionalities can be introduced to have a smart, reliable, multi-functional CD synthesis. The advancement of aggregation from nano size to macro structures and ACQ phenomenon are advantages of controlled aggregation. Hence this engineered material finds applications in almost all sectors, alone and in a matrix to enhance the property of CDs.

DOI: 10.4018/979-8-3373-4247-4.ch003

1. INTRODUCTION

Carbon Quantum Dots, or simply CDs, are known for their quantum-confined size, where these particles exhibit unique optical and electronic properties. These spherical particles are known for their luminescent properties, biocompatibility, and non-toxic nature, which distinguishes them from conventional semiconductor quantum dots. Scientists have extensively explored the electronic, magnetic, and thermal properties of CDs. The evolution of these particles, following their discovery in 2007, during the purification of carbon nanotubes (CNTs) as fluorescent impurity particles, sparked significant advancements in nanoscience and technological applications. Later, scientists thoroughly studied the structure and properties of these particles to improve their characteristics, and a proposed model design is presented in Figure *1*.

Figure 1. Structural design of CDs

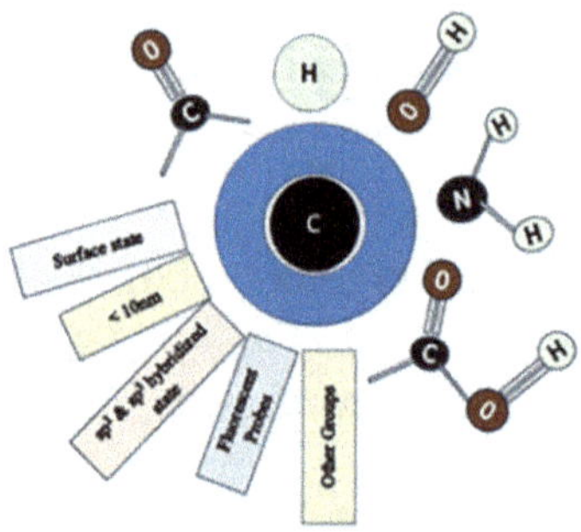

(Sen & Bose, 2025).

CDs have also been successfully produced using nanomaterial synthesis methodologies, including top-down and bottom-up strategies, such as hydrothermal, microwave irradiation, pyrolysis, ultrasonic synthesis, electrochemical methods, arc discharge, and chemical exfoliation. The top-down method yielded more crystalline and stronger core-structured CDs, which were obtained through the breakdown of bulky carbonaceous materials via electrochemical reduction or arc discharge. The bottom-up method demonstrated a wide variety of precursor material selections, including biomass waste and organic materials, and enabled a greater presence of biomolecular species (Ge et al., 2021; Junjun Liu, Li, & Yang, 2020; Sen & Bose, 2025). The different synthesis methodologies and specific precursor materials used for these syntheses are illustrated in *Figure 2*.

Figure 2. Different synthesis methodologies of CDs and corresponding precursor materials

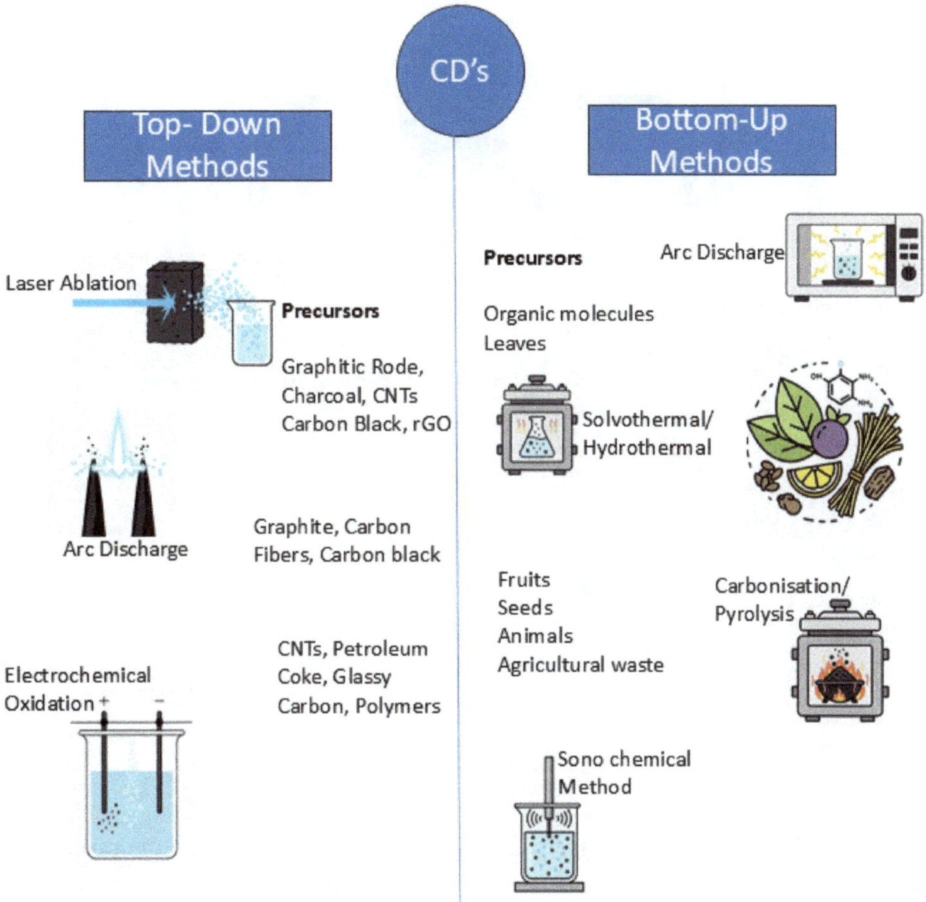

(Kumar, Kumar, & Gedanken, 2020; Sen & Bose, 2025)

As a result, the surface functional groups attached to the core structure became diverse, leading to modifications in their properties. This bottom-up method enabled the enhanced functionalisation of synthesised materials by varying functional groups, such as -OH, -COOH, and $-NH_2$, through post-synthesis treatments. These surface passivation or functionalisation modifications can be classified, basically, into covalent or non-covalent modifications, which in turn can be further classified according to the chemical reaction they undergo, as shown in the *Figure 3.*

Figure 3. Different chemical reactions used for passivation and functionalisation – a classification

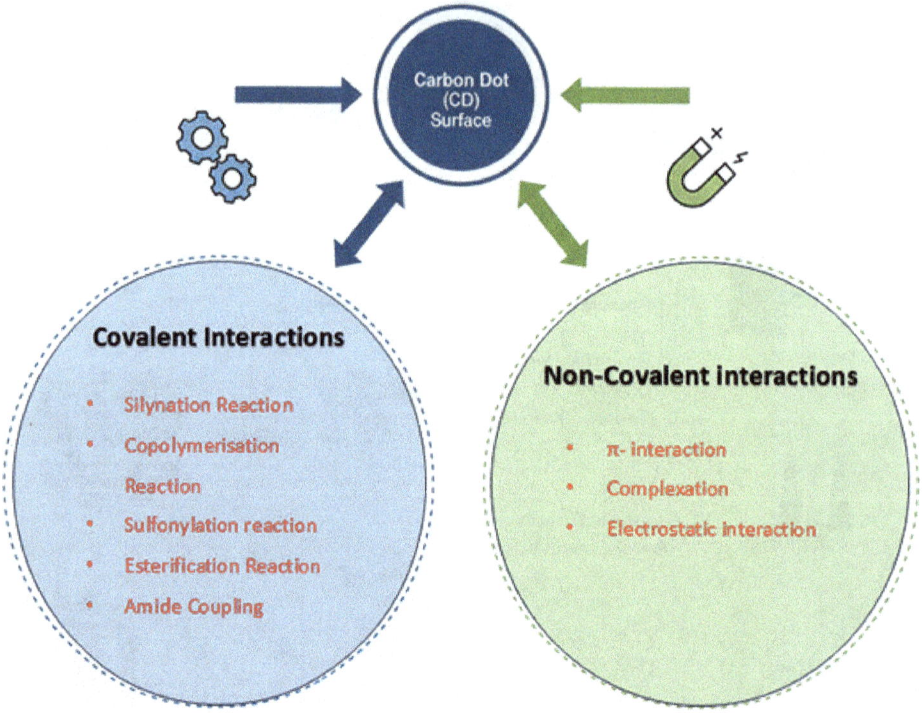

(Yan et al., 2018)

This will reduce non-radiative recombination and help enhance the properties via reversible or irreversible changes due to electrostatic interactions or hydrogen bonding. Surface functionalisation of these synthesised particles is an essential aspect, as it enables us to achieve the desired size, shape, and stability required for applications. To retain these properties, a capping layer is implemented over the core and functional group clustered CD structure. Polymer encapsulation, or the use of small molecular penetrations, can be employed to achieve this. Another way to customise the properties is by achieving surface doping with different metallic, non-metallic, transition metal, and noble gases, or with their combinations, as required for the specific application, or by aggregating CDs. Doping with heteroatoms is widespread and retains a homogeneous distribution due to its relatively small size and ability to form covalent bonds with the core carbon atom. The uncontrolled aggregation can alter the CDs' properties by causing quenching (Jorns & Pappas, 2021; Pundi & Chang, 2022).

In contrast, controlled aggregation can lead to restricted intramolecular motions, causing the fluorophore to lock into more rigid vibrational and rotational states, thereby increasing fluorescence and quantum Yield (QY). Hence, the property modification of CDs can be achieved through any surface passivation, functionalisation, doping, or aggregation. This chapter outlines the various properties of CDs and their modification strategies, highlighting their importance in diverse applications.

2. UNIQUE PROPERTIES OF CDs

The properties and their modification through functionalisation and surface passivation, as well as the variations that occur for different doping requirements, require an understanding of the properties of CQDs and what influences their fundamental properties. This section provides an overall sense of how properties arise and what contributes to a specific change in that property. The CDs possess unique properties, including quantum confinement, fluorescence, non-toxicity, eco-friendliness, reliable precursor availability, and surface functionality and tunability. The size and optical properties of CDs make them captivating among researchers and industrialists, who seek to replace conventional fluorescent materials in a wide range of cutting-edge applications. This section discusses how various modification strategies can easily tune the properties of CDs. These characteristic properties can be well identified with the existing characterisation techniques. Specific functionalisation and doping can be determined using sophisticated techniques such as XPS, NMR, and HPLC, among others. This is explained in a separate section, beyond the existing methods, under a separate subtitle.

2.1. Size, Shape and Crystallinity

After synthesis, CDs need to be characterised to confirm their structure and properties, in which size, shape, and crystallinity will be key features to confirm their formation. Transmission electron microscopy (TEM), X-ray diffraction (XRD) analysis, and Raman analysis are key tools used for such analysis. The enhancement in crystallinity or amorphous nature, as well as the ratio of core hybridisation and carbon dot size, are all interrelated and require special attention to be fine-tuned. TEM analysis is typically used to evaluate the shape and size, which are usually spherical or quasi-spherical, with diameters of less than 10nm, as calculated from size distribution histograms. High-resolution (HR)-TEM SAED (selected area electron diffraction) patterns, along with the XRD peak positions, can be compared to evaluate the interplanar spacing in the carbon lattice and hence to identify the crystallinity of the sample by placing the rings present. TEM image *Figure 4 (a) and (b)* (for samples prepared at two different temperatures) and XPS survey spectrum and deconvoluted C1s peaks for synthesised CDs is given in the *Figure 4 (c)* and *(d)* respectively. The XRD spectrum of CDs varies from amorphous to crystalline,

*Figure 4. TEM images of the sample prepared by the hydrothermal method. (**a**) TEM image showing particle size above 10nm, at 200°C of processing temperature (**b**) TEM image for CDs prepared at 240°C (**c**) XPS survey scan for the CDs sample and (d) C1s deconvoluted spectrum of the sample*

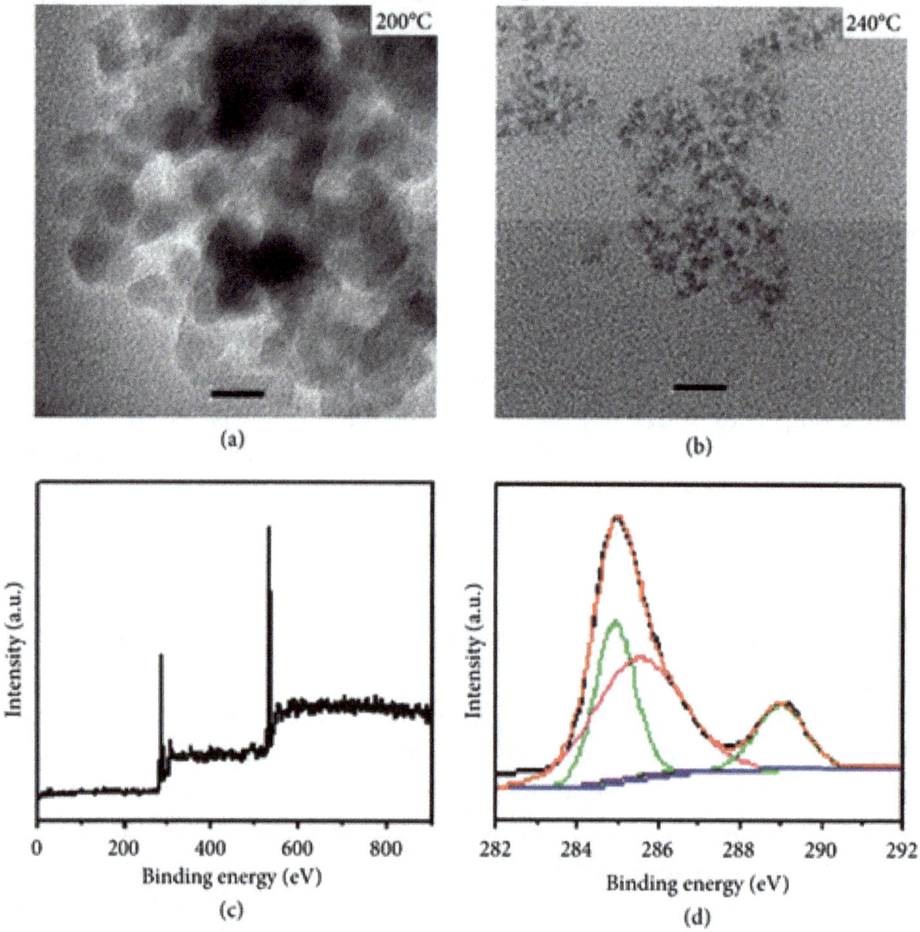

(Hoan, Tam, & Pham, 2019).

The crystalline, sharp peaks in XRD are a sign of the formation of graphene quantum dots of the (002) plane of graphitic carbon. The amorphous peaks, with a broader peak centred around 25° to 29 °, are most common, accompanied by a shoulder peak at 40 ° to 42°, indicating a short-range order of lattice points. *Figure 5 (a)* and *(b)* well depict the TEM image, particle distribution histogram and XRD pattern *Figure 5 (c)* along with FTIR analysis results, *Figure 5 (c)*, to confirm surface functional groups present(Zhong, Sun, Zhao, & Zhao, 2025).

Figure 5. (a) TEM image of Peperomia tetraphylla based CDs (PT-CDs) (b) Particle distribution histogram with diameter of spherical particles given (c) XRD spectrum of CDs (d) FTIR spectrum of raw material is compared with that of CDs after formation

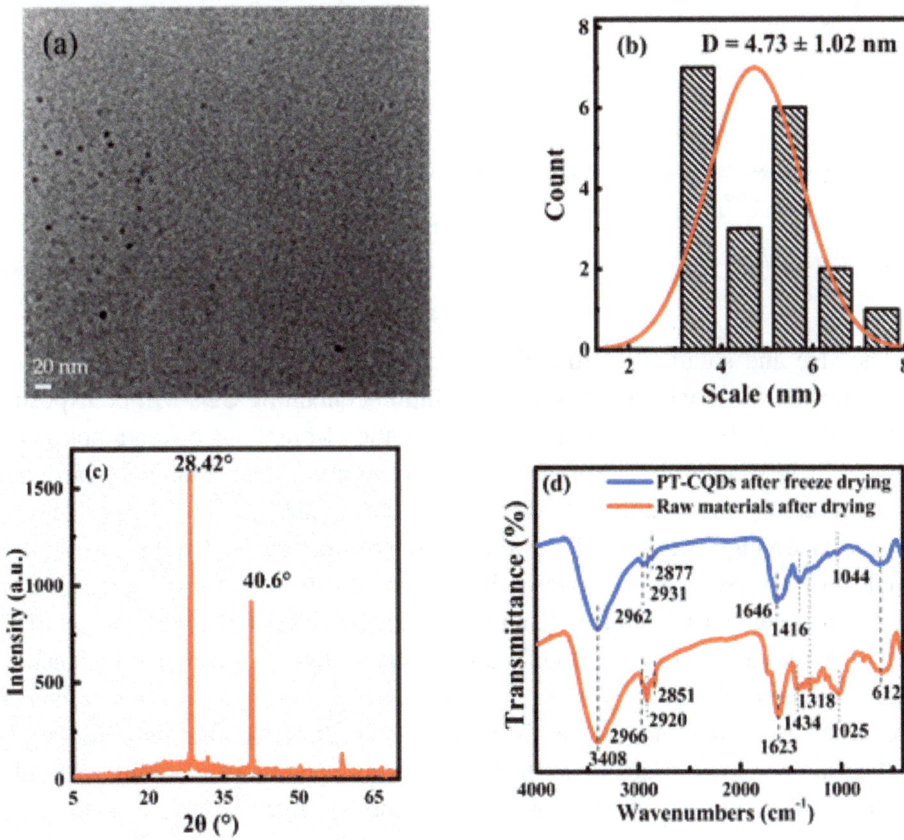

(Zhong et al., 2025).

The disordered graphitic carbon structure range can be analysed using the Raman peaks at 1350 cm^{-1} and 1600 cm^{-1} by calculating the I_d/I_g ratio(Mintz et al., 2021). *Figure 6* is an example of the obtained Raman spectrum of CDs synthesised with graphene band(I_g) at 1610cm^{-1} and defect band (I_d) at 1330cm^{-1}, I_d/I_g ratio 0.69 and 0.94 for c-dot 2000 (CDs purified with 2000kDa dialysis membrane) and c-dot 2000 (CDs purified with 500kDa dialysis membrane)(Fawaz, Hasian, & Alghoraibi, 2023).

Figure 6. Raman spectrum of CDs synthesized

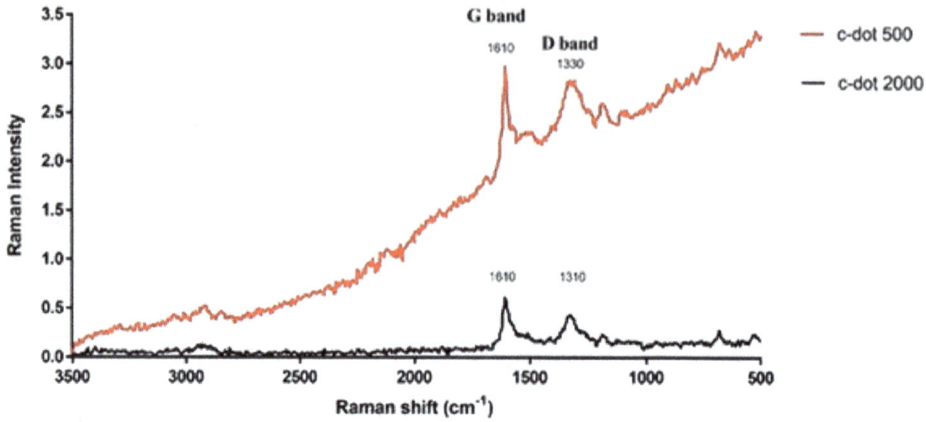

(Fawaz et al., 2023).

The size and solution-based stability can easily be found from diffuse light scattering (DLS), in which the colloidal solution containing CDs will be exposed to light of a specific wavelength and get information from the Brownian motion of particles in the solution, which provides size information and how stable the particles are distributed in the solution(Mansuriya & Altintas, 2021).

CDs have a higher density of active sites on their surface, leading to easier functionalisation and passivation, as well as heteroatom doping. In contrast, modification will be challenging as it can lead to an increased size of the CDs, with more active sites or better quantum yield. Modification of crystalline CDs is due to the anchor points for chemical groups, whereas for amorphous groups, it occurs via structural defects. At the same time, the passivation agents or functionalisation units might result in an increase in size due to the presence of an outer layer. Doping with different moieties, such as heteroatoms, metallic, and non-metallic atoms, is common in CDs structures, resulting in alterations to the electronic structure and the creation of active sites to produce a novel material with high tunability and even multiple photon emission characteristics due to its quantum confined size. Defects can lead to an increase in D-band intensity, and doping can result in a new spectrum in the CD structure, facilitating the successful incorporation of the dopant. Another peculiar aspect of the future of CDs is that, if they are of higher crystallinity, they may undergo π-π stacking due to planar graphitic layers stacking on top of each other, leading to aggregation and quenching effects.

2.2. Photoluminescence

CDs possess excellent photoluminescent properties and tunability over optical properties, making them an excellent candidate for imaging, ink, and optoelectronic applications. The comparatively smaller size of CDs leads to the quantum confinement effect of sp^2 hybridised carbon dots, where surface defects attached with functional groups and the presence of molecular fluorophores contribute to this effect. The mechanism of CD luminescent emission is still a controversial topic with these three proposed mechanisms, among which the quantum confinement effect (QCE) is most accepted, suggests as energy levels for electronic transition gets discrete due to smaller size, energy gap between the highest occupied molecular orbital (HOMO) and lowest occupied molecular orbit (LUMO) will get further widen and causes blue shift in the light emitted. The presence of numerous functional groups throughout the surface layer of CDs can also create trapping sites, resulting in radiative recombination and fluorescence. In molecular fluorophores, the presence of small organic fragments can also occur, leading to this mechanism. Functionalisation or passivation with different materials can alter the emission characteristics of CDs, indicating that the order of π electronic structure in the system changes with the addition of other groups to the structure (Mintz et al., 2021). The PL spectrum of CDs mainly exhibits excitation-dependent features and hence a shift in the wavelength corresponding to emission can occur, as shown in the *Figure 7 (a)*, whereas *Figure 7 (b)* represents the corresponding excitation and emission wavelengths and UV-Visible spectrum (Zhong et al., 2025).

Figure 7. (a) PL- excitation spectrum and (b) UV-Visible spectrum of CDs synthesized

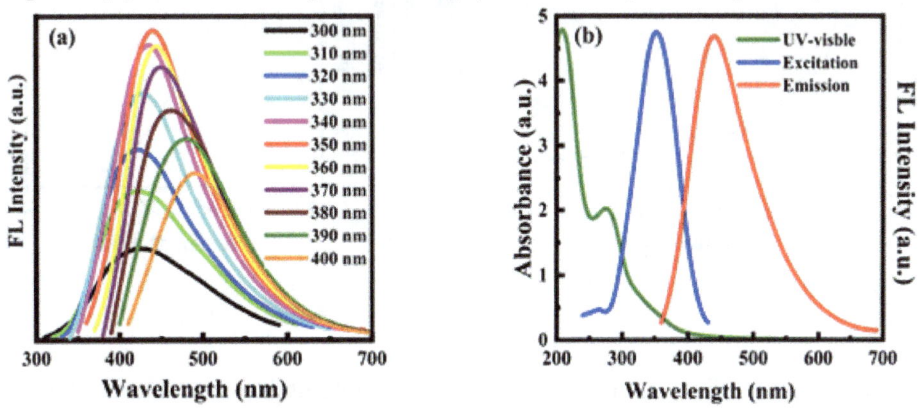

(Zhong et al., 2025).

During the passivation reaction for enhanced PL emission, surface defects are eliminated, preventing excited electrons from being trapped in non-radiative emissive sites and thereby enhancing fluorescence emission. In contrast, functionalisation will tune the CDs' responsiveness to specific pH levels, particular ions, or particular environments, making the response easier for detection, imaging, or biosensing. The doping of heteroatoms, such as nitrogen, sulfur, boron, and phosphorus, leads to electronic structural modifications and creates new emissive sites, increasing quantum yield. Aggregation of CD structures in turn leads to fluorescent quenching by creating new non-radiative pathways for energy dissipation, aggregation-induced emission due to π-π stacking mechanism and sometimes certain luminophores in higher concentrations known as aggregation-induced emission can also occur, which is opposite to that of the commonly called aggregation-caused quenching mechanism of fluorescence loss on aggregation (L. Li & Dong, 2018; Sen & Bose, 2025). The UV-Visible spectrum Figure 8 and associated PL spectrum of CDs with and without PEG functionalisation are given in Figure 8 (b), which has been optimised for drug nanocarrier application (Arsalani, Nezhad-Mokhtari, & Jabbari, 2019).

Figure 8.(a) UV–vis absorption spectra and excitation-dependent photoluminescence (PL) spectra of the PEG passivated carbon dots (CDs-PEG) (b) and the PL spectra of the CDs with and without the addition of PEG

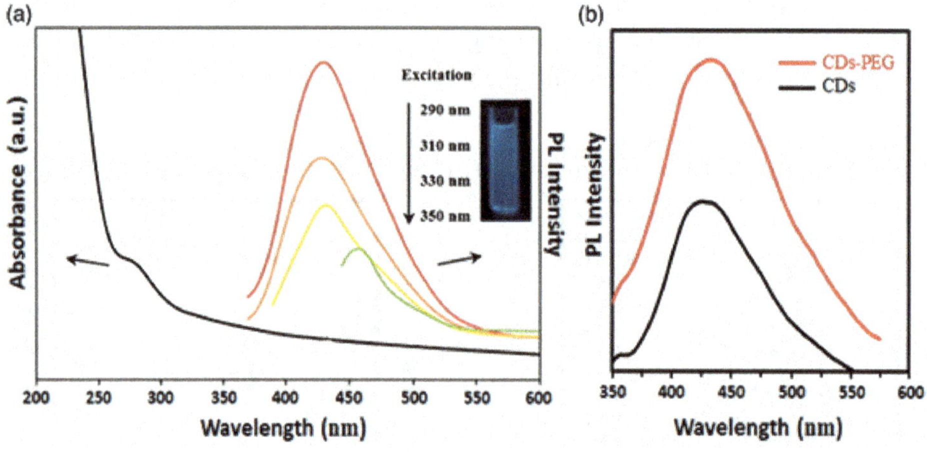

(Arsalani et al., 2019).

Up-conversion photoluminescence/anti-Stokes photoluminescence is a fascinating phenomenon observed in CD species, where they absorb multiple low-energy photons and emit a single, high-energy photon. Here, on passivation, attaching densely branched polymers or effective protective layers to the CD surface through passivation will decrease the chances of sealing surface defects and increase the availability of excited electrons. Whereas, in functionalisation, functional groups act as emissive traps, capturing electrons to create new radiative pathways and alter the energy levels by creating new energy states, leading to a red shift in the emission or an overall enhancement in fluorescence intensity. Doping or incorporation of foreign atoms into the CDs' surface can also alter the electronic structure as it creates intermediate energy levels within the bandgap of CDs. Nitrogen-, sulfur-, and phosphorus-doped CDs have been studied for up-conversion/down-conversion mechanisms, showing sequential two-photon absorption. Room-temperature fluorescence, resulting from intersystem crossing and boric acid or rigid matrix encapsulation, is enhanced in CDs with certain doping, leading to restricted vibrational states. When doped with metallic nanoparticles and specific metal ions, surface plasmon resonance (SPR) states may occur, and localisation of these states and plasmon-exciton coupling can lead to plasmon-enhanced effects, resulting in improved fluorescence, catalysis, energy conversion properties, and photostability.

2.3. Other Optical Properties

UV-Visible absorption spectra for unmodified CDs will show strong absorption peaks at 230-280nm, corresponding to the π-π^* transition of the sp^2 hybridised carbon core, and a shoulder peak attributed to the n-π^* transition of heteroatoms bonded functional groups, such as C=O or C=N, which reveals the CDs' surface chemistry. The size of synthesised CDs has a significant influence on their absorption peak positions, as a reduced size can cause a larger band gap and a resultant blue-shifted wavelength. Surface functionalisation, passivation, or doping can, on the other hand, give rise to new absorption shoulder peaks and a red shift in the system due to narrowed band gaps. Passivation, on the other hand, can increase absorbance by sealing defects and preventing aggregation, thereby improving UV-light exposure stability.

Photothermal conversion of absorbed light into heat and its applications in photothermal therapies, solar water evaporation and anti-bacterial applications. The most advanced studies on nonlinear optical properties of CDs to high-intensity laser lights have been reported to have two-photon absorption (TPA) and optical limiting, which is highly sensitive according to the surface states and doping, which is widely used in optical limiters and saturable absorbers for lasers to prevent tissue damage and sensitive equipment. In addition to these most common optical properties, rewritable displays and dynamic anti-counterfeiting, driven by photo-induced electron transfer or free radical-driven photochromism, in which the reversible optical properties are utilised, are also studied for CDs with modified properties. The use of specific chiral precursors, such as L-Cysteine, or the attachment of chiral ligands during the functionalisation and passivation process, can lead to optical responses that differ for left-handed and right-handed circularly polarised light. Hence, when these CDs with chiral groups interact, they exhibit circular dichroism and circularly polarised luminescence (CPL), and are a novel tool in asymmetric synthesis and chiral sensing.

Identification of doping or surface functionalisation requires sophisticated analytical instrumentation facilities such as X-Ray photoelectron spectroscopy (XPS), Nuclear Magnetic Resonance (NMR) spectroscopy, High Performance Liquid Chromatography (HPLC), along with shift or peak broadening in the existing peaks of FTIR, XRD or UV-Visible spectroscopy as compared to the base compound has to be studied.

2.4. Water solubility

The water solubility of CDs is mainly due to the oxygen-containing functional groups, such as -COOH and -OH, that are attached to the structure. To improve solubility in water, introducing more hydrophilic -OH functional groups will be inherent in creating a dense hydrophilic shell around the carbon core. Surface charge-induced CDs during the passivation mechanism can create electrostatic repulsion among themselves, preventing aggregation and increasing dispersion stability, as reported for ethylenediamine molecules. Hydrogen bonding due to attached functional groups may also lead to solvent-solute interactions, thereby preventing aggregation. Hydroxyl, amine, and carboxyl groups enable extensive hydrogen bonding with surrounding water molecules, creating a hydration shell around the CDs. In neutral or basic solutions, amine groups become protonated, and the resulting electrostatic repulsion can enhance colloidal stability. The functionalisation of CDs using Polyethylene Glycol (PEG) known as PEGylation with long ether linkages attached, is rich in hydrogen bonds, and a hydration layer forms, making them suitable for in vivo applications (D'souza & Shegokar, 2016).

Most likely used passivation agents and functionalisation agents to improve water solubility are Polyethylenimine, Polyethene Glycol, and amine passivation, which leads to protonated amine groups in the aqueous solution and excellent electrostatic repulsion due to the presence of highly polar groups.

The water-loving nature may not always be preferred, as it is shifting towards optoelectronics and antimicrobial coatings. In such cases, doped CDs will be selected, for which the degree of doping with heteroatoms is high. These CDs will have non-polar surfaces and be stable and soluble in organic solvents rather than water, due to the dominance of non-polar surface groups. The Hydrophobic CDs (HCDs) will be based on partial carbonisation of organic materials, in which the reaction is ceased at an extent where hydrophilic products might form. Here, partial carbonisation will lead to a non-polar, hydrophobic surface. Some post-synthesis modifications have also been proposed for HCD preparation using hydrophobic molecules, such as long-chain alkyl amines or polymers, to facilitate reactions with hydrophilic carboxyl groups. Doping of heteroatoms can control the hydrophobic core and residual polar surface groups, helping to tune hydrophobicity. The hydrophilicity or hydrophobicity of the sheets prepared or coated with CDs is studied for corrosion-resistant antimicrobial coatings using contact angle measurements.

2.5. Biocompatibility and Low-Toxicity

The biocompatibility and low toxicity of CDs make them a suitable candidate for interacting with biological systems, replacing conventional, toxic, heavy metal-based quantum dots. The properties of these CDs can be easily tuned due to their surface, which is rich in functional groups, allowing for further modification and targeted interactions with control over other biological interactions. Cell viability studies using the MTT assay, as well as live or dead staining, are employed to determine the IC50 value. Additionally, flow cytometry and genotoxicity tests, such as the comet test and micronucleus test, are performed to assess the feasibility of CDs for in vivo studies. In vivo studies, which examined living systems with excretion and clearance, pharmacokinetics, biodistribution, and histopathology, revealed the potency of CDs as a biocompatible system.

For example, PEG passivation creates a stealth layer that prevents protein corona formation and prevents CDs from being recognised by antibodies generated in the immune system, making them suitable for use as nanocarriers in anti-cancer drug delivery systems. Natural and biopolymer-based polymers, such as alginate, pectin, or chitin, are used for stabilised CDs, which contain anti-inflammatory and anti-tumour agents. EDA or PEI are mainly utilised as passivation agents, which provide amine-passivated CDs and are widely used in fluorescent probes for medical diagnostics and therapeutic applications, due to their non-disruptive nature in biological systems, smaller size, and low cytotoxicity.

Rather than passivation, functionalisation by creating stable surface sites for conjugating CDs with specific antibodies has also been explored for targeted anti-cancer drug delivery. In this approach, the real-time monitoring of drug transportation can be observed, and the visualisation of the drug release mechanism is also possible. Specific targeting ligands can be implemented for the targeted delivery of drugs, whereas targeting ligands can be eliminated for the removal of CDs from the body via the activation of protein corona. The CDs synthesised have improved colloidal stability, due to hydrophilic substitutions, which will prevent agglomeration and ensure dispersed stability in in vivo applications. The abundance of folate receptors on the surfaces of inflamed cells facilitates endocytosis, and changes in the atmospheric pH lead to the release of drugs to the targeted body, resulting in effective treatment without harming healthy cells. This can be effectively done with CD-mediated systems, along with monitoring. The functionalised or non-functionalised CDs can hence be selected according to the treatment given. Doping with Nitrogen or sulfur atoms has also been identified as improving biocompatibility due to hydrophilic substitutions, which induce cellular uptake and antioxidant properties by reducing oxidative stress in cells. ROS generation and cytotoxic effects can be reduced by eliminating specific doping sites, passivation agents, controlling concentration and dosage, preventing aggregation, and minimising byproducts in the CDs. The clumping of CDs should be controlled when used for biological applications, as it can cause ACQ or AIE emissions. Tunability over this biological activity and nano size makes this field of research interesting, with potential for greater contributions to the medical sector.

2.6. Charged CDs

Charged CDs are generally classified based on the surface charge, whether it is positive or negative, due to the presence of functional groups that protonate via functionalisation, passivation, or doping, which prevents aggregation by creating electrostatically repellent CD particles in the solution. The carboxyl groups within an aqueous system with a pH above the isoelectric point will give rise to deprotonation, forming carboxylate ions that offer a net negative charge. Citric acid, sodium citrate, and tribasic citrate are the primary precursor materials used, which undergo deprotonation in water, resulting in a negative charge. Urea, PEI, or EDA passivation, with amine groups interacting with the surrounding medium, will prevent aggregation and increase solubility in the solution phase. The folic acid attachment to negatively charged CDs and neutralisation of charge is effectively used in targeted drug delivery. Doping of heteroatoms like nitrogen can induce a positive charge on the CD surface, whereas the sulfur group will give sulfonate groups with a negative charge due to the altered electron density of the carbon lattice.

Negatively charged CDs are widely used in applications involving cationic gene delivery, as gene complexes are typically positively charged, and in fluorescent on-off sensors. Positively charged CDs, on the other hand, are used as efficient anti-bacterial agents, having strong electrostatic interaction mechanisms with the negatively charged bacterial cell membrane, and as drug or gene delivery agents.

2.7. Electrical, Thermal and Magnetic Properties

The electron transfer ability and semiconducting behaviour of CDs make them a substitute for conventional carbon quantum dots, with the least modification possibilities and toxicity. The electrical properties can be improved by covalent functionalisation with electron-donating groups, or by creating hybrid compositions with conductive materials like CNT or graphite, in which CD acts as a bridge to prevent aggregation and facilitate smooth electron transport. Piezoelectric and triboelectric effects occur due to the inherent structural asymmetry of CDs, resulting from their functional groups and structural defects. This asymmetry leads to charge separation under mechanical strain or slight friction, allowing for the generation of a small voltage. Thus, the addition of these materials can enhance their properties.

The phonon- or electron-controlled thermal conductivity in solid and nanofluidic forms, in different solutions, is the dominant mechanism that depends on the form of the surrounding mechanism of CDs. In solid form, the conductivity is highly dependent on quantised phonon lattice vibrations. As phonon-dominant transport will occur due to the graphic core and functional group interactions, CDs have been identified to have applications as a coolant in some liquid systems, providing effective thermal conductivity via convective and diffusion motion in nanofluid systems with excellent solubility in a variety of liquids. Modifications of CDs according to these applications are an emerging area of research for computational fluid dynamics researchers. In addition to this, the thermal properties of CDs have been widely utilised in photothermal studies, where the incorporation of graphitised domains within the carbon core or the addition of functional groups to this domain can modify near-infrared (NIR) absorption, leading to a photothermal effect. This effect will be directly proportional to the aggregation ability of CDs and inversely proportional to the density of the aggregate, as the collective system will exhibit more light absorption, trapping heat and localised overheating.

CDs, even though they do not have any metallic composition, fascinate in showing paramagnetism or ferromagnetism due to defects and unpaired electrons that have been trapped along with the crystalline magnetic structure. Heteroatom doping into these structures, hence, points to its application in metal-free Magnetic Resonance Imaging (MRI). The intrinsic electronic spin states, which are studied under spintronics for qubit storage, are also being focused on due to their natural spin states (Jian Liu et al., 2017).

3. SURFACE PASSIVATION

Surface passivation occurs during the bottom-up synthesis of CDs, which is a one-pot synthesis strategy, on the addition of specific compounds of interest or due to the in-situ passivation, in which the precursor material itself, with different compositions, can lead to various functional groups' caping' into the CD surface. Precursor materials, such as citric acid, glucose, and sucrose, are excellent capping agents for incorporating -OH groups into the CDs' core structure. Hydrothermal, solvothermal, or microwave-assisted synthesis strategies, as well as one-pot rapid methods of incorporation, will be suitable in such cases. In the case of top-down approaches, such as electrochemical exfoliation or chemical reduction processes, the purity of the precursor material yields highly crystalline structures with the fewest capillary cavities remaining, to which the functional groups become attached. In such cases, if we need to tune properties such as fluorescence, QY, solubility or stability, specific modifications can be done after the synthesis procedures, known as post-synthesis passivation of some chemicals, mixing or grafting into the CD surface. These two processes, with specific examples and the passivation mechanism in some biological systems, are explained in the subsections.

According to its structure, the mechanism associated with CDs passivation is due to dangling bonds and surface defects in the CD surface. The surface will have trapping sites that trap electrons during the luminescent process, causing fluorescent quenching. Hence, the stability and enhanced properties can only be attained by incorporating an additional layer of surface-active groups onto the core structure to fill the gaps and control the surface properties. The central carbon lattice core nuclear structure is surrounded by outer attached layers of functional groups, which serve as the defect-resolving layer with key functional groups such as -COOH, -OH, and-NH_2, using suitable precursor or pre-treatment agents. The depiction of the core structure of this CD, including defects and attached functional groups, is shown in *Figure 9.*

Figure 9. CDs with surface functional groups of hydroxyl, amino and carboxyl groups in the defect sites, a representation.

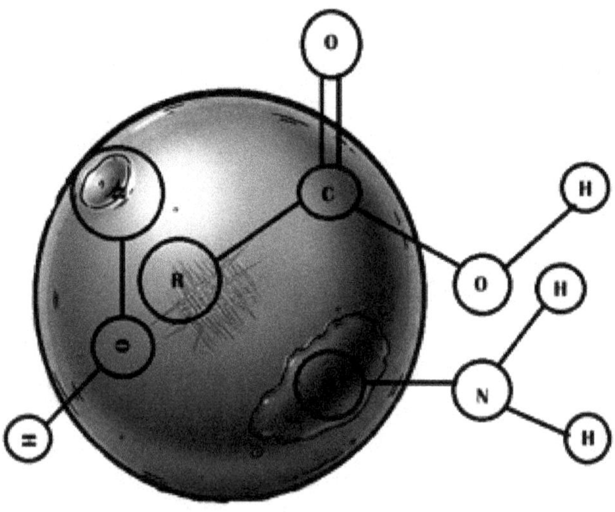

3.1. Passivation Mechanism

The overall passivation mechanism relies on the adaptability of the passivation agent, which is used to anchor functional groups in the CDs. These anchoring sites are essentially pre-existing moieties, along with carbonised carbon structures. During synthesis, active dangling bonds from these surfaces will accept the functional groups to create more stable structures.

The passivation mechanism of incorporating -OH functional groups involves the thermal decomposition of oxygen-rich precursor materials, resulting in thermal dehydration and polymerisation during a series of complex reactions inside the reaction chamber. This process leads to the formation of aromatised products, which further nucleate and grow through the creation of covalent bonds with the core structures. This is a surface healing mechanism that allows defects to be resolved.

Whereas, coming to functionalisation of carboxyl-rich CDs when treated with polymers such as PEI, with an amine group can modify CDs synthesised due to the rich -COOH groups present in the surface via carbodiimide (EDC/NHS) mechanism to form stable covalent amide (-CONH) bonds, which is crucial for drug delivery applications. This reaction mechanism involves three steps: 1. Carboxyl group activation, by forming a highly unstable and reactive intermediate of O-acylisourea formation, 2. Leading to activity with NHS (N-hydroxy succinimide) and forms an amine-reactive NHS-ester, which controls immediate hydrolysis of the activated group, 3. As the passivation agent is added to the solution, and followed by nucleophilic attack on these esters, it displaces NHS and forms covalently linking -CONH bonds on the CD surface (Dimos, 2016; Latiza, Olay, Eguico, Yan, & Rubi, 2025).

Sometimes, the passivation mechanism can be a non-covalent mechanism, such as electrostatic interactions or van der Waals forces, which are less stable than those of covalent interactions due to the surface charges of the synthesised CDs. Here, stable shells can be formed through the use of a passivation agent, such as a cationic polymer like chitosan, and can trigger luminescent mechanisms. Hydrogen bonding, or π-π stacking due to the aromatic sp3 hybridised carbon core with aromatic functional groups, can lead to a stable passivation layer where molecules are weakly bonded to the carbon core. After modification with these molecules, biocompatibility, optical properties, and physical properties improve.

3.2. In-Situ Passivation

The mechanism of one-pot passivation, which enables passivation at the time of synthesis, involves mixing specific passivation agents with the precursor materials during the synthesis process, thereby achieving simultaneous carbonisation and passivation. This approach aims to enhance particular properties by incorporating specific functional groups.

Citric acid and urea are the main precursors for CD synthesis, where citric acid serves as an efficient carbon source and urea as a nitrogen source, effectively generating more -NH_2 groups on the defect surface of the carbon network formed. Sugar-based carbohydrates will be used for the more hydrophilic CDs, as they can be carbonised, and the remaining oxygen groups can occupy the defects in the structure to form -OH functional groups rich CDs. This group's encapsulation will contribute towards the enhancement of the properties.

The primary objective of in situ passivation of CDs is to enhance their fluorescence efficiency. The fluorescent quenching in the PL emission mechanism should be minimised to achieve maximum emission and broader tunability in optical properties, such as colour purity, emission colour intensity, or quantum yield value. Non-radiative recombination sites should be eliminated to achieve a higher quantum yield, which can be achieved by passivation during synthesis. Rather than selecting a suitable precursor material for synthesis, the control over the surface chemistry of in situ passivation is comparatively less, as it will further depend on the CD size, structure, and degree of carbonisation, as well as the resultant dangling bonds formed on the surface. Therefore, during synthesis, parameters such as temperature, time, and even the ratio of precursors to passivation agent should be optimised for maximum fluorescence. The quenching will be reduced as the defect sites on the surface are reduced due to the saturation of dangling bonds, which are effectively healed by the surface functional groups. This results in a smooth surface, leading to higher efficiency in the conversion of absorbed light to emitted light without energy loss due to electron vibrations.

The -OH group incorporation will mainly achieved by oxygen rich precursor materials usage, and this will result in more enhanced solubility in water, high photoluminescent properties and antioxidant properties, which makes them a suitable candidate in fluorescent imaging, biomolecular detection, drug release studies, therapeutics etc. whereas, the incorporation of -NH2 groups into the matrix results pH sensitive CD surface, and -COOH incorporation allows further functionalisation, more stability in solution phase and helps enhancement in sensing and catalysis applications (Ayaz, Alas, & Genc, 2020).

The surface passivation agents are usually mainly organic molecules or polymers. It isn't easy to distinguish the role of the passivation agent and a carbon precursor separately in a reaction mixture using a single ratio of both. It also needs further trials. Almost all organic solvents have been used as passivation agents. In contrast, polymers such as PEG1500N, PEI, and PPEI-EI (propionylethylenimine co-ethylenimine) are the most commonly used agents for surface passivation (L.

Li & Dong, 2018). Some widely used precursor materials, passivation agents, and incorporation of functional groups are tabulated in *Table 3.1*.

Table 1. Different passivation precursor materials, agents, and functional groups are incorporated.

Precursor material	Passivation Agent	Functional groups incorporating	Reference
Citric Acid	Urea, Ethylenediamine	$-NH_2$,-CONH	(Sendão, Yuso, Algarra, Esteves da Silva, & Pinto da Silva, 2020)
Citric Acid	Polyethylenimine (PEI)	$-NH_2$	(C. Wang, Xu, & Zhang, 2015)
Sugar, Molasses	EDTA	$-NH_2$,-COOH	(Lou et al., 2020)
Citric Acid	Arginine (Amino acids)	B-C,-OH	(Iqbal et al., 2018)
Glycerol	H_3BO_4	B-C,-OH	(Shen, Chen, Cheng, & Qi, 2023)
Organic materials	No addition	-OH,-COOH, Epoxy	(Qian et al., 2013)
Chitosan	Ethylene diamine	$-NH_2$	(Maruthapandi, Sharma, Luong, & Gedanken, 2020)
Soybean Phospholipids	H_3PO_4	$-PO_4$,$-NH_2$	(R. Ding et al., 2023)
o-Phenylenediamine	Ionic Liquid	$-NH_2$,-C=O	(B. Wang et al., 2022)
Chitosan	H_3PO_4	$-PO_4$,-OH	(Zhao et al., 2025)

Surface passivation of CDs was achieved in 2011 using a single-step process, where glycerol served as the necessary carbon source and 4,7,10-trioxa-1,13-tridecanediamine (TTDDA) acted as the surface passivation agent, synthesised via a microwave method (C. Liu et al., 2011). Surface passivation by microwave irradiation using PEG and gelatin is given in *Figure 10.*

Figure 10. A schematic illustration of the preparation procedure of PEG passivated CDs (CDs-PEG) through microwave carbonisation

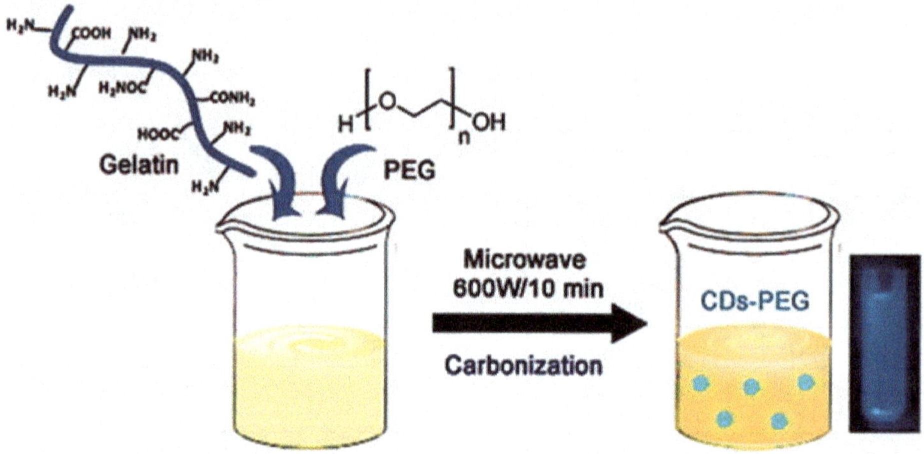

(Arsalani et al., 2019).

3.3. Post-synthesis Passivation

Direct mixing, grafting, or layer-by-layer assembly methods of post-treatment are available, and the CDs will be tuned after synthesis from the precursor material. Most top-down approaches utilise bulk carbon precursors to reduce them into CQD structures; consequently, the likelihood of surface defects is higher, resulting in a reduction in light-emitting capacity. Thus, tuning them with a suitable chemical agent and incorporating desired functional groups can be achieved for better performance. A passivation agent should be selected carefully, as its reactivity should match that of pre-existing functional groups. The agent should be soluble and stable in the same system as CD, with control over layer thickness increment after passivation, which can lead to an increase in size. Specific passivation agents, such as HNO3, EDA, PEG, PEI, folic acid, and even particular aptamers, are used to incorporate different functional groups and tune properties according to specific applications.

Incorporation of these co-precursor materials or passivation agents can influence the size of carbon dots synthesised, even though they will result in more defect-free and more fluorescent particles. For example, commonly used water-soluble PEG polymer functionalisation may lead to better bonding with the targeted ligands or peptides, resulting in improved photothermal and colloidal stability (Miruschenko et al., 2025). Whereas, suppose the proportion of PEG used is not optimised to retain fluorescence. In that case, it will result in greater quenching, and the increase in size will lead to steric hindrance and reduced stability over time*(Peng, Ji, Zhou, Zhao, & Leblanc, 2020)*.

Other than PEG, commonly used Primary precursor materials, surface passivation agents used, and the incorporated functional groups are tabulated in *Table 3.2.*

Table 2. Post-synthesis passivation agents are used to impart specific properties to precursor materials and functional groups.

Precursor material	Passivation Agent	Functional groups incorporating	References
Citric Acid	PEI	$-NH_2$, Bio conjugation	(Al Ragib et al., 2023)
Grapite (Laser ablation)	L-cystine	-SH (Thio) and $-NH_2$	(Strickland, Fourroux, & Pappas, 2025)
Graphite (Arc Discharge)	Glycolic acid	-OH	(Shafiee, Iravani, & Varma, 2022)
Graphite (Electrochemical)	PEG	-OH, Ether	(C. Ding, Zhu, & Tian, 2014)
Carbon soot (Oxidative treatment)	EDA via EDC/NHS	-CONH, $-NH_2$	(Tian & Yin, 2019)
Ascorbic acid (Hydrothermal)	Phenyl boronic acid (PBA)	Boronate group	(Anjali Devi, Anulekshmi, Salini, Aparna, & George, 2017)
L-Lysine (Hydrothermal)	H_2O_2 & H_2SO_4	-COOH,-OH	(Liao et al., 2023)
Carbon rod (Laser Ablation)	Folic acid	-COOH,-CONH	(Kim et al., 2021)
Glucose (Hydrothermal)	Chitosan	$-NH_2$	(Ahmed, El-Shahat, Allayeh, & Emam, 2023)

4. FUNCTIONALISATION

Functionalisation is also a surface modification technique, in which specific molecules are attached to the CD surface to improve its chemical and biological activities by tailoring their properties. Carboxylic acid groups, amines, or targeting ligands will be attached to the surface to enhance specific properties. It isn't easy to differentiate between passivation and functionalisation, in which functionalisation emphasises the covalent and non-covalent attachment of functional groups to improve specific biological applications. In contrast, passivation focuses mainly on enhancing the quantum yield of the synthesised CDs. The functionalisation mechanism will be the same as that of passivation, allowing for both in situ functionalisation and post-synthesis functionalisation, with a focus on specific applications. The modifications in the functional groups, upon incorporation of different functionalities, can be identified in the FTIR spectrum by analysing the changes in the functional groups (Mintz et al., 2021). Variation in the functionalisation by Tetraethylenepentamine (TEPA) functionalisation at different temperature and time interval is given in the *Figure 11*, which showing changes in the peak positions on addition

Figure 11. FTIR spectra of CQDs in the following conditions: (a) 12.5% TEPA in glycerol under 180°C for 4 hours and (b) 25% TEPA in glycerol under 180°C for 4 hours

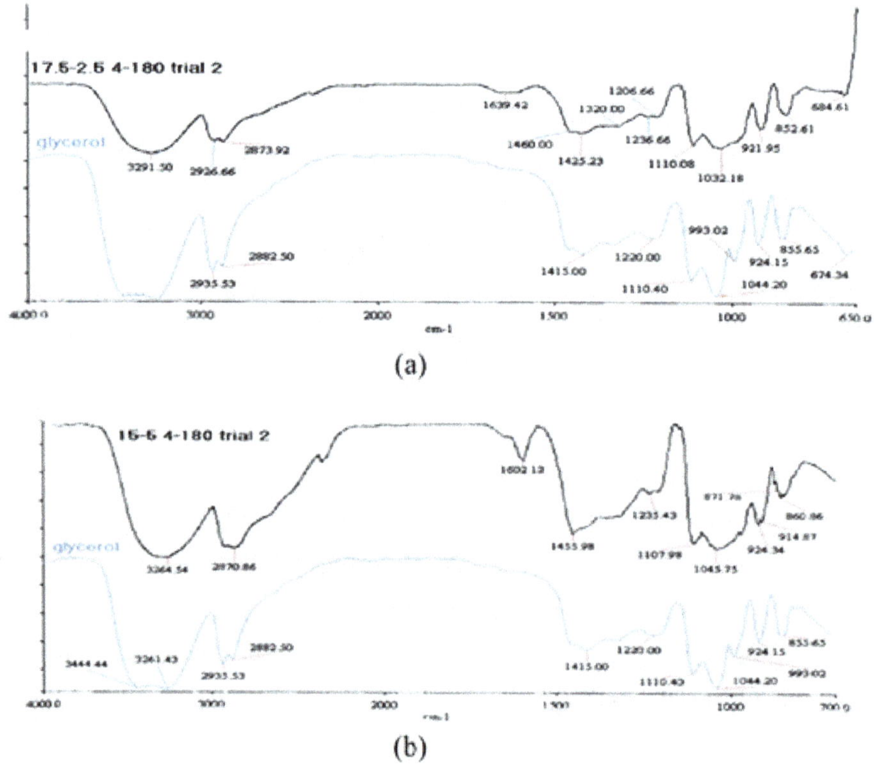

(Cruz et al., 2019)

The in-situ functionalisation and post-synthesis functionalised carbon dots, examples, and application perspectives are discussed in detail here.

4.1. In-Situ Functionalisation (One-Pot strategies)

Here, instead of focusing solely on certain functional groups, specific ligands, drugs, or chemical groups are introduced to the CD surface during a one-pot synthesis method. For example, particular proportions of amino acids are typically used in the reaction mixture to enhance the attachment of carboxyl groups and amine groups to the surface, thereby improving the biocompatibility of the prepared sample. Here, a post-synthesis possibility for specific and more complex ligand attachment will be facilitated to the synthesised CDs. Here, the docking sites will be ready to accept antibodies, peptides, folic acid, enzymes, or DNA for targeted drug delivery or to quantify the targeted molecule. Mainly hydrothermal, solvothermal, and microwave-assisted synthesis methodologies, along with appropriate co-precursor materials or chemical agents, will be employed. The first step of in situ synthesis is the carbonisation or polymerisation of the precursor materials, which leads to the elimination of water from the precursor and the formation of C=C bonds, resulting in carbonisation and subsequent polymerisation. Here, the functionalising agent will be co-carbonised and will act as a surface passivation agent after being integrated into carbon lattices (Pandit, Behera, Sahoo, & De, 2019; Ru et al., 2020).

In the case of hydrothermal synthesis, when molecules are heated inside a reaction chamber, a supercritical water environment is created with the pressure and temperature supplied. Amidation or esterification creates active sites, which in turn create further active sites. Here, an intrinsically integrated CD structure with covalent or non-covalent active sites will be made for further attachment of targeted molecules to its surface.

For example, CDs synthesised from citric acid and ethylenediamine exhibit antibacterial activity due to protonated amine groups present on their surface. In the case of CDs with charged surfaces due to protonated groups, they will be easily attracted electrostatically towards negatively charged microorganisms, causing cytotoxicity in certain microorganisms, which can be positively utilised for bacterial growth inhibition. On the other hand, reactive oxygen species (ROS) can cause superoxide radicals, which in turn lead to oxidative stress in proteins, lipids, and DNA (Latiza et al., 2025; Ru et al., 2020).

4.2. Post-synthesis Functionalisation

Here, pre-existing CDs are modified to tailor their properties for specific applications without altering the core structure, allowing for more precise control over properties than one-pot strategies. These strategies include amidation and esterification mechanisms, primarily as they utilise the embedding of negatively charged functional groups, such as those of PEG or certain drugs, to enhance biocompatibility and improve drug delivery applications. In salination, which is another covalent functionalisation mechanism, silyl groups that can react with hydroxyl groups are added using silane coupling agents like (3-Aminopropyl) triethoxysilane (APTES) to form a stable siloxane bond (Si-O-Si). Non-covalent functionalisation, such as

π-π stacking, hydrogen bonding, or electrostatic interactions, is intended to minimise disturbance to the electronic structure of the CDs synthesised and to enhance the reversibility of the system. Here, positively charged CDs are utilised so that negatively charged DNA or RNA can bind to the graphitic carbon core. *Table 4.2.* represents the primary functionalisation methods, materials used and applications.

Table 3. Functionalisation methods, precursor materials and applications that focus

Functionalisation method	Materials used	Applications
Esterification	PEG, or alcohols, with sulfuric acid catalyst	Biocompatibility, water solubility, on-off sensors
Amidation	Polyethylenimine, antibodies, Carbodiimide reagent, Doxorubicin	Drug delivery, stability improvement, bioimaging
Silanisation	APTES/ other silane coupling agents	Improved dispersibility, surface encapsulation, and improved functionalised surface for ligand attachments
Metal-ion coordination	Metal ions fe3+,cu2+,Ag+	Ion sensing
Non-covalent	DNA/RNA (Aromatic molecules), charged molecules, polymers	Biosensing, Gene delivery, Controlled drug release.

5. DOPING

Foreign atoms will be embedded in the CD surface instead of functional groups, thereby improving the existing characteristics, which is what is achieved through doping. Doping can significantly reduce the biocompatibility and structural stability of CDs, as it can lead to the formation of inevitable byproducts or degradation over time, resulting in adverse effects on their properties. In comparison, the complexity and control over the dopant atom will be a challenging factor to consider. Increased catalytic activity, enhanced water solubility and QY are certain advantages over doping. The semiconductor industry, seeking an alternative to conventional quantum dots, can utilise them as a replacement. Doping can be achieved with a single atom or by adding more co-dopants to improve the properties. The effectiveness of heteroatom doping lies in its ability to alter the electronic structure of the carbon dots. When a heteroatom with a different number of valence electrons or a different atomic size is introduced, it creates defects and changes the electron density within the carbon lattice. When more co-dopants are added to this matrix, the synthesis strategies and defect sites become more complex to handle in terms of structural explanation. Metal oxide doping and other carbon materials, such as carbon soot, carbon nanotubes, and specific metallic complexes, are also used to enhance properties like conductivity, stability, sensing, catalysis, and energy applications. The heteroatom doping, Co-dopants used, and different metal oxide composites and carbon materials used for doping are explained in the subsections (Ghosh, Das, Das, Banerji, & Das, 2022).

5.1. Hetero Atom Doping

The typical heteroatoms used for doping are Nitrogen, Sulfur, Phosphorus, and Boron. These are of a size comparable to carbon atom and are rich in electron donors. The doping with nitrogen often introduces electron-rich sites, enhancing electrical conductivity and making them ideal for biosensing, catalysis, and energy conversion devices with excitation-independent characteristics. When it comes to Sulfur doping, Sulfur doping introduces a new absorption band in the visible spectrum and creates new surface states that can enhance both photoluminescence and catalytic activity. When sulfur and nitrogen are co-doped, they often exhibit a synergistic effect, resulting in even higher quantum yields and improved performance in sensing and bioimaging. Phosphorus is particularly effective at altering the electronic structure of the carbon core and has shown promise in photocatalysis, where it can enhance charge transfer processes and improve fluorescence effects. Boron doping can induce p-type semiconductor behaviour, which helps create p-n junctions in photovoltaic devices. It primarily affects surface functionalities and can alter the absorption spectrum (Miao et al., 2020). Common atoms in doping and co-doping, along with their applications, are given in Figure *12*.

Figure 12. Doped and Co-doped CDs in a nutshell

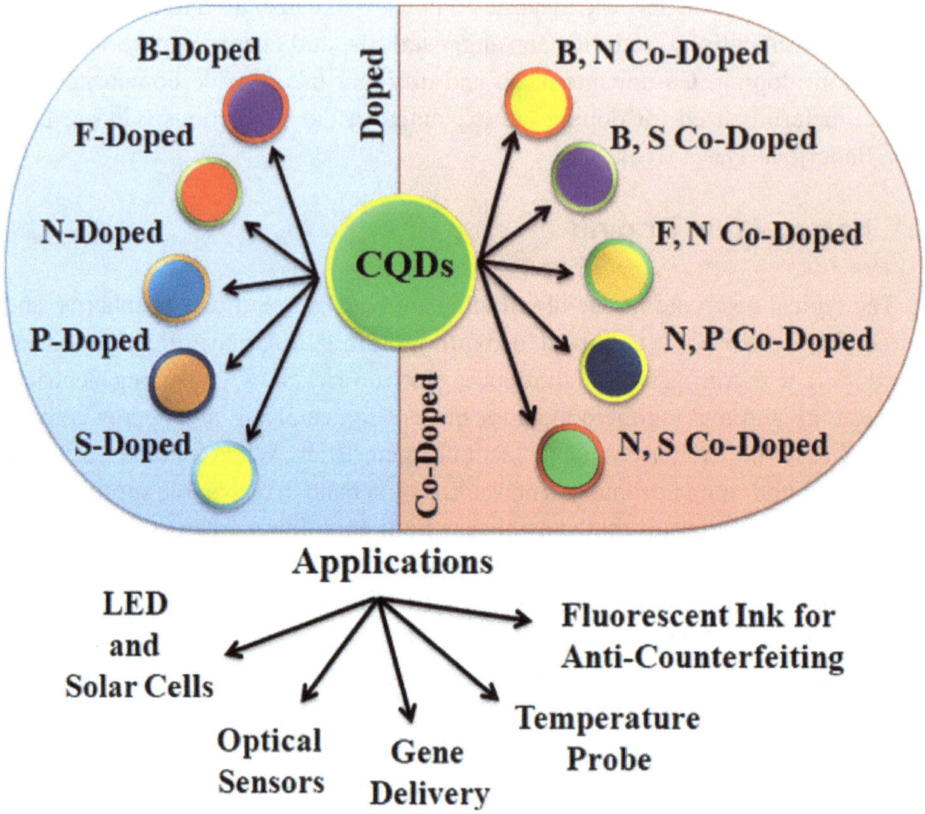

(Kandasamy, 2019)

Table 4. Different heteroatoms and their commonly used precursor materials, along with synthesis methodologies for the synthesis of CDs, are tabulated in the Table.5.1

Heteroatom	Doping precursor	Synthesis methods
Nitrogen	Urea, Ethylenediamine, L-Cysteine, Guanidine Hydrochloride, Amino Acids	Hydrothermal, solvothermal, Pyrolysis
Sulfur	Thiourea, Methionine, Cysteine, Sodium thiosulfate	Hydrothermal, Microwave
Phosphorus	Phosphoric acid, Sodium phosphate salts, phosphorous tribromide	Hydrothermal, Solvothermal
Boron	Boric acid, Phenylboronic acid	Hydrothermal, Pyrolysis

5.2. Co-Doped CDs

Co-doping involves incorporating two or more different heteroatoms into the carbon dot lattice during the synthesis process. This is a significant advancement over single-element doping, as the combined effects of the dopants can lead to superior performance. In which the mechanism involves, the presence of multiple heteroatoms creates a unique electronic environment. For example, a standard co-dopant pair is nitrogen (N) and sulfur (S). While nitrogen introduces electron-rich sites, sulfur can alter the electron density and create additional defects. This synergistic interaction can lead to a more uniform electronic structure, resulting in a higher fluorescence quantum yield and a more precise emission wavelength. The Co-doped CDs have an advantage in often exhibiting more stable and tunable photoluminescence, improved electrical conductivity, and enhanced catalytic activity compared to their mono-doped counterparts. For instance, N, S-co-doped CDs have shown superior performance in bio-sensing and photocatalysis. The combined effect of the dopants can also increase resistance to photobleaching and environmental degradation(Fu, Li, Lu, Dong, & Yang, 2024).

5.3. CD- Metal oxide Composites

The metal oxide compositions of CDs are combinations with semiconductor metal oxides, such as titanium dioxide (TiO2) or zinc oxide (ZnO). The CDs typically act as sensitisers, improving the metal oxide's ability to absorb light and separate charge carriers. Metal oxides often have a wide bandgap, which limits their photocatalytic activity to the ultraviolet (UV) region. Carbon dots, with their broad absorption spectrum in the UV-Vis range, can absorb visible light and transfer the absorbed energy or photoexcited electrons to the metal oxide. This process is known as photosensitisation. The CDs act as a "light-harvesting antenna" and a "charge carrier reservoir," preventing the rapid recombination of electron-hole pairs in the metal oxide, which is a significant limitation of these materials. Finds a wide range of applications in composites in photocatalysis for environmental remediation, such as the degradation of organic pollutants in wastewater. They are also used in dye-sensitised solar cells to improve light absorption and charge separation efficiency, as well as in gas sensors where the presence of CDs enhances the metal oxide's sensitivity and response(Chu, Lee, Chang, & Liu, 2019).

5.4. CD- Carbon material Composites

This class of composites combines carbon dots with other carbon-based nanomaterials, such as graphene, graphene quantum dots (GQDs), or CNTs. The goal is to create a hybrid material that leverages the unique properties of each component. Graphene and CNTs are known for their high electrical conductivity, large surface area, and mechanical strength. Carbon dots, in this context, act as spacers and electron transfer bridges. By attaching CDs to the surface of graphene sheets, they prevent the sheets from restacking, which preserves the high surface area crucial for energy storage. The CDs also facilitate rapid electron transport and provide additional active sites. The synergistic properties of these composites are particularly beneficial for energy-related applications such as. Supercapacitors CD-graphene and CD-CNT composites are used as electrode materials, offering high specific capacitance, excellent rate capability, and long-term cycling stability (Gurushankar et al., 2025; Kannan et al., 2024; Kannan et al., 2023; Kannan et al., 2022; Kannan et al., 2021). The CDs increase the number of active sites for charge storage and enhance ion diffusion. They are also used as anode materials in lithium-ion and sodium-ion batteries, where CDs help buffer volume changes during cycling and improve charge/discharge rates (Y. Liu et al., 2021).

6. AGGREGATION

Aggregation is crucial for the tunability of physical and optical properties. This can cause increase or decrease in the fluorescence due to aggregation-induced emission AIE or aggregation – caused quenching (ACQ). AIE can be achieved in many ways, where aggregation limits intramolecular rotation or vibration, or weakens the π-π stacking to improve the optical properties. Here, restricted intramolecular motion (RIM) in the solid state of CDs enhances the luminescent properties, enabling applications such as solid-state lighting displays, sensing probes, and anti-counterfeiting devices. The special morphology of CDs, achieved through aggregation, enables the engineering of new properties. Aggregation can otherwise cause fluorescent quenching in dilute solutions, due to steric hindrance or through molecular conformational control. When embedded in some polymeric matrices, they prevent aggregation (Ru, Waterhouse, & Lu, 2022). Hence the aggregation can be classified as either controlled aggregation or uncontrolled aggregation, where, PL property influence and shifts in wavelength, intermolecular forces influence, controlling factors, influence of solvent polarity, pH, and aggregation induced emission, different mechanisms explaining reasons, solvent engineering, matrix embedding, applications such as bioimaging, sensing, drug delivery, LEDs, aggregation caused quenching and mechanism are discussed here. The hydrophobicity of CDs can induce more aggregation-induced emission characteristics. The ACQ, AIE phenomenon and other applications are discussed with different methods to control aggregation states.

6.1. Aggregation-Caused Quenching (ACQ) Mechanisms

ACQ is a phenomenon of strong fluorescence in the solution or isolated state, and reduces fluorescence in the solid aggregates or crystals. π – π stacking interactions, excitonic coupling, solvent polarity conditions or energy transfer pathways. The π–π stacking of carbon dots, which contain π-conjugated domains, occurs as they stack and leads to exciton trapping due to the close alignment of planar aromatic rings. The formation of excimers or aggregates leads to non-emissive sites, which enhance non-radiative decay and thus quench fluorescence. Some of the aggregation can be by excitonic coupling, in which, excited states of adjacent CDs interact is where, This mechanism can be utilised in quenching based sensing applications, in which loss of fluorescence property is using or has to be avoided in solid state lighting to ensure efficient OLED/LED performance (Chen, Lam, Kwok, Liu, & Tang, 2019; Ru et al., 2022). The mechanism behind this phenomenon is the restriction in intermolecular motion (RIM) in the solid state, as compared to the vigorous, non-radiative intramolecular motion that consumes this energy and restricts the RIM, leading to radiative decay domains. This is now being extensively utilised in chemical sensing, environmental monitoring, medical diagnosis, optoelectronics, and biological imaging (Z. Li, Ji, Xie, & Tang, 2021).

6.2. Aggregation-Induced Emission (AIE)

In AIE CDs, the fluorescence state gets a significant enhancement, opposite to that of the ACQ mechanism. Here, in the solution state, CDs exhibit weak emission. In contrast, in the solid form, they exhibit higher emission, caused by the loss of excited-state energy due to vigorous internal rotation and vibration, or due to the higher concentration or solvent polarity-induced restricted motion of surface functional groups. Mechanism of AIE and its details are given as a table (*Table.6.2*) a nutshell.

Table 5. Different AIE mechanisms in CDs

Mechanism	Description	Result of aggregation
Restriction of Intra-dot Vibration/Rotation (RIV/RIR)	This is the primary and most accepted mechanism, analogous to AIE in organic luminogens. CDs have flexible surface functional groups (e.g., $-NH_2$, -COOH, phenyl rings) that rotate and vibrate freely in solution	Aggregation and close packing physically restricts this vigorous movement, blocking the non-radiative decay pathway (energy loss as heat), forcing the excited state to decay via the radiative pathway (light emission), thus turning the fluorescence ON.
Inter-dot Excitonic/ Charge Transfer Coupling	In some structured CDs, aggregation creates a new electronic environment due to the alignment of the core structures or surface states.	The new coupling leads to the formation of J-aggregates or new charge transfer states that are radiatively efficient and often result in a significant redshift (longer wavelength) of the emission.
Surface/Core Rigidification	Aggregation can induce a structural phase change or a more ordered state, rigidifying the entire CD core and surface.	This rigidification protects the emissive centers from external non-radiative interactions (like solvent effects), stabilising the excited state and boosting the quantum yield.

This is having application in high- contrast bioimaging and cell sensing, environmental sensing, optoelectronics or solid-state lightings, anticounterfeiting and information storage(Z. Li et al., 2021; Tang & Qin, 2013; H. Wang et al., 2023)

CONCLUSION

Precise modifications in the CD structures can be induced to enhance their properties. The Luminescent emission characteristics, as well as all other properties, rely on the morphology and functional groups attached to CDs. The functional groups can be modified according to the application or the property that needs to be modified. This requires a need to concentrate on the concentrations and proportions of materials chosen for selection. The non-radiative pathway reductions and targeted functionalities can be introduced to achieve an innovative, reliable, multi-functional CD synthesis. The advancement of aggregation from nano size to macro structures and the ACQ phenomenon are advantages of controlled aggregation. Hence, this engineered material finds applications in almost all sectors, both alone and in a matrix, to enhance the properties of CDs.

REFERENCES

Ahmed, H. B., El-Shahat, M., Allayeh, A. K., & Emam, H. E. (2023). Maillard reaction for nucleation of polymer quantum dots from chitosan-glucose conjugate: Antagonistic for cancer and viral diseases. *International Journal of Biological Macromolecules, 224*, 858–870. DOI: 10.1016/j.ijbiomac.2022.10.172 PMID: 36306909

Al Ragib, A., Al Amin, A., Alanazi, Y. M., Kormoker, T., Uddin, M., Siddique, M. A. B., & Barai, H. R. (2023). Multifunctional carbon dots in nanomaterial surface modification: A descriptive review. *Carbon Research, 2*(1), 37. DOI: 10.1007/s44246-023-00069-x

Anjali Devi, J., Anulekshmi, A., Salini, S., Aparna, R., & George, S. (2017). Boronic acid functionalized nitrogen doped carbon dots for fluorescent turn-on detection of dopamine. *Microchimica Acta, 184*(10), 4081–4090. DOI: 10.1007/s00604-017-2433-7

Arsalani, N., Nezhad-Mokhtari, P., & Jabbari, E. (2019). Microwave-assisted and one-step synthesis of PEG passivated fluorescent carbon dots from gelatin as an efficient nanocarrier for methotrexate delivery. *Artificial Cells, Nanomedicine, and Biotechnology, 47*(1), 540–547. DOI: 10.1080/21691401.2018.1562460 PMID: 30829085

Ayaz, F., Alas, M. O., & Genc, R. (2020). Differential immunomodulatory effect of carbon dots influenced by the type of surface passivation agent. *Inflammation, 43*(2), 777–783. DOI: 10.1007/s10753-019-01165-0 PMID: 31873835

Chen, Y., Lam, J. W., Kwok, R. T., Liu, B., & Tang, B. Z. (2019). Aggregation-induced emission: Fundamental understanding and future developments. *Materials Horizons, 6*(3), 428–433. DOI: 10.1039/C8MH01331D

Chu, K.-W., Lee, S. L., Chang, C.-J., & Liu, L. (2019). Recent progress of carbon dot precursors and photocatalysis applications. *Polymers, 11*(4), 689. DOI: 10.3390/polym11040689 PMID: 30995724

Cruz, R. A. T., Soriano, A. N., de Yro, P. A. N., Quiachon, G. M. O., Emolaga, C. S., Ysulat, M. L. M., Bigol, U. G., & Basilia, B. A. (2019). Functionalized Carbon-based Quantum Dots: Optical Characterization and Potential Application as Bio-fluorophore. *IOP Conference Series. Materials Science and Engineering, 559*(1), 012003. DOI: 10.1088/1757-899X/559/1/012003

D'souza, A. A., & Shegokar, R. (2016). Polyethylene glycol (PEG): A versatile polymer for pharmaceutical applications. *Expert Opinion on Drug Delivery, 13*(9), 1257–1275. DOI: 10.1080/17425247.2016.1182485 PMID: 27116988

Dimos, K. (2016). Carbon quantum dots: Surface passivation and functionalization. *Current Organic Chemistry*, *20*(6), 682–695. DOI: 10.2174/13852728196661507 30220948

Ding, C., Zhu, A., & Tian, Y. (2014). Functional surface engineering of C-dots for fluorescent biosensing and in vivo bioimaging. *Accounts of Chemical Research*, *47*(1), 20–30. DOI: 10.1021/ar400023s PMID: 23911118

Ding, R., Li, Z., Xiong, Y., Wu, W., Yang, Q., & Hou, X. (2023). Electrochemical (bio) sensors for the detection of organophosphorus pesticides based on nanomaterial-modified electrodes: A review. *Critical Reviews in Analytical Chemistry*, *53*(8), 1766–1791. DOI: 10.1080/10408347.2022.2041391 PMID: 35235478

Fawaz, W., Hasian, J., & Alghoraibi, I. (2023). Synthesis and physicochemical characterization of carbon quantum dots produced from folic acid. *Scientific Reports*, *13*(1), 18641. DOI: 10.1038/s41598-023-46084-1 PMID: 37903841

Fu, Q., Li, N., Lu, K., Dong, Z., & Yang, Y. (2024). The emerging development of nitrogen and sulfur co-doped carbon dots: Synthesis methods, influencing factors and applications. *Materials Today. Chemistry*, *37*, 102032. DOI: 10.1016/j.mtchem.2024.102032

Ge, G., Li, L., Wang, D., Chen, M., Zeng, Z., Xiong, W., & Guo, C. (2021). Carbon dots: Synthesis, properties and biomedical applications. *Journal of Materials Chemistry. B*, *9*(33), 6553–6575. DOI: 10.1039/D1TB01077H PMID: 34328147

Ghosh, T., Das, T. K., Das, P., Banerji, P., & Das, N. C. (2022). Current scenario and recent advancement of doped carbon dots: A short review scientocracy update (2013–2022). *Carbon Letters*, *32*(4), 953–977. DOI: 10.1007/s42823-022-00339-5

Gurushankar, K., Karuppaiya, C., Tari, V., & Kannan, K. (2025). Exploring MXenes and MBenes in Metal Sulfur and Selenium Battery Technologies. In *Novel Energy Storage and Conversion Technologies for Two-Dimensional MXenes and MBenes* (pp. 105–120). IGI Global Scientific Publishing. DOI: 10.4018/979-8-3693-9546-2.ch005

Hoan, B. T., Tam, P. D., & Pham, V.-H. (2019). Green Synthesis of Highly Luminescent Carbon Quantum Dots from Lemon Juice. *Journal of Nanotechnology*, *2019*(1), 2852816. DOI: 10.1155/2019/2852816

Iqbal, A., Iqbal, K., Xu, L., Li, B., Gong, D., Liu, X., Guo, Y., Liu, W., Qin, W., & Guo, H. (2018). Heterogeneous synthesis of nitrogen-doped carbon dots prepared via anhydrous citric acid and melamine for selective and sensitive turn on-off-on detection of Hg (II), glutathione and its cellular imaging. *Sensors and Actuators. B, Chemical*, *255*, 1130–1138. DOI: 10.1016/j.snb.2017.08.130

Jorns, M., & Pappas, D. (2021). A review of fluorescent carbon dots, their synthesis, physical and chemical characteristics, and applications. *Nanomaterials (Basel, Switzerland), 11*(6), 1448. DOI: 10.3390/nano11061448 PMID: 34070762

Kandasamy, G. (2019). Recent Advancements in Doped/Co-Doped Carbon Quantum Dots for Multi-Potential Applications. *C, 5*(2), 24.

Kannan, K., Chanda, D., Gautam, J., Behera, A., Meshesha, M. M., Jang, S. G., & Yang, B. (2023). Hydrothermally synthesized mixed metal oxide nanocomposites for electrochemical water splitting and photocatalytic hydrogen production. *International Journal of Hydrogen Energy, 48*(93), 36412–36426. DOI: 10.1016/j.ijhydene.2023.06.042

Kannan, K., Chanda, D., Gautam, J., Meshesha, M. M., Jang, S. G., & Yang, B. L. (2023). Facial synthesis of pp heterojunction composites: Evaluation of their electrochemical properties with photovoltaics-electrolyzer water splitting using two-electrode system. *International Journal of Hydrogen Energy, 48*(37), 13814–13826. DOI: 10.1016/j.ijhydene.2022.12.235

Kannan, K., Chinnaiah, K., Gurushankar, K., Krishnamoorthi, R., Chen, Y. S., Murphin Kumar, P. S., & Li, Y. Y. (2024). Investigation of the electrochemical behavior of CuO-NiO-Co3O4 nanocomposites for enhanced supercapacitor applications. *Materials (Basel), 17*(16), 3976. DOI: 10.3390/ma17163976 PMID: 39203154

Kannan, K., Radhika, D., Gnanasangeetha, D., Krishna, L. S., & Gurushankar, K. (2021). Y3+ and Sm3+ co-doped mixed metal oxide nanocomposite: Structural, electrochemical, photocatalytic, and antibacterial properties. *Applied Surface Science Advances, 4*, 100085. DOI: 10.1016/j.apsadv.2021.100085

Kannan, K., Radhika, D., Kasai, R. D., Gnanasangeetha, D., Palani, G., Gurushankar, K., Koutavarapu, R., Lee, D.-Y., & Shim, J. (2022). Facile fabrication of novel ceria-based nanocomposite (CYO-CSO) via co-precipitation: Electrochemical, photocatalytic and antibacterial performances. *Journal of Molecular Structure, 1256*, 132519. DOI: 10.1016/j.molstruc.2022.132519

Kannan, K., & Tari, V. (Eds.). (2024). Novel energy storage and conversion technologies for two-dimensional MXenes and MBenes. ISBN: 9798369395462, eISBN: 9798369395486DOI: 10.4018/979-8-3693-9546-2

Kim, T. E., Jang, H. J., Park, S. W., Wei, J., Cho, S., Park, W. I., & Jung, Y. K. (2021). Folic acid functionalized carbon dot/polypyrrole nanoparticles for specific bioimaging and photothermal therapy. *ACS Applied Bio Materials, 4*(4), 3453–3461. DOI: 10.1021/acsabm.1c00018 PMID: 35014429

Kumar, R., Kumar, V. B., & Gedanken, A. (2020). Sonochemical synthesis of carbon dots, mechanism, effect of parameters, and catalytic, energy, biomedical and tissue engineering applications. *Ultrasonics Sonochemistry*, *64*, 105009. DOI: 10.1016/j.ultsonch.2020.105009 PMID: 32106066

Latiza, R. J. P., Olay, J., Eguico, C., Yan, R. J., & Rubi, R. V. (2025). Surface matters: Passivation in carbon dot synthesis—a critical review. *Journal of Materials Science*, *60*(11), 1–13. DOI: 10.1007/s10853-025-10764-2

Li, L., & Dong, T. (2018). Photoluminescence tuning in carbon dots: Surface passivation or/and functionalization, heteroatom doping. *Journal of Materials Chemistry. C, Materials for Optical and Electronic Devices*, *6*(30), 7944–7970. DOI: 10.1039/C7TC05878K

Li, Z., Ji, X., Xie, H., & Tang, B. Z. (2021). Aggregation-induced emission-active gels: Fabrications, functions, and applications. *Advanced Materials*, *33*(33), 2100021. DOI: 10.1002/adma.202100021 PMID: 34216407

Liao, X., Wu, B., Li, H., Zhang, M., Cai, M., Lang, B., & Zhou, P. (2023). Fluorescent/colorimetric dual-mode discriminating Gln and Val enantiomers based on carbon dots. *Analytical Chemistry*, *95*(39), 14573–14581. DOI: 10.1021/acs.analchem.3c01854 PMID: 37729469

Liu, C., Zhang, P., Tian, F., Li, W., Li, F., & Liu, W. (2011). One-step synthesis of surface passivated carbon nanodots by microwave assisted pyrolysis for enhanced multicolor photoluminescence and bioimaging. *Journal of Materials Chemistry*, *21*(35), 13163–13167. DOI: 10.1039/c1jm12744f

Liu, J., Bi, H., Cesar Morais, P., Zhang, X., Zhang, F., & Hu, L. (2017). Room-temperature Magnetism in Carbon Dots and Enhanced Ferromagnetism in Carbon Dots-Polyaniline Nanocomposite. *Scientific Reports*, *7*(1), 2165. DOI: 10.1038/s41598-017-01350-x PMID: 28526848

Liu, J., Li, R., & Yang, B. (2020). Carbon Dots: A New Type of Carbon-Based Nanomaterial with Wide Applications. *ACS Central Science*, *6*(12), 2179–2195. DOI: 10.1021/acscentsci.0c01306 PMID: 33376780

Liu, Y., Roy, S., Sarkar, S., Xu, J., Zhao, Y., & Zhang, J. (2021). A review of carbon dots and their composite materials for electrochemical energy technologies. *Carbon Energy*, *3*(5), 795–826. DOI: 10.1002/cey2.134

Lou, Y., Ji, J., Qin, A., Liao, L., Li, Z., Chen, S., Zhang, K., & Ou, J. (2020). Cane molasses graphene quantum dots passivated by PEG functionalization for detection of metal ions. *ACS Omega*, *5*(12), 6763–6772. DOI: 10.1021/acsomega.0c00098 PMID: 32258911

Mansuriya, B. D., & Altintas, Z. (2021). Carbon dots: Classification, properties, synthesis, characterization, and applications in health care—an updated review (2018–2021). *Nanomaterials (Basel, Switzerland)*, *11*(10), 2525. DOI: 10.3390/nano11102525 PMID: 34684966

Maruthapandi, M., Sharma, K., Luong, J. H., & Gedanken, A. (2020). Antibacterial activities of microwave-assisted synthesized polypyrrole/chitosan and poly (pyrrole-N-(1-naphthyl) ethylenediamine) stimulated by C-dots. *Carbohydrate Polymers*, *243*, 116474. DOI: 10.1016/j.carbpol.2020.116474 PMID: 32532398

Miao, S., Liang, K., Zhu, J., Yang, B., Zhao, D., & Kong, B. (2020). Hetero-atom-doped carbon dots: Doping strategies, properties and applications. *Nano Today*, *33*, 100879. DOI: 10.1016/j.nantod.2020.100879

Mintz, K. J., Bartoli, M., Rovere, M., Zhou, Y., Hettiarachchi, S. D., Paudyal, S., Chen, J., Domena, J. B., Liyanage, P. Y., Sampson, R., Khadka, D., Pandey, R. R., Huang, S., Chusuei, C. C., Tagliaferro, A., & Leblanc, R. M. (2021). A deep investigation into the structure of carbon dots. *Carbon*, *173*, 433–447. DOI: 10.1016/j.carbon.2020.11.017

Miruschenko, M. D., Kosolapova, K. D., Aleinik, I. A., Borodina, L. N., Vedernikova, A. A., Sokolova, A. V., Sandzhieva, M. A., Mitroshin, A. M., Yakimansky, A. V., Koroleva, A. V., Zhizhin, E. V., Cherevkov, S. A., Langer, M., Otyepka, M., Ushakova, E. V., & Rogach, A. L. (2025). Functionalization of Hydrophilic and Amphiphilic Carbon Dots with Polyethylene Glycol for Electroluminescent Devices. *Small Structures*, *6*(3), 2400528. DOI: 10.1002/sstr.202400528

Pandit, S., Behera, P., Sahoo, J., & De, M. (2019). In situ synthesis of amino acid functionalized carbon dots with tunable properties and their biological applications. *ACS Applied Bio Materials*, *2*(8), 3393–3403. DOI: 10.1021/acsabm.9b00374 PMID: 35030781

Peng, Z., Ji, C., Zhou, Y., Zhao, T., & Leblanc, R. M. (2020). Polyethylene glycol (PEG) derived carbon dots: Preparation and applications. *Applied Materials Today*, *20*, 100677. DOI: 10.1016/j.apmt.2020.100677

Pundi, A., & Chang, C.-J. (2022). Recent advances in synthesis, modification, characterization, and applications of carbon dots. *Polymers*, *14*(11), 2153. DOI: 10.3390/polym14112153 PMID: 35683827

Qian, Z., Ma, J., Shan, X., Shao, L., Zhou, J., Chen, J., & Feng, H. (2013). Surface functionalization of graphene quantum dots with small organic molecules from photoluminescence modulation to bioimaging applications: An experimental and theoretical investigation. *RSC Advances*, *3*(34), 14571–14579. DOI: 10.1039/c3ra42066c

Ru, Y., Ai, L., Jia, T., Liu, X., Lu, S., Tang, Z., & Yang, B. (2020). Recent advances in chiral carbonized polymer dots: From synthesis and properties to applications. *Nano Today*, *34*, 100953. DOI: 10.1016/j.nantod.2020.100953

Ru, Y., Waterhouse, G. I. N., & Lu, S. (2022). Aggregation in carbon dots. *Aggregate*, *3*(6), e296. DOI: 10.1002/agt2.296

Sen, S., & Bose, A. (2025). Carbon dots: A review of innovations, applications, challenges, and future prospects. *Inorganic Chemistry Communications*, *173*, 113852. DOI: 10.1016/j.inoche.2024.113852

Sendão, R., Yuso, M. V. M., Algarra, M., Esteves da Silva, J. C. G., & Pinto da Silva, L. (2020). Comparative life cycle assessment of bottom-up synthesis routes for carbon dots derived from citric acid and urea. *Journal of Cleaner Production*, *254*, 120080. DOI: 10.1016/j.jclepro.2020.120080

Shafiee, A., Iravani, S., & Varma, R. S. (2022). Graphene and graphene oxide with anticancer applications: Challenges and future perspectives. *MedComm*, *3*(1), e118. DOI: 10.1002/mco2.118 PMID: 35281783

Shen, Z., Chen, L., Cheng, H., & Qi, Z. (2023). Selective synthesis of glyceryl monolaurate intensified by boric acid based deep eutectic solvent. *Catalysis Today*, *407*, 291–300. DOI: 10.1016/j.cattod.2021.05.018

Strickland, S., Fourroux, L., & Pappas, D. (2025). Effect of precursors on carbon dot functionalization and applications: A review. *Analyst*, *150*(8), 1448–1469. DOI: 10.1039/D4AN01554A PMID: 40105280

Tang, B. Z., & Qin, A. (2013). *Aggregation-induced emission: Applications*. John Wiley & Sons.

Tian, X. T., & Yin, X. B. (2019). Carbon dots, unconventional preparation strategies, and applications beyond photoluminescence. *Small*, *15*(48), 1901803. DOI: 10.1002/smll.201901803 PMID: 31240816

Wang, B., Wei, Z., Sui, L., Yu, J., Zhang, B., Wang, X., & Tian, Y. (2022). Electron–phonon coupling-assisted universal red luminescence of o-phenylenediamine-based carbon dots. *Light, Science & Applications*, *11*(1), 172. DOI: 10.1038/s41377-022-00865-x PMID: 35668065

Wang, C., Xu, Z., & Zhang, C. (2015). Polyethyleneimine-Functionalized Fluorescent Carbon Dots: Water Stability, pH Sensing, and Cellular Imaging. *ChemNanoMat : Chemistry of Nanomaterials for Energy, Biology and More*, *1*(2), 122–127. DOI: 10.1002/cnma.201500009

Wang, H., Li, Q., Alam, P., Bai, H., Bhalla, V., Bryce, M. R., & Chen, X. (2023). Aggregation-induced emission (AIE), life and health. *ACS Nano*, *17*(15), 14347–14405. DOI: 10.1021/acsnano.3c03925 PMID: 37486125

Yan, F., Jiang, Y., Sun, X., Bai, Z., Zhang, Y., & Zhou, X. (2018). Surface modification and chemical functionalization of carbon dots: A review. *Microchimica Acta*, *185*(9), 424. DOI: 10.1007/s00604-018-2953-9 PMID: 30128831

Zhao, X., Lei, Z., Hou, H., Lin, G., Sun, Y., Li, H., Yang, X., Su, Y., Ali, K. M. Y., Hu, E., Wang, H., Wang, Q., & Hu, E. (2025). Purification of high–toxicity beryllium–containing water resources using carbon dot–enhanced phosphoric acid/chitosan composite gels. *Separation and Purification Technology*, *358*, 130299. DOI: 10.1016/j.seppur.2024.130299

Zhong, L., Sun, C., Zhao, X., & Zhao, Q. (2025). Dual Roles of Carbon Quantum Dots from Green Carbon Sources: A Fluorescence Sensor for Fe3+ Ions, UV and High-Energy Blue Light Screening. *Nanomaterials (Basel, Switzerland)*, *15*(6), 436. DOI: 10.3390/nano15060436 PMID: 40137609

Chapter 4
Spectroscopic, Microscopic, and Surface Analytical Characterization Techniques for Carbon Dots

Julin Joseph
https://orcid.org/0000-0001-7846-0670
St. Josephs College for Women, Alappuzha, India

Cicy Varghese
St. Josephs College for Women, Alappuzha, India

Bincy John
St. Josephs College for Women, Alappuzha, India

ABSTRACT

Characterisation of carbon dots involves the process of determining their physical, chemical and optical properties using different spectroscopic, microscopic and surface analytical approaches. Characterization of carbon dots (CDs) is crucial for understanding their properties, behavior, and potential applications. Different spectroscopic techniques used in the study of carbon dots are UV-Visible Spectroscopy, Photoluminescence (PL) Spectroscopy, Fourier Transform Infrared Spectroscopy (FTIR), Raman Spectroscopy, X-ray Photoelectron Spectroscopy (XPS), and Nuclear Magnetic Resonance (NMR) Spectroscopy. Microscopic techniques

DOI: 10.4018/979-8-3373-4247-4.ch004

used for characterisation of carbon dots include Transmission Electron Microscopy (TEM), Scanning Electron Microscopy (SEM), and Atomic Force Microscopy (AFM). Moreover, different kinds of surface analytical techniques are also used for the study of carbon dots.

1. INTRODUCTION: SPECTROSCOPIC TECHNIQUES

These techniques help to determine optical properties, electronic structure, and chemical bonding.

(a) UV-Visible Spectroscopy

UV-visible region is a part of electromagnetic spectrum that spans wavelengths from 190 nm to 800 nm. Across this wavelength range, the absorption of the electromagnetic radiation is caused by the excitation (transition to higher energy level) of bonding and non-bonding electrons of molecules. σ, π and n are the different types of molecular orbitals. The energy required to excite an electron in a bonding s orbital to the corresponding antibonding orbital is high and the transition is represented as σ-σ*. Saturated compounds containing atoms with lone pairs (non-bonding electrons) are capable of n-σ* transition. Most absorption spectrum of organic compounds is based on transitions of n or π electrons to the π* excited state. The energy of electronic transitions generally follows the order n-π*< π-π*<n-σ*<σ-π*<σ-σ* (Valeur & Berberan-Santo, 2012). UV-visible spectroscopy is a widely used technique to analyze the optical absorption properties of carbon dots (CDs), providing valuable insights into their electronic structure and surface functionalities. Most carbon dots exhibit characteristic absorption bands attributed to electronic transitions (Surendran et al., 2024). The UV–Vis spectra of carbon dots typically show two main types of absorption bands. One band arises from electronic transitions in the conjugated carbon core while the other is associated with the surface functional groups. Transition between bonding (π) and antibonding (π*) orbitals of conjugated carbon atoms can be indicated as π → π* transition. In carbon dots, it arises from C=C (sp^2) bonds in aromatic or graphitic domains. Transitions based on surface functional groups involve electron transfer from a nonbonding electron pair (n) on oxygen or nitrogen to an antibonding π* orbital of a carbonyl or similar group represented as n → π* transition. It is usually observed beyond 300-380 nm. The appearance of peaks near 300 nm indicates surface defects or functional groups such as C=O, C–O, C–N. Sometimes, a broad tail appears for wavelengths >400 nm results due to surface defect states. A prominent π–π* transition typically appears around 250 nm, originating from C=C bonds within the carbon core (Joseph &

Anappara, 2016a). Corresponding figure is given in Figure 1a. In contrast, an n–π* transition, is usually observed near 300 nm, and is associated with non-bonding electrons on heteroatoms, such as oxygen, in surface functional groups like C=O as shown in Figure 1b.These absorption features not only reveal the structural nature of CDs but also help in understanding surface modifications, including metal ion doping. Changes in the UV-visible spectrum such as shifts in peak positions or intensity can indicate successful doping or interaction with other chemical species. UV-visible spectroscopy also plays a crucial role in fluorescence quenching studies, which help determine the interaction mechanisms between CDs and quenchers. Two primary types of quenching mechanisms exist: static and dynamic quenching. In static quenching, quencher forms a non-fluorescent ground-state complex with fluorophore and this complex absorbs without emission. The new or altered UV–Vis absorption band is a characteristic of complex formation. In dynamic quenching excited fluorophore collides with quencher leading to a nonradiative deactivation. This does not result any change in ground-state absorption. Static quenching involves the formation of a non-fluorescent ground-state complex between surface functional groups of the CDs and the quencher. This interaction can lead to changes in the UV-visible absorption spectrum, often seen as new peaks or shifts in existing ones(Demchenko, 2010). Figure 1c shows the UV visible spectrum of CDs in the presence of various concentrations of Hg^{2+} ions (Joseph & Anappara, 2017a). To confirm the occurrence of static quenching, temperature-dependent studies can be performed. As temperature increases, complex formation is typically reduced due to thermal dissociation, and the original absorption spectrum of the CDs is restored. This thermal reversibility is a strong indication of static quenching. In summary, UV-visible spectroscopy serves as a powerful tool to characterise CDs, monitor surface interactions, confirm metal ion doping, and elucidate quenching mechanisms in fluorescence studies.

Figure 1. (a) UV-Visible spectrum of CDS showing π-π transition (b) π-π* transition and n-π* transition (c) UV-Visible spectrum of CDs in the presence of various concentrations of Hg^{2+} ions.*

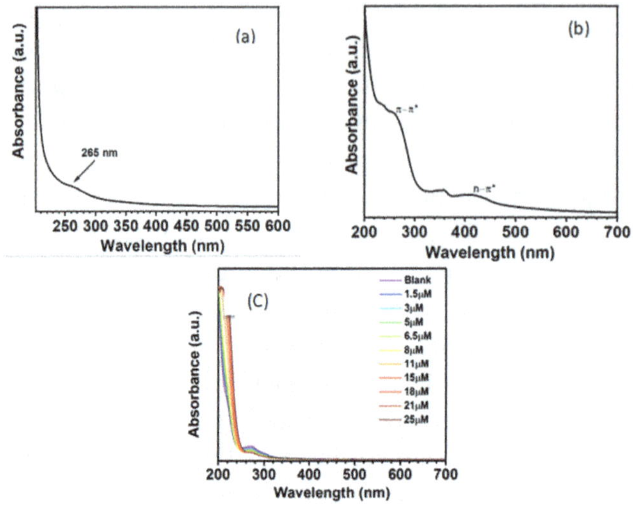

(b) Photoluminescence (PL) Spectroscopy

Photoluminescence (PL) spectroscopy is a contactless, non-destructive method to analyse the radiative electronic transitions of materials. There are two types of photoluminescence measurements (a) steady-state (b) time resolved measurements (Valeur & Berberan-Santo, 2012). Steady state photoluminescence spectroscopy monitors the time average of radiative emission events in the sample, which enable the determination of parameters such as spectral position, spectral width and spectral intensity. In most of the steady-state experimental setup, a xenon lamp is used as the excitation source and is equipped with two monochromators to select both the excitation and emission wavelengths. To record an excitation spectrum, the emission monochromator is fixed at the desired wavelength and excitation monochromator is scanned through the excitation bands of the luminescence centre (Valeur & Berberan-Santo, 2012). The emission spectrum is measured by choosing an appropriate excitation wavelength and scanning wavelength in the emission monochromator. Two types of emission can be observed in the case of CDs: Excitation dependent emission and Excitation independent emission. Excitation-dependent emission means that the emission color (wavelength) of carbon dots (CDs) changes when they are excited with different wavelengths of light. This happens because each excitation wavelength preferentially excites a different emissive site or domain within the CDs. In carbon dots, excitation-dependent emission arises due to: Variations in dot size, leading to different band gaps, and the presence of multiple surface functional groups, defects, and heterogeneous energy traps on the surface. As a result, different excitation energies produce emissions from different localized states, causing the emission peak to shift with the excitation wavelength. Excitation-independent emission, on the other hand, means that the emission color remains constant regardless of the excitation wavelength. This occurs when there is only one dominant emissive state. In this case, all absorbed excitation energy rapidly relaxes to the same lowest-energy emissive level before emission. Carbon dots exhibit excitation-independent emission when they have a uniform carbon core, minimal variation in size, and homogeneous surface chemistry, ensuring a single, well-defined emissive pathway.

Compared to organic fluorophores, emission spectra of CDs are broad. Most of the carbon dots emits light in bluish green region. Therefore, adjusting the reaction parameters could help scientists to tune emission up to NIR region. In some CDs, excitation-dependent PL can be observed due to the heterogeneity in size, surface states, and emissive traps. Excitation independent PL are also observed in some other CDs. Examples for excitation dependent and independent emission spectra of CDs are illustrated in Figure 2.

Figure 2. (a)Excitation independent photoluminescence spectra of CDs(Joseph & Anappara, 2016b) (b) excitation dependent emission spectra of CDs from benzene tricarboxylic and formamide

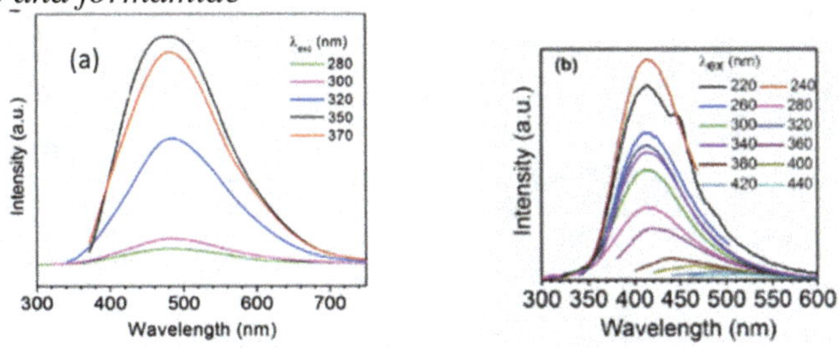

(Joseph & Anappara, 2017b)

Quantum yield of CDs varies with synthesis method and surface passivation. The main factors affecting photoluminescence are size of the carbon core, surface passivation, heteroatom doping, synthesis method and pH. The characteristics of CDs include strong and tuneable photoluminescence, high photostability, low toxicity, high biocompatibility, water solubility, cost effective and green synthesis strategies suitable in bioimaging applications. Chemical sensing of CDs is based on their ability to respond to various analytes. In the presence of certain analytes, there is a chance of change in fluorescence. This makes them ideal for detecting ions, small molecules and environmental toxins. There are different kinds of fluorescence sensing mechanisms. In fluorescence turn-off sensing, analyte interaction leads to a decrease in fluorescence (Dong et al., 2015; Shahbaz et al., 2025; Tian et al., 2023; Zheng et al., 2016). In fluorescence turn-on sensing, weakly fluorescent CDs or quenched CDs become fluorescent in the presence of the target analyte such as dopamine, glucose etc. In Photoinduced Electron Transfer (PET), electron transfer between CDs and analyte leads to changes in emission. In inner filter effect, absorption of excitation/emission light by the analyte causes a change in observed fluorescence. Another sensing mechanism is Fluorescence Resonance Energy Transfer (FRET)(Demchenko, 2010; Dong et al., 2015; Joseph & Anappara, 2016b, 2017a, 2017b; Lakowicz, 2006; Shahbaz et al., 2025; Tian et al., 2023; Zheng et al., 2016), where the energy tranfer from donor CDs to acceptor molecule quenches or alters emission. Carbon dots are particularly sensitive to metal ions due to their interaction with surface functional groups. For example, Fe^{3+} ions have strong binding to -OH/-COOH groups. Different fluorescence sensing mechanism using CDs are demonstrated in Figure 3.

Figure 3. (a) Photoluminescence intensity variation of the quantum dots in the presence of different analytes (fluorescence turn off) sensor (b) FRET based fluorescence quenching (Chawre et al., 2023) (c) PET based fluorescence quenching of carbon dots.

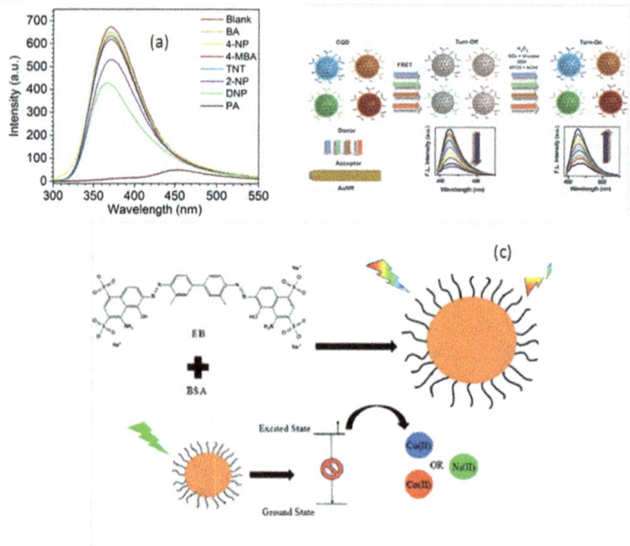

(Mintz et al., 2018)

Traditionally, carbon dots (CDs) are well known for their strong fluorescence; however, recent advances have revealed that some CDs can also exhibit room-temperature phosphorescence (RTP). This development opens exciting possibilities for applications in bioimaging, anti-counterfeiting, and optoelectronic devices. To enable RTP in CDs, several strategies have been employed. Heteroatom doping (e.g., with nitrogen, sulfur, phosphorus, or boron) introduces additional energy levels and promotes intersystem crossing (ISC), facilitating the population of the triplet excited state. Embedding CDs into matrices—such as silica, polymers, or crystalline hosts—helps to suppress non-radiative relaxation pathways and protects the triplet states from quenching by oxygen. Rigid matrices in particular restrict molecular motion, further stabilizing triplet excitons and thereby enhancing RTP efficiency. Phosphorescence life time and emission spectra of hetero atom doped CDs are demonstrated in Figure 4.

Figure 4. (a) Fluorescence spectrum of different carbon dots samples, after incorporating in the potash alum matrix (b) phosphorescence emission spectrum of the CDs/potash alum matrix (c) phosphorescence excitation spectra of the samples (d) phosphorescence life-time measurements of CDs/potash alum matrix.

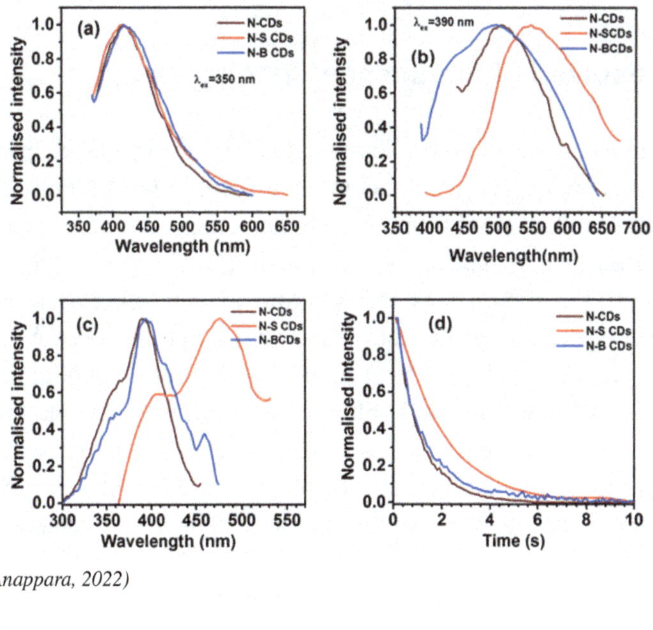

(Joseph & Anappara, 2022)

Fluorescence and phosphorescence can be visually observed under 365 nm UV lamp as shown in Figure 5.

Figure 5. (a) digital photographs of CDs under UV lamp N-B CDs represented in A, N-S CDs represented in B, N-CDs represented in C (b) CDs dispersed in potash alum under 365 nm UV excitation (c) afterglow of CDs dispersed in potash alum for few seconds after the UV light is turned off

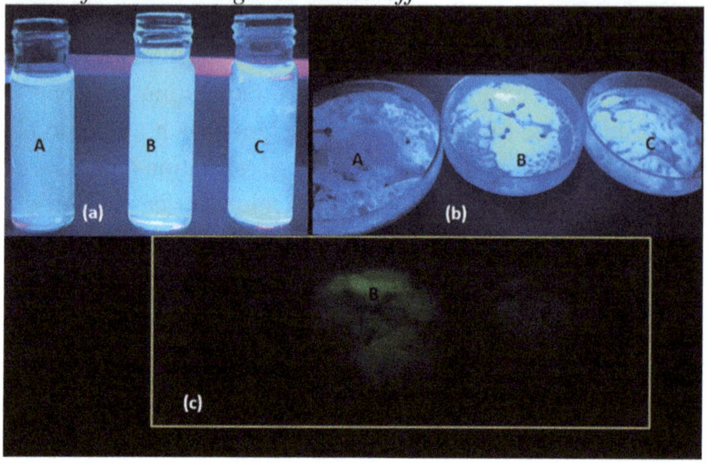

(Joseph & Anappara, 2022).

(c) Time-Resolved Fluorescence Spectroscopy

Time-Correlated Single Photon Counting (TCSPC) is a high-precision technique used to measure how long a material continues to emit light (fluoresce) after being excited by a short pulse of light. In this technique, the sample (e.g., carbon dots) is excited by an ultrashort light pulse, typically in the picosecond or nanosecond range. After absorption, the carbon dots emit photons as they relax back to the ground state, and each emitted photon appears after a short, random delay. A single-photon detector then records the exact arrival time of each emitted photon relative to the excitation pulse. To maintain single-photon statistics, only one photon is counted per excitation cycle. By repeating this process over millions of excitation cycles, the instrument constructs a histogram of photon arrival times, which represents the fluorescence decay curve — a plot of emission intensity versus time after excitation. This decay curve reveals the fluorescence lifetime, which is the average time the excited electrons remain before returning to the ground state. Analysis of the fluorescence lifetime provides valuable insights into the electronic structure, energy relaxation pathways, and nonradiative processes occurring in carbon dots. For instance, multi-exponential decay behavior often indicates the presence of multiple emissive sites, such as core states and surface defect states, each contributing differently to the overall photoluminescence. Time-resolved photoluminescence measurement gives information about the temporal evolution of the emission of a sample. For these measurements, the sample is exposed to a pulse of light. This short irradiance generates electron-hole pairs that decay to lower energy levels of the sample. The measurement of emission intensity as a function of time provides the life-time of the sample. Fluorescence life-time of a fluorophore is very sensitive to molecular environment. Many macromolecular events, such as resonance-energy transfer and dynamic quenching can be confirmed by fluorescence decay measurements. Time-correlated single-photon counting (TCSPC) is the most widely used technique for fluorescence life-time measurements. In TCSPC, the sample is excited with a pulsed laser and the time between the sample excitation and arrival of the emitted photon at the detector is measured (delay time). The condition of TCSPC is adjusted so that one photon is detected per laser pulse. The recording of single photons is repeated many times and the delay time is sorted using a histogram analysis. Since the number of photons is proportional to fluorescence intensity, the resultant histogram represents fluorescence intensity decay curve. The x-axis of the decay curve represents delay time and y-axis represents intensity. The fluorophores show an exponential decay curve and can be represented as $I(t)=I_0\, exp^{\frac{t}{\tau}}$ where I_0 is the intensity at time zero and τ is the time for the intensity to drop by or to 37% and is called life-time of the sample(Becker et al., 2004; Guilbault, 2020; Marcu et al., 2015). The fluorescence decay curve is shown in Figure 6.

Figure 6. (a) A typical fluorescent decay curve of CDs(b) Fluorescence decay measurements without and with (0.1 mM Fe^{3+} ions) of N-CDs (c) Time-resolved decay curves of carbon dots collected in the presence of different Cu^{2+} ion concentration.

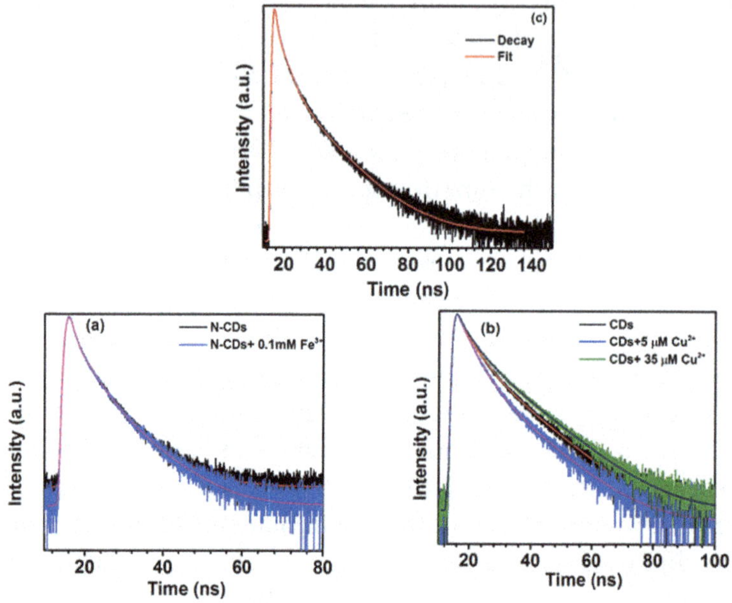

(d) Fourier Transform Infrared Spectroscopy

Fourier transform infrared spectroscopy (FTIR) is used to analyse the functional groups present on the organic and inorganic samples. In FTIR spectroscopy a beam containing many frequencies of infrared radiation are passed through the sample. Different bonds in the sample absorb radiations at specific frequencies and the remaining radiation is transmitted; and the resulting spectrum represents molecular absorption and transmission of the sample (Gupta et al., 2024; Kannan et al., 2024; Kannan et al., 2023; Kannan et al., 2021; Pavia et al., 2015) Since the natural frequencies of different bonds are different, the IR spectrum represents the fingerprints of molecules. In the case of carbon dots distinct functional groups can be identified through recording of typical vibration bands. FT-IR spectroscopy is a powerful technique used to identify and characterize various functional groups present on the surface of synthesized carbon quantum dots (CQDs).

By analysing the FTIR spectrum of the carbon dots, the presence of particular bonds can be determined based on the positions of these peaks. Typically, the O–H stretching vibration appears as a broad band around 3200–3600 cm^{-1}, the

C=O stretching vibration is observed at 1650–1750 cm^{-1}, and the C=C stretching vibration appears in the range of 1500–1650 cm^{-1}.(Joseph & Anappara, 2022). In addition to these, heteroatom doping with elements such as nitrogen, sulphur, and boron significantly alters the surface chemistry and can also be identified using FT-IR spectroscopy. When atoms such as nitrogen (N), sulfur (S), or boron (B) are introduced into the carbon dot structure, they form new chemical bonds with carbon, oxygen, or hydrogen atoms either on the surface or within the lattice. These structural modifications can be detected using FTIR spectroscopy, which reveals new absorption peaks or shifts in existing ones corresponding to the newly formed bonds. In nitrogen-doped CDs, N–H stretching vibrations typically appear in the range of 3200–3400 cm^{-1}, while C–N stretching occurs at 1200–1350 cm^{-1}, and C=N stretching is observed around 1550–1650 cm^{-1}. For sulfur-doped CDs, characteristic peaks include C–S stretching at 600–800 cm^{-1}, S=O stretching at 1000–1200 cm^{-1}, and S–H stretching around 2550 cm^{-1}. In the case of boron-doped CDs, B–C stretching bands are usually observed at 1300–1450 cm^{-1}, and B–O stretching appears around 1100–1200 cm^{-1}. These distinct vibrational features confirm the successful incorporation of heteroatoms into the carbon dot framework (Joseph & Anappara, 2022). Figure 7 shows FTIR spectra of CDs.

Figure 7. ATR-FTIR spectra of (a) N-CDs (b) N-S CDs (c) N-B CDs (Becker, 2005)

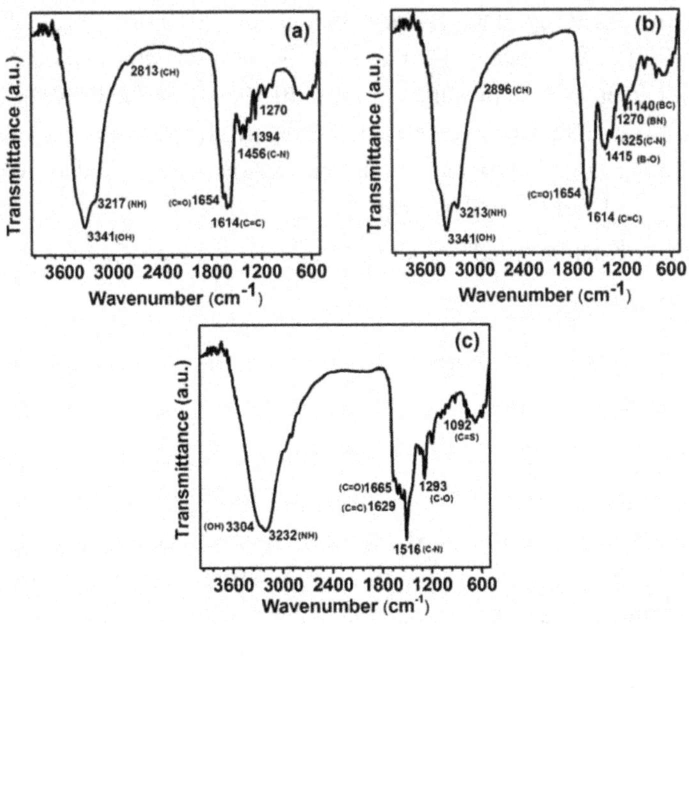

(e) Raman Spectroscopy

Raman spectroscopy provides information about molecular vibrations, and it gives structural features of carbon atoms with in the CDs. Raman effect is the basic principle used in Raman spectroscopy. The technique involves irradiation of a monochromatic light source (Source: Nd: YVO4 diode laser operating at 532 nm) on a sample and detecting the scattered light. Most of the scattered lights have the same frequency as the excitation source (Rayleigh or elastic scattering). Due to interaction between the incident electromagnetic waves and the vibrational energy levels of the molecules, the energy of a portion of scattered light is shifted from the laser frequency(Jorio et al., 2011; Lewis & Edwards, 2001). When the frequency of incident radiation is lower than frequency of scattered radiation, anti-Stokes lines appear in the Raman spectrum. But when the frequency of incident radiation is higher than frequency of scattered radiation, Stokes lines appear in the Raman spectrum. Raman spectrum is a plot of the intensity of Raman scattered radiation as a function of its frequency difference from the incident radiation (represented in cm^{-1}). In some cases, Raman spectra of CDs appear on fluorescence back ground as in Figure 8 a. The main features in the Raman spectra of graphitic carbon nanomaterials are G band(~1600 cm^{-1}) and D band (~1350 cm^{-1}).(Figure 8 b and c) G band arises from the in-plane stretching vibration mode (E_{2G}) of crystalline graphite carbons and D-band arises from defects or disorder (impurities, edges, functional groups) that breaks the translational symmetry(Dresselhaus et al., 2010; Jorio et al., 2011; Lewis & Edwards, 2001). Presence of D and G bands are ascribed to sp^3 and sp^2 carbon respectively. Depending on presence of graphitic layers, some CDs show 2D band (~2650–2700 cm^{-1}). This can be seen in CDs synthesized from graphite (Figure 8 d).The ratio of sp^2 to sp^3 carbons in nanomaterial can be calculated by comparing the area of G-band and D-band as $(\frac{ID}{IG})$

Tuinstra and Koenig noted that the crystal size is inversely related to the $\left(\frac{I_D}{I_G}\right)$ ratio as

$$\frac{I_D}{I_G} = \frac{C(\lambda)}{L_a}$$

where L_a is the cluster size, C(λ) is an empirical constant that depends on the excitation laser energy, where C(λ) at 532 nm is around 4.96 nm (Caņado et al., 2006). The intensity ratio of the two bands (I/I) serves as an indicator of the relative degree of disorder or crystallinity in the carbon structure. This ratio reveals how ordered or disordered the carbon framework is, indicating the proportion of graphitic (sp^2) versus amorphous or defective (sp^3) carbon, and provides insight into the size of the sp^2 domains within the carbon dots. A low I/I ratio (< 0.5) implies larger sp^2

domains and fewer defects, corresponding to a more graphitic structure. A moderate ratio (≈ 0.8–1.2) indicates a partially graphitic structure, while a high ratio (> 1.5) suggests a highly defective or amorphous nature with small or distorted sp^2 domains.

Figure 8. Raman spectra of (a) N-CDs (b) N-S CDs (c) N-B CDs (d) CDs from graphite.

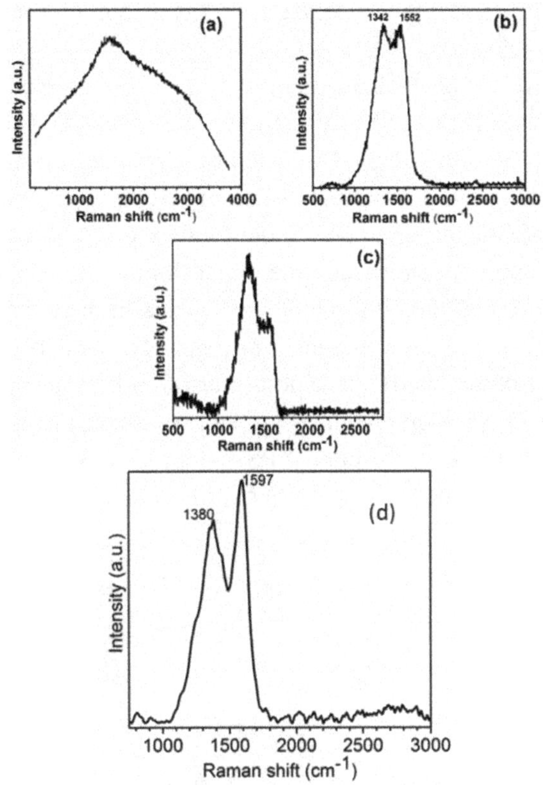

(f) X-ray Photoelectron Spectroscopy (XPS)

X-ray photoelectron spectroscopy (XPS) is a surface-sensitive quantitative spectroscopic technique. X-ray photoelectron spectroscopy (XPS) is a powerful surface analysis technique used to investigate the elemental composition, chemical states, and electronic environment of materials. A monochromatic source of X-ray is used to irradiate the sample in ultrahigh vacuum conditions. Due to interaction of the X-ray with the samples, the core-level electrons get ejected. The energy of a photo emitted core electron is a function of its binding energy and is the characteristic of the element from which it was emitted. XPS is used to analyse the surface chemical composition, depth profile of the composition and bonding (Watts & Wolstenholme, 2003).For carbon dots (CDs), which are a type of zero-dimensional carbon nanomaterial, XPS can be particularly useful for analysing their surface chemistry, which plays a significant role in their optical, electronic, and catalytic properties. After obtaining XPS data, it can be deconvoluted using Gaussian or Lorentzian fitting methods. The elemental percentage of CDs is found by relative areas under the peaks in wide scan spectrum.

All the CDs have C_{1s} peak as a signature of C=C bonds around 284.5 eV. Depending on the nature of surface functionalities, presence of C–O bonds around 286 eV and C=O bonds around 287 eV, O–C=O or C–N bonds around 288-290 eV are very likely to occur. Oxygen is another prominent element present in all CDs. Here we can see peaks of C–O around 532 eV and C=O or O–C=O around 530-531 eV. Nitrogen, sulphur and phosphorus are the major doping elements present in CDs. If the CDs have nitrogen, then we can observe peaks in the range of 399–401 eV, depending on the chemical environment of nitrogen. The peak observed at 168–170 eV correspond to sulphur doped CDs. If phosphorus is used as a dopant, peak appears typically around 133 eV. Deconvoluted C_{1s} and N_{1s} spectrum is shown in Figure 9. Figure 10 implies the deconvoluted XPS spectra of NS doped CDs. Figure 11 demonstrates deconvoluted XPS spectrum of N,B doped CDs.

Figure 9. (a) C_{1s} XPS spectrum of N-CDs (b) N_{1s} XPS spectrum of N-CDs.

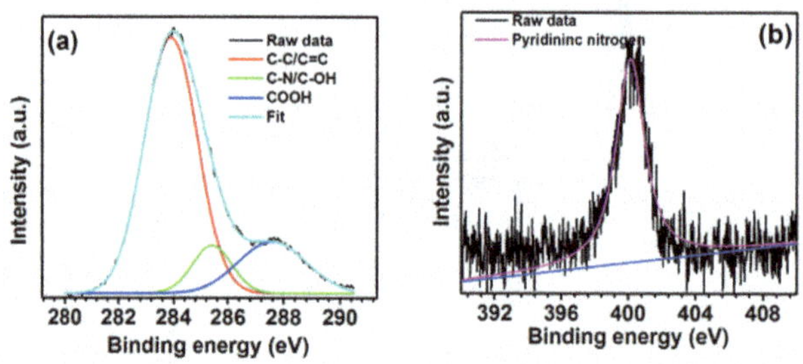

(Joseph & Anappara, 2022)

Figure 10. (a) XPS survey spectrum of N-S doped CDs (b) C_{1s} XPS spectrum of CDs. (c) N_{1s} XPS spectrum of CDs. (d) S_{2p} XPS spectrum of CDs.

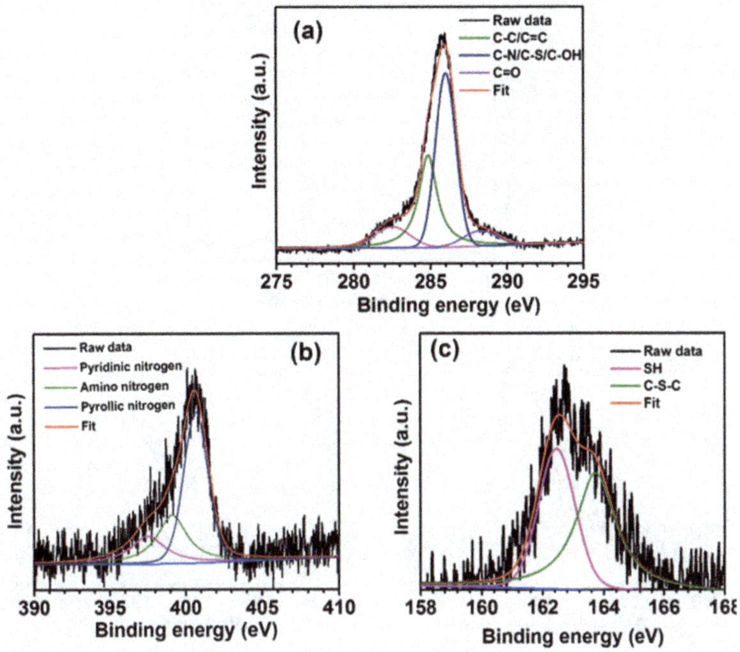

(Joseph & Anappara, 2022)

In the case of carbon dots, XPS analysis reveals distinct nitrogen, oxygen, carbon and heteroatom bonded units. The binding energy of carbon C_{1s} is about 285 eV, nitrogen N_{1s} is around 400 eV and oxygen O_{1s} is about 530 eV. On de-convoluting the C_{1s} and O_{1s} different bonds such as C=C, C=O, C-N, C-OH, C-O-C can be identified and by analysing nitrogen N_{1s}, presence of pyrrolic, pyridinic, graphitic nitrogen can be identified. Presence of other heteroatoms in carbon dots such as sulphur, phosphorous and boron can be confirmed using XPS (Joseph & Anappara, 2022).

Figure 11. XPS high resolution survey scans of (a) C_{1s}, (b) N_{1s} and (c) B_{1s} region of CDs. The sample was prepared by coating on a glass plate

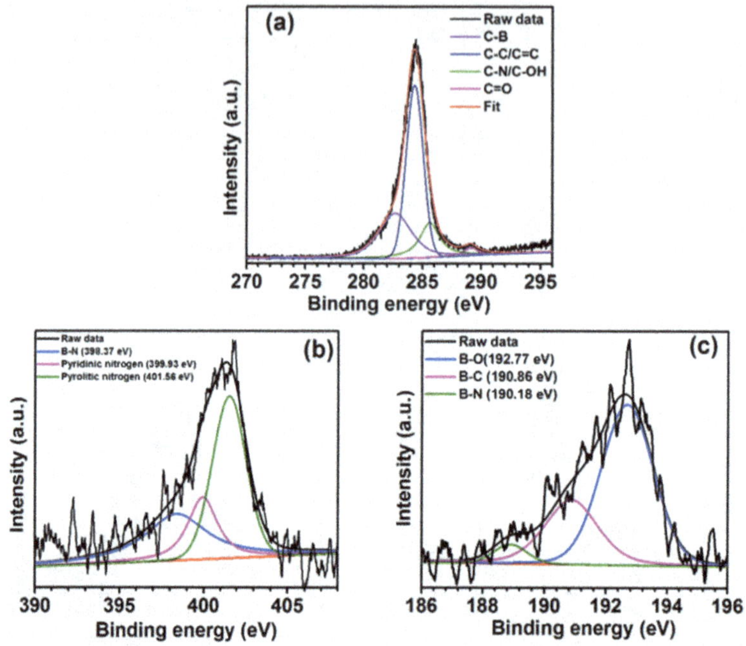

(Joseph & Anappara, 2022)

(g) Nuclear Magnetic Resonance (NMR) Spectroscopy

Nuclear magnetic resonance spectroscopy is a research technique to determine the structure, dynamics, reaction state, and chemical environment of molecules. Here the sample is excited with a radio wave (60-1000 MHz) pulse, and the nuclear magnetic resonance response obtained as time domain data (FID). The Fourier transform of the FID data gives frequency-domain spectrum. There are two types of NMR spectrum carbon-13 NMR (^{13}C NMR) and proton nuclear magnetic resonance (^{1}H NMR). ^{13}C NMR determines the type and number of carbon atoms in the molecule and ^{1}H NMR determine the types and number of hydrogen atoms in the molecules. In a fully decoupled ^{13}C NMR spectrum, the number of peaks indicates the number of types of carbon present in the sample (Pavia et al., 2015). The chemical shift gives information about the hybridization of each carbon (e.g., sp, sp^2,sp^3); electronegative atoms bonded to carbon, and pi bonds. In ^{1}H NMR, the number of signals equals to the number of sets of equivalent protons; integration of the signals yields the relative numbers of non-equivalent protons; the chemical shift (position of the peaks relative to a reference) gives information about the proton environment; splitting of the signals give information about neighbouring protons. In the case of CDs the ^{13}C NMR spectra show three types of peaks: around 20-30 ppm (residual sp3 carbon atoms), 100-150 ppm (sp^2 carbon atoms C=C) and 150-210 ppm (sp^2 carbon atoms C=O) (Nagaiah et al., 2012). In the ^{1}H NMR spectra show peaks in the region 6-8 ppm due to the presence of sp^2 hybridized aromatic protons; 3-6 ppm peaks are due to the protons attached with hydroxyl and carbonyl groups; carboxylic acid protons of carbon dot exhibit peaks in the range of 9-9.5 ppm (De et al., 2015). Figure 12 implies ^{1}H NMR spectrum of CDS synthesized from tannic acid. Analyte (picric acid) interaction study can also be visualised using NMR as in Figure 12. The spectra of PA alone, is represented by the blue curve, and the CDs - PA mixture (with picric acid concentration of about 5 μM), is represented by the red curve. The measurement was done after dispersing the solid sample in DMSO (Joseph & Anappara, 2016b).

Figure 12. ^{1}H NMR of carbon dots (black spectra).

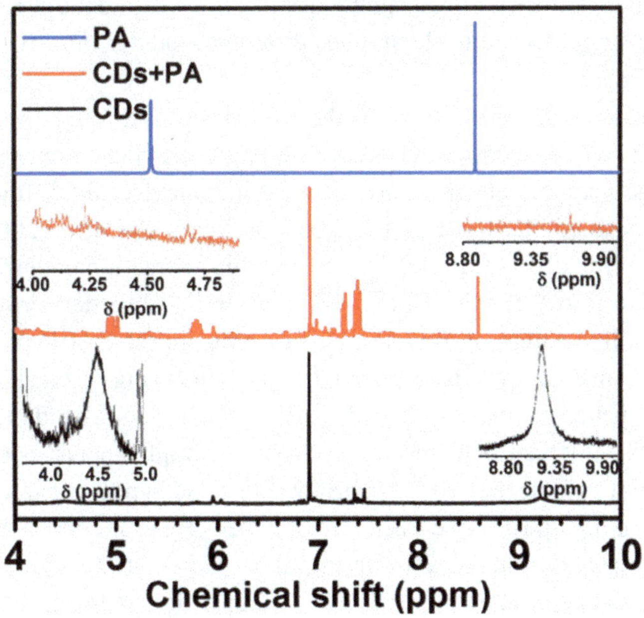

2. MICROSCOPIC TECHNIQUES

Microscopic Techniques are performed for direct visualization of size, shape, and structural organization.

(a) Transmission Electron Microscopy (TEM)

Transmission electron microscopy is one of the main instrumentation technique to analyse the morphology of nanoparticles. Transmission Electron Microscopy (TEM) is a powerful technique widely used to characterize carbon dots (CDs) at the atomic or nanometre scale. It provides essential information about the morphology, particle size, and degree of crystallinity of the CDs. A beam of electrons is transmitted through ultra-thin sample and the interactions between electrons and the atoms in the specimen are used to observe the structure of the nanoparticles. Imaging mode and diffraction mode are two modes of operations in TEM. When an electron beam passes through a thin sample, part of the electrons get transmitted and another part gets scattered. In the imaging mode, by using electromagnetic lenses, the transmitted electrons can be focussed into an image on a charge coupled device (CCD) camera or a phosphor screen. Typically, carbon dots exhibit a spherical or quasi-spherical morphology. The TEM images can be analysed using software such as ImageJ, which enables the determination of particle size and the generation of size distribution histograms. Figure13 implies different kinds of information obtained from TEM analysis.

Whereas in the diffraction mode, as the electrons pass through the sample, they are scattered by the electrostatic potential set up by the constituent element in the specimen. The scattered electrons are focussed into an electron diffraction pattern using electromagnetic lenses (Egerton, 2005; Williams & Carter, 2009). The area chosen for electron diffraction is 0.5-1 µm and this diffraction pattern is known as selected area electron diffraction pattern (SAED). The SAED pattern provides insights into the crystalline nature of the CDs. Diffuse rings in the SAED pattern suggest an amorphous structure, whereas sharp or well-defined rings indicate a crystalline nature. SAED patterns can be used to identify crystal structures and measure lattice parameters. The diffraction pattern of amorphous sample appear as diffuse rings, while crystalline sample patterns have bright spots; and in polycrystalline materials small spots making up ring-like traces are obtained (Egerton, 2005). Every ring in SAED pattern corresponds to a group of lattice planes of the same Miller index family in the polycrystalline sample. The radii of the rings are inversely proportional to the interplanar spacings (d_{hkl}) of the lattice planes of crystals. Diffraction pattern reveals that CDs prepared from graphite are more crystalline, than CDs synthesized from molecular precursors. SAED of CDs from graphite appears as single spots with rings (Figure14a.) and for CDs from other precursors is usually obtained as rings as shown in Figure 14b. By measuring the radius of the ring the interplanar distance (d) can be calculated.

SAED pattern of CDs demonstrated in Figure 14.

Figure 13. (a) Representative TEM images of the CDs (b) Size-histogram of CDs obtained from the TEM data (c) The high resolution TEM images of an individual carbon dot.

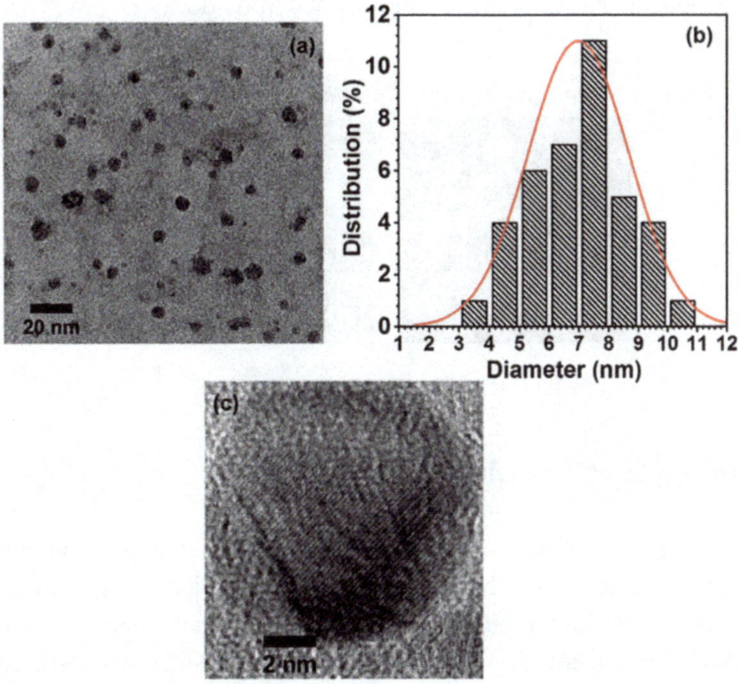

High-resolution transmission electron microscopy.

The image formation in TEM and HRTEM is identical, but HRTEM gives high-resolution images in the atomic-scale. This technique is also known as phase-contrast imaging, since the phase of the diffracted electron wave is preserved and interferes constructively or destructively with the phase of the transmitted wave. The highest point of the resolution obtained in HRTEM is about 0.5 Å. Using HRTEM individual atoms of a crystal and its defects can be resolved (Egerton, 2005; Hu et al., 2015; Ngafwan et al., 2022; Williams & Carter, 2009; Zhu et al., 2015). HR-TEM images often reveal lattice fringes within the carbon dots, especially when crystalline or graphitic domains are present. A commonly observed interlayer spacing (d-spacing) is ~0.21 nm, which corresponds to the (100) plane of graphitic carbon. In some cases, a spacing of ~0.34 nm may be observed, which corresponds to the (002) plane of graphite. The presence of these lattice fringes indicates the crystalline domains embedded within an amorphous carbon matrix.

Furthermore, Energy-Dispersive X-ray Spectroscopy (EDS or EDX) is often conducted alongside HR-TEM to determine the elemental composition of the CDs. EDS provides qualitative and semi-quantitative analysis, typically detecting elements such as carbon (C), oxygen (O), and in some cases, nitrogen (N) or other dopants, depending on the synthesis route. This technique is essential for confirming the presence of functional groups or heteroatom doping, which has a significant influence on the optical and electronic properties of carbon dots.

Figure 14. A typical selected area electron diffraction (SAED) pattern of CDs

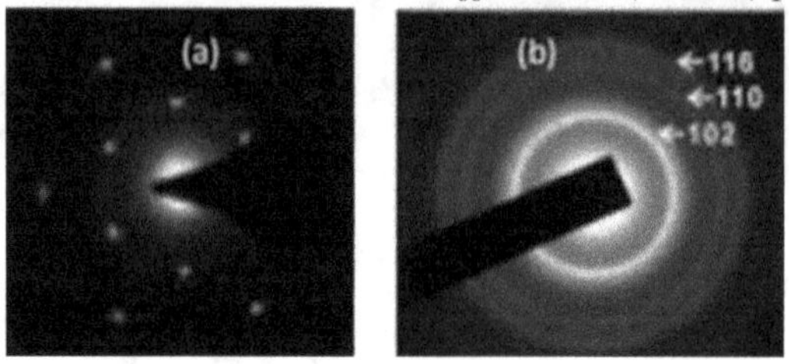

(b) Scanning Electron Microscopy (SEM)

Scanning Electron Microscopy (SEM) is a powerful imaging technique used in nanomaterials research, but it has limitations when applied to carbon dots (CDs), which are extremely small (typically <10 nm). But in certain other cases it is useful. CDs have a tendency to aggregate into clusters due to their small size and high surface energy. SEM can be used to visualize these aggregated clusters or thin films of CDs rather than individual dot. SEM is useful when CDs are deposited on a substrate (e.g., glass, silicon wafer, polymer). Using the image, uniformity, coverage, and distribution of CDs on the surface can be analysed. SEM combined with Energy-Dispersive X-ray Spectroscopy (EDX/EDS) can provide elemental composition (e.g., C, N, O) and it confirms the presence of dopants (like N, S, P) in doped CDs. There are certain limitations on SEM images of CDs. Resolution is not sufficient to get clear image of individual carbon dots. CDs are usually non-conductive, so they require metal coating (e.g., gold or platinum), which can obscure fine details. Figure 15 provides SEM images of CDs hematite composite.

Figure 15. Carbon quantum dots -SEM image of mesoporous hematite composite

(Ngafwan et al., 2022)

(c) Atomic Force Microscopy (AFM)

Atomic force microscope (AFM) is a type of scanning probe microscope to measure the topography of the sample. The AFM instrument uses a cantilever associated with a sharp tip (probe) made up of Si or Si_3N_4 to scan over the sample surface. The van der Waals force between the sample and probe is used to image the samples. When the tip approaches the sample, the attractive force between the sample surface and the tip causes the cantilever to deflect towards the surface. But when the tip makes contact with a sample, the repulsive forces takes over and causes the cantilever to deflect away from the surface. A laser beam deflection system is set to detect cantilever deflections towards or away from the surface. A position-sensitive photodiode can be used to track these changes in the deflections. By using a feedback loop to control the height of the tip above the surface-thus maintaining constant laser position-the AFM can generate an accurate topographic map of the surface features. There are three types of operating modes in the AFM (a) contact mode (b) tapping mode (c) and non-contact mode (Haugstad, 2012; Yao & Wang, 2005). Contact mode is known as static mode whereas non-contact and tapping mode are known as dynamic mode due to the involvement of oscillations of a cantilever. Atomic Force Microscopy (AFM) is a powerful technique for characterizing nanoscale surface morphology and height profiles of carbon dots (CDs). Unlike SEM or TEM, AFM provides topographical information in real space and does not require conductive samples, making it highly suitable for studying individual carbon dots, especially in the 1–10 nm range. AFM images of CDs provide information about height and size distribution, aggregation behaviour and surface roughness. CDs typically appear as nanodots with heights of 1–10 nm, depending on synthesis and aggregation. AFM can be employed to investigate whether CDs are well-dispersed as individual dots, or aggregated into clusters or forming continuous films or islands. Figure 16 implies a representative image of AFM of CDs.

Figure 16. AFM image of CDs

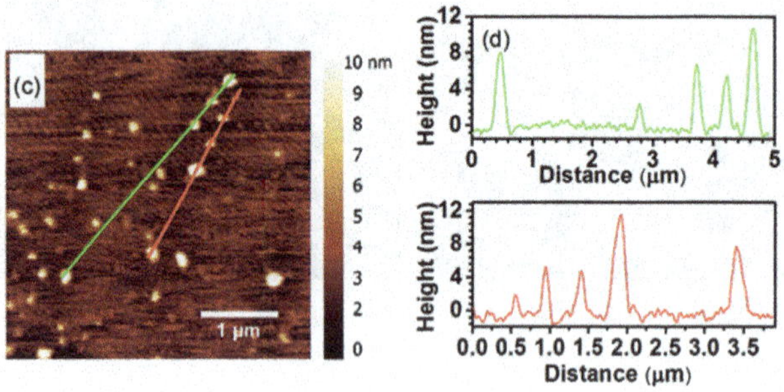

3. SURFACE ANALYTICAL TECHNIQUES

These techniques analyse surface functionality, elemental composition, and chemical bonding.

(b) X-ray Diffraction (XRD)

X-ray powder diffraction is an analytical technique used for identification of shape, number of atoms per unit cell and position of the atoms in crystalline materials. In the case of nanoparticles, approximate particle size can be obtained from the XRD data. X-ray diffraction is based on the constructive interference of monochromatic X-rays and a crystalline sample. The X-rays are generated from a cathode ray tube, filtered to produce monochromatic radiation and directed to sample. When a beam of X-rays is incident on the atom there is an absorption and re-emission of electromagnetic radiations known as scattering. Two or more scattered waves get interfered. The interaction of the incident rays with the sample produces constructive interference when the conditions satisfy Bragg's law(Reifenberger, 2016). This law relates the wavelength of electromagnetic radiation to the diffraction angle and the lattice spacing in a crystalline sample. These diffracted X-rays are then detected, processed and counted. By scanning the sample through a range of angles, all possible diffraction directions of the lattice should be attained due to the random orientation of the powdered material (Jeffery, 1957).In pure crystalline material, XRD peak appears as sharp peak, whereas in amorphous and nanoscale samples exhibit broadening of peaks. With help of this broadening the size of nanoparticles can be determined. X-ray diffraction (XRD) is a technique used to study the crystallinity and structural order of materials. For carbon dots (CDs) which are often amorphous or partially graphitic, XRD provides important information about their internal structure, especially the presence (or absence) of graphitic domains. XRD can reveal degree of crystallinity of carbon Dots. XRD typically shows broad peaks instead of sharp diffraction lines, indicating disordered carbon structures. XRD spectrum of CDs having graphitic structure shows a broad peak around 2θ ≈ 20–26°. This corresponds to the (002) plane of graphite, but the peak is broader and shifted compared to pure graphite (~26.5°). This indicates disordered graphitic layers or turbostratic carbon. If there's partial crystallinity, crystallite size (L_A) can be estimated using the Scherrer equation. Figure 17 implies XRD of CDs.

Figure 17. XRD of carbon dots

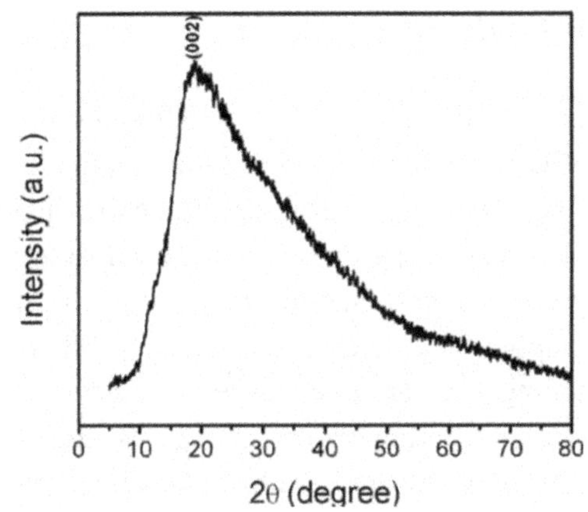

(Shaikh et al., 2019)

(c) Dynamic Light Scattering (DLS)

Dynamic Light Scattering (DLS) is a technique used to measure the hydrodynamic diameter of particles in suspension. For carbon dots (CDs), DLS gives insight into their size distribution in solution, aggregation behaviour, and colloidal stability.DLS measures the size of the particle along with its hydration shell. The reported size is usually larger than the core size measured by TEM or AFM. The common size range for CDs via DLS is 5–50 nm, (Figure 18) depending on surface functionalization and dispersion. Colloidal stability can be analysed using DLS. If CDs are stable, DLS shows consistent particle size over time. An increase in size suggests aggregation or poor dispersion. There are some limitations for DLS such as overestimation of size, high sensitivity to impurities, and it cannot distinguish aggregates from large dots (Dager et al., 2019).

Figure 18. DLS of carbon dots

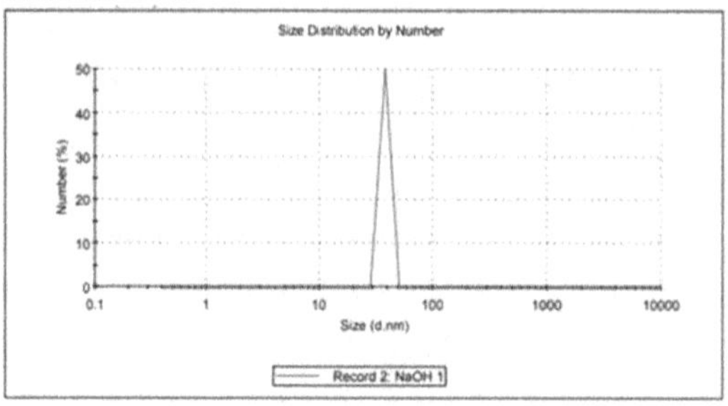

(d) Zeta Potential Analysis

Zeta potential is the electric potential at the slipping plane of a particle in a colloidal system. It reflects the degree of repulsion or attraction between adjacent, similarly charged particles in a dispersion. Zeta potential analysis of carbon dots (CDs) is an important characterization technique used to understand the surface charge and colloidal stability of carbon dots in a suspension. High negative/positive zeta potential of CDs implies they are electrostatically stabilized. Low zeta potential suggests the risk of aggregation and poor dispersion. Zeta potential values can be used to optimize the synthesis parameters (e.g., dopants, precursors, pH) for desired stability or functionality. Most of the CDs have negative surface charge which is attributed to the presence of carboxyl and hydroxyl groups. Figure 19 implies zeta potential analysis of CDs.

Figure 19. Zeta potential analysis of CDs

1 zeta.nzt

Measurement Results

Date	: Friday, May 13, 2022 6:17:20 PM
Measurement Type	: Zeta Potential
Sample Name	: 1 zeta
Temperature of the Holder	: 25.1 °C
Dispersion Medium Viscosity	: 0.893 mPa·s
Conductivity	: 0.205 mS/cm
Electrode Voltage	: 3.3 V

Calculation Results

Peak No.	Zeta Potential	Electrophoretic Mobility
1	-41.9 mV	-0.000325 cm2/Vs
2	--- mV	--- cm2/Vs
3	--- mV	--- cm2/Vs

Zeta Potential (Mean)	: -41.9 mV
Electrophoretic Mobility Mean	: -0.000325 cm²/Vs

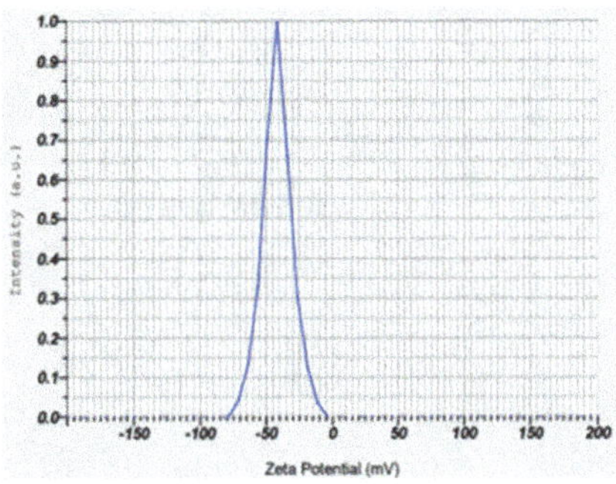

(Joseph & Anappara, 2022)

(e) Thermogravimetric Analysis (TGA)

TGA was conducted to assess the thermal stability and compositional characteristics of the synthesized carbon dots. The TGA curve revealed an initial weight loss of approximately 7% below 150 °C, attributed to the evaporation of adsorbed moisture and residual solvents. A more significant weight loss (~25%) occurred between 200–350 °C, corresponding to the decomposition of oxygen- and nitrogen-containing surface functional groups. A gradual decline in weight was observed from 350–550 °C, indicating the thermal breakdown of the carbon core structure. The final residue at 800 °C was around 15%, suggesting partial carbonisation or the presence of thermally stable components. These results confirm the multi-component nature of the CDs and their moderate thermal stability, suitable for applications in bioimaging and optoelectronic devices. Figure 20 implies TGA analysis of CDs.

Figure 20. TGA analysis of A- solid part of CDs, B- aqueous CDs, C- oleic acid, D- anhydrous CDs

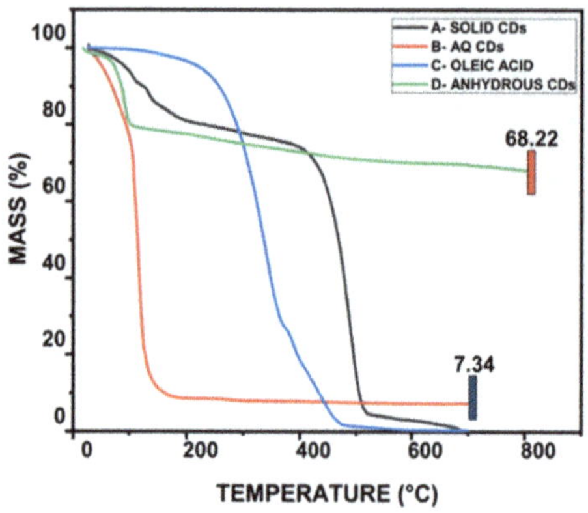

(Rimal et al., 2020)

4. CONCLUSION

The detailed characterization of carbon dots (CDs) using spectroscopic and analytical methods valuable insights into their structural, optical, and chemical properties. The combination of methods such as UV-Vis absorption, photoluminescence (PL) spectroscopy, X-ray diffraction (XRD), Fourier-transform infrared (FTIR) spectroscopy, and transmission electron microscopy (TEM) facilitates a detailed understanding of the synthesis, functionalization, and application potential of CDs. Spectroscopic data reveals key features such as the excitation-dependent emission behavior, tunable fluorescence, and surface functional groups, which are crucial for tailoring CDs for specific applications, including bioimaging, sensing, and energy conversion. Analytical techniques, on the other hand, help confirm the presence of elemental composition, crystallinity, and the morphology of the carbon dots, providing essential information for optimizing their properties. Overall, these characterization methods not only validate the structural integrity and functionality of CDs but also enable the development of more efficient and versatile nanomaterials for diverse scientific applications.

ACKNOWLEDGEMENT

We deeply acknowledge the research funding provided under the Consolidation of University Research for Innovation and Excellence (CURIE) program (DST/CURIE-PG/ 2023/9) of the Department of Science and Technology (DST), Government of India, New Delhi for the completion of this work. The authors are additionally thankful for the financial assistance received through the Anweshan Faculty Research Grant 2022-2023, (Ref. No. 134/2022, dated 28-2-2022), sanctioned by Management, St. Josephs College for Women, Alappuzha, Kerala. We also acknowledge the financial assistance provided by the Kerala State Council for Science Technology and Environment (under the student project programme (File No:00711/SPS65/2021/KSCSTE).

REFERENCES

Becker, W., Bergmann, A., Biscotti, G. L., & Rueck, A. (2004). Advanced time-correlated single photon counting techniques for spectroscopy and imaging in biomedical systems. *Commercial and Biomedical Applications of Ultrafast Lasers IV*, *5340*, 104. DOI: 10.1117/12.529143

Caņado, L. G., Takai, K., Enoki, T., Endo, M., Kim, Y. A., Mizusaki, H., Jorio, A., Coelho, L. N., Magalhães-Paniago, R., & Pimenta, M. A. (2006). General equation for the determination of the crystallite size la of nanographite by Raman spectroscopy. *Applied Physics Letters*, *88*(16), 163106. Advance online publication. DOI: 10.1063/1.2196057

Chawre, Y., Satnami, M. L., Kujur, A. B., Ghosh, K. K., Nagwanshi, R., Karbhal, I., Pervez, S., & Deb, M. K. (2023). Förster Resonance Energy Transfer between Multicolor Emissive N-Doped Carbon Quantum Dots and Gold Nanorods for the Detection of H2O2, Glucose, Glutathione, and Acetylcholinesterase. *ACS Applied Nano Materials*, *6*(9), 8046–8058. DOI: 10.1021/acsanm.3c01518

Dager, A., Uchida, T., Maekawa, T., & Tachibana, M. (2019). Synthesis and characterization of Mono-disperse Carbon Quantum Dots from Fennel Seeds: Photoluminescence analysis using Machine Learning. *Scientific Reports 2019 9:1*, *9*(1), 1–12. DOI: 10.1038/s41598-019-50397-5

De, B., Kumar, M., Mandal, B. B., & Karak, N. (2015). An in situ prepared photoluminescent transparent biocompatible hyperbranched epoxy/carbon dot nanocomposite. *RSC Advances*, *5*(91), 74692–74704. DOI: 10.1039/C5RA12131K

Demchenko, A. P. (Ed.). (2010). *Advanced Fluorescence Reporters in Chemistry and Biology II. 9*. DOI: 10.1007/978-3-642-04701-5

Dong, Y., Cai, J., You, X., & Chi, Y. (2015). Sensing applications of luminescent carbon based dots. *Analyst, 140*(22), 7468–7486. DOI: 10.1039/C5AN01487E PMID: 26396996

Dresselhaus, M. S., Jorio, A., Hofmann, M., Dresselhaus, G., & Saito, R. (2010). Perspectives on Carbon Nanotubes and Graphene Raman Spectroscopy. *Nano Letters, 10*(3), 751–758. DOI: 10.1021/nl904286r PMID: 20085345

Egerton, R. F. (2005). Physical principles of electron microscopy: An introduction to TEM, SEM, and AEM. *Physical Principles of Electron Microscopy: An Introduction to TEM, SEM, and AEM*, 1–202. https://doi.org/DOI: 10.1007/B136495/COVER

Guilbault, G. (2020). Practical fluorescence. In *Practical Fluorescence* (2nd ed.). CRC Press., DOI: 10.1201/9781003066514

Gupta, V., Kannan, K., Tari, V., & Chandra, N. (2024). Bioengineered silver nanoparticles using Brassica oleracea sub sp. botrytis (L.) for enhanced antibacterial activity. *Physics and Chemistry of Solid State, 25*(2), 303–310. DOI: 10.15330/pcss.25.2.303-310

Haugstad, G. (2012). Atomic Force Microscopy: Understanding Basic Modes and Advanced Applications. *Atomic Force Microscopy: Understanding Basic Modes and Advanced Applications*. DOI: 10.1002/9781118360668

Hu, Y., Yang, J., Jia, L., & Yu, J. S. (2015). Ethanol in aqueous hydrogen peroxide solution: Hydrothermal synthesis of highly photoluminescent carbon dots as multifunctional nanosensors. *Carbon, 93*, 999–1007. DOI: 10.1016/j.carbon.2015.06.018

Jeffery, G. A. (1957). Elements of x-ray diffraction (Cullity, BD). *Journal of Chemical Education, 34*(4), A178. DOI: 10.1021/ed034pA178

Jorio, A., Saito, R., Dresselhaus, G., & Dresselhaus, M. S. (2011). *Raman Spectroscopy in Graphene Related Systems*. Raman Spectroscopy in Graphene Related Systems., DOI: 10.1002/9783527632695

Joseph, J., & Anappara, A. A. (2016a). Microwave-assisted hydrothermal synthesis of UV-emitting carbon dots from tannic acid. *New Journal of Chemistry, 40*(9), 8110–8117. DOI: 10.1039/C6NJ02107G

Joseph, J., & Anappara, A. A. (2016b). White light emission of carbon dots by creating different emissive traps. *Journal of Luminescence, 178*, 128–133. DOI: 10.1016/j.jlumin.2016.05.051

Joseph, J., & Anappara, A. A. (2017a). Ellagic acid-functionalized fluorescent carbon dots for ultrasensitive and selective detection of mercuric ions via quenching. *Journal of Luminescence*, *192*, 761–766. DOI: 10.1016/j.jlumin.2017.08.008

Joseph, J., & Anappara, A. A. (2017b). Long Life–time Room–temperature Phosphorescence of Carbon Dots in Aluminum Sulfate. *ChemistrySelect*, *2*(14), 4058–4062. DOI: 10.1002/slct.201700786

Joseph, J., & Anappara, A. A. (2022). Comparative photoluminescence study of nitrogen-doped carbon dots co-doped with boron and sulphur. *Luminescence*, *37*(9), 1475–1481. DOI: 10.1002/bio.4320 PMID: 35797229

Kannan, K., Radhika, D., Gnanasangeetha, D., Krishna, L. S., & Gurushankar, K. (2021). Y3+ and Sm3+ co-doped mixed metal oxide nanocomposite: Structural, electrochemical, photocatalytic, and antibacterial properties. *Applied Surface Science Advances*, *4*, 100085. DOI: 10.1016/j.apsadv.2021.100085

Kannan, K., Radhika, D., Gnanasangeetha, D., Lakkaboyana, S. K., Sadasivuni, K. K., Gurushankar, K., & Hanafiah, M. M. (2021). Photocatalytic and antimicrobial properties of microwave synthesized mixed metal oxide nanocomposite. *Inorganic Chemistry Communications*, *125*, 108429. DOI: 10.1016/j.inoche.2020.108429

Kannan, K., Radhika, D., Kasai, R. D., Gnanasangeetha, D., Palani, G., Gurushankar, K., Koutavarapu, R., Lee, D.-Y., & Shim, J. (2022). Facile fabrication of novel ceria-based nanocomposite (CYO-CSO) via co-precipitation: Electrochemical, photocatalytic and antibacterial performances. *Journal of Molecular Structure*, *1256*, 132519. DOI: 10.1016/j.molstruc.2022.132519

Kannan, K., Radhika, D., Reddy, K. R., Raghu, A. V., Sadasivuni, K. K., Palani, G., & Gurushankar, K. (2021). Gd3+ and Y3+ co-doped mixed metal oxide nanohybrids for photocatalytic and antibacterial applications. Nano express, 2(1), 010014.

Lakowicz, J. (2006). Principles of fluorescence spectroscopy. In *Principles of Fluorescence Spectroscopy*. Springer US., DOI: 10.1007/978-0-387-46312-4

Lewis, I. R., & Edwards, H. (2001). Handbook of Raman Spectroscopy : From the Research Laboratory to the Process Line. *Handbook of Raman Spectroscopy*. DOI: 10.1201/9781420029253

Marcu, L., French, P. M. W., & Elson, D. S. (2015). *Fluorescence lifetime spectroscopy and imaging : principles and applications in biomedical diagnostics*. 554. https://books.google.com/books/about/Fluorescence_Lifetime_Spectroscopy_and_I.html?id=6xvcBQAAQBAJ

Mintz, K. J., Guerrero, B., & Leblanc, R. M. (2018). Photoinduced Electron Transfer in Carbon Dots with Long-Wavelength Photoluminescence. *The Journal of Physical Chemistry. C, Nanomaterials and Interfaces*, *122*(51), 29507–29515. DOI: 10.1021/acs.jpcc.8b06868

Nagaiah, T. C., Bordoloi, A., Sánchez, M. D., Muhler, M., & Schuhmann, W. (2012). Mesoporous nitrogen-rich carbon materials as catalysts for the oxygen reduction reaction in alkaline solution. *ChemSusChem*, *5*(4), 637–641. DOI: 10.1002/cssc.201100284 PMID: 22431491

Ngafwan, N., Rasyid, H., Salaam Abood, E., Kamal Abdelbasset, W., Al-Shawi, S. G., Bokov, D., & Jalil, A. T. (2022). Study on novel fluorescent carbon nanomaterials in food analysis. *Food Science and Technology (Brazil), 42*. https://doi.org/DOI: 10.1590/FST.37821

Pavia, D. L., Lampman, G. M., Kriz, G. S., & Vyvyan, J. R. (2015). Introduction to Spectrometry. *Cengage Learning*, 1–690. https://www.researchgate.net/publication/44372742_Introduction_to_spectroscopy_Donald_L_Pavia_Gary_M_Lampman_George_S_Kriz_and_James_R_Vyvyan

Reifenberger, R. (2016). *Fundamentals of atomic force microscopy: part I: foundations*. https://doi.org/DOI: 10.1557/MRS

Rimal, V., Shishodia, S., & Srivastava, P. K. (2020). Novel synthesis of high-thermal stability carbon dots and nanocomposites from oleic acid as an organic substrate. *Applied Nanoscience*, *10*(2), 455–464. DOI: 10.1007/s13204-019-01178-z

Shahbaz, M., Salma, U., Alam, M. Z., Mujeeb, M. A., Ansari, R., Basha, H. A., Alimuddin, , & Khan, S. A. (2025, February 11). ul H., Basha, H. A., Alimuddin, & Khan, S. A. (2025). Fluorescent/ Photoluminescent Carbon Dots as a Sensor for the Selective and Sensitive Detection of Fe3+/Fe2+ Metal Ions. A Review of the Last Decade. *Journal of Fluorescence*, *35*(9), 7451–7474. Advance online publication. DOI: 10.1007/s10895-025-04166-5 PMID: 39932667

Shaikh, A. F., Tamboli, M. S., Patil, R. H., Bhan, A., Ambekar, J. D., & Kale, B. B. (2019). Bioinspired Carbon Quantum Dots: An Antibiofilm Agents. *Journal of Nanoscience and Nanotechnology*, *19*(4), 2339–2345. DOI: 10.1166/jnn.2019.16537 PMID: 30486995

Surendran, P., Lakshmanan, A., Priya, S. S., Balakrishnan, K., Rameshkumar, P., Kannan, K., & Vinitha, G. (2024). Synthesis of fluorescent carbon quantum dots from Manihot esculenta waste peels for nonlinear optical and biological applications. *Chemical Physics Impact*, *8*, 100515. DOI: 10.1016/j.chphi.2024.100515

Tian, J., An, M., Zhao, X., Wang, Y., & Hasan, M. (2023). Advances in Fluorescent Sensing Carbon Dots: An Account of Food Analysis. *ACS Omega*, *8*(10), 9031–9039. DOI: 10.1021/acsomega.2c07986 PMID: 36936334

Valeur, B., & Berberan-Santo, M. N. (2012). *Molecular Fluorescence - Principal and Applications*. 331.

Watts, J. F., & Wolstenholme, J. (2003). An Introduction to Surface Analysis by XPS and AES. *An Introduction to Surface Analysis by XPS and AES*. DOI: 10.1002/0470867930

Williams, D. B., & Carter, C. B. (2009). The Transmission Electron Microscope. *Transmission Electron Microscopy*, 3–22. DOI: 10.1007/978-0-387-76501-3_1

Yao, N., & Wang, Z. L. (2005). *Handbook of microscopy for nanotechnology*. 731. DOI: 10.1007/1-4020-8006-9

Zheng, M., Li, Y., Zhang, Y., & Xie, Z. (2016). Solvatochromic fluorescent carbon dots as optic noses for sensing volatile organic compounds. *RSC Advances*, *6*(86), 83501–83504. DOI: 10.1039/C6RA16055G

Zhu, H., Liu, A., Xu, Y., Shan, F., Li, A., Wang, J., Yang, W., Barrow, C., & Liu, J. (2015). Graphene quantum dots directly generated from graphite via magnetron sputtering and the application in thin-film transistors. *Carbon*, *88*, 225–232. DOI: 10.1016/j.carbon.2015.03.007

Chapter 5
Carbon Quantum Nanostructures With Theranostic Applications in Cancer

Tanmay J. Urs
https://orcid.org/0009-0001-6822-5738
Amity Institute of Biotechnology, India

Mayanglambam Princy Devi
Amity Institute of Biotechnology, India

Shreya Gupta
Amity Institute of Biotechnology, India

Navya Aggarwal
Amity Institute of Biotechnology, India

Shinjini Sen
Amity Institute of Biotechnology, India

Banashree Bondhopadhyay
https://orcid.org/0000-0002-6679-7791
Amity University, India

ABSTRACT

Numerous studies have previously reported the applications of carbon-based materials, in diagnostic imaging and therapies as standalone treatments. This chapter specifically explores the capable role of carbon nanostructures, to synergistically integrate both the platforms, for a combined treatment. The chapter classifies the five structural classes of carbon, and their respective physiochemical characteristics such as size, morphology, surface and optical properties. It further explores the top-down and bottom-up fabrication methods, like arc discharge, CVD, hydrothermal and detonation methods. Several studies are reported to support successful theragnostic applications such as photothermal, photodynamic and chemodynamic treatments, synergistically enabling MRI, fluorescence and photoacoustic imaging. Mechanistic

DOI: 10.4018/979-8-3373-4247-4.ch005

roles such as EPR effect, ROS-generation, are elaborately discussed. Limitations around lack of clinical studies, long-time exposure and cost-effective functionalization are challenges to be critically evaluated to accelerate translational research.

1. INTRODUCTION

The integration of the diagnostic imaging and therapies, commonly called as "theranostics", has emerged on-demand as a wholesome strategy to manage and treat cancer. The conventional chemotherapy and radiotherapy treatments are subject to criticism, despite being effective, due to their challenges such as low specificity, systemic toxicity in organs, and sub-optimal tumor targeting (Roszkowski et al., 2025). In real circumstances, doctors initially diagnose cancer through imaging techniques and then introduce treatment into the body to tackle the tumor. This conventional approach creates an uncertainty with the amount of therapeutics, actually reaching the site of action (Gupta et al., 2024). This indicates a clear disconnect between seeing the problem and actually treating it. Therefore, this creates a need for smarter and combined approaches that can simultaneously tackle both the objectives.

Modern nanomaterials, especially carbon-based nanostructures, are of particular interest in this context to close the existing gap. The carbon nanostructures offer vivid physiochemical properties such as tunable surface chemistry for functionalization, large surface area for drug-loading and intrinsic optical properties for imaging purposes (Sindhwani et al., 2020; Mohammadpour et al., 2024; Gupta et al., 2024; Surendran et al., 2024; Kannan et al., 2024). These multi-functional characteristics approve their eligibility to be engineered for combined roles such as drug/gene delivery, imaging and therapeutic intervention (Gao et al., 2024). Carbon-based nanomaterials are preferred over other existing nanoparticle systems which typically specialize in either therapy (e.g., liposomes and polymeric NPs) (Dristant et al., 2023; Hamad et al., 2024) or diagnostics (e.g., metallic NPs)(Khorasani et al., 2023), but rarely both with equal efficacy.

Among the various carbon nanostructures, nanotubes (CNTs), graphene/graphene oxide (GO), fullerenes, nanodiamonds (NDs), carbon quantum dots (CQDs) have been utilised mainly to precisely deliver drugs and generate therapeutic effects such as photodynamic, photothermal or chemodynamic therapy in tumors (Wang et al., 2023). The CNTs, which are folded graphene sheets, can penetrate cell membranes (Papi et al., 2024), act as MRI contrast agents (D. Liu et al., 2022), and serve as effective drug carrier (Kostarelos et al., 2007). Graphene, a single layer of carbon atoms in a honey comb lattice, is strong flexible and electric conductor (Tao & Rouhi, 2023), while its derivatives like GOs are used for drug delivery and biosensing (Goenka et al., 2014). Fullerenes are spherical carbon cages capable of carrying drugs or atoms, generating reactive oxygen species for cancer therapy (Sushko et al., 2022), and showing low toxicity when used at controlled doses (T. Li et al., 2015). CQDs are very small fluorescent nanoparticles with excellent biocompatibility, suitable for imaging, biosensing and targeted therapy(Sun et al., 2006). NDs are useful in cancer therapy and bioimaging, due to their small size (4-8 nm) (Patil et al., 2022), biocompatibility (Leung et al., 2023) and large surface area that can be functionalised for drug delivery (Priyadarshni et al., 2024).

In their classification, it is useful to differentiate them based on their dimensionality and structural foundation. The distinction includes one-dimensional nanotubes (CNTs), two-dimensional graphene derivatives, and zero-dimensional structures such as fullerenes, carbon quantum dots (CQDs), and nanodiamonds (NDs) (Z. Liu & Liang, 2012). Their fabrication methodologies are diverse and assorted, these nanostructures fall within two broad categories: top-down and bottom-up methodologies (Abid et al., 2022). Top-down methods involve the reduction of bulk sources of carbon into nanoscale structures, for example, oxidative cutting of graphene oxide (Wen et al., 2016) and detonation synthesis of nanodiamonds (Kaur & Badea, 2013). On the other hand, while bottom-up methods include chemical vapor deposition of CNTs (Pant et al., 2021) or hydrothermal carbonization of CQDs (Z. Wang et al., 2025), which involve fabrication from molecular or atomic precursors.

2. TYPES OF CARBON NANOSTRUCTURES AND THEIR PHYSIOCHEMICAL CHARACTERISTICS

2.1 Carbon Nanotubes (CNTs)

CNTs are cylindrical-shaped nanostructures which are exclusively comprised of carbon atoms arranged in a hexagonal lattice. These structures closely resemble to rolled graphene sheets. CNTs are primarily classified into single-walled carbon nanotubes (SWCNTs) and multi-walled carbon nanotubes (MWCNTs). SWCNTs consists of a single graphene cylinder, with diameters ranging from 0.5 to 2nm. On the other hand, MWCNTs are composed of multiple concentric graphene cylinders with diameters upto 100nm (Ali et al., 2023).

CNTs possess a high surface area due to their elongated tubular architecture, which offers ample sites for the adsorption and conjugation of therapeutic and diagnostic agents. CNTs efficiently converts near-infrared (NIR) light rays into heat during photothermal therapy. This phenomenon enables precise and localized tumor removal. Additionally, CNTs can be subjected to covalent functionalization to introduce reactive groups such as -carboxyl and -amino groups. This improves solubility, biocompatibility, and targeting specificity of administered agents (Ijaz et al., 2023).

2.2 Graphene Oxide Nanostructures (GOs)

Graphene oxide (GO) is a 2-D carbon nanomaterial, produced from graphene by the process of chemical oxidation. GO contains a single layer of sp^2-hybridized structure of carbon atoms. These are arranged in a hexagonal lattice similar to graphene (Shafiee et al., 2022). However, they contain many oxygen-functional groups such as -hydroxyl, -epoxy, -carbonyl, and -carboxyl groups adsorbed on the basal plane and edges. These oxygen functionalities break the extended π-conjugation which is present in pure graphene and introduces hydrophilicity (Muazim & Hussain, 2017).

GO possesses intrinsic luminescent properties including NIR, which is very useful for deep-tissue tumor imaging as it is capable of producing clear signals with minimal background noise. The surface of GO contains oxygen groups that can bind with metal ions such as gadolinium (Gd^{3+}) or manganese (Mn^{2+}). This binding allows GO to produce MRI contrast images (Orecchioni et al., 2015).

2.3 Fullerenes

Fullerenes (C_{60}) are cage-like, hollow carbon nanostructures. The carbon atoms bond covalently to yield a truncated icosahedron structure, often compared to a soccer ball-like appearance. Fullerenes have a remarkable chemical and thermal stability, this allows C_{60} to remain stable under physiological conditions. The electron-rich surface around the curvature, provides them high electron affinity and fast-energy transfer processes, which are exploited in photodynamic applications. Structurally, the fullerenes possess excellent tensile strength which ensures the structural integrity of fullerenes during drug loading or surface modification (Fernandes et al., 2022).

Fullerenes potentially can carry out dual functions as photosensitizers and delivery vehicles. Upon light excitation, pristine and derivative fullerenes induce apoptotic cell death in the tumor tissues, while sparing neighbouring healthy cells. Their intrinsic near-infrared fluorescence provides optical imaging of deep-tissues with low background interference (Bolshakova et al., 2025; Zhen et al., 2024). The metal-functionalized derivatives of fullerene such as $Gd@C_{82}(OH)_{40}$ are used as MRI contrast agents, that permit concurrent structural imaging and therapy monitoring (H. Zhang, 2008). Fullerenes also possess a hollow core that can encapsulate chemotherapeutic agents for controlled release and multimodal imaging.

2.4 Nanodiamonds

Nanodiamonds (NDs) are very small carbon particles with a diamond-like core. They contain oxygen-rich surface groups and typically 2-8 nm in diameter. They have good mechanical strength, stability in circulation and biocompatibility. Their surface chemistry (carboxyl, hydroxyl, and amine groups) makes it convenient to attach drugs, targeting molecules and imaging agents. NDs are inherently fluorescent due to the presence of nitrogen-vacancy centres. They emit stable red light lasting significantly longer than typical dyes, to allow long-term cell tracking and clear imaging. With a very large surface area, NDs are capable of encapsulating very high doses of drugs (Gismondi et al., 2015; Mumtaz et al., 2022).

2.5 Carbon Quantum Dots (QCDs)

CQDs are small, spherical NPs which are smaller than 10nm. They contain a crystalline carbon core and a richly functionalize-able surface. The carbon core contains sp^2-hybridized domains as a result, CQDs absorb and emit light in a size-dependent manner. CQDs surfaces are densely covered with oxygen- and nitrogen-bearing groups (Ozyurt et al., 2023). This property makes them highly soluble in water and create "trap" states that can further modulate fluorescence. CQDs are capable of resisting photobleaching and can maintain luminescence for extended illumination. Their small size and surface charges prevent aggregation in solution, while robust core–shell structures withstand chemical and thermal stress (Yang et al., 2023).

3. FABRICATION AND SYNTHESIS OF CARBON NANOSTRUCTURES

3.1 Carbon Nanotubes (CNTs)

Several methods have been developed in the recent years, for the mass production of high quality CNTs. The frequently used approaches include, chemical vapour deposition (CVD), arc discharge, and laser ablation, meanwhile, the microwave irradiation technique is lately being preferred for fabrication of highly pure CNTs (Baghel et al., 2022).

(Yilmaz et al., 2017) demonstrated in a study, the synthesis of vertically aligned CNT arrays on aluminium substrates utilizing the hot-filament CVD method. The process involved the breakdown of methane (CH_4), a precursor carbon source gas, into smaller, reactive carbon pieces. Simultaneously, these pieces were subjected to land on cobalt NPs for absorption. Upon maximum absorption, the excess carbon was forced to eject out in a controlled manner. The carbon arranged themselves into rolled-up graphite sheets, leading to the formation of cylindrical nanotube forests.

Another study synthesized multiwalled CNTs by arc discharge method, supplying direct current. The approach created an intense electric spark between two graphite electrodes, in a chamber filled with argon gas. The anode consisted of nickel/yttrium gas which acted as catalysts. However, the cathode was a solid graphite rod. High current and voltage were supplied to generate extreme heat conditions. This phenomenon vapourized anode and released carbon atoms into the hot plasma. The carbon atoms cooled in the surrounding argon gas chamber and condensed around the catalysts to assemble as cylindrical, graphene nanotubes (Tepale-Cortés et al., 2021).

A few researchers developed an innovative, straight-forward method to synthesize SWCNTs, by pulse laser ablation of a coal-cobalt-nickel composite target. A powerful laser beam is employed to vapourize the coal pellet. This pellet is further subjected to rest inside a sealed quartz tube furnace, which is maintained at 1200 °C filled with argon gas. The laser fire creates a small, hot plasma of carbon atoms of coal and metal clusters from nickel-cobalt. Later on, argon carries the vapours to the cooler sections of the tube, where nickel-cobalt particles form a site of nucleation. The catalytic action of cobalt and nickel, along with the high-energy environment, results in the self-assembly of vapourised carbon atoms into the shape of cylindrical sheets, leading to the formation of carbon nanotubes (Chen et al., 2023)(*(PDF) Carbon Nanotubes: Methods, Achievements, Challenges, and Future Directions*, n.d.)(Ghosh, 2022).

3.2 Graphene Oxide (GOs)

A study reported a green, safe and low-cost approach to synthesize GO nanosheets substituting hydrogen peroxide (H_2O_2) instead of heavy metal oxidizers, traditionally used for fabrication. The process involves stirring of graphite powder in an acidic solution with the slow addition of H_2O_2. The mixture is subsequently heated as peroxide gradually oxidizes graphite, by adding oxygen-rich groups without the need of volatile compounds such as $KMnO_4$ or $NaNO_3$. Upon cooling, the suspension is washed thoroughly to neutralize the pH and subjected to ultrasonic water treatment. This peels the oxidized graphene into thin individual sheets of GO. This approach eliminates the risks of explosion, toxic residue accumulation as H_2O_2 is a safe and only oxidant and the acid can be recycled (Cao et al., 2023).

3.3 Carbon Quantum Dots (CQDs)

A study used a simple, effective and eco-friendly method to synthesize CQDs using kitchen waste. Mango peels were collected from a house in Bangladesh and was washed and dried for seven days, later crushed into powder. The peel powder was treated with sodium hypochlorite, a bleaching solution, to purify and neutralize the pH. The powder was further placed in an autoclave at 180 degrees for 12 hours to convert the mango peel powder into small dots of carbon. This was achieved by breaking down the components of the powder hydrothermally, under optimal pressure and heat. The impurities were removed from the product and was freeze-dried (lyophilized) to obtain solid carbon-quantum dots (Malitha et al., 2024).

(Nazar et al., 2024) successfully synthesized QCDs from arabica coffee ground by once again using a simple, cost-effective strategy of microwave-assisted irradiation. The coffee powder was utilized as a precursor, and was carbonized at 500 degrees. The activated carbon molecules were added into deionized water and vigorously stirred to form uniform suspension. The solution was exposed to microwave radiation, which assisted in breaking down the acti9vated carbon into quantum-sized carbon dots. Further, impurities were removed several times by centrifugation and the final product was dried to obtain brownish, thermally stable CQDs.

CQDs were fabricated from hazelnut husk waste, using hydrothermal-assisted pyrolysis approach in a recent study (Yuncu et al., 2024). The husks were washed, dried and frozen with liquid nitrogen and simultaneously, crushed into powder. The powder was mixed with ammonium citrate and urea, the sources for nitrogen and carbon, and heated at 200 °C for 25 minutes to initiate pyrolysis. The pyrolysis phenomenon is effectively utilised to break complex organic materials into quantum-sized carbon dots.

3.4 Nanodiamonds (NDs)

The primary and most abundantly occupied approach to synthesize nanodiamonds include the detonation of explosives (Pichot et al., 2013). In this method, explosives with negative oxygen balance such as mixture of 60% TNT and 40% hexogen is chosen to induce incomplete combustion, to produce products other than non-toxic gases. The explosives are subjected to detonation inside a sealed chamber containing cooling gases such as nitrogen and argon. The carbon precursor atoms are availed either from the explosive molecules itself or from the graphite, additionally kept in the chamber. During detonation, the explosive breaks apart releasing carbon atoms that are further condensed by the coolant inside the chamber. The condensed carbon atoms crystallize when the pressure and temperature inside the chamber, reach the levels sufficient for diamond formation. The final product undergoes purification to remove non-diamond carbon and metal contaminants from the chamber. Additionally, the particles are subjected to milling process or ultrasonic treatment which is applied to break aggregates or clusters into individual nano-sized carbon diamonds.

Another strategy to synthesize nanodiamonds resembles the natural process of formation of the diamonds. The high-pressure high-temperature (HPHT) milling technique is an approach where a carbon precursor, often graphite, is subjected to very high temperatures (2000 degree) and pressures (GPa) inside a chamber. The technique crushes the carbon source into nanoclusters, forming bulk or microdiamonds. The clustered aggregate is further milled to obtain nanodiamonds with impurities. The impurities are further washed off using acid treatment to obtain pure nanodiamonds (Basso et al., 2020).

3.5 Fullerenes

Fullerene nanostructures are typically assembled from fullerene molecules which are synthesized mainly by arc discharge method. In this approach, a very hot electric arc is created between 2 graphite electrodes inside a chamber filled with water. Initially, the chamber is made vacuum and consequently filled with helium gas at low pressures. Large amounts of current are made to jump between electrodes, while anode is electrically heated to produce an arc with a temperature around 2000°C. The intense heart produced vapourizes carbon atoms from the electrode into gaseous phase. Upon cooling and condensation, the atoms begin to stick together to form clusters of carbon. This leads to the gradual formation of fullerene molecules (mainly C_{60} and C_{70}) which are hollow and spherical in morphology (Kyesmen et al., 2016). These fullerene molecules are further utilized for assembly of nanostructures by drop-drying process or template method (*Sci-Hub : {title} [{doi}]*, n.d.-a).

4. THERANOSTIC APPLICATIONS OF CARBON NANOSTRUCTURES IN CANCER TREATMENT

A recent study revealed a one-step strategy to utilise MWCNTs into a single nanoplatform, that can locate and destroy breast cancer cells. Initially, raw MWCNTs were chemically functionalized with butyric acid for hydrophilicity and to introduce carboxyl groups on their surface. These –COOH groups were later activated and linked to a short DNA aptamer that specifically binds to the MUC-1 protein, a hallmark of aggressive breast tumors. The aptamer-coated MWCNTs circulated harmlessly inside the body. The cellular uptake studies and MRI imaging verified its successful attachment to the MUC-1^{+} MDA-MB-231 cells. Furthermore, the exposure of trapped aptamer-labelled MWCNTs to the heat emitted by NIR light, assisted photothermal cell death of MDA-MB-231 cells *in vitro*, without harming the neighbouring tissues (Khajuria et al., 2024).

In another original study, researchers created a magnetic theranostic system by integrating MWCNTs with FeO NPs and characterizing with polyethylene glycol (PEG) to obtain MWCNT–Fe_3O_4–PEG hybrids. The PEGylation made the tubes biocompatible and water-soluble, prolonging their circulation time in vivo. The hybrids additionally served as efficient T_2-weighted MRI contrast agents, darkening tumor tissues proportionally to Fe concentration in gel. Simultaneously, the anti-cancer drug doxorubicin (DOX) was non-covalently loaded onto the nanotubes via π–π stacking. In vitro release assays showed minimal DOX leakage at physiological pH (7.4) but accelerated release under acidic conditions (pH 6.3 and 5.2) in HeLA human cervical cancer cell line. Furthermore, NIR irradiation of DOX-loaded MWCNT–Fe_3O_4–PEG significantly increased drug release due to localized heating. This dual-responsive platform thus integrates MRI diagnosis, pH-triggered drug delivery, and photothermal-enhanced chemotherapy into a single nanomaterial for targeted cancer theranostics (Alkattan, 2014).

In a study, iron oxide nanoparticles (FeONPs) and reduced graphene oxide (rGO) were combined via self-assembly to create a multifunctional nanocomposite. The hybrid was then coated with PEG–NH_2 for biocompatibility. The FNPs/rGO-PEG quickly heated to 60 °C when it was exposed to near-infrared laser light, killing roughly 80% of 4T1 metastatic breast cancer cells in vitro. The heated nanocomposite directly destroyed the tumour and simultaneously released signals that stimulated the production of cytokines and matured dendritic immune cells. This converts the destroyed tumour into an in-situ vaccine. In BALB/c mice model, a single treatment with FNPs/rGO-PEG with laser eliminated the primary tumours and produced systemic antitumor immunity, which prolonged survival and stopped distant tumours from growing. Strong MRI contrast was another benefit of the iron oxide component, which made image-guided photothermal immunotherapy possible (L. Wang et al., 2019).

A group of researchers designed a platform using PEG-FA functionalized GO nanocomposites, that combined real-time fluorescence imaging with dual therapeutics. GO nanosheets were chosen for their 2D carbon scaffold, beneficial for loading molecules. They possess strong NIR absorption useful for heating purposes and contain oxygen groups at edges for effective chemical functionalization. PEG improved water dispersibility and reduced rapid clearance by MPS, meanwhile, FA functionalization enabled active targeting of tumor, typically overexpressed in cancer cells. Rhodamine B, was incorporated additionally as a fluorescent tracker to allow imaging. The conjugation of Indocyanine green (ICG) served as active photothermal and photosensitizer agent as it can strongly absorb NIR and generate ROS. The complex was intraperitoneally injected into mice bearing Ehrlich tumors, which successfully generated localised hyperthermia for photothermal therapy, upon NIR irradiation in tumors. This phenomenon induces hyperthermic cytotoxicity, protein denaturation, membrane disruption and enzyme inactivation, causing irreversible damage due to thermal ablation. The complex simultaneously activates ICG for photodynamic therapy, generating ROS that causes oxidation of lipids, proteins and nucleic acids, leading to oxidative stress, mitochondrial damage and apoptosis selectively in tumors (Romero et al., 2021).

In a study, a few scientists prepared magnetic nanodiamonds (FeNDs) by coating the multifunctional, carbon-rich surface NDs with Fe_3O_4 nanoparticles. The conjugate was subjected to physically adsorb DOX onto with FeND surface, due to the presence of abundant functional groups on the surface with characterization capabilities. Raman spectroscopy and XPS were used to verify the element composition. These analytical techniques confirmed the presence of Fe_3O_4 deposition on NDs with no loss of fluorescence, allowing for real-time cellular imaging and biodistribution tracking within A549 lung cancer cells. Following the uptake by the cancer cells, it was shown that FeND could release DOX into the cytoplasm of tumor cells successfully. Simultaneously, it catalyzed the Fenton reaction ($Fe^{2+} + H_2O_2$) to produce reactive oxygen species (ROS). In addition, NIR laser irradiation encouraged ROS production and show dual mode, combined photothermal-chemodynamic therapy (CDT) (Selvam et al., 2024).

Heterofunctional NDs were engineered in a study to combine targeted delivery system and optical imaging platform to treat breast cancer cell lines (MDA-MB-231 & MCF-7). The nanostructure was conjugated covalently with fluorescent labeled paclitaxel (PTX)-DNA conjugates and anti-EGFR monoclonal antibodies (mAbs) to enable a simultaneous system of therapy and visualization. Further, the incorporation of fluorescent oligonucleotide linkers enabled tracking of cellular uptake by confocal microscopy and flow cytometry. Upon exposure, the anti-EGFR mAbs directed the NDs specifically to the EGFR-overexpressing MDA-MB-231 cells, where EGFR is highly expressed. The conjugates were largely internalized due to multivalent interactions, while minimal uptake was observed in the minimally expressing MCF7 cells. The binding to free EGF by MDA-MB-231, confirmed that uptake was EGFR-targeting dependent. The PTX functionalization achieved increased therapeutic potential, with IC_{50} levels lower than free drug in MCF7 cells, and the anti-EGFR mAbs doubled efficacy as observed in MDA-MB-231 cells (*Sci-Hub : {title} [{doi}]*, n.d.-b; X. Q. Zhang et al., 2011).

The authors of a study, synthesised and tested a nanocomposite (UCNP–PEG–FA/PC70) in cervical cancer model (HeLa-luc), both *in vitro* and in mice bearing HeLa-luc tumors. As per the procedure, the team first synthesized β-$NaGdF_4$:Yb, Tm@$NaGdF_4$ core-shell UCNPs by thermal decomposition in organic solvent. Later on, the surface ligands were exchanged for citric acid to make them water soluble.

On the other hand, they electrostatically loaded PC70 onto the UCNP surface and incorporated folate modified polyethylene glycol (PEG-FA) to increase circulation time in physiology and promote targeting of folate-receptor$^+$ tumor cells. This complex enabled a two-step mechanism. The complex enabled the upconversion core to convert deeply penetrating 980nm NIR light to UV-blue emissions. Simultaneously, the UV-blue photons activated the attached fullerene-porphyrin dyad (PC70) core to generate singlet oxygen ($^1O_2$), a lethal reactive oxygen species, under hypoxic conditions to activate PDT (Guan et al., 2015).

A study could successfully fabricate a special large-amino-acid-mimicking CQDs (LAAM TC-CQDs) by hydrothermal approach. The synthesis was prepared using tetra-aminoanthraquinone, a large aromatic molecule with -amino groups, and citric acid as precursors. As a result, it produced a multi-layered graphene crystal, which contained paired α-carboxyl and amino groups sticking to the edges. The engineered CQDs were capable of emitting deep-tissue fluorescence imaging under NIR light, as NIR could penetrate deep into tissues than visible light. Additionally, they served as photoacoustic agents, which could generate sound waves when hit with a laser light, detectable by ultrasound. The conjugate was intravenously injected into xenograft mice bearing U87 gliomas of human brain. The LAAM-TC CQDs selectively accumulated in the tumors due to LAT1 transporter, typically overexpressed by cancer cells. This was achieved as CQDs could mimic amino acids which was selectively picked by cancer cells and ignored by healthy ones. After reaching the tumor site, the -carboxyl and -amino groups trigger clathrin-mediated endocytosis, a natural process where cells wrap and swallow substances. The healthy cells were spared as the uptake was tumor-selective. The CQDs which were loaded with topotecan by π–π stacking released their drug cargo intracellularly in the tumor, dropping tumor growth and increasing mice survival (S. Li et al., n.d.).

Folate-conjugated, reducible polyethyleneimine-passivated CQDs (fc-rPEI-CQDs) were fabricated by microwave pyrolysis approach using glycerol and branched PEI. The complex was designed to work as a smart gene delivery and imaging system for lung cancers. The conjugate carried siRNA cargo to attack EGFR and cyclin B1 through electrostatic complexation. EGFR blockage could reduce cancer growth signals, whereas, cyclin B1 restriction would arrest the tumor's cell division, leading to programmed cell death. Upon inhalation and tumor entry, the high levels of glutathione broke the disulfide bonds of the conjugate, releasing loaded siRNA selectively to the target. The CQDs glowed upon UV light exposure allowing tracking of the treatment. In lung cancer cells, the approach shut down tumor growth signals for 48 hours and reduced the cell survival significantly. The comprehensive results were similarly obtained in mouse models as well, with slowed tumor growth and direct imaging of the mice's lungs (Wu et al., 2016) (Table 1).

Table 1. A comprehensive table summarizing the theranostic applications of various carbon-based nanotubes, fullerenes, graphene oxides, quantum dots and nanodiamonds in the combined treatment of different cancers in vitro and in vivo

Nanocarbon Platform	Function--alization	Cancer Model	Mechanism	Theranostic Applications
MWCNTs	-COOH groups & MUC1 DNA Aptamer	MDA-MB-231 (breast cancer line)	Aptamer directs the MWCNTs to the MUC-1 receptors. NIR produces heat for photo ablation	Photothermal Therapy + MRI-assisted tumor localization
MWCNTs	Fe_3O_4 NPs + PEG + DOX through π–π stacking	HeLa (cervical cancer model)	Fe_3O_4 provides T_2 MRI contrast. The acidic pH triggers DOX release. NIR heating accelerates DOX release and create photothermal damage	MRI diagnosis + pH-responsive chemo + photothermal therapy
GO	Fe_3O_4 + rGO self-assembly + PEG–NH_2 coating	4T1 tumors in BALB/c mice	NIR irradiation heats rGO for PTT. Fe_3O_4 enables T_2 MRI guidance.	Image-guided PT immunotherapy, local tumor ablation MRI tracking.
Fullerene	PEG-FA + Rhodamine B + ICG	HeLa (mouse model)	FA targets folate receptors. ICG absorbs NIR to generate heat and ROS. GO supports PT and PD conversion. RhB enables fluorescence tracking.	Dual fluorescence/ photothermal–PDT for targeted tumor ablation and real-time imaging.
NDs	Fe_3O_4 on ND surface + DOX adsorption	A549 (lung cancer cells)	FeND catalyzes Fenton reaction to generate ROS (CDT). NIR boosts ROS. DOX release adds chemotherapy. ND fluorescence tracks biodistribution. Fe_3O_4 provides MRI contrast.	Combined fluorescent imaging, MRI guidance, chemodynamic and chemotherapy for hypoxia-amplified tumor killing.
NDs	Fluorescent PTX-DNA + anti-EGFR mAb	MDA-MB-231 & MCF7 (breast cells)	Anti-EGFR mAb directs NDs to EGFR+ cells. DNA linker releases PTX intracellularly. Fluorescein tag monitors uptake via microscopy and flow cytometry.	Receptor-targeted chemotherapy with real-time fluorescence imaging of drug delivery and uptake.
Fullerene	PC70 dyad + PEG-FA on UCNP	HeLa-luc (cervical cancer model)	UCNPs upconvert NIR to UV/blue to activate PC70 for singlet oxygen generation under hypoxia. PC70 and UCNPs emit fluorescence/UCL. Gd^{3+} in UCNPs provides T_1 MRI. FA enhances tumor targeting.	Trimodal FL/UCL/MR imaging-guided PDT for deep-tissue and hypoxic limitations.

continued on following page

Table 1. Continued

Nanocarbon Platform	Function- -alization	Cancer Model	Mechanism	Theranostic Applications
CQDs	Edge α-carboxyl & amino groups from tetra-aminoanthraquinone	U87 glioma in mice	CQDs mimic large amino acids to enter via LAT1 transporter. Deep-tissue NIR fluorescence/ photoacoustic imaging. π–π stacking loads topotecan. Intracellular release kills tumor.	Dual fluorescence/ photoacoustic imaging with targeted chemotherapy.
CQDs	Reducible PEI + folate conjugation	Lung cancer models	Folate targets folate receptors. Disulfide bonds cleaved by high GSH in tumor cells to release siRNA against EGFR and cyclin B1. CQD fluorescence monitors delivery.	Folate-targeted, reduction-responsive gene therapy with real-time imaging of siRNA delivery.

5. MECHANISMS OF ACTION IN THERANOSTICS

The theranostic applications of carbon nanostructures in cancer is maintained by their ability to accumulate within tumor and support therapy and diagnosis through distinct mechanisms ("Journal of Materials Chemistry B," n.d.). In cancer theranostic, diagnosis is an important step where carbon-based nanomaterial, has provided great advancements. The early and precise detection of tumours is often limited in conventional methods due to low sensitivity high cost and delayed results (Tang et al., 2021). Their nanoscale size, high surface to volume ratio and excellent conductivity enables them to interact efficiently with biomolecule and enhance the sensitivity of detection systems (Sheikhpour et al., 2020).

Functionalisation of this nanomaterial with biological molecules molecule like antibodies, peptide, aptamers or nucleic acids provides selectivity toward specific cancer biomarkers, ensuring targeted and reliable diagnostics outcomes (Safari et al., 2023). Studies have shown the use of carbon nanostructure in biosensing platforms designed to detect well-known cancer biomarkers, including HER2, CEA, CA125, VEGF, PSA, alpha-fetoprotein, and miRNA 21 (Safari et al., 2023). For instance, electrochemical biosensors developed using CNT or graphene electrodes significantly improve the detection limits for circulating tumour cells and genetic mutations due to their high conductivity and responsiveness.

Similarly, graphene oxide and reduced graphene oxide can enhance multimodal imaging systems, contributing to fluorescence, Raman and photoacoustic imaging signal (Keramat et al., 2022). Carbon quantum dots due to their flexible photoluminescence, allow real-time fluorescence imaging of single cancer cells, offering the possible of detecting tumor at extremely early stages(Chung et al., 2021; Tade & Patil, 2020).

Furthermore, carbon nanostructures can be used for designing hybrid diagnostic devices that integrate biosensing with therapeutic monitoring. Graphene- based electrochemical biosensors, for example, provide not only ultrasensitive detection but also versatile in integrating with portable point of care devices for real time analysis (Fu et al., 2023). Nanodiamonds and fullerenes, due to their biocompatibility and surface chemistry, have been used as contrast agents in imaging modalities, enabling improved visualisation of tumor at both cellular and tissue levels. Importantly, these diagnostics mechanism are not limited to single techniques but can be combined with therapeutic approaches, supporting the development of integrated theranostic systems(Ahmadian et al., 2022; Hosseini et al., 2023a).

Diagnostic agents like contrast dyes, MRI agents, or radioactive tracers can be encapsulated in nanoparticles that, due to the EPR effect, naturally gather in tumours (Kush et al., 2024). This allows clinicians to achieve better imaging contrast and more accurate tumor localization (Figure 1).

On the therapeutic side, anti-cancer drugs carried within nanoparticles take advantage of the same mechanism, leading to increase local concentrations at the tumor site and lower systemic toxicity (Singh & Kumar, 2022).

The EPR effect occurs because of structural and functional issues in tumor blood vessels and lymphatic systems(Mauro et al., 2021). These issues create conditions that differ greatly from normal tissues. Tumors, in their effort to support uncontrolled growth, stimulate the rapid formation of new blood vessels, a process known as angiogenesis (Bonferoni et al., 2021). The combined effect of increased permeability and poor clearance leads to a selective accumulation of nanoparticles in tumor tissue, creating a natural way for targeted delivery (Gou et al., 2024). Highly vascularized tumors tend to show stronger EPR activity, while small or poorly vascularized tumors may not (Elabed et al., 2025).

Figure 1. Graphical representation of the major mechanism through which carbon nanostructures interaction with cancer cells. These include enhanced permeability and EPR effect for passive targeting and ligand-receptor binding for active targeting

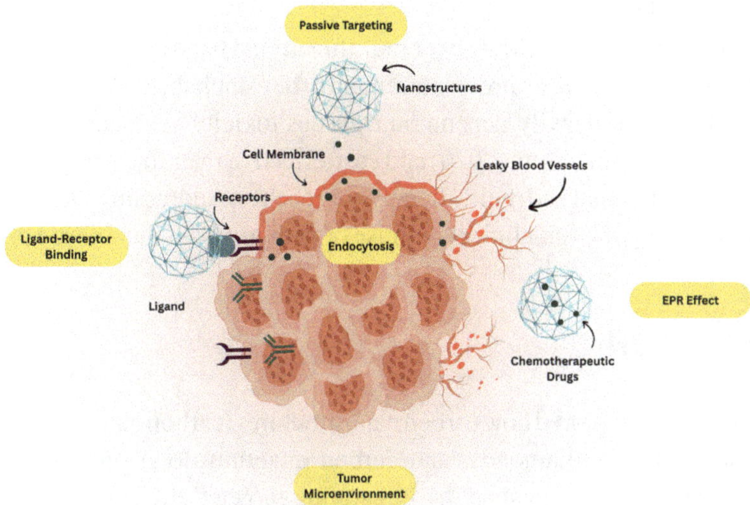

Due to the leaky blood vessels and insufficient drainage of tumors, nanostructures like carbon-based systems can accumulate in malignant tissues (Hosseini et al., 2023b). This selective localization is crucial for both diagnostic accuracy and treatment effectiveness, providing the foundation for more advanced targeting strategies(Alfei et al., 2025; Sharma, 2025). Ligand binding mechanism process starts when ligand-decorated nanoparticles circulate in the bloodstream and reach the tumor environment(Moghimi et al., 2023). When they encounter cancer cells with the matching receptor, the ligand on the nanoparticle surface binds to the receptor with strong affinity (Semenov et al., 2024).

6. FUTURE PERSPECTIVES

Carbon nanostructures have capably shown immense advantages in cancer theranostics. However, the existing studies are limited to only laboratory and early animal models. There is a clear lack of understanding of long-term safety of these nanostructures upon prolonged exposure in the physiology. Additionally, these challenges are coupled with biodistribution and clearance by MPS (Y. Li et al., 2024). These nanostructures currently face a critical challenge of reliable, large-scale production, as achieving particle quality and controlled surface-functionalization is resource intense (Kashyap et al., 2023). On the other hand, most research existing today, focus on photothermal and photodynamic therapies, limited by shallow light penetration into tumors and less effectiveness in hypoxic condition. Extensive research can be carried out to introduce new therapies for combination treatment (Deng et al., 2021).

In addition, regulatory guidance for multifunctional nanomaterials is vague, and standardized characterization protocols remain to be established. The future progress of these systems will heavily depend on rigorous toxicity evaluation, advances in reproducible (Rajakumar et al., 2020) and cost-effective manufacturing, exploration of robust multifunctional and multi-therapy approaches. Addressing these issues will be vital for realizing the full clinical promise of carbon-based theranostic platforms.

7. CONCLUSION

This chapter has explored how carbon nanostructures (carbon nanotubes, graphene oxide, fullerenes, nanodiamonds, and carbon quantum dots) provide integrated platforms for cancer theranostics, by integrating targeted drug delivery and multimodal imaging with stimulus-responsive therapies within individual materials. It investigated their unique physicochemical properties (top-down and bottom-up fabrication routes), surface functionalization routes, and exemplary in vitro and in vivo experimental works showing photothermal, photodynamic, chemodynamic, and immunomodulatory activities under real-time imaging guidance. Although these platforms overcome key dilemmas of specificity, resistance, and monitoring, practical issues remain, ensuring long-term biocompatibility and scalable and reproducible fabrication and clear regulatory routes. For clinical translation forward, future work must emphasize severe safety assessment, standardized syntheses of higher quality, and multifunctional combination therapies development (e.g., combining carbon platforms and immunotherapy or gene editing) to overcome hypoxia and heterogeneity of tumors. By matching up innovations at the material science level with realistic fabrication and regulatory paradigms, carbon nanostructure–based theranostics can escape proof-of-concept into personalized effective care for cancer.

REFERENCES

Abid, N., Khan, A. M., Shujait, S., Chaudhary, K., Ikram, M., Imran, M., Haider, J., Khan, M., Khan, Q., & Maqbool, M. (2022). Synthesis of nanomaterials using various top-down and bottom-up approaches, influencing factors, advantages, and disadvantages: A review. *Advances in Colloid and Interface Science*, *300*, 102597. DOI: 10.1016/j.cis.2021.102597 PMID: 34979471

Ahmadian, E., Janas, D., Eftekhari, A., & Zare, N. (2022). Application of carbon nanotubes in sensing/monitoring of pancreas and liver cancer. *Chemosphere*, *302*, 134826. Advance online publication. DOI: 10.1016/j.chemosphere.2022.134826 PMID: 35525455

Alfei, S., Reggio, C., & Zuccari, G. (2025). Carbon Nanotubes as Excellent Adjuvants for Anticancer Therapeutics and Cancer Diagnosis: A Plethora of Laboratory Studies Versus Few Clinical Trials. *Cells*, *14*(14), 1052. DOI: 10.3390/cells14141052 PMID: 40710305

Ali, A., Rahimian Koloor, S. S., Alshehri, A. H., & Arockiarajan, A. (2023). Carbon nanotube characteristics and enhancement effects on the mechanical features of polymer-based materials and structures – A review. *Journal of Materials Research and Technology*, *24*, 6495–6521. DOI: 10.1016/j.jmrt.2023.04.072

Alkattan, N. (2014). *Magnetic Carbon Nanotubes as a Theranostic Platform for Drug Delivery and Magnetic Resonance Imaging*.

Baghel, P., Sakhiya, A. K., & Kaushal, P. (2022). Ultrafast growth of carbon nanotubes using microwave irradiation: Characterization and its potential applications. *Heliyon*, *8*(10), e10943. DOI: 10.1016/j.heliyon.2022.e10943 PMID: 36276756

Basso, L., Cazzanelli, M., Orlandi, M., & Miotello, A. (2020). Nanodiamonds: Synthesis and application in sensing, catalysis, and the possible connection with some processes occurring in space. *Applied Sciences (Basel, Switzerland)*, *10*(12), 1–28. DOI: 10.3390/app10124094

Bolshakova, O., Zherebyatieva, O., & Sarantseva, S. V. (2025). Fullerenes in vivo. Toxicity and protective effects. *Nanotoxicology*, *19*(3), 233–258. DOI: 10.1080/17435390.2025.2471273 PMID: 40015266

Bonferoni, M. C., Rassu, G., Gavini, E., Sorrenti, M., Catenacci, L., Torre, M. L., Perteghella, S., Ansaloni, L., Maestri, M., & Giunchedi, P. (2021). Electrochemotherapy of Deep-Seated Tumors: State of Art and Perspectives as Possible "EPR Effect Enhancer" to Improve Cancer Nanomedicine Efficacy. *Cancers 2021, Vol. 13, Page 4437, 13*(17), 4437. DOI: 10.3390/cancers13174437

Cao, K., Tian, Z., Zhang, X., Wang, Y., & Zhu, Q. (2023). Green preparation of graphene oxide nanosheets as adsorbent. *Scientific Reports*, *13*(1), 1–11. DOI: 10.1038/s41598-023-36595-2 PMID: 37291198

Chen, S., Chen, Y., Xu, H., Lyu, M., Zhang, X., Han, Z., Liu, H., Yao, Y., Xu, C., Sheng, J., Xu, Y., Gao, L., Gao, N., Zhang, Z., Peng, L. M., & Li, Y. (2023). Single-walled carbon nanotubes synthesized by laser ablation from coal for field-effect transistors. *Materials Horizons*, *10*(11), 5185–5191. DOI: 10.1039/D3MH01053H PMID: 37724683

Chung, S., Revia, R. A., & Zhang, M. (2021). Graphene Quantum Dots and Their Applications in Bioimaging, Biosensing, and Therapy. *Advanced Materials*, *33*(22), 1904362. Advance online publication. DOI: 10.1002/adma.201904362 PMID: 31833101

Deng, X., Shao, Z., & Zhao, Y. (2021). Solutions to the Drawbacks of Photothermal and Photodynamic Cancer Therapy. *Advancement of Science*, *8*(3), 2002504. DOI: 10.1002/advs.202002504 PMID: 33552860

Dristant, U., Mukherjee, K., Saha, S., & Maity, D. (2023). An Overview of Polymeric Nanoparticles-Based Drug Delivery System in Cancer Treatment. *Technology in Cancer Research & Treatment*, *22*, 15330338231152084. DOI: 10.1177/15330338231152083 PMID: 36718541

Elabed, S., Sheirf, A., & Ali, M. (2025). Nanostructures for cancer therapeutics and diagnostics: Recent advances and future outlook. *Radiation Physics and Chemistry*, *226*, 112295. DOI: 10.1016/j.radphyschem.2024.112295

Fernandes, N. B., Shenoy, R. U. K., Kajampady, M. K., DCruz, C. E. M., Shirodkar, R. K., Kumar, L., & Verma, R. (2022). Fullerenes for the treatment of cancer: An emerging tool. *Environmental Science and Pollution Research International*, *29*(39), 58607–58627. DOI: 10.1007/s11356-022-21449-7 PMID: 35790637

Fu, L., Zheng, Y., Li, X., Liu, X., Te Lin, C., & Karimi-Maleh, H. (2023). Strategies and Applications of Graphene and Its Derivatives-Based Electrochemical Sensors in Cancer Diagnosis. *Molecules (Basel, Switzerland)*, *28*(18), 6719. Advance online publication. DOI: 10.3390/molecules28186719 PMID: 37764496

Gao, S., Xu, B., Sun, J., & Zhang, Z. (2024). Nanotechnological advances in cancer: Therapy a comprehensive review of carbon nanotube applications. *Frontiers in Bioengineering and Biotechnology*, *12*, 1351787. DOI: 10.3389/fbioe.2024.1351787 PMID: 38562672

Ghosh, R. (2022). *Carbon Nanotubes: Methods, Achievements, Challenges, and Future Directions*. DOI: 10.36227/techrxiv.19502455.v2

Gismondi, A., Reina, G., Orlanducci, S., Mizzoni, F., Gay, S., Terranova, M. L., & Canini, A. (2015). Nanodiamonds coupled with plant bioactive metabolites: A nanotech approach for cancer therapy. *Biomaterials*, *38*, 22–35. DOI: 10.1016/j.biomaterials.2014.10.057 PMID: 25457980

Goenka, S., Sant, V., & Sant, S. (2014). Graphene-based nanomaterials for drug delivery and tissue engineering. *Journal of Controlled Release*, *173*(1), 75–88. DOI: 10.1016/j.jconrel.2013.10.017 PMID: 24161530

Gou, Z., Tang, K., Zeng, C., Yuan, H., Zhang, C., Huang, Y., Qu, T., Xin, Q., Zhao, Y., Zeng, G., Yang, J., Xie, P., Yang, S. T., & Tang, X. (2024). Photothermal therapy of xenografted tumor by carbon nanoparticles-Fe(II) complex. *Colloids and Surfaces. B, Biointerfaces*, *240*, 113968. DOI: 10.1016/j.colsurfb.2024.113968 PMID: 38788472

Guan, M., Dong, H., Ge, J., Chen, D., Sun, L., Li, S., Wang, C., Yan, C., Wang, P., & Shu, C. (2015). Multifunctional upconversion-nanoparticles-trismethylpyridylporphyrin-fullerene nanocomposite: A near-infrared light-triggered theranostic platform for imaging-guided photodynamic therapy. *NPG Asia Materials*, *7*(7), e205–e205. DOI: 10.1038/am.2015.82

Gupta, D., Roy, P., Sharma, R., Kasana, R., Rathore, P., & Gupta, T. K. (2024). Recent nanotheranostic approaches in cancer research. *Clinical and Experimental Medicine*, *24*(1), 8. DOI: 10.1007/s10238-023-01262-3 PMID: 38240834

Gupta, V., Kannan, K., Tari, V., & Chandra, N. (2024). Bioengineered silver nanoparticles using Brassica oleracea sub sp. botrytis (L.) for enhanced antibacterial activity. *Physics and Chemistry of Solid State*, *25*(2), 303–310. DOI: 10.15330/pcss.25.2.303-310

Hamad, I., Harb, A. A., & Bustanji, Y. (2024). Liposome-Based Drug Delivery Systems in Cancer Research: An Analysis of Global Landscape Efforts and Achievements. *Pharmaceutics*, *16*(3), 400. DOI: 10.3390/pharmaceutics16030400 PMID: 38543294

Hosseini, S. M., Mohammadnejad, J., Najafi-Taher, R., Zadeh, Z. B., Tanhaei, M., & Ramakrishna, S. (2023a). Multifunctional Carbon-Based Nanoparticles: Theranostic Applications in Cancer Therapy and Diagnosis. *ACS Applied Bio Materials*, *6*(4), 1323–1338. DOI: 10.1021/acsabm.2c01000 PMID: 36921253

Hosseini, S. M., Mohammadnejad, J., Najafi-Taher, R., Zadeh, Z. B., Tanhaei, M., & Ramakrishna, S. (2023b). Multifunctional Carbon-Based Nanoparticles: Theranostic Applications in Cancer Therapy and Diagnosis. *ACS Applied Bio Materials*, *6*(4), 1323–1338. DOI: 10.1021/acsabm.2c01000 PMID: 36921253

Ijaz, H., Mahmood, A., Abdel-Daim, M. M., Sarfraz, R. M., Zaman, M., Zafar, N., Alshehery, S., Salem-Bekhit, M. M., Ali, M. A., Eltayeb, L. B., & Benguerba, Y. (2023). Review on carbon nanotubes (CNTs) and their chemical and physical characteristics, with particular emphasis on potential applications in biomedicine. *Inorganic Chemistry Communications*, *155*, 111020. DOI: 10.1016/j.inoche.2023.111020

Kannan, K., Radhika, D., Gnanasangeetha, D., Krishna, L. S., & Gurushankar, K. (2021). Y3+ and Sm3+ co-doped mixed metal oxide nanocomposite: Structural, electrochemical, photocatalytic, and antibacterial properties. *Applied Surface Science Advances*, *4*, 100085. DOI: 10.1016/j.apsadv.2021.100085

Kannan, K., Radhika, D., Kasai, R. D., Gnanasangeetha, D., Palani, G., Gurushankar, K., Koutavarapu, R., Lee, D.-Y., & Shim, J. (2022). Facile fabrication of novel ceria-based nanocomposite (CYO-CSO) via co-precipitation: Electrochemical, photocatalytic and antibacterial performances. *Journal of Molecular Structure*, *1256*, 132519. DOI: 10.1016/j.molstruc.2022.132519

Kannan, K., Radhika, D., Reddy, K. R., Raghu, A. V., Sadasivuni, K. K., Palani, G., & Gurushankar, K. (2021). Gd3+ and Y3+ co-doped mixed metal oxide nanohybrids for photocatalytic and antibacterial applications. Nano express, 2(1), 010014.

Kashyap, B. K., Singh, V. V., Solanki, M. K., Kumar, A., Ruokolainen, J., & Kesari, K. K. (2023). Smart Nanomaterials in Cancer Theranostics: Challenges and Opportunities. *ACS Omega*, *8*(16), 14290–14320. DOI: 10.1021/acsomega.2c07840 PMID: 37125102

Kaur, R., & Badea, I. (2013). Nanodiamonds as novel nanomaterials for biomedical applications: Drug delivery and imaging systems. *International Journal of Nanomedicine*, *8*, 203. DOI: 10.2147/IJN.S37348 PMID: 23326195

Keramat, A., Kadkhoda, J., Farahzadi, R., Fathi, E., & Davaran, S. (2022). The Potential of Graphene Oxide and Reduced Graphene Oxide in Diagnosis and Treatment of Cancer. *Current Medicinal Chemistry*, *29*(26), 4529–4546. DOI: 10.2174/0929867329666220208092157 PMID: 35135444

Khajuria, A., Alajangi, H. K., Sharma, A., Kaur, H., Sharma, P., Negi, S., Kumari, L., Trivedi, M., Yadav, A. K., Kumar, R., Raghuvanshi, R. S., Kaur, I. P., Tyagi, R. K., Jaiswal, P. K., Lim, Y. B., Barnwal, R. P., & Singh, G. (2024). Theranostics: Aptamer-assisted carbon nanotubes as MRI contrast and photothermal agent for breast cancer therapy. *Discover Nano*, *19*(1), 145. DOI: 10.1186/s11671-024-04095-5 PMID: 39256285

Khorasani, A., Shahbazi-Gahrouei, D., & Safari, A. (2023). Recent Metal Nanotheranostics for Cancer Diagnosis and Therapy: A Review. *Diagnostics (Basel)*, *13*(5), 833. DOI: 10.3390/diagnostics13050833 PMID: 36899980

Kostarelos, K., Lacerda, L., Pastorin, G., Wu, W., Wieckowski, S., Luangsivilay, J., Godefroy, S., Pantarotto, D., Briand, J. P., Muller, S., Prato, M., & Bianco, A. (2007). Cellular uptake of functionalized carbon nanotubes is independent of functional group and cell type. *Nature Nanotechnology*, *2*(2), 108–113. DOI: 10.1038/nnano.2006.209 PMID: 18654229

Kush, P., Kumar, P., & Singh, R. (2024). Functionalized Carbon Nanostructures in Cancer Diagnosis and Therapy. *Handbook of Functionalized Carbon Nanostructures: From Synthesis Methods to Applications*, 2091–2130. DOI: 10.1007/978-3-031-32150-4_60

Kyesmen, P. I., Onoja, A., & Amah, A. N. (2016). Fullerenes synthesis by combined resistive heating and arc discharge techniques. *SpringerPlus*, *5*(1), 1323. DOI: 10.1186/s40064-016-2994-7 PMID: 27563518

Leung, H. M., Chu, H. C., Mao, Z. W., & Lo, P. K. (2023). Versatile nanodiamond-based tools for therapeutics and bioimaging. *Chemical Communications*, *59*(15), 2039–2055. DOI: 10.1039/D2CC06495B PMID: 36723092

Li, S., Su, W., & Wu, H. (n.d.). *Targeted tumour theranostics in mice via carbon quantum dots structurally mimicking large amino acids*. DOI: 10.1038/s41551-020-0540-y

Li, T., Murphy, S., Kiselev, B., Bakshi, K. S., Zhang, J., Eltahir, A., Zhang, Y., Chen, Y., Zhu, J., Davis, R. M., Madsen, L. A., Morris, J. R., Karolyi, D. R., LaConte, S. M., Sheng, Z., & Dorn, H. C. (2015). A New Interleukin-13 Amino-Coated Gadolinium Metallofullerene Nanoparticle for Targeted MRI Detection of Glioblastoma Tumor Cells. *Journal of the American Chemical Society*, *137*(24), 7881–7888. DOI: 10.1021/jacs.5b03991 PMID: 26022213

Li, Y., Xu, Z., Qi, Z., Huang, X., Li, M., Liu, S., Yan, Y., & Gao, M. (2024). Application of Carbon Nanomaterials to Enhancing Tumor Immunotherapy: Current Advances and Prospects. *International Journal of Nanomedicine*, *19*, 10899–10915. DOI: 10.2147/IJN.S480799 PMID: 39479174

Liu, D., Pan, X., Zhao, X., & Zhang, H. (2022). Preparation and In Vitro and In Vivo Evaluation of Lung Cancer-Targeted Carbon Nanotube Ultrasound Contrast Agents. *Journal of Biomedical Nanotechnology*, *18*(6), 1684–1691. DOI: 10.1166/jbn.2022.3366

Liu, Z., & Liang, X. J. (2012). Nano-Carbons as Theranostics. *Theranostics*, *2*(3), 235–237. DOI: 10.7150/thno.4156 PMID: 22448193

Malitha, M. D., Molla, M. T. H., Bashar, M. A., Chandra, D., & Ahsan, M. S. (2024). Fabrication of a reusable carbon quantum dots (CQDs) modified nanocomposite with enhanced visible light photocatalytic activity. *Scientific Reports*, *14*(1), 1–18. DOI: 10.1038/s41598-024-66046-5 PMID: 39095409

Mauro, N., Utzeri, M. A., Varvarà, P., & Cavallaro, G. (2021). Functionalization of Metal and Carbon Nanoparticles with Potential in Cancer Theranostics. *Molecules 2021, Vol. 26, Page 3085, 26*(11), 3085. DOI: 10.3390/molecules26113085

Moghimi, M., Akpa, P. A., Peter, I. E., Onwuka, A. M., Obi, B. C., Akunne, M. O., Nworu, C. S., Ejikeme, P. M., Akunne, T. C., Attama, A. A., Akah, P. A., & Ng, A. A. A. (2023). Nanotheranostics: Platforms, Current Applications, and Mechanisms of Targeting in Breast and Prostate Cancers. *Journal of Nanotheranostics 2023, Vol. 4, Pages 346-383, 4*(3), 346–383. DOI: 10.3390/jnt4030016

Mohammadpour, A., Kazemi, A., Baghapour, M. A., Samaei, M. R., Kannan, K., & Khaneghah, A. M. (2024). Bioengineered FeZn/GA@ Cu nanocomposite utilizing spent coffee ground extract and gum arabic: Enhanced nitrate removal via (RSM) and machine learning optimization. *International Journal of Biological Macromolecules*, *277*, 134060. DOI: 10.1016/j.ijbiomac.2024.134060 PMID: 39097464

Muazim, K., & Hussain, Z. (2017). Graphene oxide — A platform towards theranostics. *Materials Science and Engineering C*, *76*, 1274–1288. DOI: 10.1016/j.msec.2017.02.121 PMID: 28482495

Mumtaz, M., Hussain, N., Salam, S., & Bilal, M. (2022). Multifunctional nanodiamonds as emerging platforms for cancer treatment, and targeted delivery of genetic factors and protein medications—a review. *Journal of Materials Science 2022 57:17, 57*(17), 8064–8099. DOI: 10.1007/s10853-022-07168-x

Nazar, M., Hasan, M., Wirjosentono, B., Gani, B. A., & Nada, C. E. (2024). Microwave Synthesis of Carbon Quantum Dots from Arabica Coffee Ground for Fluorescence Detection of Fe3+, Pb2+, and Cr3+. *ACS Omega*, *9*(18), 20571–20581. DOI: 10.1021/acsomega.4c02254 PMID: 38737012

Orecchioni, M., Cabizza, R., Bianco, A., & Delogu, L. G. (2015). Graphene as Cancer Theranostic Tool: Progress and Future Challenges. *Theranostics*, *5*(7), 710–723. DOI: 10.7150/thno.11387 PMID: 25897336

Ozyurt, D., Al Kobaisi, M., Hocking, R. K., & Fox, B. (2023). Properties, synthesis, and applications of carbon dots: A review. *Carbon Trends*, *12*, 100276. DOI: 10.1016/j.cartre.2023.100276

Pant, M., Singh, R., Negi, P., Tiwari, K., & Singh, Y. (2021). A comprehensive review on carbon nano-tube synthesis using chemical vapor deposition. *Materials Today: Proceedings*, *46*, 11250–11253. DOI: 10.1016/j.matpr.2021.02.646

Papi, R. M., Tasioulis, K. S., Kechagioglou, P. V., Papaioannou, M. A., Andriotis, E. G., & Kyriakidis, D. A. (2024). Carbon Nanotube-Mediated Delivery of PTEN Variants: In Vitro Antitumor Activity in Breast Cancer Cells. *Molecules (Basel, Switzerland)*, *29*(12), 2785. DOI: 10.3390/molecules29122785 PMID: 38930850

Patil, S., Mishra, V. S., Yadav, N., Reddy, P. C., & Lochab, B. (2022). Dendrimer-Functionalized Nanodiamonds as Safe and Efficient Drug Carriers for Cancer Therapy: Nucleus Penetrating Nanoparticles. *ACS Applied Bio Materials*, *5*(7), 3438–3451. DOI: 10.1021/acsabm.2c00373 PMID: 35754387

(PDF) Carbon Nanotubes: Methods, Achievements, Challenges, and Future Directions. (n.d.). Retrieved August 30, 2025, from https://www.researchgate.net/publication/360022778_Carbon_Nanotubes_Methods_Achievements_Challenges_and_Future_Directions

Pichot, V., Risse, B., Schnell, F., Mory, J., & Spitzer, D. (2013). Understanding ultrafine nanodiamond formation using nanostructured explosives. *Scientific Reports*, *3*(1), 1–6. DOI: 10.1038/srep02159 PMID: 23831716

Priyadarshni, N., Singh, R., & Mishra, M. K. (2024). Nanodiamonds: Next generation nano-theranostics for cancer therapy. *Cancer Letters*, *587*, 216710. DOI: 10.1016/j.canlet.2024.216710 PMID: 38369006

Rajakumar, G., Zhang, X. H., Gomathi, T., Wang, S. F., Ansari, M. A., Mydhili, G., Nirmala, G., Alzohairy, M. A., & Chung, I. M. (2020). Current Use of Carbon-Based Materials for Biomedical Applications—A Prospective and Review. *Processes 2020, Vol. 8, Page 355*, *8*(3), 355. DOI: 10.3390/pr8030355

Romero, M. P., Buzza, H. H., Stringasci, M. D., Estevão, B. M., Silva, C. C., Pereira-Da-Silva, M. A., Inada, N. M., & Bagnato, V. S. (2021). *Graphene Oxide Theranostic Effect: Conjugation of Photothermal and Photodynamic Therapies Based on an in vivo Demonstration*. DOI: 10.2147/IJN.S287415

Roszkowski, S., Durczyńska, Z., & Szablewska, S. (2025). Targeted nanodelivery systems for personalized cancer therapy. *Reports of Practical Oncology and Radiotherapy : Journal of Greatpoland Cancer Center in Poznan and Polish Society of Radiation Oncology*, *29*(6), 776–788. DOI: 10.5603/rpor.103524 PMID: 40104662

Safari, M., Moghaddam, A., Salehi Moghaddam, A., Absalan, M., Kruppke, B., Ruckdäschel, H., & Khonakdar, H. A. (2023). Carbon-based biosensors from graphene family to carbon dots: A viewpoint in cancer detection. *Talanta*, *258*, 124399. Advance online publication. DOI: 10.1016/j.talanta.2023.124399 PMID: 36870153

Selvam, R., Pearl, W. G., Perevedentseva, E., Karmenyan, A., & Cheng, C. L. (2024). Fe-doped nanodiamond-based photo-Fenton catalyst for dual-modal fluorescence imaging and improved chemotherapeutic efficacy against tumor hypoxia. *RSC Advances*, *14*(6), 4285–4300. DOI: 10.1039/D3RA08465E PMID: 38298935

Semenov, K. N., Shemchuk, O. S., Ageev, S. V., Andoskin, P. A., Iurev, G. O., Murin, I. V., Kozhukhov, P. K., Maystrenko, D. N., Molchanov, O. E., Kholmurodova, D. K., Rizaev, J. A., & Sharoyko, V. V. (2024). Development of Graphene-Based Materials with the Targeted Action for Cancer Theranostics. *Biochemistry (Moscow) 2024 89:8*, *89*(8), 1362–1391. DOI: 10.1134/S0006297924080029

Shafiee, A., Iravani, S., & Varma, R. S. (2022). Graphene and graphene oxide with anticancer applications: Challenges and future perspectives. *MedComm*, *3*(1), e118. DOI: 10.1002/mco2.118 PMID: 35281783

Sharma, D. K. (2025). Recent advancements in nanoparticles for cancer treatment. *Medical Oncology 2025 42:3*, *42*(3), 1–17. DOI: 10.1007/s12032-025-02609-4

Sheikhpour, M., Naghinejad, M., Kasaeian, A., Lohrasbi, A., Shahraeini, S. S., & Zomorodbakhsh, S. (2020). The applications of carbon nanotubes in the diagnosis and treatment of lung cancer: A critical review. *International Journal of Nanomedicine*, *15*, 7063–7078. DOI: 10.2147/IJN.S263238 PMID: 33061368

Sindhwani, S., Syed, A. M., Ngai, J., Kingston, B. R., Maiorino, L., Rothschild, J., MacMillan, P., Zhang, Y., Rajesh, N. U., Hoang, T., Wu, J. L. Y., Wilhelm, S., Zilman, A., Gadde, S., Sulaiman, A., Ouyang, B., Lin, Z., Wang, L., Egeblad, M., & Chan, W. C. W. (2020). The entry of nanoparticles into solid tumours. *Nature Materials 2020 19:5*, *19*(5), 566–575. DOI: 10.1038/s41563-019-0566-2

Singh, R., & Kumar, S. (2022). Cancer Targeting and Diagnosis: Recent Trends with Carbon Nanotubes. *Nanomaterials 2022, Vol. 12, Page 2283, 12*(13), 2283. DOI: 10.3390/nano12132283

Sun, Y. P., Zhou, B., Lin, Y., Wang, W., Fernando, K. A. S., Pathak, P., Meziani, M. J., Harruff, B. A., Wang, X., Wang, H., Luo, P. G., Yang, H., Kose, M. E., Chen, B., Veca, L. M., & Xie, S. Y. (2006). Quantum-sized carbon dots for bright and colorful photoluminescence. *Journal of the American Chemical Society, 128*(24), 7756–7757. DOI: 10.1021/ja062677d PMID: 16771487

Surendran, P., Lakshmanan, A., Priya, S. S., Balakrishnan, K., Rameshkumar, P., Kannan, K., & Vinitha, G. (2024). Synthesis of fluorescent carbon quantum dots from Manihot esculenta waste peels for nonlinear optical and biological applications. *Chemical Physics Impact, 8*, 100515. DOI: 10.1016/j.chphi.2024.100515

Sushko, E. S., Vnukova, N. G., Churilov, G. N., & Kudryasheva, N. S. (2022). Endohedral Gd-Containing Fullerenol: Toxicity, Antioxidant Activity and Regulation of Reactive Oxygen Species in Cellular and Enzymatic Systems. *International Journal of Molecular Sciences, 23*(9), 5152. DOI: 10.3390/ijms23095152 PMID: 35563539

Tade, R. S., & Patil, P. O. (2020). Theranostic Prospects of Graphene Quantum Dots in Breast Cancer. *ACS Biomaterials Science & Engineering, 6*(11), 5987–6008. DOI: 10.1021/acsbiomaterials.0c01045 PMID: 33449670

Tang, L., Xiao, Q., Mei, Y., He, S., Zhang, Z., Wang, R., & Wang, W. (2021). Insights on functionalized carbon nanotubes for cancer theranostics. *Journal of Nanobiotechnology, 19*(1), 423. Advance online publication. DOI: 10.1186/s12951-021-01174-y PMID: 34915901

Tao, C., & Rouhi, J. (2023). A biosensor based on graphene oxide nanocomposite for determination of carcinoembryonic antigen in colorectal cancer biomarker. *Environmental Research, 238*, 117113. DOI: 10.1016/j.envres.2023.117113 PMID: 37696325

Tepale-Cortés, A., Moreno-Saavedra, H., Hernandez-Tenorio, C., Rojas-Ramírez, T., Illescas, J., Tepale-Cortés, A., Moreno-Saavedra, H., Hernandez-Tenorio, C., Rojas-Ramírez, T., & Illescas, J. (2021). Multi-walled Carbon Nanotubes Synthesis by Arc Discharge Method in a Glass Chamber. *Journal of the Mexican Chemical Society, 65*(4), 480–490. DOI: 10.29356/jmcs.v65i4.1486

Wang, L., Wang, M., Zhou, B., Zhou, F., Murray, C., Towner, R. A., Smith, N., Saunders, D., Xie, G., & Chen, W. R. (2019). PEGylated reduced-graphene oxide hybridized with Fe3O4 nanoparticles for cancer photothermal-immunotherapy. *Journal of Materials Chemistry. B*, *7*(46), 7406–7414. DOI: 10.1039/C9TB00630C PMID: 31710067

Wang, Z., Changotra, R., Dasog, M., Singh Selopal, G., Yang, J., & He, Q. S. (2025). Carbon quantum dots: Synthesis via hydrothermal processing, doping strategies, integration with photocatalysts, and their application in photocatalytic hydrogen production. *Sustainable Materials and Technologies*, *44*, e01386. DOI: 10.1016/j.susmat.2025.e01386

Wen, J., Li, M., Xiao, J., Liu, C., Li, Z., Xie, Y., Ning, P., Cao, H., & Zhang, Y. (2016). Novel oxidative cutting graphene oxide to graphene quantum dots for electrochemical sensing application. *Materials Today. Communications*, *8*, 127–133. DOI: 10.1016/j.mtcomm.2016.07.006

Wu, Y. F., Wu, H. C., Kuan, C. H., Lin, C. J., Wang, L. W., Chang, C. W., & Wang, T. W. (2016). Multi-functionalized carbon dots as theranostic nanoagent for gene delivery in lung cancer therapy. *Scientific Reports 2016 6:1, 6*(1), 1–12. DOI: 10.1038/srep21170

Yang, H. L., Bai, L. F., Geng, Z. R., Chen, H., Xu, L. T., Xie, Y. C., Wang, D. J., Gu, H. W., & Wang, X. M. (2023). Carbon quantum dots: Preparation, optical properties, and biomedical applications. *Materials Today. Advances*, *18*, 100376. DOI: 10.1016/j.mtadv.2023.100376

Yilmaz, M., Raina, S., Hsu, S. H., & Kang, W. P. (2017). Growing micropatterned CNT arrays on aluminum substrates using hot-filament CVD process. *Materials Letters*, *209*, 376–378. DOI: 10.1016/j.matlet.2017.08.061

Yuncu, H., Nadaroglu, H., & Bozkurt, E. (2024). Eco-friendly synthesis of Carbon Quantum Dots (CQDs) from hazelnut husk for sensitive Aflatoxin B1 (AFB1) detection. *Toxicology Reports*, *13*, 101824. DOI: 10.1016/j.toxrep.2024.101824 PMID: 39649378

Zhang, H. (2008). Gd@C82 Fullerenol. *Molecular Imaging and Contrast Agent Database (MICAD)*. https://www.ncbi.nlm.nih.gov/books/NBK23539/

Zhang, X. Q., Lam, R., Xu, X., Chow, E. K., Kim, H. J., & Ho, D. (2011). Multimodal Nanodiamond Drug Delivery Carriers for Selective Targeting, Imaging, and Enhanced Chemotherapeutic Efficacy. *Advanced Materials*, *23*(41), 4770–4775. DOI: 10.1002/adma.201102263 PMID: 21932280

Zhen, M., Xu, Y., Wang, C., & Bai, C. (2024). Fullerene-Based Immunoregulatory Nanomaterials for Immunotherapy of Tumor and Immune-Related Inflammatory Diseases. *Advanced Functional Materials*, *34*(49), 2409319. DOI: 10.1002/adfm.202409319

Chapter 6
Green Synthesis of Quantum Dots for Sustainable Drug Delivery

T. Gayathri
https://orcid.org/0000-0002-1079-4057
KPR Institute of Engineering and Technology, India

G. V. Aishwarya
KPR Institute of Engineering and Technology, India

R. K. Harshavardhini
https://orcid.org/0009-0009-6270-2159
KPR Institute of Engineering and Technology, India

R. Rekha
KPR Institute of Engineering and Technology, India

ABSTRACT

Green synthesis of quantum dots (QDs) uses environmentally benign, low-toxicity, and energy efficient routes plant extracts, agricultural residues and harmless solvents to prepare nanomaterials designed for future biomedical applications. Carbon- and graphene-based QDs provide excellent fluorescence, photostability, and surface functionalities with varied versatility, enabling the synthesis of multifunctional drug delivery vehicles. Their inherent biocompatibility and low cytotoxicity render them suitable for deep-tissue bioimaging, targeted therapy of cancer and gene delivery, with real-time monitoring of therapy. Renewable biomass valorization to produce QDs advocates the concepts of circular economy, reducing both environmental

DOI: 10.4018/979-8-3373-4247-4.ch006

impact and industrial dependency on toxic precursors. Green-synthesized QDs are a viable approach for the provision of high-performance and eco-friendly drug delivery systems for human health and environmental stewardship.

INTRODUCTION

Quantum dots (QDs) are semiconductor nanocrystals ranging from 10 nm or smaller in diameter. They emit bright colors that change with size. Small dots are blue, larger red due to quantum confinement effects (Yu & Schanze, 2023). They are used extensively in electronics displays, bioimaging, sensors, and solar cells. The Nobel Prize was awarded to Brus, Ekimov, and Bawendi for breaking ground on this technology in the 1980s and early 1990s. Conventional quantum dot fabrication has included poisonous precursors (such as cadmium telluride or dimethyl cadmium), elevated temperatures, and fossil fuels, but hardly green.

Green synthesis involves environmentally friendly, low-toxicity, low-energy techniques employing water, plant waste, and benign reagents. For carbon quantum dots (CQDs) or graphene quantum dots (GQDs), it involves avoiding strong acids and heavy metals. Benefits are:

1. Natural, renewable precursors (e.g. plant extracts, agro-waste)
2. Eco-friendly solvents such as water
3. Low cost, low waste, scalable processes
4. Biocompatible and high photostability product

There are three categories of green carbon dots: plant-based, microorganism-based, and natural polymer-based carbon quantum dots.

1. Plant-Based Carbon Dots

They are derived from leaves, fruit peels, seeds, and other biomass of plants. They are generally synthesized through hydrothermal or microwave routes. Bamboo stems, eucalyptus twigs, lemon/orange peels, black sesame seeds, and lotus leaves are some examples. CDs are frequently used as sensors for Fe^{3+}, Cr^{6+}, Hg^{2+} ions, or bio-imaging probes. Quantum yields vary extensively e.g. cucumber peel CDs more stable than pineapple peel ones; pineapple ones degrade sooner. Nitrogen doped CDs prepared from bamboo exhibit intense fluorescence and Fe^{3+} selectivity. Pomelo juice-derived Up conversion or Nitrogen doped QDs have an average size of 3 nm, and N-doping improves photocatalytic dye degradation under solar light (Navarro-Badilla et al., 2023). Few advantages of these plant-based CDs are abundantly found, environment-friendly, low cost and tunable photoluminescence. Few listable disadvantages are size control and reproducibility is variable; yields are plant- and protocol-dependent.

2. Microorganism-Based Carbon Dots

CDs are produced from algae, microbial biomass, fungal residues, or even bacterial cell walls. For example, *Dunaliella salina* (a green algae) hydrothermal CDs have been exploited for the detection of heavy metals such as Hg (II) and Cr (VI), and live-cell bioimaging (Khairol Anuar et al., 2021). Also, onion waste, silkworm chrysalis, crab shell biomass, all microbe or insect-derivative are processed to form bioactive CDs. Silk fibroin (silkworm cocoons' protein) produces SF-CQDs through microwave-induced synthesis (20 min at 200 °C) with water solubility and photostability . Few advantages of these microorganism-based carbon dots are composition (proteins, chitin) being unique permits intrinsic N, S doping, biocompatible and functionalizable. Few listable disadvantages are purity of biomass is crucial; batch-to-batch inconsistency unless highly standardized.

3. Natural Polymer-Based Carbon Dots

This classification is based on biopolymers such as polysaccharides like cellulose, starch, pectin, chitin or proteins. They are not raw biomass from plants, but polymers that have been extracted. CQDs from xylan (a hemicellulose) using microwave in ammonia solution: rapid (10 min), green synthesized, N-doped, high crystallinity and tolerance towards salt/metal. CQDs from chitosan, cellulose, pectin through hydrothermal carbonization provides sustainability, biodegradability, and multifunctional application in textiles, environmental, and biomedical applications. Few advantages of these natural polymer-based carbon dots structured design is its consistent size/emission, functional polymer chemistry enables doping and surface passivation. Few listable disadvantages are extraction/purification of polymer is required and not as low-cost or waste-valorization focused as raw biomass.

Important observations under consideration:

1. Size control is more difficult with bottom-up approaches; dots can be large in size distribution unless precisely adjusted.
2. Surface imperfections or by-products from biomass pathways may require additional purification (dialysis or chromatography) to prevent fluorescent impurities.
3. While they look green, scalability and yield still differs traditional approaches can be better in industrial yield unless optimized.

SYNTHESIS METHODS

1. Plant-Based Carbon Quantum Dots (CQDs)

Synthesis of carbon quantum dots (CQDs) from plants is a green and ecofriendly process that makes use of several carbon-rich plant sources such as lemon peel, orange peel, banana peel, eucalyptus leaves, aloe vera, or even waste materials like coconut husk and sugarcane bagasse. These precursors are initially harvested and washed extensively with deionized water to eliminate surface impurities. The pure biomass is subsequently oven-dried or shade dried at 60–80 °C for 24–48 hours to eliminate moisture content. After full dryness, the biomass is powdered finely with the aid of a mechanical grinder to enhance the surface area and allow uniform thermal degradation.

In hydrothermal synthesis, this powder is suspended in deionized water with a 1:10 w/v ratio and stirred intensely to prepare a homogenous suspension. The mixture is placed in a Teflon-lined stainless-steel autoclave and is heated at 180–220 °C for 6–12 hours. With this, the carbonization of biomass happens, leading to nucleation and growth of CQDs. The autoclave, after the reaction, is cooled down to room temperature naturally. The dark brown to black suspension obtained is filtered through Whatman filter paper to remove particulates and centrifuged at 12,000 rpm for 20 minutes.

The supernatant, in which CQDs are present, is then purified. Membrane filtration (0.22 µm) assists in the removal of colloidal debris, while dialysis against distilled water for 48–72 hours removes low-molecular-weight impurities. Finally, the solution is lyophilized to get dry powder of CQD. This process is scalable and toxin-free, making it suitable for biomedical purposes. Characterization involves TEM for morphology, FTIR for functional groups, and PL for fluorescent behavior. Experiments demonstrated high quantum yields (10–25%) with orange peel and lemon juice precursors, respectively.

2. Microorganism-Based Carbon Quantum Dots (CQDs)

Microorganism-based synthesis utilizes the natural constitution and biosynthetic mechanisms of microorganisms like bacteria (*Escherichia coli*), fungi (*Aspergillus niger*), and algae (*Chlorella vulgaris*). The process begins with cultivation of the desired microorganism in nutrient broth under optimized conditions usually 37 °C and pH 7.0 for bacterial growth, and 28 °C with pH 5.0–6.0 for fungi. After 24–72 hours of incubation, the microbial biomass is recovered through centrifugation at

10,000 rpm and washed several times with sterile deionized water to eliminate culture media residues.

The purified biomass is resuspended in distilled water and pre-treated with dilute nitric acid (0.1 M) or hydrogen peroxide (3%) in order to improve carbonization efficiency. This pre-treated biomass is either utilized straight or homogenized through ultrasonication (20 kHz, 30 min) to lyse cell walls. The slurry is then moved to a hydrothermal reactor and heated at 200–240 °C for 12–24 hours. With this interval, carbonization, dehydration, and passivation take place, resulting in CQDs formation. The mixture is cooled, filtered, and centrifuged to eliminate unreacted biomass.

In the microwave-assisted synthesis method, microbial biomass slurry is microwaved at 700–900 W for 5–10 minutes. The swift heating is responsible for efficient conversion into CQDs. After synthesis, the solution is filtered and dialyzed against a 1 kDa membrane for 48–72 hours. Freeze drying provides solid CQD powder. Analytical characterization involves UV-Vis and PL spectroscopy to evaluate excitation-emission characteristics, TEM to measure size distribution (usually 2–6 nm), and XPS to validate elemental composition. CQDs prepared from chitosan and algal derivatives with quantum yields of more than 30%, which are helpful in sensing and imaging.

3. Natural Polymer-Based Carbon Quantum Dots (CQDs)

Natural polymer-based CQDs are prepared from biopolymers including starch, cellulose, gelatin, pectin, gum arabic, and carboxymethyl cellulose (CMC). They possess numerous hydroxyl and carboxyl functional groups, hence making them good precursors for the synthesis of CQDs. For example, starch or cellulose is dissolved in deionized water at a concentration of 1–5% w/v and stirred at mild heating (50–60 °C) under stirring for dissolution. Urea or ethylenediamine can be added to act as nitrogen dopant and passivating agents.

In hydrothermal synthesis, polymer solution is loaded into a 100 mL Teflon-lined autoclave and kept at 200 °C for 6–12 hours. Dehydration and carbonization of the polymer chains are enhanced through reaction, producing CQDs with trapped functional groups. On cooling, the dark solution is centrifuged and filtered to eliminate large aggregates. Purification involves membrane filtration (0.22 μm) and dialysis against distilled water for 48–72 hours. Lastly, the CQDs are concentrated through rotary evaporation or freeze-dried into a powder.

Microwave-assisted synthesis entails suspending the polymer solution in a microwave-reactive container and irradiating at 700–900 W for 8–12 minutes. The merit of this process is fast synthesis and increased yield with reproducible particle size. The resulting CQDs are typically 3–8 nm in diameter and fluorescence in the blue-green region. Structural characterization instruments like FTIR, XRD, and

Raman spectroscopy validate the structural aspects, whereas PL and quantum yield analysis assess the photonic performance. CQDs obtained from sugarcane bagasse CMC exhibited intense photoluminescence and high water dispersibility.

This process is especially effective for the tuning of optical properties through doping with changes in precursor composition. Introduction of nitrogen, sulfur, or phosphorus groups during synthesis increases the stability and intensity of photoluminescence. The adaptability of the process makes it applicable across many disciplines ranging from bioimaging to drug delivery and environmental monitoring.

The steps presented for plant-based, microorganism-based, and natural polymer-based CQD synthesis demonstrate a range of green synthesis approaches. Every approach offers an avenue to CQD synthesis that is renewable, economical, and clean. Plant based routes are preferred for simplicity and scalability, routes based on microorganisms have the advantage of intrinsic nitrogen doping and functionalization and routes based on natural polymers offer structural diversity and process regulation and allow for the fine-tuning of the optical properties for the requirements of the targeted application (Zhao et al., 2022). Through the integration of these green synthesis methods, scientists make contributions towards sustainable nanomaterial production with immense applications in medicine, the environment, and industry.

4. Surface Functionalization and Modifications

Plant-based CQDs contain inherent surface functionalities like hydroxyl, carboxyl, and carbonyl groups, making them amenable for further chemical modifications. Surface passivation is typically done using polymers like polyethylene glycol (PEG) or polyvinylpyrrolidone (PVP) to make them more biocompatible and stable (Zhang & Han, 2012). Amino functionalization with ethylenediamine or polyethylenimine (PEI) imparts positive surface charges and enhances cellular uptake in biomedical applications. Thiol-functionalized probes such as L-cysteine can also be grafted for selective detection of heavy metals. In addition, conjugation with targeting ligands (e.g., folic acid) or drugs is established through 1-Ethyl-3-(3-dimethylaminopropyl) carbodiimide/N-Hydroxy succinimide coupling chemistry, which allows targeted delivery. Surface-engineered CQDs derived from lemon peels had higher quantum yields and enhanced bioimaging performance after PEGylation.

Microbial CQDs are rich in intrinsic doping with heteroatoms such as nitrogen and sulfur originated from the biomass. Nevertheless, post-synthetic adjustment is necessary in order to further optimize their applications. Doping with biocompatible substances such as dextran or PEG improves dispersion in biological media (Parashar et al., 2024). Surface oxidation with hydrogen peroxide or nitric acid adds carboxyl groups, and this can further be amidated with amines for targeted purposes. In addition, enzymatic coupling reactions permit conjugation with peptides or antibodies for biosensing purposes. CQDs derived from yeast biomass exhibited greater photoluminescence and cytocompatibility upon functionalization with chitosan (Mirseyed et al., 2024).

CQDs derived from biomimetic polymers are preserved with functional groups that render them extremely reactive and flexible. Surface functionalization with carboxyl or amino can enable coupling with fluorophores, metal ions, or nanoparticles to enhance sensitivity in detection. For biomedical applications, PEGylation can be employed to minimize cytotoxicity and prolong circulation time. In environmental sensing, functionalization with targeted ligands such as EDTA or thiol-containing species increases the selectivity towards heavy metal ions. Furthermore, gelatin CQDs can be crosslinked with molecules containing aldehyde groups for stability improvement. Drug loading capacity and sustained release characteristics are greatly enhanced by functionalized gelatin-CQDs (Ding et al., 2024).

5. Surface Coating and Drug Delivery Mechanism

Plant quantum dots (QDs), especially carbon quantum dots (CQDs), have been considered a cutting-edge and environmentally friendly platform for drug delivery systems because they are biocompatible, green synthesis-friendly, and inherently photoluminescent. These nanoscale carbon materials are normally produced from renewable biomass sources such as fruit peels, leaf extracts, seeds, and even whole plant tissues by green synthetic processes like hydrothermal treatment, microwave-assisted synthesis, or pyrolysis (Usman & Cheng, 2024). These syntheses produce CQDs decorated with rich surface functional groups such as hydroxyl (–OH), carboxyl (–COOH), and amine ($–NH_2$) that play a key role in their surface chemistry and interaction with therapeutic compounds. Not only do plant-based QDs differ from others by virtue of their intrinsic fluorescence, but also by virtue of their functional versatility and biocompatibility that comes from the plant phytochemicals involved in their synthesis.

These CQDs, having been synthesized, present several routes for drug loading that are primarily fueled by their extensive surface chemistry. The simplest one is physical adsorption or electrostatic interaction. Drugs having oppositely charged functionalities or polar moieties can be attached to the CQDs through non-covalent interactions like hydrogen bonding, Van der Waals forces, or ionic interactions (Bao et al., 2019). This approach is widely applied to reversible loading of small molecules with controlled release upon physiological stimuli like pH or temperature. Covalent conjugation approaches, however, are utilized when a more permanent drug-carrier interaction is required. During this process, surface carboxyl groups on the CQD are activated with EDC (1-ethyl-3-(3-dimethylaminopropyl) carbodiimide) and NHS (N-hydroxy succinimide) chemistry for the formation of amide bonds with amine-bearing drugs or targeting ligands like folic acid, hyaluronic acid, or peptides. Covalent attachment keeps the drug attached in circulation but releasable under intracellular conditions like the acidic pH of lysosomes or the high glutathione content of the cytosol.

In addition to direct attachment, CQDs can also be incorporated into or coated by biopolymeric layers with core-shell or nanogel structures. Such systems use polymers such as polyethylene glycol (PEG), chitosan, alginate, or hyaluronic acid (HA) as coating materials. These coatings have several purposes: enhancing colloidal stability in physiological fluids, boosting biocompatibility, hiding toxicity, and minimizing immunogenic recognition. For example, Hydroxyapatite coatings facilitate active targeting of tumor cells with CD44 receptor overexpression, dramatically enhancing uptake efficiency and biodistribution PEGylation, which involves grafting PEG chains onto the surface of CQDs, extends systemic circulation by inhibiting opsonization and Reticuloendothelial System clearance (RES) (Bao et al., 2019). These coatings may be deposited either through physical adsorption or covalent attachment based on the desired strength of the system.

Surface coating is initiated even from the synthesis phase, in which plant extracts serve not just as carbon sources but also as natural capping agents (Kumar et al., 2023). These agents bring hydrophilic functionalities that facilitate water solubility and surface passivation. After synthesizing the CQDs, they are purified (in most cases through dialysis or ultrafiltration) to get rid of unreacted precursors and remaining plant biomolecules. Subsequently, surface modification is carried out. For coating with PEG or HA, carbodiimide chemistry is once used to attach polymer chains onto the CQD surface. In advanced systems, linker molecules such as bifunctional PEG (bearing terminal $-NH_2$ or $-COOH$ groups) are employed to bridge between the CQD and the drug molecule, enabling concomitant functionalization and drug loading.

Some important physicochemical phenomena govern the drug delivery mechanism of plant-derived CQDs. One of the most significant is the particle size. Smaller CQDs (<10 nm) have better permeability across biological barriers and are typically cleared efficiently by the kidneys. But for drug delivery, somewhat larger particles (~10–20 nm) with suitable coating can take advantage of the Enhanced Permeability and Retention effect (EPR) in tumor tissue for passive targeting. Surface charge, as evidenced by the zeta potential, also plays a decisive role. Positively charged CQDs (e.g., chitosan-coated or amino group-rich) enable cellular uptake through electrostatic attraction with negatively charged cell membranes. Yet they can also be associated with greater cytotoxicity and serum protein adsorption. Neutral or weakly negative particles (usually obtained through PEGylation) show low non-specific interactions and improved in vivo stability.

Another crucial aspect is the colloidal stability of QDs in biological fluids. Without proper coating, QDs aggregate with salts or proteins and reduce their functional surface area and also elicit immune responses. Hydrophilic coatings such as PEG, alginate, or HA inhibit aggregation and improve dispersion, which is critical in ensuring uniform biodistribution and pharmacokinetics. The hydrophobic or hydrophilic character of both the drug and the CQD surface also controls drug loading efficiency. Hydrophobic drugs are directed to non-polar zones or inner micelle-type centers if available, whereas hydrophilic drugs adhere to polar groups at the periphery. This selective binding will affect both the loading as well as the release kinetics since hydrophobic interactions tend to be slower to reverse.

The drug release mechanism from CQDs is typically controlled by environmental stimuli. pH-responsive systems, for instance, take advantage of the tumor's acidic microenvironment (pH ~6.5) or endosomes/lysosomes (pH ~5.5) to release the drug. CQDs with pH-cleavable linkers (e.g., hydrazones or esters) cleave their bonds under acidic conditions for targeted release (Bao et al., 2019). Redox-responsive systems leverage disulfide bonds that cleave upon exposure to intracellular glutathione (GSH), a higher concentration of which exists in cancer cells than in normal cells. This facilitates triggered release only after cellular internalization, minimizing systemic toxicity. In addition, some coatings are enzymatically degradable, allowing site-specific drug release where certain enzymes are expressed in tissues.

Density and arrangement of ligands in space on the CQD surface also adjust targeting efficiency. For instance, folic acid ligands must be spaced well with flexible linkers to avoid steric hindrance that may detract from receptor binding. Too much PEG may obscure such ligands; too little may reveal the core and signal immune detection. These balance issues need to be optimized upon formulation. Furthermore, cellular internalization mechanism whether it is clathrin-mediated endocytosis, caveolae-mediated endocytosis, macropinocytosis, or direct membrane translocation is influenced by particle size, surface chemistry, and the inclusion of cell-penetrating peptides (CPPs). Plant-derived CQDs with CPPs or ligand-targeting receptors can fundamentally redirect the uptake pathway, changing intracellular trafficking and final release site.

There is also added complexity in the fluorescence behavior of CQDs. Although their native luminescence allows real-time imaging of drug delivery and biodistribution, loading drugs or surface coating may quench or alter this fluorescence. Thus, maintaining the optical signature and designing the surface for functional purpose is a fine balance act. Additionally, although plant-derived CQDs exhibit lower cytotoxicity than conventional metal-based QDs (such as CdSe or ZnS), the effect of remaining phytochemicals or surface contaminants on immune interactions should be thoroughly assessed, particularly for systemic applications (Cai et al., 2021).

In general mechanism of drug delivery by plant-derived CQDs is one of a multifaceted interaction of synthesis process, surface chemistry, functionalization strategy, physicochemical properties, and biological context. Drugs may be loaded by either physical or covalent means, and biopolymer or targeting ligand-based surface coatings increase stability, targeting, and release control. The primary importance are phenomena such as size, charge, coating density, hydrophilicity, and stimuli-responsiveness, which need to be optimized for optimal therapeutic performance. Consequently, plant-derived CQDs are extremely promising as a next-generation nanomedicine platform that couples diagnosis and therapy into one highly tunable and sustainable system. Nevertheless, current challenges persist in large-scale reproducibility, long-term toxicity profiling, and clinical translation, but the future outlook is unequivocally bright.

Microorganism-based CQDs are an intriguing new area in green nanomedicine, using bacteria, fungi, yeast, or algae as living "factories" to biosynthesize fluorescent carbon nanodots under gentle green conditions (Wang et al., 2025). Unlike plant-mediated CQDs, microbial protocols offer "intrinsic capping layers" formed by secreted biomolecules during synthesis of proteins, polysaccharides, enzymes that inherently become part of the surface coating and confer colloidal stability and abundant functional groups (e.g., amide I/II bonds, carboxyl, hydroxyl moieties). These CQDs derived from microorganisms are usually products of intracellular or extracellular processes: in extracellular metabolism, microbial enzymes lower carbon precursors outside of cells, producing nanoparticles naturally covered with extracellular proteins or polysaccharides; in intracellular pathways, ions are brought into the microbial cytoplasm and enzymatically transformed into QDs, which subsequently need cell-lysis for harvesting and purification for the elimination of native biomaterials .

The intrinsic surface covering obtained through microbial synthesis performs a double function: it stabilizes CQDs in water media by steric or electrostatic repulsion and offers reactive groups (–NH, –COOH, –OH) to be used for additional drug loading or conjugation of ligands. For instance, E. coli-mediated CdTe QDs had zeta potential of approximately 36 mV, much lower than hydrothermal counterparts, evidence of strong biomolecular capping through biosynthesis (MDPI). Fourier-transform infrared spectroscopy of these microbial CQDs identified distinct amide I and II peaks, establishing proteinaceous coverings from microbial enzymes or secreted proteins.

Loading of drugs onto microbial CQDs utilizes comparable channels as plant-based systems but with microbial context-dependent characteristics (Giordano et al., 2023). Physical adsorption and electrostatic binding continue as major avenues: drugs with opposite charges or those that can engage in hydrogen bonding bind to CQD surfaces with proteins and polysaccharides. For antimicrobial purposes, positively charged CQDs (prepared through microbial modification or doping) bind strongly to the negatively charged bacterial cell wall particularly Gram-negative lipopolysaccharide-rich walls causing disruption of the cell wall and inactivation of microorganisms; the same binding mechanism also allows for drug delivery targeted against bacteria (Wang et al., 2025), where enhanced antimicrobial activity is required, functionalization with quaternizing agents forms ultra-positive surface charges (e.g., zeta +33 mV), allowing strong electrostatic and hydrophobic interactions with bacterial membranes and membrane insertion and disruption the process investigated in quaternized nitrogen-doped CQDs (Jiang et al., 2023).

Covalent conjugation is also exploited, especially when drugs or target ligands require stable coupling. The original surface biomolecules of microbial-derived CQDs (e.g., amine group-containing proteins) can be activated through carbodiimide chemistry for conjugation of drug molecules or targeting groups like folic acid. Such biopolymers are particularly abundant in those microbial CQDs derived from fungi or yeast and offer multiple reactive sites for functionalization (Maity et al., 2023). Although direct literature references are still fewer for plant-CQDs, the identical 1-Ethyl-3-(3-dimethylaminopropyl) carbodiimide/N-Hydroxy succinimide approach can be used for microbial derived CQDs after identification of their surface groups.

In addition, microorganism-derived CQDs can be inserted into polymeric matrices e.g., chitosan, hydrogels or antimicrobial wound dressing films in which the CQDs are responsible for fluorescence and antimicrobial activity and the host polymer dictates drug release and mechanical properties. A prime example is gentamicin-loaded chitosan/folic acid CQD nanocomposite hydrogel films exhibited intense H-bonding interactions demonstrated through FTIR (wide O–H peaks, aromatic skeleton shift), verifying incorporation of CQDs into the CS network; the CQD enhanced tensile strength and altered swelling behavior, facilitating sustained drug release for wound dressing application.

Various phenomena decisively regulate the efficacy of microorganism-based CQDs for drug delivery. Initial size and surface charge determine biodistribution, cell uptake, and microbial targeting (Giordano et al., 2023). The microbial synthesis pathway tends to produce CQDs of 2–10 nm size, which is optimal for penetration into tissues and clearance by the kidneys, but thickness of coating or embedding in polymer matrices can drive sizes into 10–20 nm improving retention in inflamed or tumorous tissues through the EPR effect (Hettiarachchi et al., 2019). Positively charged CQDs, either through microbial surface capping or deliberate doping (e.g. nitrogen or quaternary ammonium groups), are bound avidly by negatively charged bacterial or mammalian cell membranes (Wang et al., 2025).

Second, colloidal stability and aggregation are influenced by the microbial biomolecular shell. Aggregation can suppress fluorescence, interfere with targeting ability, and result in rapid clearance or immune activation (Giordano et al., 2023).

The protein/polysaccharide capping inherently enhances dispersion and minimizes nonspecific adsorption in physiological media.

Third, the hydrophilic and hydrophobic balance of microbial CQDs affects drug affinity: hydrophobic binding sites within coating proteins aid in loading hydrophobic drugs, whereas polar surface groups draw in hydrophilic drugs (Giordano et al., 2023). For antimicrobial payloads, numerous loaded drugs are hydrophilic (e.g., gentamicin), promoting adsorption onto polar surfaces and allowing sustained release in the hydrogel matrix. Fourth, microbial CQDs are generally photodynamic antimicrobial agents; upon exposure to light, they produce reactive oxygen species (ROS) that destroy microbial membranes and DNA (Wang et al., 2024; Kannan et al., 2024; Mohammadpour et al., 2024; Gupta et al., 2024; Kannan et al., 2022; Kannan et al., 2021). The two-pronged mechanism drug cargo and photodynamic activity enhances antibacterial potency, especially against drug-resistant bacteria and biofilms (Wang et al., 2025).

Fifth, stimuli-responsive release can be designed. While less frequently documented for microbial CQDs than for plant CQDs, pH or redox-responsive systems are possible: Microbial capping with disulfide linkages or acid-cleavable linkers can be prepared to release drugs within intracellular acidic or reductive milieu (Giordano et al., 2023).

Sixth, ligand density and steric arrangements are important where targeting is wanted (e.g., folate receptor targeting) (Karakoçak et al., 2020). Microbial CQDs with indigenous surface proteins offer many sites for functionalization, but excessive coatings can occlude ligand access. The art is to have stealth (through PEG or polysaccharide coatings) balanced with exposure of targeting ligands (Tegafaw et al., 2025). Seventh, fluorescence properties vs. coating/drug loading also involve a compromise. Although microbial CQDs are naturally fluorescent, excessive drug loading or dense coatings will quench emission. Optimization is important to maintain functionality for theranostic tracking e.g., minimal but adequate coating, or the use of linkers that maintain electron states. Eighth, immunogenicity and cytotoxicity are a function of both the core CQD and surface coating material. Microbial capping tends to diminish acute toxicity but may trigger immune reactions through residual microbial endotoxins or foreign proteins. Thorough purification and characterization (e.g., endotoxin assay, protein analysis) are necessary for in vivo use with safety (Mangini et al., 2021).

Mechanistically, a prototype microbial CQD drug delivery route would proceed as follows: initially, an enzyme-secreting fungus or bacterium releases enzymes that convert carbon precursors outside the cell to form ~5 nm CQDs in a proteinaceous coat. Surface functionalized CQDs are purified through ultrafiltration and dialysis. Subsequently, surface amine or carboxyl groups are activated through 1-Ethyl-3-(3-dimethylaminopropyl) carbodiimide/N-Hydroxy succinimide (EDC/NHS) to bind a drug molecule (e.g. antibiotic or anticancer agent) and/or a targeting ligand such as folic acid (Maity et al., 2023). Optionally, the CQDs are loaded into a chitosan hydrogel film to release gentamicin. The nanofilm produced shows improved mechanical strength, controlled release, antimicrobial activity under fluorescent observation, and potential photodynamic synergism upon illumination (Wang et al., 2024). When applied, CQDs loaded film binds to wound surfaces; drug diffusion follows due to moisture or pH, while fluorescent emission tracks distribution; positively charged CQDs also kill remaining microbial cells through electrostatic and ROS-based mechanisms (Knoblauch & Geddes, 2020). This multi-functional design integrates microbial-based green synthesis benefit with surface-engineered delivery capability.

In short, microorganism-based CQDs utilize the native biochemical machinery of microbes to synthesize inherently coated, functionalized nanodots that are particularly suitable for drug delivery and antimicrobial applications. Their strengths include environment friendly fabrication, inbuilt surface capping, intrinsic fluorescence, and compatibility with both passive and active drug delivery approaches. Pivotal phenomena like size, zeta potential, colloidal stability, hydrophobic affinity, stimuli responsiveness, ligand architecture, fluorescence integrity, and immunogenic safety need to be all finely balanced. Though clinical translation remains to overcome challenges regarding scalability, reproducibility, and regulatory clearances, microorganism-based CQDs offer a radical, futuristic tip of the hat to a world in which smart nanomedicines are being produced by biological systems themselves.

Polymer-cloaked CQDs are designed nanometer-sized carbon centers whose surfaces are intentionally wrapped, grafted, or passivated by synthetic or biogenic polymers (PEG, PEI, PAA, chitosan, polysaccharides, block copolymers), resulting in multifunctional carriers that integrate the intrinsic fluorescence and nanoscale size of CQDs with the colloidal stability, stealth, and chemistries that can be tuned with polymer coats. This hybrid strategy is widely documented as a successful method for enhancing circulation half-life, minimizing immunogenicity, and facilitating controlled drug release and targeting (Hailing et al., 2020). Polymer coatings are deposited in several technologically different manners: (1) physical adsorption or entanglement wherein amphiphilic polymers or block copolymers self-assemble onto hydrophobic CQD surfaces or intercalate through hydrophobic/hydrophilic interactions; (2) covalent grafting wherein polymer chains with terminal functional groups ($-NH_2$, $-COOH$) are linked to CQD surface groups using carbodiimide chemistries or click reactions to form stable shell architectures; and (3) in-situ polymerization wherein monomers are polymerized on the CQD surface to produce crosslinked shells or nanogel crowns that encapsulate payloads and leakage is avoided (Alkahtani et al., 2024).

These coating methods modify critical physical properties hydrodynamic diameter, zeta-potential, steric layer, hydrophilicity/hydrophobicity balance and thus modulate biodistribution (renal clearance vs. EPR accumulation), opsonization, and cellular uptake mechanisms (Alkahtani et al., 2024). Drug loading onto polymer-based CQDs capitalizes on three main mechanisms selected based on the chemistry of the therapeutic and release kinetics desired: (a) non-covalent adsorption (electrostatic binding, hydrogen bonding, π–π stacking) ideal for reversible, stimulus-enabled release; (b) covalent pro-drug conjugation (e.g., amide or hydrazone linkages introduced by EDC/NHS or aldehyde chemistry) for circulation stability with programmable cleavage in tumoral/lysosomal milieus; (Hailing et al., 2020) and (c) interior encapsulation when CQDs are designed as hollow or core–shell nanostructures and subsequently capped with polymeric gates allowing increased loading of hydrophobic drugs such as doxorubicin (DOX) and controlled leakage through gate opening (Alkahtani et al., 2024). The literature provides specific examples: PEI-passivated Carbon dots complexed with DOX demonstrate how a cationic polymer shell can increase loading and cellular delivery while also facilitating imaging through CQD fluorescence. The key to this is strategic polymer choice with PEGylation. PEGylation is the paradigm stealth coat (minimizing serum protein adsorption and RES clearance), whereas PEI or other polyamines are specifically selected for complexation of nucleic acid or anionic drugs and for endosomal escape through the "proton sponge" mechanism in which polymer buffering leads to endosome swelling and burst release of cargo intracellularly.

Stimuli-responsive release motifs underlie polymer-CQD designs targeting spatially selective therapy: pH-sensitive polymers (polyacrylic acid, hydrazone linkers, acid-labile esters) release in acidic tumors or endo/lysosomes; redox-sensitive disulfide linkers take advantage of high intracellular glutathione to break bonds preferentially inside cells; enzyme-sensitive peptides or polysaccharide shells (e.g., hyaluronic acid cleavable by hyaluronidase) mediate receptor-targeted release; and photo- or thermo-responsive polymers allow external regulation of release by light or heat. Engineering these stimuli into the polymer layer transforms CQDs from passive carriers to "smart" theranostic agents that only release payloads upon arriving in the target microenvironment, reducing systemic toxicity and enhancing therapeutic index. Several recent communications show pH-responsive CQD-polymer hydrogels and polymer-gated mesoporous systems releasing model drugs in acidic environments that simulate tumor microenvironments (Das et al., 2020).

Many interdependent factors decide if a polymer-CQD system works in reality particle size and hydrodynamic profile (narrow size distributions centered around 5–20 nm are preferable for tissue penetration but larger coated particles take advantage of EPR retention) (Alkahtani et al., 2024); surface charge (posh shells enhance uptake but increase risk of cytotoxicity; neutral PEG minimizes nonspecific binding);

grafting density and spacer length of the polymer (over dense PEG conceals targeting ligands; insufficient spacing leads to steric hindrance and suboptimal receptor interaction); colloidal stability in proteinaceous environments (aggregation silences fluorescence, distorts biodistribution and can induce complement activation); and preservation of optical properties (surface passivation must not quench fluorescence by dense polymer brushes or excessive drug loading).

For instance, DOX–PEI-CD complexes exhibit improved intracellular delivery but need to be optimized in PEI molecular weight and grafting density in order to reduce cytotoxicity without compromising endosomal escape. At a mechanistic pharmacokinetics and cell biology level, polymer-CQDs follow a predetermined route: following systemic delivery the polymer coat regulates opsonization and renal clearance; targeted ligands (antibodies, peptides, folic acid) displayed on the polymer surface target receptor-mediated endocytosis, whereas non-specific uptake takes macropinocytosis or clathrin/caveolae routes depending on size and charge; internalized particles are subjected to endosomal acidification and redox gradients that induce polymer bond cleavage or swelling, allowing attached or encapsulated drugs to be released into the cytosol; and lastly, the carbon core and byproducts of polymer degradation are metabolized or excreted based on size and biodegradability.

Real-time tracking is possible because the CQD core retains fluorescence if coatings and loading are tuned to avoid absolute quenching enabling true theranostic readouts of delivery and effect. Safety, scalability, and maintenance of theranostic function are pragmatic limitations: high-molecular-weight PEI enhances transfection efficiency but cytotoxicity (requiring partial PEGylation or low-MW PEI substitutes), whereas polymer crosslinking can enhance drug retention but hinder biodegradation and clearance; platform reproducibility in polymer grafting and CQD core size are necessary for batch-to-batch reproducibility and clinical translation; and polymer or monomer impurities create regulatory challenges demanding extensive characterization (FT-IR, XPS, DLS, TEM, zeta potential, in vitro/in vivo toxicity assays). Reviews highlight that polymer-CQD systems have been highly preclinically promising in chemotherapy, gene delivery, and imaging, but that translation will depend on resolving long-term toxicity and manufacturing reproducibility issues. Overall, polymer-based CQDs are modular, tunable nanocarriers: polymer shells offer the levers (stealth, targeting, stimuli-sensitivity, loading mode) and the CQD core offers the imaging handle and ultrasmall footprint. Careful polymer chemistry (polymer selection, grafting strategy, stimuli-responsive linkers, ligand density) combined with judicious control of size and surface charge allows designs that offer high loading capacity, pH/redox/enzyme-controlled release, effective cellular uptake, and concurrent optical readout — a recipe consistently proven in the literature.

APPLICATIONS

CQDs find application in various fields such as bioimaging, targeted cancer therapy, gene delivery and RNA therapy, anti-microbial and anti-bacterial applications, agriculture, environment, nanocarrier for brain drug delivery and other applications which is depicted in the figure 1.

Figure 1. Applications of Quantum Dots

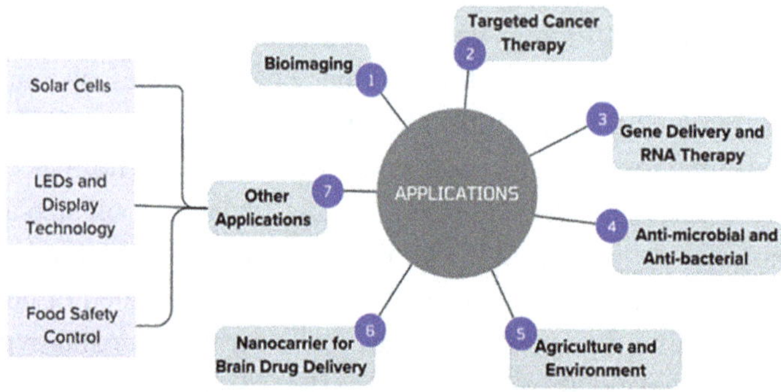

1. Bioimaging

Bioimaging is defined as imaging of living cells, tissues, or whole organisms, generally at the cellular or below-cellular level. Quantum dots synthesized using green methods are very appealing for bioimaging because:

1. Bright, tuneable fluorescence: Emission may be designed throughout visible and near-infrared (NIR) wavelength ranges to carry out deep-tissue imaging.
2. Photostability: Withstand photobleaching more effectively than organic dyes.
3. Low cytotoxicity: Non-toxic to live cell and in vivo imaging.

a) Cell Marking and Tracking Based on Plant Waste-Derived Quantum Dots

Plant waste-derived QDs provide a green and low-cost bioimaging product with great biocompatibility. These QDs exhibit the capability of penetrative entry into living cells, spreading across both cytoplasm and nuclei and giving off bright, stable fluorescence. Their photostability and high resolution render them particularly well-suited for live-cell microscopy over prolonged periods, allowing researchers to observe cellular processes and morphological events in real time without substantial photobleaching. This technique not only benefits cell biology studies but also demonstrates a green strategy for the production of advanced nanomaterials from crop by-products.

b) Multicolour Imaging through Atom Doping

By doping quantum dots with heteroatoms like nitrogen or boron, it is possible to tailor their optical properties for emission in multiple wavelengths. This multicolour fluorescence ability allows simultaneous imaging of various biological targets in one sample, making complex cellular and molecular investigations feasible (Ren et al., 2023). Such a system facilitates the application of separate colour channels for various biomarkers or cellular compartments, enhancing the accuracy of multiplexed imaging and minimizing cross-interference between the signals. This function is particularly useful in high-content screening, tissue mapping, and examining interactions among many cell types within the same environment.

c) Real-Time In Vivo Imaging for Disease Diagnosis and Monitoring

Quantum dots administered intravenously have shown excellent performance in detecting tumours or target tissues in living animal models during preclinical research. Utilizing strong near-infrared (NIR) fluorescence, these QDs provide deep-tissue penetration and high-contrast imaging, which is essential for non-invasive disease diagnosis (Das et al., 2024). Biodistribution can also be followed in real time for monitoring therapeutic delivery, confirming that drug carriers are delivered to the targeted location while avoiding off-target accumulation. This simultaneous diagnostic and monitoring capacity holds great promise for the precision medicine and enhanced assessment of treatment effectiveness in cancer and other diseases.

2. Targeted Cancer Therapy

a) Drug Delivery and Tumour Targeting

- Increased Permeability & Retention (EPR) effect: Because of their tiny size (typically <10nm), QDs tend to be localized in tumour tissues via the leaky vasculature, facilitating passive targeting.
- Surface modification: QDs may be coupled with antibodies, peptides, or aptamers specific for binding and recognizing tumour cell markers for active targeting of drug delivery (Zhao et al., 2020).

- Controlled drug delivery: Surface-functionalized QDs can be stimulated by tumour microenvironment signals (such as acidic pH) to achieve site-specific drug release.
- Theragnostic: Green QDs serve as drug carriers and imaging probes simultaneously, permitting real-time tracking of drug biodistribution and tumour response.

b) Multifunctional Cancer Therapy Modalities

During Photothermal and photodynamic therapy certain QDs can take up near-infrared radiation and convert it into heat (photothermal effect) or produce reactive oxygen species (photodynamic effect) to ablate cancer cells without harming normal tissue (Sarkar et al 2024).

- Nitrogen-doped graphene quantum dots (N-GQDs) encapsulating methotrexate (MTX) are more cytotoxic to breast cancer cells than free methotrexate(MTX). This is attributed to enhanced cellular uptake and a pH-responsive drug release profile, where increased drug release takes place in the acidic tumour microenvironment and slower release in healthy tissues. This site-specific, sustained release augments therapeutic efficacy while minimizing off-target toxicity.
- Quantum dots (QDs) attached to anti-EGFR or anti-IGFR antibodies bind specifically to breast, liver, and brain tumours. They provide both high-resolution imaging and targeted drug delivery, enhancing the localization of the tumour and the therapeutic outcome. Their dual functionality enables their application in "theragnostic," uniting diagnosis and therapy into one platform.

Advantages

- Low toxicity: Green-synthesized QDs, particularly carbon or graphene-based, exhibit lower toxicity compared to heavy-metal QDs, with greater potential for clinical translation.
- High drug loading: The high surface area-to-volume ratio and rich surface chemistry of QDs facilitate high loading and controlled delivery of chemotherapeutics.
- Enhanced therapeutic indices: By saturating the tumour site and limiting off-target effects, QDs decrease doses needed and alleviate side effects, resulting in enhanced patient safety and outcomes.

3. Gene Delivery and RNA Therapies

Green-synthesized QDs are proving themselves to be efficient vectors for gene delivery due to their size tunability, photostability of fluorescence, and broad surface chemistry.

- High Efficiency in Nucleic Acid Binding: The large number of functional groups (such as amine and carboxyl) on QDs facilitates efficient conjugation or electrostatic binding of gene molecules, leading to a high gene load.
- Prevention of Degradation: QDs protect DNA and RNA payloads against enzymatic degradation, increasing their stability and enhancing the chances of successful transfection.
- Cellular Internalization and Targeting: Targeting ligand-modified surfaces (such as aptamers or antibodies) enable receptor-mediated uptake into specific cell types, enhancing gene delivery specificity (Won et al., 2025).

QDs prepared through green methods are ideal platforms for RNA-based medicines such as small interfering RNA (siRNA), microRNA (miRNA), and messenger RNA (mRNA).

- Effective RNA Loading and Delivery: The cationic nature of modified QDs (e.g., polyethyleneimine-coated) enables complexation with anionic RNA molecules, resulting in high payload and improved cellular uptake.
- Protection and Release: QDs shield RNA from nuclease degradation. pH or redox-induced release mechanisms allow for controlled delivery of RNA in target cells.
- Gene Silencing/Messaging: RNA delivery with QDs enhances RNA stability and bioavailability, which facilitates effective gene silencing (siRNA/miRNA) or protein synthesis (mRNA) to generate therapeutic effect in conditions like cancer, genetic diseases, and inflammation.

Green-synthesized carbon quantum dots (CQDs), surface-functionalized with the cationic polymer polyethyleneimine (PEI), have been used successfully to deliver microRNA-10a (miR-10a) (Liu et al., 2021) into macrophages for the regulation of immune responses in models of inflammatory disease. The PEI functionalization enhances tight electrostatic attraction between the CQDs and the negatively charged miRNA strands, with efficient loading capacity and protection from degradation. When taken up by cells, the CQDs effectively release miR-10a, causing gene expression-targeted regulation and inhibition of inflammation signalling pathways. Besides their therapeutic action, the native fluorescence of CQDs can be used to track cellular uptake in real time and intracellular distribution, enabling researchers to monitor delivery efficiency and therapeutic effect simultaneously. This synergy of low toxicity and high biocompatibility along with dual imaging–delivery capability underscores the potential of green-synthesized CQDs as next-generation gene therapy platforms (Won et al., 2025).

4. Antimicrobial and Antibacterial Applications

Plant extract and natural biomaterial mediated green-synthesized quantum dots (QDs), particularly those from carbon and graphene, have shown remarkable antibacterial and more general antimicrobial activity, and thus are very promising for future sustainable drug delivery. These QDs possess potent inhibitory activities against a broad range of pathogens, such as Gram-positive and Gram-negative bacteria and fungi, by the generation of reactive oxygen species (ROS) that trigger oxidative stress and cause cellular membrane, protein, and nucleic acid damage. The green QDs surface chemistry, which is rich in functional groups, allows for close contact with microbial cell walls destroying their integrity, inducing leakage of intracellular content, and culminating in cell death (Dong et al., 2020).

These mechanisms not only allow for efficient killing of free-living bacteria and fungi but also disrupt the process of biofilm formation, further increasing their potential as therapeutic agents against resistant or recurrent infections. Moreover, the biocompatible synthesis pathways of green QDs provide a safety profile for targeted drug delivery with low cell toxicity against human cells. This is augmented by their inherent fluorescence, allowing real-time imaging and monitoring of the treatment. Studies have demonstrated that QDs synthesized from medicinal plant sources like neem and mango leaves and functionalized derivatives like amine-functionalized graphene QDs exhibit spectacular antibacterial and antifungal activity at low concentrations without being cytotoxic to mammalian tissues (Dong et al., 2017).

5. Nanocarrier for Brain Drug Delivery

Green-synthesized QDs are becoming increasingly promising nanocarriers for brain-targeted drug delivery because they are produced in an eco-friendly manner, have tuneable surface chemistry, ultra-minimum size, and superior biocompatibility. Drug delivery to the brain appears to be a daunting task since the BBB (Blood Brain Barrier) prevents most therapeutic drugs from entering the brain, thereby affecting effective treatment of neurological disorders such as brain tumours, Alzheimer's disease, and other CNS diseases. Green QDs like carbon or graphene quantum dots derived from plant-based or biowaste precursors have specific advantages here (Liu et al., 2021). Their surface can be modified with certain ligands, antibodies, or peptides to enable receptor-mediated delivery across the BBB through processes like endocytosis or efflux pump inhibition. Furthermore, green QDs are naturally small enough to penetrate the BBB through passive diffusion pathways and can be developed to take advantage of multiple transport mechanisms while having low toxicity and little risk of immune response.

After passing through the BBB, green QDs function as novel nanocarriers for targeted delivery and regulated release of drugs, genes (e.g., siRNA), or imaging probes directly into the brain tissue. These QDs can be loaded with anti-tumour agents or neuroprotective drugs, providing enhanced therapeutic efficacy at lower doses and thereby minimizing peripheral side effects. Their intrinsic fluorescence allows real-time tracking and monitoring of nanocarrier movement, cargo release, and drug distribution in the brain, facilitating personalized and adaptive therapies (Alotaibi et al., 2021). Moreover, most recently, studies have shown that green QDs can be designed to allow stimuli-responsive release e.g., enzyme or pH triggered maximizing drug activity directly within diseased neural microenvironments. Illustrative examples are QDs-based delivery systems for chemotherapeutic drugs (such as temozolomide or epirubicin) in glioblastoma treatment, where surface modification facilitates BBB crossing, targeted tumour uptake, and concomitant therapeutic and diagnostic functions.

The environmental sustainability of their synthesis, using no heavy metals or toxic chemicals, renders green QDs extremely promising for clinical translation with low toxicity for repeated or chronic treatment. In the end, green QDs are a revolutionary leap ahead for brain nanomedicine, combining precision targeting, non-invasive imaging, and eco-safety, which are critical for the future of CNS drug delivery and neurotherapeutics.

LIMITATIONS AND CHALLENGES

1. Reproducibility Problems: Differences in natural precursors and synthesis processes typically result in non-uniform QD size, shape, and properties, impairing their function in drug delivery.
2. Scalability Limitations: Numerous green synthesis methods are challenging to scale from lab to industrial scale production without compromising quality and environmental friendliness.
3. Multistep Purification and Characterization: Natural precursors can present impurities or heterogeneous surface chemistries, making purification and exact characterization of QDs difficult.
4. Limited Standardization: Absence of standard procedures impairs direct comparison and regulatory certification of green QD products.

5. Toxicity and Biocompatibility Uncertainties: Long-term, complete toxicology studies on biodistribution, metabolism, and environmental fate remain to be conducted to confirm safety.
6. Surface Functionalization Challenges: Obtaining reproducible, stable, and controlled drug attachment or release with biological surface groups is challenging.
7. Regulatory Barriers: Drug and nanomaterial regulations are still difficult to navigate with the absence of complete safety data and harmonized guidelines.
8. Batch-to-Batch Variability: Fluctuations in natural sources may lead to inconsistency in QD properties, affecting reproducibility and therapeutic efficacy (Kohli & Parab, 2025).

RECENT ADVANCEMENTS

1. Biomass and Waste-Derived Quantum Dots for Green Manufacturing

A paradigm shift towards sustainable nanotechnology lies in the employment of renewable biomass sources like agricultural waste, fruit peels, leaves, or herbal plant extracts for the eco-friendly synthesis of quantum dots. Not only does this technology solve issues related to waste valorization and environmental sustainability, but also imbues QDs with novel surface chemistries from their biological sources—leading to improved biocompatibility, antioxidant, and even inbuilt antimicrobial activity. Plant extract or food waste-based methods can produce carbon or graphene quantum dots with fluorescent properties tuneable through the incorporation of chemical functionalities, enhancing their interaction with biological molecules for high drug loading and targeting efficacy. The method avoids dangerous solvents and precursors, simplifies synthetic routes, and enables cost-efficient, large-scale fabrication. The obtained QDs are usually better tolerated by biological systems, less toxic, and environmentally friendly in terms of circular economy. These green materials can be designed to have high surface functional groups (e.g., carboxyl, hydroxyl, amine), which enhance the follow-up drug conjugation and modification (Ren et al., 2023). Such QDs therefore have potential as the nucleus of new nanocarrier systems that not only provide therapeutics but do so with minimal environmental impact, speeding up the applications of green chemistry principles in biomedicine.

2. Mild, Energy-Efficient, and Scalable Synthesis Techniques

Current technological advancements have highlighted the need to develop synthesis routes that are not only environmentally sustainable but also energy-efficient, fast, and scale-up amenable for industrial applications. Hydrothermal, microwave radiation, and ultrasonic syntheses enable the synthesis of quantum dots at reduced temperatures and pressures than traditional synthetic routes. These methodologies minimize energy usage and reaction times, eliminate toxic reagent usage, and preserve the structural integrity of sensitive biomolecules from natural starting materials. Microwave-assisted synthesis, for instance, can quickly yield highly fluorescent, monodispersed QDs in minutes, not hours. The ease of one-pot processes assures facile purification and minimizes waste, which is beneficial for potential pharmaceutical-grade manufacture (Dhariwal et al., 2024). These techniques have shown high reproducibility, product yield, and optical property tunability key performance metrics for biomedical applications. Additionally, the versatility of these eco-synthesis protocols allows for QD tailoring for individual applications, e.g., size-tuned nanocarriers for passive tumour targeting or surface engineered particles for active targeting. The combination of energy efficiency, process safety, and scalability makes these technologies the foundation for commercially practical, sustainable quantum dot-based drug delivery systems.

3. Advanced Surface Functionalization and Stimuli-Responsive Drug Release

A paradigm-breaking trend involves the engineering of quantum dots with advanced surface chemistries that support high drug loading, controlled release, and targeted delivery. Heteroatom doping (e.g., nitrogen, sulphur, or phosphorus) and conjugation of stimuli-responsive moieties (e.g., pH-, enzyme-, or redox-sensitive groups) equip QDs with responsiveness to modify their behaviour based on precise physiological environments. For instance, in cancer microenvironments, which are acidic and possess specific enzyme profiles, these QDs release their payload only at the site of disease, while avoiding side effects in healthy tissues. The surface-functionalization also enables conjugation with antibodies, aptamers, peptides, or small molecules to active target diseased cells or tissues (Jaldurgam et al., 2021). The surface can also be engineered to improve biocompatibility and avoid rapid immune-mediated clearance. This accuracy in drug release and targeting translates into better therapy, reduced systemic toxicity, and potential for personalized medicine. Since green synthesis methods offer QDs with inherently rich surface functionalities, the combination of smart release mechanisms is synergistic augmenting their use in next-generation, highly selective drug delivery (Zhao et al., 2020).

4. Integration as Multifunctional Theragnostic Platforms

One of the most advanced recent advancements in the green QD area is their use within multifunctional nanoplatforms for the conjugated therapy and diagnostics—termed theragnostic. By merging quantum dots with other nanomaterials, like metals (e.g., gold or silver nanoparticles) or polymers, scientists have made hybrid systems that can be used to deliver drugs, conduct enhanced fluorescence or photoacoustic imaging, and carry out photothermal or photodynamic therapy (using light to destroy cancer cells or disease agents) in a single device. This permits real-time monitoring of drug delivery and treatment response, allowing for optimal treatment regimens tailored to individual patient requirements. In addition, low toxicity and high bio-imaging capabilities of green-synthesized QDs allow accurate, safe biodistribution mapping of drugs and disease locations. Hybrid platforms can react to stimuli (such as light or magnetism), offering therapy with spatial and temporal control. Recent studies show that they can have integrated delivery, hyperthermia therapy, and on-demand "smart" therapies, bringing nanomedicine even closer to actual precision and personalized medicine. As these platforms mature, their sustainability, increased effectiveness, and multi-functionality place them at the cutting edge of novel drug delivery and diagnostic technology.

FUTURE DIRECTIONS AND PERSPECTIVES

The future direction and vision of green synthesis of quantum dots (QDs) for sustainable drug delivery are at an interesting juncture where multidisciplinary innovation, environmental responsibility, and clinical requirements meet. With the increased need for safer, more efficient, and environmentally friendly nanomedicine, green-synthesized QDs provide a strong platform for revolutionary innovation in targeted and personalized drug delivery. The future is based on several interconnected pillars that, as a whole, address the challenges of today while unfolding new therapeutic avenues. One main direction is the optimization of synthesis protocols to improve reproducibility, scalability, and quality control. Traditional green synthesis, although eco-friendly, tends to have batch-to-batch inconsistency caused by variations in biological precursors and process conditions. To address this, future work will leverage automation and artificial intelligence to regulate reaction kinetics, precursor concentrations, and environmental conditions in real time. Integration of biotechnology strategies like microbial or plant cell factories designed for uniform precursor production will normalize raw material quality. Such technologies will facilitate industrial-scale production of uniform, high-purity QDs designed for pharmaceutical applications, scaling up from laboratory feasibility to clinical and commercial translation.

At the same time, the marriage of green synthesis with advanced molecular engineering portends next-generation "designer" QDs that engage biological systems on the molecular level. By employing genomics, proteomics, and metabolomics-based insights, precursors and synthesis conditions can be optimized to produce QDs with ideal surface chemistries, sizes, and electronic properties for specific drug binding, tissue targeting, and cellular uptake. This tailored design will enable real personalized medicine, wherein nanocarriers are made for individual patients or for particular disease phenotypes, thereby optimizing therapeutic efficacy with the least adverse effects (Kohli & Parab, 2025).

Another frontier involves intelligent, stimuli-responsive QDs with the ability to manage complex drug release. Future QDs will be equipped with dynamic molecular switches that sense subtle physiological signals such as pH gradients, enzymatic activity, or redox states found in tumour microenvironment or inflamed tissues. These responsive carriers will modulate spatiotemporal delivery such that drugs are released only at the target site, thus significantly enhancing safety profiles and therapeutic efficacy. Multiplexed sensitivity can combine many stimuli within a single platform, clearing the way for multistage therapies that take on complex disease pathways.

Multifunctionality is poised to characterize the second generation of green QD platforms, combining diagnostic imaging, targeted treatment, and environmental sensitivity into single, biocompatible entities. Hybrid nanostructures of green QDs with metal nanoparticles, polymers, or biomolecules will facilitate synchronized drug delivery and real-time tracking through intrinsic fluorescence or photoacoustic signals. In addition, externally activatable therapies like photothermal or photodynamic therapy will be seamlessly incorporated so that physicians can trigger therapy on demand with spatial control. Such theragnostic systems are a paradigm shift away from the classical one-size-fits-all treatments toward dynamic, tailored intervention strategies that can change over the progression of disease. Sustainability will continue to be the top priority, but with greater focus on lifecycle analysis and environmental impact assessments during QD design, synthesis, use, and disposal. Leveraging waste streams of agro-industries and renewable biowaste as feedstock is in line with circular economy principles and minimizes the ecological footprint of nanomedicine manufacture.

Lastly, stringent biosafety and regulatory infrastructures will be the backbone of the successful clinical translation of green-synthesized QDs. Detailed toxicological assessments including acute and chronic exposure, immunogenicity, metabolism, and clearance are required. Future studies will work closely with regulatory bodies to establish standardized protocols for characterization, safety testing, and quality assurance, enabling early-stage approvals and patient confidence. These advances will bring forth a new era of safe, responsive, and green therapeutics that respond to global healthcare needs and environmental demands both.

CONCLUSION

Green synthesis of quantum dots appears as a revolutionary method in nanomedicine, which allows for the creation of stable, harmless, and sustainable drug delivery systems. By moving away from the traditional approach based on toxic heavy metals and energy expensive reactions, green synthesis exploits renewable raw materials such as plant extracts and agricultural residues along with benign solvents like water. Not only is this in accordance with the global sustainability agenda, but it also yields nanomaterials with enhanced biocompatibility and reduced environmental footprint. Carbon and graphene quantum dots, isolated through such green pathways, retain critical optical properties fluorescent tunability, photostability, and functionalize able surfaces without presenting serious toxicity concerns.

These attributes make them eligible for multifaceted use in drug delivery: quantum dots can be designed to guide therapeutic agents straight to ailing tissues, using surface manipulations and ligand conjugations for targeted cancer therapy and site-release. Their fluorescence enables concurrent imaging and real-time drug biodistribution tracking, creating avenues for integrated theragnostic. QDs are also safeguarding vehicles for sensitive molecules such as RNA and DNA, enabling sophisticated gene therapies by controlled cellular delivery and release.

However, the field has significant challenges to overcome before broad clinical use can be achieved. These include scaling up synthesis processes without compromising consistency of particle size and functionalization, high-level analysis of long-term toxicity and biodegradability, and the implementation of universally recognized safety guidelines. Future research needs to focus on high-throughput manufacturing strategies, standardized quality assurance, and holistic life-cycle assessment, while broadening toxicological analyses to accommodate realistic exposure conditions.

Through methodically solving these challenges, quantum dots of green synthesis are poised to transform sustainable drug delivery throughout healthcare and biotechnology. Their distinct integration of performance, safety, and environmental sustainability sets them as foundation materials for nanotechnologies of the future, enabling a greener and healthier world for humans and the environment.

REFERENCES

Alkahtani, M. E., Elbadawi, M., Chapman, C. A., Green, R. A., Gaisford, S., Orlu, M., & Basit, A. W. (2024). Electroactive polymers for on-demand drug release. *Advanced Healthcare Materials, 13*(3), 2301759. DOI: 10.1002/adhm.202301759 PMID: 37861058

Alotaibi, B. S., Buabeid, M., Ibrahim, N. A., Kharaba, Z. J., Ijaz, M., Noreen, S., & Murtaza, G. (2021). Potential of nanocarrier-based drug delivery systems for brain targeting: A current review of literature. *International Journal of Nanomedicine, 16*, 7517–7533. DOI: 10.2147/IJN.S333657 PMID: 34795481

Bao, W., Ma, H., Wang, N., & He, Z. (2019). pH-sensitive carbon quantum dots–doxorubicin nanoparticles for tumor cellular targeted drug delivery. *Polymers for Advanced Technologies, 30*(11), 2664–2673. DOI: 10.1002/pat.4696

Cai, R., Xiao, L., Liu, M., Du, F., & Wang, Z. (2021). Recent advances in functional carbon quantum dots for antitumour. *International Journal of Nanomedicine, 16*, 7195–7229. DOI: 10.2147/IJN.S334012 PMID: 34720582

Das, S., Mondal, S., & Ghosh, D. (2024). Carbon quantum dots in bioimaging and biomedicines. *Frontiers in Bioengineering and Biotechnology, 11*, 1333752. DOI: 10.3389/fbioe.2023.1333752 PMID: 38318419

Das, S. S., Bharadwaj, P., Bilal, M., Barani, M., Rahdar, A., Taboada, P., & Kyzas, G. Z. (2020). Stimuli-responsive polymeric nanocarriers for drug delivery, imaging, and theragnosis. *Polymers, 12*(6), 1397. DOI: 10.3390/polym12061397 PMID: 32580366

Dhariwal, J., Rao, G. K., & Vaya, D. (2024). Recent advancements towards the green synthesis of carbon quantum dots as an innovative and eco-friendly solution for metal ion sensing and monitoring. *RSC Sustainability, 2*(1), 11–36. DOI: 10.1039/D3SU00375B

Ding, C., Tong, L., Feng, J., & Fu, J. (2016). Recent advances in stimuli-responsive treatment. *Molecules (Basel, Switzerland), 21*(12), 1715. DOI: 10.3390/molecules21121715 PMID: 27999414

Ding, L., Agrawal, P., Singh, S. K., Chhonker, Y. S., Sun, J., & Murry, D. J. (2024). Polymer-based drug delivery systems for cancer therapeutics. *Polymers, 16*(6), 843. DOI: 10.3390/polym16060843 PMID: 38543448

Dong, X., Awak, M. A., Tomlinson, N., Tang, Y., Sun, Y. P., & Yang, L. (2017). Antibacterial effects of carbon dots in combination with other antimicrobial reagents. *PLoS One, 12*(9), e0185324. DOI: 10.1371/journal.pone.0185324 PMID: 28934346

Dong, X., Liang, W., Meziani, M. J., Sun, Y. P., & Yang, L. (2020). Carbon dots as potent antimicrobial agents. *Theranostics*, *10*(2), 671–686. DOI: 10.7150/thno.39863 PMID: 31903144

Giordano, M. G., Seganti, G., Bartoli, M., & Tagliaferro, A. (2023). An overview on carbon quantum dots optical and chemical features. *Molecules (Basel, Switzerland)*, *28*(6), 2772. DOI: 10.3390/molecules28062772 PMID: 36985743

Gupta, V., Kannan, K., Tari, V., & Chandra, N. (2024). Bioengineered silver nanoparticles using Brassica oleracea sub sp. botrytis (L.) for enhanced antibacterial activity. *Physics and Chemistry of Solid State*, *25*(2), 303–310. DOI: 10.15330/pcss.25.2.303-310

Hailing, Y., Xiufang, L., Lili, W., Baoqiang, L., Kaichen, H., Yongquan, H., & Hong, S. (2020). Doxorubicin-loaded fluorescent carbon dots with PEI passivation as a drug delivery system for cancer therapy. *Nanoscale*, *12*(33), 17222–17237. DOI: 10.1039/D0NR01236J PMID: 32671377

Hettiarachchi, S. D., Graham, R. M., Mintz, K. J., Zhou, Y., Vanni, S., Peng, Z., & Leblanc, R. M. (2019). Triple conjugated carbon dots as a nano-drug delivery model for glioblastoma brain tumors. *Nanoscale*, *11*(13), 6192–6205. DOI: 10.1039/C8NR08970A PMID: 30874284

Jaldurgam, F. F., Ahmad, Z., & Touati, F. (2021). Synthesis and performance of large-scale cost-effective environment-friendly nanostructured thermoelectric materials. *Nanomaterials (Basel, Switzerland)*, *11*(5), 1091. DOI: 10.3390/nano11051091 PMID: 33922455

Jiang, Y., Yin, C., Mo, J., Wang, X., Wang, T., Li, G., & Zhou, Q. (2023). Recent progress in carbon dots for anti-pathogen applications in oral cavity. *Frontiers in Cellular and Infection Microbiology*, *13*, 1251309. DOI: 10.3389/fcimb.2023.1251309 PMID: 37780847

Kannan, K., Radhika, D., Gnanasangeetha, D., Krishna, L. S., & Gurushankar, K. (2021). Y3+ and Sm3+ co-doped mixed metal oxide nanocomposite: Structural, electrochemical, photocatalytic, and antibacterial properties. *Applied Surface Science Advances*, *4*, 100085. DOI: 10.1016/j.apsadv.2021.100085

Kannan, K., Radhika, D., Kasai, R. D., Gnanasangeetha, D., Palani, G., Gurushankar, K., Koutavarapu, R., Lee, D.-Y., & Shim, J. (2022). Facile fabrication of novel ceria-based nanocomposite (CYO-CSO) via co-precipitation: Electrochemical, photocatalytic and antibacterial performances. *Journal of Molecular Structure*, *1256*, 132519. DOI: 10.1016/j.molstruc.2022.132519

Kannan, K., Radhika, D., Reddy, K. R., Raghu, A. V., Sadasivuni, K. K., Palani, G., & Gurushankar, K. (2021). Gd3+ and Y3+ co-doped mixed metal oxide nanohybrids for photocatalytic and antibacterial applications. Nano express, 2(1), 010014.

Kannan, K., & Tari, V. (Eds.). (2024). Novel energy storage and conversion technologies for two-dimensional MXenes and MBenes. ISBN: 9798369395462, eISBN: 9798369395486DOI: 10.4018/979-8-3693-9546-2

Karakoçak, B. B., Laradji, A., Primeau, T., Berezin, M. Y., Li, S., & Ravi, N. (2020). Hyaluronan-conjugated carbon quantum dots for bioimaging use. *ACS Applied Materials & Interfaces*, *13*(1), 277–286. DOI: 10.1021/acsami.0c20088 PMID: 33355448

Khairol Anuar, N. K., Tan, H. L., Lim, Y. P., So'aib, M. S., & Abu Bakar, N. F. (2021). A review on multifunctional carbon-dots synthesized from biomass waste: Design/fabrication, characterization and applications. *Frontiers in Energy Research*, *9*, 626549. DOI: 10.3389/fenrg.2021.626549

Knoblauch, R., & Geddes, C. D. (2020). Carbon nanodots in photodynamic antimicrobial therapy: A review. *Materials (Basel)*, *13*(18), 4004. DOI: 10.3390/ma13184004 PMID: 32927673

Kohli, H. K., & Parab, D. (2025). Green synthesis of carbon quantum dots and applications: An insight. *Next Materials*, *8*, 100527. DOI: 10.1016/j.nxmate.2025.100527

Kumar, A., Kumar, D., & Saikia, M. (2023). A review on plant derived carbon quantum dots for bio-imaging. *Materials Advances*, *4*(18), 3951–3966. DOI: 10.1039/D3MA00254C

Liu, Y., Zhao, C., Sabirsh, A., Ye, L., Wu, X., Lu, H., & Liu, J. (2021). A novel graphene quantum dot-based mRNA delivery platform. *ChemistryOpen*, *10*(7), 666–671. DOI: 10.1002/open.202000200 PMID: 33829677

Maity, P. P., Kapat, K., Poddar, P., Bora, H., Das, C. K., Das, P., & Dhara, S. (2023). Capra cartilage-derived peptide delivery via carbon nano-dots for cartilage regeneration. *Frontiers in Bioengineering and Biotechnology*, *11*, 1213932. DOI: 10.3389/fbioe.2023.1213932 PMID: 37701494

Mangini, M., Verde, A., Boraschi, D., Puntes, V. F., Italiani, P., & De Luca, A. C. (2021). Interaction of nanoparticles with endotoxin: Importance in nanosafety testing and exploitation for endotoxin binding. *Nanotoxicology*, *15*(4), 558–576. DOI: 10.1080/17435390.2021.1898690 PMID: 33784953

Mirseyed, P. S., Arjmand, S., Rahmandoust, M., Kheirabadi, S., & Anbarteh, R. (2024). Green synthesis of yeast cell wall-derived carbon quantum dots with multiple biological activities. *Heliyon*, *10*(9), e24827. DOI: 10.1016/j.heliyon.2024.e29440 PMID: 38699041

Mohammadpour, A., Kazemi, A., Baghapour, M. A., Samaei, M. R., Kannan, K., & Khaneghah, A. M. (2024). Bioengineered FeZn/GA@ Cu nanocomposite utilizing spent coffee ground extract and gum arabic: Enhanced nitrate removal via (RSM) and machine learning optimization. *International Journal of Biological Macromolecules*, *277*, 134060. DOI: 10.1016/j.ijbiomac.2024.134060 PMID: 39097464

Pandiyan, S., Arumugam, L., Srirengan, S. P., Pitchan, R., Sevugan, P., Kannan, K., & Gandhirajan, V. (2020). Biocompatible carbon quantum dots derived from sugarcane industrial wastes for effective nonlinear optical behavior and antimicrobial activity applications. *ACS Omega*, *5*(47), 30363–30372. DOI: 10.1021/acsomega.0c03290 PMID: 33283084

Parashar, A. K., Saraogi, G. K., Jain, P. K., Kurmi, B., Shrivastava, V., & Arora, V. (2024). Polymer-drug conjugates: Revolutionizing nanotheranostic agents for diagnosis and therapy. *Discover. Oncology*, *15*(1), 641. DOI: 10.1007/s12672-024-01509-9 PMID: 39527173

Ren, H., Yuan, Y., Labidi, A., Dong, Q., Zhang, K., Lichtfouse, E., & Wang, C. (2024). Biological imaging in vitro and in vivo. *Chinese Chemical Letters*, *34*(6), 107998. DOI: 10.1016/j.cclet.2022.107998

Sarkar, S., Srivastava, T. P., Sahoo, O. S., Shankar, A., Rai, A., Pethusamy, K., & Karmakar, S. (2024). Applications of quantum dots in preventive oncology. *Asian Pacific Journal of Cancer Prevention : APJCP*, *25*(3), 747–756. DOI: 10.31557/APJCP.2024.25.3.747 PMID: 38546057

Surendran, P., Lakshmanan, A., Priya, S. S., Balakrishnan, K., Rameshkumar, P., Kannan, K., & Vinitha, G. (2020). Bioinspired fluorescence carbon quantum dots extracted from natural honey: Efficient material for photonic and antibacterial applications. *Nano-Structures & Nano-Objects*, *24*, 100589. DOI: 10.1016/j.nanoso.2020.100589

Surendran, P., Lakshmanan, A., Priya, S. S., Geetha, P., Rameshkumar, P., Kannan, K., & Vinitha, G. (2021). Fluorescent carbon quantum dots from Ananas comosus waste peels: A promising material for NLO behaviour, antibacterial, and antioxidant activities. *Inorganic Chemistry Communications*, *124*, 108397. DOI: 10.1016/j.inoche.2020.108397

Surendran, P., Lakshmanan, A., Vinitha, G., Ramalingam, G., & Rameshkumar, P. (2020). Facile preparation of high fluorescent carbon quantum dots from orange waste peels for nonlinear optical applications. *Luminescence*, *35*(2), 196–202. DOI: 10.1002/bio.3713 PMID: 31591819

Tegafaw, T., Mulugeta, E., Zhao, D., Liu, Y., Chen, X., Baek, A., & Lee, G. H. (2025). Surface modification, toxicity, and applications of carbon dots to cancer theranosis: A review. *Nanomaterials (Basel, Switzerland)*, *15*(11), 781. DOI: 10.3390/nano15110781 PMID: 40497830

Usman, M., & Cheng, S. (2024). Recent trends and advancements in green synthesis of biomass-derived carbon dots. *Eng*, *5*(3), 2223–2263. DOI: 10.3390/eng5030116

Wang, S., McCoy, C. P., Li, P., Li, Y., Zhao, Y., Andrews, G. P., & Ge, Y. (2024). Carbon dots in photodynamic/photothermal antimicrobial therapy. *Nanomaterials (Basel, Switzerland)*, *14*(15), 1250. DOI: 10.3390/nano14151250 PMID: 39120355

Wang, S., Wang, D., Wang, G., Zhang, M., Sun, Y., & Ding, J. (2025). Antibacterial carbon dots. *Materials Today. Bio*, *30*, 101383. DOI: 10.1016/j.mtbio.2024.101383 PMID: 39811607

Won, J. E., Park, M., Hong, S. H., Kim, Y. S., & Song, H. (2025). Quantum dots as biocompatible small RNA nanocarriers modulating macrophage polarization to treat Asherman's syndrome. *NPJ Regenerative Medicine*, *10*(1), 15. DOI: 10.1038/s41536-025-00403-4 PMID: 40133362

Yu, K., & Schanze, K. S. (2023). Commemorating the Nobel Prize in Chemistry 2023 for the discovery and synthesis of quantum dots. *ACS Central Science*, *9*(11), 1989–1992. DOI: 10.1021/acscentsci.3c01296 PMID: 38033796

Zhang, P., & Han, H. (2012). Compact PEGylated polymer-caged quantum dots with improved stability. *Colloids and Surfaces. A, Physicochemical and Engineering Aspects*, *402*, 72–79. DOI: 10.1016/j.colsurfa.2012.03.022

Zhao, C., Song, X., Liu, Y., Fu, Y., Ye, L., Wang, N., & Liu, J. (2020). Synthesis of graphene quantum dots and their applications in drug delivery. *Journal of Nanobiotechnology*, *18*(1), 142. DOI: 10.1186/s12951-020-00698-z PMID: 33008457

Zhao, P., Zhang, Q., Cao, J., Qian, C., Ye, J., Xu, S., & Li, Y. (2022). Facile and green synthesis of highly fluorescent carbon quantum dots from water hyacinth for the detection of ferric iron and cellular imaging. *Nanomaterials (Basel, Switzerland)*, *12*(9), 1528. DOI: 10.3390/nano12091528 PMID: 35564237

Chapter 7
Carbon Quantum Dots in Energy Storage: Supercapacitors, Batteries, and Hybrid Systems

Sathish Kumar Ponnaiah
https://orcid.org/0000-0002-5572-7683
Pohang University of Science and Technology, South Korea

Balakumar Vellaichamy
https://orcid.org/0000-0002-7673-767X
Sri Ramakrishna College of Arts and Science, India

Sungwon Lee
Daegu Gyeongbuk Institute of Science and Technology, South Korea

ABSTRACT

Zero-dimensional (0D) carbon quantum dots (CQDs) have attracted significant attention due to their nontoxicity, tunable photoluminescence, chemical stability, biocompatibility, low cost, and ease of surface functionalization. These properties enable applications in biomedicine, drug delivery, sensing, optoelectronics, photocatalysis, and energy storage. In supercapacitors, batteries, and hybrid systems, CQDs contribute to enhance charge storage by improving ion transport, increasing active sites, and boosting conductivity. In hybrid systems, CQDs help achieve both high energy and power densities, with minimal environmental impact. This chapter explores on the role of CQDs-based composites in electrochemical systems, highlighting their mechanisms, contributions to energy storage devices, and the challenges and future prospects for real-world application.

DOI: 10.4018/979-8-3373-4247-4.ch007

1 INTRODUCTION

Global energy demand is rising rapidly due to population growth and technological progress, while fossil fuel reserves continue to decline. Renewable sources such as solar, wave, and wind are expanding but remain insufficient and intermittent, underscoring the urgent need for efficient energy storage (Kumar et al., 2025). Electrochemical systems, including batteries and supercapacitors (SCs), have therefore attracted major attention. Conventional batteries—such as lithium-ion batteries (LIBs), nickel-cadmium (Ni-Cd), nickel–metal hydride (Ni-MH), and lead-acid, and—offer high energy densities but are limited in power output and cycle life, whereas SCs deliver ultrahigh power densities and long durability but suffer from low energy capacity (Gopi and Ramesh, 2024). Over the past thirty years, the development of next-generation batteries such as aqueous metal-ion, sodium/aluminum-ion, and lithium–air/sulfur, variants has been accompanied by intensive research aimed at optimizing their performance for practical applications (Muthukutty et al., 2024). This complementarity makes LIBs suitable for sustained energy output and SCs for rapid charge–discharge applications such as regenerative braking, communication devices, and hybrid vehicles. To meet both high energy and high-power demands, battery-supercapacitor hybrid systems (BSHSs) integrate battery-like and capacitor-like electrodes. LIBs, first commercialized in 1991, have achieved ~200 Wh kg^{-1}, while SCs, introduced in 1978, deliver up to 1000 kW kg^{-1} with cycle lives exceeding 100000 cycles, although their energy storage capacity remains relatively low (<10 Wh kg^{-1}) (Huan et al., 2026). This contrast in performance can be attributed to their distinct operating mechanisms: batteries store energy through diffusion-limited faradaic reactions, providing high energy but slower kinetics, whereas SCs rely on interfacial charge transfer, allowing for ultrafast cycling with stable performance (Zhang et al., 2013). BSHSs aim to combine the advantages of both, addressing the challenge of achieving both high energy density and fast cycling rates. This goal is driving ongoing research in advanced materials, electrode–electrolyte compatibility, and nanoscale engineering, all of which are crucial for the development of next-generation energy storage technologies.

Understanding these hybrids begins with a comparison of conventional devices. The lead–acid battery, introduced by Gaston Planté in 1859, was the first rechargeable battery and dominated for decades before being replaced by Ni-Cd and Ni-MH batteries (30–70 Wh kg^{-1}) (Zuo et al., 2017). Later advances produced Ni–Fe alkaline batteries with hybrid electrodes exceeding 100 Wh kg^{-1}. LIBs, first commercialized by Sony in 1991 from $LiCoO_2$/graphite chemistry, now reach ~200 Wh kg^{-1}. In contrast, SCs, first demonstrated by Becker in 1957 and commercialized by NEC Japan in 1978, are recognized for their exceptional power densities, often above 10 kW kg^{-1} (Phogat et al., 2025). The fundamental difference between batteries and SCs lies in their energy storage mechanisms. Batteries rely on diffusion-controlled faradaic redox reactions in bulk electrodes, enabling high energy densities (30–200 Wh kg^{-1}) but limiting power output (~500 W kg^{-1}) and cycle life (500-2000 cycles) (Rohith et al., 2024). These reactions produce distinct redox peaks in cyclic voltammetry (CV) and flat charge–discharge plateaus. SCs, in contrast, store energy through interfacial charge transfer at the electrolyte-electrode interface, achieving ultrahigh power densities (up to 1000 kW kg^{-1}), extended cycle lifetimes (>1000000 cycles), and minimal degradation. Their nearly rectangular CV curves and linear voltage–time profiles reflect rapid charge–discharge behavior (Kumar et al., 2025). Despite these differences, both devices share core components—anode, cathode, electrolyte, separator, and current collector enabling hybrid systems that combine the high power and durability of SCs with the high energy density of batteries. Current research focuses on advanced materials, electrode–electrolyte compatibility, and nanoscale engineering to further enhance hybrid performance (Muthukutty et al., 2024).

Carbon nanomaterials are promising candidates for advancing energy storage due to their chemical stability, large surface area, and high conductivity. Among them, carbon quantum dots (CQDs) have attracted particular interest for batteries and SCs owing to their tunable electrochemical properties, excellent conductivity, and high surface-to-volume ratio (Arora et al., 2024). First observed in 2004 during single-walled carbon nanotube synthesis, CQDs are quasi-spherical nanoparticles (<10–20 nm) with mixed sp^2/sp^3 hybridizations, giving rise to unique electrochemical behaviors (Murakami et al., 2004). The term "carbon quantum dots" was introduced in 2006 when Sun et al. synthesized them via laser ablation of graphite, enhancing fluorescence through surface passivation. Functional groups such as aldehyde, hydroxyl, and carboxyl, improve dispersibility and electron transport, while their small size and modifiable chemistry enable high charge storage and cycling stability (Hoang et al., 2019). As highlighted in **Figure. 1**, these zero-dimensional carbon nanomaterials are well suited for future-generation energy storage. Building on this foundation, the chapter focuses on internally connected serial BSHS devices that combine one battery-type and one capacitive electrode. Unlike external or parallel designs, this section emphasizes recent progress in CQDs-based hybrids, covering synthesis, surface modification, heteroatom doping, and integration into high-performance systems, before concluding with multifunctional applications, current challenges, and future prospects.

Figure 1. Key properties of zero-dimensional CQDs, including their high conductivity, biocompatibility, tunable photoluminescence, rich hetero atom doping, eco-friendly synthesis and surface reactivity, which enhance their potential for energy storage applications.

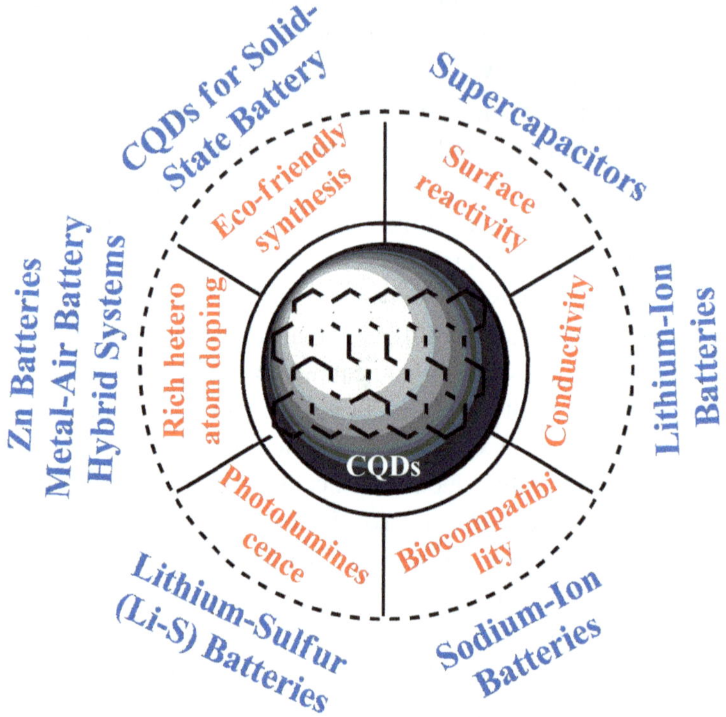

2 SYNTHESIS OF CQDS

Various methods have been reported for synthesizing CQDs from a wide range of complex raw materials. However, unlike true graphene-based structures, CQDs lack a well-defined graphitic domain. Interestingly, the physicochemical properties of CQDs are strongly influenced by the chosen synthesis method, which enables their use in diverse applications (Yang et al., 2023). In general, CQDs synthesis approaches can be classified into two main categories, as shown in: **Figure 2**: top-down and bottom-up strategies. Recently, the preparation of CQDs from waste materials has also emerged as a promising alternative approach (Jiang et al., 2015).

Figure 2. Schematically presents an overview of CQDs synthesis via top-down (common precursor materials such as graphene, carbon nanotube and fullerene), bottom-up (common precursor materials such as glucose and citric acid), and waste-derived methods (common precursor materials from various forms of plant and animal waste).

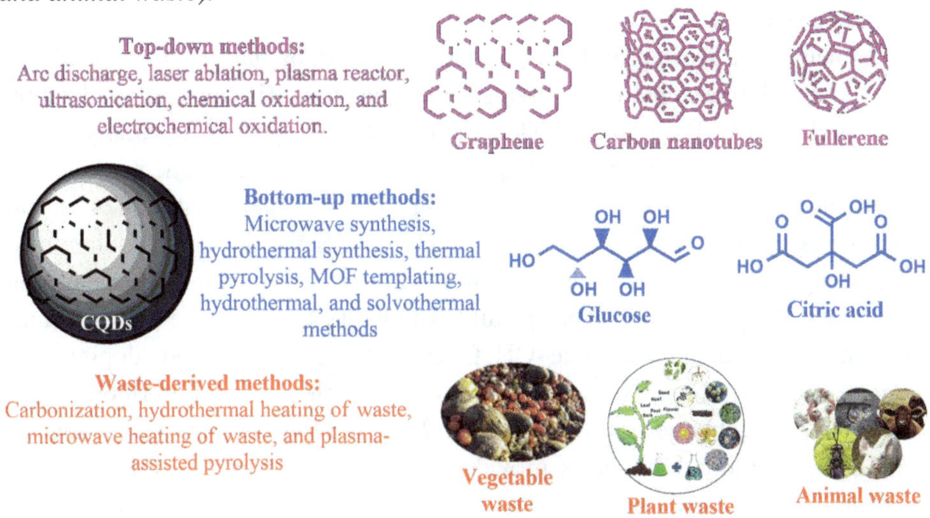

2.1 Top-down approach

The top-down method employs various techniques to produce CQDs from substances derived from carbon, like carbon fibers, graphite, graphene, carbon nanotubes, and activated carbon (Peng et al., 2012). The main top-down methods for producing CQDs are briefly outlined below. One of the earliest top-down techniques is arc-discharge, a high-energy method in which carbon material at the anode is broken down using plasma generated between two electrodes within a sealed reactor under high voltage. At temperatures reaching up to 3727 °C, carbon atoms evaporate from the anode and reassemble as CQDs at the cathode (Wang et al., 2019). These oxygen-rich CQDs exhibit strong fluorescence even without surface treatment. However, the presence of contaminants complicates purification, and the broad particle size distribution decreases the number of electrocatalytic active sites and surface area (Talat et al., 2020). To address some of these issues, laser ablation offers a more controlled alternative. First introduced by Sun et al., this method involves irradiating a carbon target (e.g., graphite) with a pulsed Nd:YAG laser, followed by acid treatment. Surface functionalization with polymers like PEG or poly(propionylethylenimine-co-ethylenimine) enhances the photoluminescence of CQDs, even if the initial particles are non-luminescent (Sun et al., 2006). The laser pulse width can be tuned to control nucleation and particle size. The process creates plasma at the solid-liquid interface, leading to bubble formation and collapse under pressure -producing CQDs that are water-soluble, highly fluorescent, and size-uniform. Despite these advantages, laser ablation remains costly and technically demanding (Hu et al., 2011).

Another innovative approach was introduced in 2010 by Jiang et al., who utilized a one-step arc plasma technique involving helium plasma and in situ functionalization (Jiang et al., 2010). By adding ethylenediamine during synthesis, primary amine groups were introduced to enhance CQDs dispersibility in water. The reaction was carried out in a cylindrical Pyrex reactor using a benzene–helium gas mixture, and rapid discharge resulted in carbonization and CQDs formation. The photoluminescence, composition, and size of the CQDs were influenced by precursor concentration and applied voltage. These CQDs typically exhibit blue fluorescence (365–490 nm) and contain functional groups like–OH, C–O and C=O. Furthermore, doping with elements such as nitrogen, boron, or phosphorus enabled red-shifted fluorescence and improved quantum yields (Wang and Hu, 2014) . In contrast to these high-energy techniques, ultrasound synthesis provides a simpler, ambient-condition method by using high-frequency sound waves to induce cavitation. The collapse of sonic bubbles generates localized high temperatures and pressures, fragmenting carbon precursors into CQDs (Li et al., 2012). The morphology and optical properties of the resulting CQDs depend on the sonication conditions (power, frequency, duration). For example, Kumar et al. successfully used this technique to fabricate Sn@ CQDs@Sn hybrid anodes (Kumar et al., 2016). Ultrasound-synthesized CQDs are typically more biocompatible and safer, making them ideal for biomedical applications. Doping with metals or heteroatoms can further enhance their photoluminescence and quantum yield (Kumar et al., 2020). These versatile CQDs find applications in water purification, energy storage, catalysis, cosmetics, bioimaging, and antimicrobial systems.

Another prominent top-down technique is electrochemical synthesis, which involves breaking down carbon sources like graphite, graphene, reduced graphene oxide (rGO), activated carbon and carbon nanotubes (CNTs) using reactive species generated during water oxidation. Hydroxyl and oxygen radicals act as "scissors" to cleave carbon frameworks at defect sites and edges. Anion intercalation in organic solvents assists exfoliation under applied voltage. In parallel, combustion-based methods offer a low-cost alternative using waste-derived carbon, such as candle soot or paraffin (Liu et al., 2007). For instance, HNO_3 reflux and electrophoresis were used to convert candle soot into ~2 nm CQDs with strong multicolor fluorescence and high water solubility. Oxidative acid treatment not only disintegrates aggregates but also introduces surface functional groups, enhancing dispersibility and tunable photoluminescence. Furthermore, RasulKhan et al. successfully used this technique to fabricate CQDs@functionalized multi-walled carbon nanotubes (f-MWCNTs) hybrid nanocomposite. Subsequently, a non-enzymatic electrochemical sensor was developed using the synthesized CQDs@f-MWCNTs electrode material for the sensitive and selective sensing of toxic mercury (Hg^+) ions (RasulKhan et al., 2020). A more refined version of electrochemical synthesis was later reported by Zhou et al., who used a three-electrode setup with MWCNTs in degassed acetonitrile containing 0.1 M tetrabutylammonium perchlorate (Zhou et al., 2007). Platinum was used as the counter electrode, $Ag/AgClO_4$ as the reference electrode, and MWCNT-coated carbon paper as the working electrode. Building on this, Lu et al. developed an ionic liquid-assisted exfoliation method using graphite or highly oriented pyrolytic graphite as the anode in 1-methyl-3-butylimidazolium chloride-based ionic liquids (Lu et al., 2009). This approach produced CQDs with bright luminescence, excellent crystallinity, and a uniform size distribution, highlighting the scalability and efficiency of electrochemical routes.

2.2 Bottom-Up Approach

CQDs produced via the bottom-up method are synthesized by carbonizing or self-assembling tiny organic pioneers such as citric acid (CA) and saccharides. This method enables precise tuning of size and surface properties by adjusting reaction conditions. The process is scalable and cost-effective, making it suitable for large-scale production (Ozyurt et al., 2023). Common bottom-up approaches include thermal pyrolysis, hydrothermal or solvothermal treatments, microwave-assisted synthesis, and template-assisted growth using resources like metal-organic frameworks (MOFs) (Gu et al., 2017).These methods are outlined in detail below.

Among these techniques, microwave-assisted synthesis stands out for its rapidity and efficiency. It typically produces CQDs in under ten minutes by employing electromagnetic radiation, which induces rapid heating and bond cleavage (Bian et al., 2014). This fast synthesis allows for high-quality CQDs with tunable properties. For example, Zhu and colleagues used poly(ethylene glycol) and saccharides such as glucose and fructose as carbon precursors (Zhu et al., 2009). Building on this, Zhai et al. synthesized highly photoluminescent CQDs via microwave pyrolysis of CA and studied the effects of various amine groups on luminescence properties (Zhai et al., 2012). The oxygen-containing surface groups in these CQDs enhanced metal ion coordination, making them particularly promising for electrocatalyst applications. Complementing microwave synthesis, hydrothermal synthesis offers a simple and economical route for CQDs production using common organic precursors (Bian et al., 2014). This method involves thermally carbonizing materials like saccharides or organic acids in a sealed autoclave at elevated temperatures. In a foundational study, Hsu and Chang described a four-step mechanism-surface passivation, polymerization, carbonization, and dehydration-highlighting the key roles of amino and carboxylic groups in facilitating hydrogen bonding and nucleation (Hsu and Chang, 2012). By varying precursor composition, both water-soluble and insoluble CQDs can be obtained. Additionally, this method supports heteroatom doping, allowing for controlled electronic structures and the fabrication of functional electrocatalytic materials (De Yro et al., 2019). In parallel, thermal pyrolysis provides a scalable approach to CQDs synthesis by decomposing organic precursors at high temperatures in an oxygen-free environment. Giannelis and co-workers (Bourlinos et al., 2008) demonstrated this by thermally carbonizing ammonium citrate salts, yielding CQDs with tailored hydrophilicity or organophilicity depending on the precursor. Follow-up studies examined the temperature-dependent photoluminescence mechanisms, showing a shift from molecular fluorophore emission at lower temperatures to carbogenic core emission at higher ones. Given its low cost and compatibility with biocompatible polymers, this method is particularly suited for applications in drug delivery and bimodal imaging (Lai et al., 2012).

To achieve precise size control, template-assisted synthesis using MOFs has gained attention. These porous crystalline structures act as nanoreactors, enabling the growth of uniform CQDs within their defined pore spaces. In 2017, Gu et al. utilized MOFs such as MIL-101, ZIF-8, and HKUST-1 to template glucose decomposition into CQDs via mild heat treatment (Gu et al., 2017). Afterward, the MOF templates were removed using KOH washing, releasing monodisperse CQDs. This strategy enables fine control over size and morphology, making MOF-templated CQDs particularly useful for optoelectronic and catalytic applications (Amali et al., 2014). In addition to conventional precursors, biomass-derived waste offers an environmentally friendly and sustainable alternative for CQDs synthesis (Luo et al., 2013). Low-cost materials like fruit and vegetable peels are readily available and rich in carbon. For instance, Sahu et al. utilized orange juice to produce CQDs with impressive photoluminescence quantum yields (PLQY) (Sahu et al., 2012). Alam et al. demonstrated a hydrothermal approach using cabbage peels, highlighting the viability of kitchen waste as a feedstock. These green methods are not only economical and eco-friendly but also add value to agricultural waste streams (Alam et al., 2015). To further enhance CQDs performance, surface functionalization is often employed. This post-synthesis step enables fine-tuning of photophysical, chemical, and biological properties through interactions such as coordination bonding, π–π stacking, and covalent modification (Li et al., 2011). Most bottom-up-derived CQDs inherently contain -OH and -COOH groups, facilitating straightforward covalent coupling. Functionalization with amine-rich compounds can significantly enhance PLQY. For instance, CQDs produced from EDTA were further modified with spiropyran molecules, resulting in reversible fluorescence switching behavior under light stimuli (Liao et al., 2013). Post-synthetic chemical modifications can further tune CQDs properties. For example, NaOH-induced hydrolysis of alkyl-functionalized CQDs introduced -COO^- groups, red-shifting emissions. Similarly, para-substituted aniline groups formed new emissive states, producing CQDs with narrow red emission peaks. Ding et al. demonstrated band gap narrowing and red-shifted PL after separating CQDs fractions based on surface oxygen content using silica chromatography (Ding et al., 2016). In summary, surface functionalization significantly affects CQDs' photoluminescence, stability, dispersibility, and bio-compatibility, thereby expanding their potential in fields such as sensing, catalysis, energy storage, and environmental remediation. By combining bottom-up synthesis with tailored functionalization strategies, researchers can develop CQDs with highly customizable belongings for a wide range of advanced applications.

3 ELECTRICAL AND ELECTROCHEMICAL PROPERTIES OF CQDS

The effectiveness of CQDs as electrolyte additives in energy storage materials is largely determined by their electrical and electrochemical properties. Their redox activity, conductivity, and electrochemical stability contribute significantly to improving energy storage, charge transport, and electrolyte stability. Specifically, the conductivity of CQDs (10^{-2} to 10^{-3} S/cm) promotes ion diffusion and charge transfer, while their redox-active functional groups enhance pseudo-capacitive charge storage in SCs (Pholauyphon et al., 2022). Moreover, CQDs support ion transport, stabilize the solid electrolyte interphase (SEI), and can even promote redox reactions in batteries, such as zinc-ion batteries and lithium-sulfur, making them versatile in various energy storage systems.

Building on these properties, doping CQDs with heteroatoms like nitrogen, sulfur, or phosphorus further improves their conductivity and electrochemical stability. This modification reduces internal resistance and prevents electrolyte degradation, which is essential for ensuring long-term performance (Sharma et al., 2022). Additionally, CQDs naturally possess redox-active functional groups-such as –OH, C–O and C=O-which enhance Faradaic reactions. This further boosts their ability to store charge, pushing capacitance levels beyond those seen in conventional electrochemical double-layer capacitances (EDLC) and increasing overall efficiency in SCs (Li et al., 2021).

CQDs also exhibit notable electrochemical behavior, as evidenced by redox peaks in CV, ranging from 0.2 to 0.6 V vs. Ag/AgCl, depending on the electrolyte used. These redox peaks contribute to improved energy density and cycling stability, making CQDs particularly advantageous for redox-enhanced batteries and hybrid SCs

(Shaker et al., 2024). Another important factor in CQDs performance is their broad chemical stability across a pH range of 1-14, enabling them to function effectively in various electrolyte types (aqueous, organic, and ionic liquid) and in both acidic and alkaline environments (Li et al., 2021). Their resistance to oxidation ensures reliable performance in electrochemical systems, extending the lifespan of electrolytes and improving cycling durability by preventing unwanted side reactions. Additionally, the large specific surface area (SSA) of CQDs enhances ion accessibility at the electrode surface, which is crucial for EDLC SCs performance. In addition to their inherent properties, CQDs can be treated with high temperatures and activating agents (e.g., KOH, $ZnCl_2$, H_3PO_4) to further boost their surface area. For instance, one study demonstrated that KOH-activated CQDs achieved a capacitance of 0.66 F/cm^2 (Chen et al., 2016). However, while these treatments increase surface area, they may also create micropores that restrict access for larger ions, potentially compromising performance. To further fine-tune CQDs characteristics, the surface area and size can be regulated by varying precursor concentrations, temperature, and pressure during synthesis. The sp^2 hybridized network within CQDs is critical for facilitating electron transfer. It serves as both a conduction medium and as electron acceptors or donors, ensuring efficient charge transport. This network is vital for maintaining electron conductivity and preventing degradation of primary materials, especially after repeated cycling (Kakaei et al., 2021). Additionally, CQDs exhibit superior performance in both SCs and optoelectronics due to their higher electron mobility-two to four times greater than their hole mobility-likely a result of their larger conduction band. Another factor that influences CQDs performance is the length of the ligands attached to the surface, which can affect charge carrier mobility (Prabakaran et al., 2023). Longer ligands typically reduce mobility, while shorter ones may enhance it. Moreover, doping CQDs with heteroatoms like phosphorus, nitrogen, sulfur, or boron can further improve conductivity and electron transport. These modifications improve the electrical structure and generate additional charge carriers, which are beneficial for energy storage applications (Kakaei et al., 2021).

In addition to these structural and compositional modifications, CQDs also exhibit two main types of defects: point defects (vacancies and dislocations) and line defects (edges and crystal interfaces). These defects break electron-hole symmetry, which increases carrier density and chemical reactivity. While defect engineering can improve CQDs performance by increasing surface area and creating added active places for storing charges, and too many flaws might cause structural instability, ultimately degrading performance (Bao et al., 2023). Therefore, optimizing CQDs involves finding a balance between the advantages and drawbacks of these defects. The stability and surface charge of CQDs solutions are largely governed by their zeta potential. This zeta potential, which reflects the electrostatic potential at the slipping plane where particles interact with the surrounding fluid, is crucial in determining the performance of CQDs in colloidal systems. An extreme absolute zeta potential (around ±30) specifies excellent solubility and stability in polar fluids, which in turn enhances electrochemical performance and ensures consistent electrode coatings (Fu et al., 2023). This stability is crucial for ensuring the reliable and efficient operation of energy storage applications. Finally, the overall effectiveness of energy storage is influenced not only by electronic properties but also by factors such as surface functionalization, shape, compact size, quantum confinement, and the environmentally friendly nature of CQDs. These factors, combined with their versatile properties, make CQDs a promising material for advancing the performance and sustainability of energy storage systems.

4 APPLICATIONS OF CQDS IN ENERGY STORAGE

4.1 CQDs in Supercapacitors

Unlike batteries, SCs deliver high power output, and their performance can be improved with nanomaterial-based electrodes offering greater stability and capacitance (Raza et al., 2018). CQDs are particularly attractive for SCs because of their excellent conductivity and tunable structure and size. Charge accumulation is facilitated by the formation of double charge layers at the electrolyte-CQDs interface in the EDLC mechanism. Adding CQDs to SCs electrodes not only enhances electrical conductivity but also improves electron transfer from the electrode to the current collector, reduces internal resistance, and boosts charge/discharge efficiency. Combining metal oxides or sulfides with CQDs to create nano-hybrid structures can further enable the development of efficient and reliable SCs (Magesh et al., 2022). Electrodes made from nanomaterials show great potential for enhancing SCs due to their higher specific capacitance and long-term stability. Additionally, expandable electrodes contribute to SCs by enabling fast energy storage, and rapid charging and high energy densities. For example, Prasath et al. developed CQDs–MnO_2 nanostructures through a sustainable, eco-friendly approach using waste resources, aiming to achieve high specific capacitance and power density (Prasath et al., 2018). Structural analysis showed that, compared to pure MnO_2, the CQDs–MnO_2 nanostructures exhibited a larger surface area and improved conductivity. This hybrid proved excellent cycle life in symmetric SCs, achieving a specific capacitance of 189 Fg^{-1}. Electrochemical tests showed a rapid current-voltage response, with enhanced performance ascribed to its large surface area and conductive CQDs nanonetwork. These results suggest that novel CQDs–MnO_2 electrode materials for energy storage devices can be developed in a sustainable manner. Transition-metal sulfides are predominantly favored for energy storage systems because to their enhanced electrochemical activity and excellent electrical conductivity (Barik and Ingole, 2020). When combined with CQDs, sulfide-based materials show improved conductivity, increased ion accessibility, and greater stability during charge/discharge cycles, all of which significantly enhance electrochemical performance. For Quan et al. established the CuS@CQDs composite by using solvothermal and impregnation methods to incorporate CQDs into a CuS nanostructure (Quan et al., 2020). BET analysis revealed an enhanced surface area of 111.2 m^2g^{-1} for the composite, related to just 22.8 m^2g^{-1} for pure CuS. A greater specific capacitance of 920.5 Fg^{-1} and a power density of 397.75 Wkg^{-1} were demonstrated by the composite, which also demonstrated outstanding long-term stability, maintaining 92.8% of its performance after 10,000 cycles. The CuS@CQDs-ICM composite displayed a specific capacitance of 920.5 Fg^{-1}, significantly higher than the 369.9 F/g observed for pure CuS. The synergy between CuS and CQDs enhances electron transport, which is why the resulting asymmetric supercapacitor exhibited outstanding electrochemical performance. Doping CQDs with various heteroatoms enhances their specific capacitance and extends the lifespan of energy storage devices. Dopants such as nitrogen, phosphorus, iodine, oxygen, and boron have been explored for their ability to improve the photoelectronic properties of carbon-based materials. For instance, Wang et al. (2020) developed nitrogen-doped CQDs/polyaniline (NCQDs/PANI) nanocomposites, which demonstrated superior electrochemical properties and electrical conductivity compared to the individual materials (Wang et al., 2017). The flake-shaped N-CQDs/PANI-3 composite exhibited the best performance, with enhanced cycling stability and capacitance. In another study, Yu et al. incorporated carbon dots into lignin-based activated carbons (ACs) to enhance their surface by introducing CD binding sites (Yu et al., 2021). The capacitance of the modified AC (301.7 Fg^{-1}) was significantly greater than pure AC's (125.8 Fg^{-1}), suggesting that the hydrophilic nature of CDs improved electrode interaction with electrolytes and reduced ion diffusion pathways. CQDs can be combined with heteroatoms or transition-metal oxides/sulfides and polymer based materials to form nanocomposites that increase active sites and improve electrochemical performance by serving as electron donors and acceptors. This approach provides a straightforward method for producing stable, efficient electrodes (Arora et al., 2024). Future studies are probable to focus on enhancing performance and optimizing the synthesis of these nanocomposites.

4.2 Lithium-Ion Batteries (LIBs)

LIBs are generally used in renewable energy storage, portable electronics, and electric vehicles because of their long cycle life and high energy density, but challenges such as capacity fading, thermal instability, and safety risks still limit their performance and lifespan. To address these issues, scholars are wandering innovative alternatives such as solid-state electrolytes, advanced electrode materials, and novel electrolyte additives. CQDs, with their high conductivity, excellent dispersibility, and tunable surface chemistry, have developed as a promising material for improving LIB performance (Sung et al., 2023). They achieve this by creating new pathways for lithium-ion transport, enhancing ionic conductivity, which is especially beneficial for high-rate applications. Secondly, they from a stable layer between the electrode and electrolyte, which enhances cycling stability and capacity retention by minimizing harmful reactions. Additionally, CQDs reduce capacity fading by stabilizing the SEI and preventing degradation during repeated charge-discharge cycles. Furthermore, CQDs improve electrolyte wettability and promote uniform lithium-ion deposition, helping to prevent dendrite formation. This is crucial for safety, as dendrites can reduce battery lifespan and cause short circuits. Thanks to these combined mechanisms, CQDs serve as highly effective electrolyte additives, addressing key challenges in LIB performance and enhancing the safety, efficiency, and durability of energy storage solutions (Sung et al., 2023). CQDs, especially when functionalized with heteroatoms like fluorine, sulfur, and nitrogen, have demonstrated great potential as interfacial modifiers and electrolyte additives in LIBs and lithium-ion capacitors (LICs). Fluorine doped CQDs (F-CQDs) are receiving increased attention because they enhance the interface between electrodes and current collectors, thereby improving lithium-ion storage and cycling stability. A study found that electrochemically etching copper foil with F-CQDs created a stockade-like surface that improved adhesion, reaction kinetics, and interface stability (Shin et al., 2021). These F-CQDs-modified electrodes retained 94.1% of their capacity subsequently 500 high-rate cycles and revealed an ultrafast cycling capacity of 82.9 mAh g^{-1}. The fluorine atoms in F-CQDs help reduce capacity loss by enhancing electrolyte stability and preventing the formation of harmful SEI layers. In the process, copper is electrochemically etched, and F-CQDs are subsequently spray-coated to increase surface area, enhance stability, and improve electron and ion transport. With 94.3% capacity retention and a specific capacity of 297.3 mAh g^{-1}, the F-CQDs@etched -Cu electrode outperformed bare copper. At a current density of 2000 mA g^{-1}, the F-CQDs@etched-Cu electrode maintained 94.1% of its capacity over 500 cycles. These data demonstrate the role of F-CQDs in improving electrochemical stability and performance, particularly at high rates. In LICs, sulfur and nitrogen co-doped CQDs (SNCQDs) have shown great potential as interfacial modifiers, combining the high the power density of LIBs with the energy density of capacitors. One study enhanced the electrode-electrolyte interface by depositing SNCQDs onto copper foil. After 500 cycles at 3000 mA g^{-1}, the electrochemical performance improved, reaching a specific capacity of 80.8 mAh g^{-1} with 90.1% capacity retention. At a 5C rate, the electrode retained 84.7% of its capacity in full-cell LICs (Sung et al.,

2023). Sulfur and nitrogen co-doping enhanced performance and safety by improving electrolyte wettability, promoting uniform lithium-ion deposition, and suppressing dendrite formation. In a study of GaO_3@N-CQDs, nitrogen-doped CQDs (N-CQDs) enhanced the properties of both the electrolyte and the electrode material (Guo et al., 2020). Nitrogen doping boosts ionic conductivity by improving lithium-ion solvation, promoting uniform ion distribution, and creating additional active sites for ion interactions. By scavenging reactive species and preventing unwanted side reactions, this enhances electrochemical performance and stabilizes the electrolyte. Furthermore, the CQDs enhance the self-healing ability of GaO_3, helping to maintain the electrode's structural integrity during prolonged cycling. For instance, anodes based on Ga_2O_3@N-CQDs had an initial discharge capacity of 1348.5 $mAhg^{-1}$ at 0.1 Ag^{-1} and sustained 700.5 $mAhg^{-1}$ at 0.5 Ag^{-1} following 500 cycles. The incorporation of N-CQDs enhances both capacity retention and rate performance. These improvements are ascribed to the increased ionic conductivity and stability provided by the N-CQDs. Sulfur- and N-doped graphene sheets, including bio-derived graphitic CQDs, have shown exceptional performance in high-performance energy storage devices like SCs. Incorporating these materials into the electrolyte enhances supercapacitor performance by significantly increasing both ionic conductivity and electronic mobility. N- and S-doping in CQDs introduces redox-active sites, which accelerate ion transport and enhance electrochemical stability. N- and S-doped CQDs synergistically enhance ionic and electronic mobility, leading to increased specific capacitance. After 10,000 cycles, S-doped CQDs in graphene, for instance, maintain 94.5% of their capacitance, demonstrating exceptional cycling stability and minimizing energy loss with a Coulombic efficiency of 98%. These materials showcase their efficiency and rapid charge/discharge capabilities in symmetric SCs, delivering impressive power and energy densities of 700 W kg^{-1} and 15.2 Wh kg^{-1}, respectively (Pallavolu et al., 2023). CQDs derived from carrot juice biomass have been used as a coating to improve the electrochemical performance of $Li_4Ti_5O_{12}$ (LTO). The CQDs coating improves electrical conductivity and lithium-ion diffusion by shortening the Li^+ ion diffusion paths. Adding CQDs to the electrolyte enhances LIBs by improving ion diffusion, stabilizing the electrolyte, and strengthening the electrolyte/electrode interface, ensuing in better cycling stability and rate performance. With the addition of CQDs, LTO/C attains a high specific capacity of 211 mAh g^{-1} at 0.1 C and retains 53% of its capacity even at a high rate of 20 C (Krajewski et al., 2022).

In addition to improving the electrode-current collector interface, CQDs play a key role in synthesizing advanced carbon-based materials like ultrathin carbon nanosheets (CNSs) and 3D porous carbon frameworks, which significantly enhance structural stability and electrochemical performance. For example, CQDs derived from coal tar pitch are produced through mild oxidation and then assembled using an ice template to form three-dimensional porous carbon frameworks. These frameworks demonstrate outstanding electrochemical performance, achieving a reversible capacity of 273 mAh g^{-1} after 1000 cycles at 2000 mAg^{-1}—due to enhanced ionic conductivity and a stable electrode/electrolyte interface provided by oxygen-containing functional groups on the CQDs surface (Shi et al., 2023). CQDs have also been used to create ultrathin CNSs with controlled distribution through the use of NaCl-KCl templates. The resulting CNSs, approximately 1.70 nm thick, exhibited a SSA of 302 m^2g^{-1} and featured hierarchical nanopores. These structures deliver remarkable electrochemical performance, containing a reversible capacity of 453 $mAhg^{-1}$ at 50 mAg^{-1}, exhibiting excellent cycle stability, with 94.3% capacity retention after 500 cycles at 500 mA g^{-1}, and remarkable rate capability, delivering 225 mAh g^{-1} at 1000 mA g^{-1}. CQDs promote uniform deposition of active materials during cycling by enhancing ion diffusion, conductivity, and stabilizing the electrode/electrolyte interface (Shi et al., 2024). While CQDs show significant promise in enhancing the performance of LIB system, their benefits extended to other battery technologies, such as sodium-ion batteries (SIBs), which present unique advantages and challenges compared to lithium-based system.

4.3 Sodium-Ion Batteries

Sodium-ion batteries (SIBs) offer a more economical alternative to LIBs because of the widespread availability of sodium. Nevertheless, their larger ionic radius and lower electrochemical potential lead to low conductivity, limited cycle life, and restricted rate performance. These factors, along with the lack of suitable anode materials, restrict their energy density. CQDs address these issues by enhancing ion transport, stabilizing the SEI, and suppressing dendrite growth (Rüther et al., 2020). Their tunable structure, surface functionalization, and high conductivity improve Na^+ solvation, reduce migration barriers, and boost ionic conductivity. Incorporating oxygen (O)- or N-doped CQDs into electrolytes further reduces interfacial resistance and increases Na^+ diffusivity, significantly enhancing overall electrochemical performance (Yang et al., 2025). Beyond their role as electrolyte additives, CQDs also help stabilize electrode materials such as antimony (Sb), which offers high capacity but suffers from significant volume expansion during cycling. By minimizing structural degradation and alleviating mechanical stress, CQDs regulate the growth of Sb particles (Hou et al., 2015). For instance, the Sb@CQDs composite, synthesized via a one-step reduction method, delivers impressive performance with a specific capacity of 334 $mAhg^{-1}$ at 2 Ag^{-1} and 635 $mAhg^{-1}$ at 0.1 Ag^{-1}. This improvement is attributed to CQDs enhancing both ionic and electronic conductivity, thus improving the lifespan and efficiency of SIBs (Mahmood et al., 2025). Stabilizing the electrolyte is vital for SIBs, as the SEI layer governs ion transport, capacity retention, and cycling stability. CQDs help form a uniform, durable SEI, reducing interfacial resistance, limiting electrolyte decomposition, and controlling degradation. Their surface functionalization further fine-tunes the SEI, minimizing side reactions and improving sodium-ion storage efficiency. Moreover, CQDs enhance electrolyte stability by scavenging reactive intermediates and inhibiting parasitic degradation reactions. They also modify solvation structures, which enhances charge transfer and Na^+ ion transport. Research on CQDs-coated VO_2 nanowires illustrates how CQDs stabilize the electrolyte, boost charge transport, and protect the electrode surface (Balogun et al., 2016). This versatile role is reflected in the CQDs-modified VO_2 electrode delivering high capacities of 420 $mAhg^{-1}$ for lithium and 328 $mAhg^{-1}$ for sodium storage. Similarly, a study on $Na_3V_2(PO_4)_2F_3$ (NVPF@CQDs) hierarchical microspheres highlights CQDs' role in enhancing cycle stability and high-rate capability (Liu et al., 2020). After 6000 cycles at 30C, the NVPF@CQDs electrode retains 90.2% of its capacity due to CQDs stabilizing the SEI, preventing adverse reactions, and improving ion diffusion and charge transfer. Overall, as electrolyte additives, CQDs improve interface compatibility and stabilize the electrode/electrolyte interface, as demonstrated by enhanced performance at high current densities and extended cycle life.

Taken together, CQDs are vital additives in modern SIBs, as they enhance ion transport, stabilize the SEI layer, and suppress side reactions, collectively boosting the stability and performance of SIBs. Extending beyond CQDs, recent studies on other quantum dots (QDs) reveal significant impacts on the charge storage capacity of SIBs. For example, Fe_3O_4 QDs embedded in a three-dimensional carbon-coated structure enhance charge storage efficiency in ultrafast rechargeable SIBs (Qi et al., 2016). MOFs were used as templates to synthesize Fe_3O_4 QDs uniformly embedded within a protective carbon shell. Thanks to rapid ion diffusion enabled by this hierarchical conductive network, the electrode retained over 90% of its theoretical capacity while completing full charge/discharge cycles in just six minutes. The surface-induced pseudocapacitive effect of this system suggests that adding CQDs to the electrolyte may similarly enhance interfacial stability and rate performance by promoting charge redistribution mechanisms. CQDs not only enhance conductivity but also improve polyanionic electrodes by reducing interface impedance and strengthening interfacial stability. As electrolyte additives, they promote ion transport, stabilize the SEI, boost charge transfer kinetics, and improve long-term cycling performance (Zhao et al., 2024). CQDs enhance sodium-ion migration in fast-charging SIBs by shortening pathways and increasing diffusion, improving charge transfer kinetics and overcoming rate limitations for use in electric vehicles and extensive storage. In flexible SIBs, CQDs boost ionic conductivity, mechanical stability, and thermal resilience in gel electrolytes, enhancing cycling performance and rate capability (Lu et al., 2024). Overall, CQDs improve electrolyte stability and charge transport, enabling high-performance SIBs across diverse applications.

4.4 Lithium-Sulfur (Li-S) Batteries

Lithium-sulfur (Li-S) batteries are attractive due to sulfur's high theoretical energy density and low cost, but commercialization is limited by the polysulfide shuttle effect, lithium dendrite formation, and poor conductivity. CQDs offer a solution with their tunable surface chemistry, large surface area, and high conductivity. Specifically, the polysulfide shuttle effect occurs when lithium polysulfides (LiPSs) dissolve in the electrolyte and migrate to the anode, reduced cycle life and leading to capacity loss (Liu et al., 2024). CQDs mitigate this issue by chemically adsorbing polysulfides through their surface functional groups, preventing their migration. Their large surface area and tunable chemistry can be optimized for strong polysulfide binding, making them particularly effective in this role (Wang et al., 2023). Building on this, when incorporated into graphitic carbon nitride (g-C_3N_4), CQDs act as an efficient sulfur host that captures polysulfides and improves the conversion kinetics of LiPSs during charging and discharging. This enhancement in reaction kinetics leads to better sulfur utilization, reduced shuttle effect, and improved cycling stability. Batteries with this CQDs-decorated g-C_3N_4 composite demonstrate outstanding performance, maintaining a retaining 582.3 mAh g^{-1} after 200 cycles and specific capacity of 1300.0 mAh g^{-1} at 0.1C (Liu et al., 2024). Another critical challenge is the sulfur's insulating properties and its discharge products (e.g., Li_2S) limits electron transport, slowing reactions and reducing rate capability. CQDs improve sulfur cathode conductivity, enhance electron transfer, and optimize sulfur utilization, boosting overall electrochemical performance (Pan et al., 2022). They also stabilize the SEI by mitigating polysulfide-induced degradation, enhancing ionic conductivity, suppressing dendrite formation, and reducing electrolyte decomposition, thereby extending cycle life and improving battery safety. To further address sulfur electrode drawbacks, CQDs have recently been incorporated into VO_2, to accelerate the adsorption and conversion of LiPSs. Besides promoting the nucleation of Li_2S during discharge and its breakdown during charging, CQDs expressively improve the conductivity of the VO_2 surface. Li-S pouch cells using this composite exhibited exceptional performance, with a capacity retention of 88.8% at 0.5C after 100 cycles and an energy density of 386.1 Wh kg^{-1} (Yang et al., 2024).

Moreover, biomass-derived CQDs, such as those obtained from lignin, have been used to improve sulfur utilization in Li-S batteries. These CQDs offer a sustainable alternative to metal-based additives, with a lower environmental impact and smaller carbon footprint. Their production utilizes waste biomass and requires less energy compared to metal synthesis, and life-cycle analysis suggests they could contribute to a circular economy by reducing reliance on finite resources. The lignin-derived CQDs form a porous carbon structure that facilitates sulfur infiltration, enhances redox reactions, and increases sulfur loading further boosting the battery's performance. These CQDs also boost ionic conductivity, resulting in excellent cycling stability, with an initial capacity of 1066 mAh g^{-1} at 1C and minimal capacity decay of only 0.095% per cycle over 400 cycles (Xu et al., 2021). Beyond their role in cathode materials, CQDs serve as effective binder components and electrolyte additives, providing both mechanical and electrochemical benefits. When added to electrolytes, CQDs improve ionic conductivity, regulate lithium-ion diffusion, and prevent polysulfide dissolution . For example, CQDs act as binders in waterborne polyurethane-based systems, stabilizing electrodes while enhancing ionic conductivity and minimizing polysulfide migration. CQDs scavenge reactive species, stabilize redox reactions, and interact strongly with electrolyte components, reducing side reactions and enhancing long-term performance (Kim et al., 2018). They improve conductivity, sulfur utilization, SEI stability, and suppress dendrite growth, while their sustainable derivation from sources like lignin adds an eco-friendly advantage. Overall, CQDs serve as multifunctional binders, sulfur hosts, and electrolyte additives, addressing key Li–S battery challenges and paving the way for commercialization.

4.5 Zn Batteries

Zinc-ion batteries (ZIBs) are an emerging energy storage technology offering high capacity, safety, cost-effectiveness, and environmental benefits. With abundant, non-toxic zinc anodes and aqueous electrolytes, they are well suited for grid storage and portable electronics. Zinc deposits form at the anode, while Zn^{2+} reversibly intercalates into cathodes such as manganese oxides or vanadium compounds. Despite these advantages, limited cycle life, cathode degradation, and dendrite formation remain challenges, though advances in cathode design, electrolyte optimization, and interface engineering have significantly improved performance and stability (Wang et al., 2025). Among the various strategies to overcome these issues, CQDs have shown great promise in addressing electrolyte stability problems in both zinc-air (Zn-air) batteries and ZIBs. CQDs help suppress dendrite formation, enhance ion transport, and improve electrochemical stability. For example, V_2O_5 nanobelts and VO_2 nanorods treated with CQDs exhibit enhanced ion and electron transport, resulting in boosted capacity and cycling stability (Zhang et al., 2021). In addition to enhancing transport properties, CQDs regulate zinc deposition by reducing side reactions and promoting uniform Zn plating, which improves cycling efficiency and prevents dendrite growth. This effect significantly extends battery life by minimizing water-induced side reactions and enhancing the uniformity of zinc deposition.

To further support stable operation, CQDs help stabilize the anode and modify the electrolyte structure while also enhancing ionic conductivity. They improve electrolyte stability and charge transfer in hybrid systems, thereby boosting the efficiency of Zn-air batteries, nickel-zinc batteries, and zinc-ion capacitors. Moreover, by stabilizing the electrolyte, reducing impedance, and scavenging reactive species, CQDs contribute to improved long-term cycling performance and prevent dendrite formation. Overall, CQDs are vital for improving ZIB performance and safety, tackling key challenges such as electrolyte degradation, dendrite growth and low ionic conductivity. Looking ahead, CQDs could prove vital for advancing ZIBs and other energy storage technologies through improved ion transport, regulated zinc deposition, and electrolyte stabilization (Pradhan et al., 2025).

4.6 CQDs for Solid-State Battery

The outstanding electrical and ionic properties of CQDs have sparked significant interest in their use as advanced electrolyte additives for solid-state batteries (SSBs). CQDs enhance ionic conductivity in solid-state electrolytes (SSEs) by reducing energy barriers for ion migration and creating new ion-conducting pathways. Their surface functional groups (such as –COOH and –OH) and nanoscale size facilitate the solvation and desolvation of lithium ions (Li^+), boosting overall ionic conductivity. Building on their role in ion transport, CQDs also disrupt crystalline domains in ceramic or polymer SSEs, generating amorphous regions that promote faster Li^+ diffusion. In particular, CQDs used in lithium metal anodes improve interfacial contact, reduce interfacial resistance, and prevent dendritic growth at the electrode-electrolyte interface (Xu et al., 2022). They further enhance cycling stability by forming a stable, ionically conductive SEI through their redox-active surface groups. In addition to improving ionic conductivity and interface stability, CQDs contribute significantly to the mechanical and thermal robustness of SSEs. In polymer electrolytes, CQDs increase mechanical strength, which helps prevent cracks and mechanical failure. Meanwhile, in ceramic electrolytes, CQDs reduce porosity and promote densification. Their high thermal conductivity also helps mitigate the serious safety risk of thermal runaway. Moreover, the antioxidant properties of CQDs protect against reactive oxygen species, further enhancing the electrolyte's electrochemical stability (Liu et al., 2023).

Focusing more specifically on polymer electrolytes, CQDs have been shown to significantly boost lithium-ion transport and ionic conductivity in solid-state polymer electrolytes (SPEs). For instance, a study demonstrated that CQDs derived from poly(lithium 4-styrene sulfonate) and CA exhibited exceptionally high lithium-ion transport numbers, increasing the electrolyte's conductivity to 2.02×10^{-4} $S \cdot cm^{-1}$, surpassing that of conventional electrolytes (Z. Li et al., 2021). Beyond conductivity, CQDs also enhance the mechanical flexibility of SPEs by protecting against dendritic penetration and mechanical deformation. Supporting their use in practical devices, CQDs-based SPEs offer a wider electrochemical stability window, making them well-suited for high-voltage applications. Correspondingly, lithium and sodium metal cells incorporating CQDs-based electrolytes have demonstrated stable cycling performance and effective dendrite suppression. Overall, these properties highlight CQDs as a proficient component for future-generation solid-state batteries, as they enhance electrochemical performance, mechanical stability, and ionic conductivity.

4.7 Metal-Air Battery

Li-Oxygen (Li-O_2), aluminum-air (Al-air), and Zn-air batteries are categories of metal-air batteries known for their impressive theoretical energy densities-approximately 1,350 Wh kg^{-1} for Zn-air, 11,680 Wh kg^{-1} for Li-O_2, and 8,100 Wh kg^{-1} for Al-air. However, practical applications of these batteries are limited by issues such as low cycling stability, high charge-discharge overpotentials, and electrolyte degradation (Pradhan et al., 2025). To overcome these challenges, CQDs offer effective solutions owing to their customizable surface chemistry, nanoscale size and high electrical conductivity. CQDs improve electrolyte performance by enhancing ion transport, stabilizing electrode surfaces, and reducing side reactions. Focusing first on Li-O_2 batteries, CQDs boost Li^+ diffusion (10^{-8} to ~ 10^{-7} cm^2/s,) and lower charge transfer resistance, enabling high-rate operation. They also increase efficiency by improving oxygen solubility and electrolyte wettability (Sead et al., 2025).

In a similar manner, CQDs contribute significantly to Zn-air batteries by enhancing zinc deposition, extending cycle life, buffering electrolyte pH, and mitigating anode corrosion and hydrogen evolution (Pradhan et al., 2025). Likewise, in Al-air batteries, CQDs passivate the aluminum anode, resulting in enhanced discharge capacity and reduced self-corrosion. More broadly, across all metal-air battery systems, CQDs help extend cycle life and prevent dendritic growth by stabilizing the SEI (Wang et al., 2017). Their electroactive functional groups also enable reversible redox reactions, which reduce overpotentials and improve the efficiency of product decomposition. Despite these promising properties, challenges remain. CQDs aggregation in liquid electrolytes and instability under oxidative conditions hinder their catalytic efficiency. Therefore, future research must focus on improving stability, scalability, hybrid composite designs, and surface chemical customization to maximize CQDs performance. Importantly, CQDs derived from renewable resources not only enhance performance but also offer cost benefits-reducing costs by 15–20% in Zn-air systems-and provide an environmentally friendly, a less expensive substitute for rare metal catalysts. In conclusion, given their excellent electrochemical performance and sustainability, CQDs represent an extremely promising choice for next-generation metal-air batteries (Sead et al., 2025).

4.8 CQDs in Hybrid Systems

Hybrid energy storage system (HESS) combine battery-type and supercapacitor-type electrodes, offering complementary performance: high power density from the SC and high energy density from the battery. Batteries store energy via faradaic redox reactions, while supercapacitors rely on electrostatic double-layer capacitance, using materials such as porous carbons, graphene, metal oxides, conducting polymers, and lithium-based compounds. Integrating supercapacitors with LIB leverages the high power density of SCs (2–5 kW kg^{-1}) and the high energy density of Li-ion systems (120–200 Wh kg^{-1}). Passive hybrid systems demonstrate enhanced energy storage, with SCs supplying peak-load power and batteries providing sustained output, improving both performance and overall lifespan (Chen et al., 2025). Over time, HESS technologies have progressed remarkably—from pre-2000s prototype systems and basic control frameworks that established the foundation for hybrid storage concepts, to the 2000s proof-of-concept integrations validating the combination of batteries and SCs. During the 2010s, advances in materials, energy management systems, and control algorithms further enhanced performance and reliability, leading to practical implementation. In the 2020s and beyond, research focuses on integrating smart grids, IoT connectivity, AI-driven control, and next-generation materials. This continuous evolution illustrates how modern CQDs-based hybrid systems build upon decades of innovation in material design, system integration, and control optimization (Adeyinka et al., 2024).

Despite these promising developments, challenges remain-particularly in Zn-ion hybrid capacitors (ZHCs), where a lack of suitable cathode materials limits performance. A one-pot hydrothermal method was used to synthesize a 3D N-and boron (B)-dual-doped carbon quantum dots/rGO (N,B-CQDs/rGO) composite aerogel to address this issue. The dual doping of N and B into CQDs significantly enhances electrochemical performance. For the first time, this N, B-CQDs/rGO composite aerogel is applied as a cathode in both flexible and aqueous quasi-solid-state ZHCs. The aqueous N, B-CQDs/rGO//Zn ZHC achieves a notable energy density of 96.2 Wh·kg^{-1} at 80 W·kg^{-1}, while retaining 54.7 Wh·kg^{-1} at a very power density of 80 kW·kg^{-1}. In flexible quasi-solid-state configurations, the same system delivers a high energy density of 89.1 μWh·cm^{-2}, an outstanding power density of 96,000 μW·cm^{-2}, and excellent mechanical flexibility (Yao et al., 2023). In a related area of research, ZnS-QDs/graphene-based dual-anode systems have been investigated for both LIB and SIBs. These nanocomposites show strong potential for next-generation storage, with the ZnS-QDs/graphene hybrid delivering 491 mAh g^{-1} in SIBs and 759 mAh g^{-1} in LIBs after 100 cycles. The synergistic interaction between ZnS-QDs and graphene reduces the Na^+ and Li^+ ion diffusion distances, enhancing ion transport and boosting storage capacity. These results highlight the potential of ZnS-QDs/g-raphene hybrids for constructing high-capacity, high-rate electrodes in both sodium and lithium systems (Zhang et al., 2018). Extending these innovations further, Zhu and colleagues, developed a hybrid supercapacitor (HSC) using an N,S-CQDs/rGO composite as the anode and an N,S-CQDs/rGO/$NiCo_2S_4$ composite as the cathode (Zhu et al., 2021). This device demonstrated exceptional cycling performance, maintaining 82.9% capacitance after 10,000 cycles. It also delivered an ultra-high power density of 14.4 kW kg^{-1} and an impressive energy density of 51.0 Wh kg^{-1}. These findings suggest that CQDs/rGO-based aerogels, when used as both cathode and anode materials, present an effective and scalable electrodes for next-generation energy storage devices.

5 CHALLENGES IN THE CQDS AS ELECTROLYTE ADDITIVES IN BATTERIES AND SUPERCAPACITORS

CQDs face several challenges when used as electrolyte additives in batteries and SCs, primarily due to their electrochemical instability over prolonged cycling. Redox reactions involving oxygen-containing functional groups on CQDs can trigger unintended side reactions and structural degradation, ultimately leading to performance deterioration.

Various doping strategies have been investigated to overcome these challenges. For instance, fluorine-doped CQDs (F-CQDs) help stabilize the SEI in LIBs, while S and N co-doped CQDs (SNCQDs) improve cycling stability (Sung et al., 2023). However, despite these enhancements, the relatively lower electrical conductivity of CQDs-compared to graphene and carbon nanotubes-still limits their power density. To overcome these conductivity limitations, researchers have turned to elemental doping with boron and nitrogen, as well as the incorporation of CQDs into MXene-based anodes, both of which show improved performance. Furthermore, optimizing the sp^2/sp^3 hybridization ratio within CQDs has been found to enhance charge transfer capabilities, providing another route to improve overall efficiency (Guan et al., 2023). In addition to conductivity and stability issues, dispersion remains a critical concern, particularly in liquid-phase electrolytes. CQDs tend to agglomerate in these environments, reducing both their surface area and electrochemical activity. To counteract this, polymer functionalization and the use of surfactants have been investigated to improve dispersion and maintain their electrochemical properties. Beyond lithium-ion systems, CQDs have also demonstrated potential in SIBs (Balogun et al., 2016). Here, they enhance both capacity and stability by improving compatibility with the electrolyte and facilitating charge transfer. Nonetheless, CQDs may still interact with metal ions and electrolytes in undesirable ways, resulting in parasitic reactions that reduce efficiency. Functionalized CQDs-especially those doped with N and S-offer a promising solution by inhibiting dendrite formation and stabilizing the electrolyte composition (Balogun et al., 2016; Cai et al., 2023).

While these modifications improve performance, scalability remains a significant barrier. Producing uniform CQDs on a large scale is difficult due to variability in synthesis methods. In this context, biomass-derived CQDs, such as those obtained from carrot juice, represent a sustainable and environmentally friendly alternative that has shown promising performance in LIBs (Shi et al., 2023). However, to transition from laboratory-scale to commercial-scale production, further optimization of synthesis techniques is required. Microwave-assisted synthesis, for example, offers a potentially cost-effective method for scalable CQDs production, but it still demands refinement to ensure consistency and yield.

To fully understand and optimize CQDs behavior in energy systems, advanced characterization techniques are essential. Tools such as X-ray absorption spectroscopy and in situ electrochemical scanning tunneling microscopy enable researchers to explore the fundamental mechanisms behind CQDs function-such as quantum effects and defect-driven charge transfer-at a deeper level. These unique properties also open up new application areas. The mechanical flexibility and thermal stability of CQDs make them especially attractive for emerging technologies, including wearable and flexible energy storage devices (Yu et al., 2022). In addition, both hybrid energy storage systems and all- SSBs stand to benefit from the improved solid-state electrolyte performance offered by CQDs. Looking ahead, the integration of CQDs with machine learning represents an exciting frontier. This combination could lead to the development of smart electrolytes capable of real-time battery condition monitoring, which would significantly enhance safety and extend the operational life of energy storage systems (Xie et al., 2023). In conclusion, while challenges remain, ongoing research and development efforts suggest that CQDs have strong potential to enable high-performance, sustainable and flexible energy storage solutions in the near future.

The comparison of doping strategies and scalability methods for CQDs highlights key trade-offs:

1. Fluorine Doping (F-CQDs) improves stability but reduces energy density, limiting high-power applications.
2. S/N Co-Doping (SNCQDs) enhances cycling stability and capacity but suffers from low conductivity, restricting high-power performance.

3. Boron/Nitrogen Doping boosts conductivity and charge transfer, ideal for power applications, but may introduce defects that impact long-term stability.
4. Biomass-Derived CQDs are sustainable but face scalability and performance consistency issues at large scales.
5. Microwave-Assisted Synthesis offers cost-effective scalability but requires refinement to ensure uniformity and yield.

6 CONCLUSION

Thanks to their distinctive structural, optical, and electrochemical properties, CQDs are highly effective electrolyte additives that improve battery and SCs performance. CQDs significantly improve ion transport, charge storage capacity, and electrode-electrolyte interactions across various energy storage platforms. For example, they boost ionic conductivity in LIBs by up to 30%, increasing it from 5–8 mS cm^{-1} to 10–15 mS cm^{-1}. This improvement, combined with a stabilized SEI, results in energy densities of 150–180 Wh kg^{-1} and 90% capacity retention after 1,000 cycles. In SIBs, CQDs-modified electrolytes prevent dendrite formation and enhance stability, achieving over 92% capacity retention after 1,000 cycles, with energy densities ranging from 100 to 130 Wh kg^{-1}. After 10,000 cycles, CQDs enhance specific capacitance in SCs to 300–500 F g^{-1} in aqueous environments (compared to 150–250 F g^{-1} for activated carbon), with a 95% retention rate. CQD/polymer composites also outperform graphene-based systems (100–200 F g^{-1}) in solid-state SCs, delivering capacitances of 200–350 F g^{-1} with 93% retention after 8,000 cycles. Advances in CQDs synthesis, including heteroatom doping and biomass-based methods, are enhancing electrochemical performance, enabling safer, more efficient, and eco-friendly energy devices.

Key Research Priorities for the Future:

- Defect Engineering: Optimize CQDs surface chemistry and defect structures to enhance conductivity, stability, and charge transfer.
- Biomass-Derived CQDs: Develop scalable and sustainable synthesis methods for CQDs from biomass sources to reduce costs and environmental impact.
- Machine-Learning-Assisted Design: Leverage machine learning to predict and optimize CQDs properties for specific energy storage applications.
- Flexible and Solid-State Batteries: Integrate CQDs into flexible energy storage devices and solid-state batteries to improve performance and adaptability.
- Electrolyte Optimization: Further refine CQDs-modified electrolytes for broader applications in high-performance energy storage systems.

REFERENCES

Adeyinka, A. M., Esan, O. C., Ijaola, A. O., & Farayibi, P.-K. (2024). Advancements in hybrid energy storage systems for enhancing renewable energy-to-grid integration. *Sustainable Energy Research, 11*(1), 26. DOI: 10.1186/s40807-024-00120-4

Alam, A.-M., Park, B.-Y., Ghouri, Z. K., Park, M., & Kim, H.-Y. (2015). Synthesis of carbon quantum dots from cabbage with down- and up-conversion photoluminescence properties: Excellent imaging agent for biomedical applications. *Green Chemistry, 17*(7), 3791–3797. DOI: 10.1039/C5GC00686D

Amali, A. J., Hoshino, H., Wu, C., Ando, M., & Xu, Q. (2014). From Metal–Organic Framework to Intrinsically Fluorescent Carbon Nanodots. *Chemistry (Weinheim an der Bergstrasse, Germany), 20*(27), 8279–8282. DOI: 10.1002/chem.201402982 PMID: 24889063

Arora, G., Sabran, N. S., Ng, C. Y., Low, F. W., & Jun, H. K. (2024). Applications of carbon quantum dots in electrochemical energy storage devices. *Heliyon, 10*(15), e35543. DOI: 10.1016/j.heliyon.2024.e35543 PMID: 39166025

Balogun, M.-S., Luo, Y., Lyu, F., Wang, F., Yang, H., Li, H., Liang, C., Huang, M., Huang, Y., & Tong, Y. (2016). Carbon Quantum Dot Surface-Engineered VO_2 Interwoven Nanowires: A Flexible Cathode Material for Lithium and Sodium Ion Batteries. *ACS Applied Materials & Interfaces, 8*(15), 9733–9744. DOI: 10.1021/acsami.6b01305 PMID: 27028048

Bao, H., Liu, Y., Li, H., Qi, W., & Sun, K. (2023). Luminescence of carbon quantum dots and their application in biochemistry. *Heliyon, 9*(10), e20317. DOI: 10.1016/j.heliyon.2023.e20317 PMID: 37790961

Barik, R., & Ingole, P. P. (2020). Challenges and prospects of metal sulfide materials for supercapacitors. *Current Opinion in Electrochemistry, 21*, 327–334. DOI: 10.1016/j.coelec.2020.03.022

Bian, J., Huang, C., Wang, L., Hung, T., Daoud, W. A., & Zhang, R. (2014). Carbon Dot Loading and TiO_2 Nanorod Length Dependence of Photoelectrochemical Properties in Carbon Dot/TiO_2 Nanorod Array Nanocomposites. *ACS Applied Materials & Interfaces, 6*(7), 4883–4890. DOI: 10.1021/am4059183 PMID: 24601482

Bourlinos, A. B., Stassinopoulos, A., Anglos, D., Zboril, R., Karakassides, M., & Giannelis, E. P. (2008). Surface Functionalized Carbogenic Quantum Dots. *Small, 4*(4), 455–458. DOI: 10.1002/smll.200700578 PMID: 18350555

Cai, S., Chang, G., Hu, J., Wu, J., Luo, Y., Zou, G., Hou, H., & Ji, X. (2023). N, S-Doped Carbon Dots as Additive for Suppression of Zinc Dendrites in Alkaline Electrolyte †. *Chinese Journal of Chemistry*, *41*(14), 1697–1704. DOI: 10.1002/cjoc.202200799

Chen, B., Yang, Q., Yang, Y., Chen, J., Yan, B., Gu, Y., Fu, R., & Chen, S. (2025). PANI doped with carbon quantum dots for the asymmetric high-performance solid-state electrochromic supercapacitor. *Journal of Power Sources*, *633*, 236407. DOI: 10.1016/j.jpowsour.2025.236407

Chen, G., Wu, S., Hui, L., Zhao, Y., Ye, J., Tan, Z., Zeng, W., Tao, Z., Yang, L., & Zhu, Y. (2016). Assembling carbon quantum dots to a layered carbon for high-density supercapacitor electrodes. *Scientific Reports*, *6*(1), 19028. DOI: 10.1038/srep19028 PMID: 26754463

De Yro, P. A. N., Quaichon, G. M. O., Cruz, R. A. T., Emolaga, C. S., Que, M. C. O., Magdaluyo, E. R., & Basilia, B. A. (2019). *Hydrothermal synthesis of carbon quantum dots from biowaste for bio-imaging*., DOI: 10.1063/1.5094310

Ding, H., Yu, S.-B., Wei, J.-S., & Xiong, H.-M. (2016). Full-Color Light-Emitting Carbon Dots with a Surface-State-Controlled Luminescence Mechanism. *ACS Nano*, *10*(1), 484–491. DOI: 10.1021/acsnano.5b05406 PMID: 26646584

Fu, C., Qin, X., Zhang, J., Zhang, T., Song, Y., Yang, J., Wu, G., Luo, D., Jiang, N., & Bikker, F. J. (2023). In vitro and in vivo toxicological evaluation of carbon quantum dots originating from Spinacia oleracea. *Heliyon*, *9*(2), e13422. DOI: 10.1016/j.heliyon.2023.e13422 PMID: 36820041

Gopi, C. V. M., & Ramesh, R. (2024). Review of battery-supercapacitor hybrid energy storage systems for electric vehicles. *Results in Engineering*, *24*, 103598. DOI: 10.1016/j.rineng.2024.103598

Gu, Z., Li, D., Zheng, C., Kang, Y., Wöll, C., & Zhang, J. (2017). MOF-Templated Synthesis of Ultrasmall Photoluminescent Carbon-Nanodot Arrays for Optical Applications. *Angewandte Chemie International Edition*, *56*(24), 6853–6858. DOI: 10.1002/anie.201702162 PMID: 28481003

Guan, G., Lu, L., Meng, W., Ye, T., Zhang, Y., & Guo, F. (2023). Nb_2CT_x MXene Anchored with Carbon Quantum Dots for Lithium-Ion Batteries. *ACS Applied Nano Materials*, *6*(24), 23620–23629. DOI: 10.1021/acsanm.3c05424

Guo, J., Gao, F., Li, D., Luo, X., Sun, Y., Wang, X., Ran, Z., Wu, Q., & Li, S. (2020). Novel Strategy of Constructing Hollow Ga_2O_3@N-CQDs as a Self-Healing Anode Material for Lithium-Ion Batteries. *ACS Sustainable Chemistry & Engineering*, *8*(36), 13692–13700. DOI: 10.1021/acssuschemeng.0c03756

Hoang, V. C., Dave, K., & Gomes, V. G. (2019). Carbon quantum dot-based composites for energy storage and electrocatalysis: Mechanism, applications and future prospects. *Nano Energy*, *66*, 104093. DOI: 10.1016/j.nanoen.2019.104093

Hou, H., Banks, C. E., Jing, M., Zhang, Y., & Ji, X. (2015). Carbon Quantum Dots and Their Derivative 3D Porous Carbon Frameworks for Sodium-Ion Batteries with Ultralong Cycle Life. *Advanced Materials*, *27*(47), 7861–7866. DOI: 10.1002/adma.201503816 PMID: 26506218

Hsu, P.-C., & Chang, H.-T. (2012). Synthesis of high-quality carbon nanodots from hydrophilic compounds: Role of functional groups. *Chemical Communications*, *48*(33), 3984. DOI: 10.1039/c2cc30188a PMID: 22422194

Hu, S., Liu, J., Yang, J., Wang, Y., & Cao, S. (2011). Laser synthesis and size tailor of carbon quantum dots. *Journal of Nanoparticle Research*, *13*(12), 7247–7252. DOI: 10.1007/s11051-011-0638-y

Huan, H., Wu, W., Feng, Q., Tang, Y., Li, S., Deng, H., Xiang, X., Zu, X., & Li, Y. (2026). Li-ion battery based on B, N-graphene prepared by ball milling. *Journal of Physics and Chemistry of Solids*, *208*, 113002. DOI: 10.1016/j.jpcs.2025.113002

Jiang, H., Chen, F., Lagally, M. G., & Denes, F. S. (2010). New Strategy for Synthesis and Functionalization of Carbon Nanoparticles. *Langmuir*, *26*(3), 1991–1995. DOI: 10.1021/la9022163 PMID: 20038107

Jiang, K., Sun, S., Zhang, L., Lu, Y., Wu, A., Cai, C., & Lin, H. (2015). Red, Green, and Blue Luminescence by Carbon Dots: Full-Color Emission Tuning and Multicolor Cellular Imaging. *Angewandte Chemie International Edition*, *54*(18), 5360–5363. DOI: 10.1002/anie.201501193 PMID: 25832292

Kakaei, K., Khodadoost, S., Gholipour, M., & Shouraei, N. (2021). Core-shell polyaniline functionalized carbon quantum dots for supercapacitor. *Journal of Physics and Chemistry of Solids*, *148*, 109753. DOI: 10.1016/j.jpcs.2020.109753

Kannan, K., & Tari, V. (Eds.). (2024). Novel energy storage and conversion technologies for two-dimensional MXenes and MBenes. ISBN: 9798369395462, eISBN: 9798369395486DOI: 10.4018/979-8-3693-9546-2

Kim, M. J., Alvarez, S., Chen, Z., Fichthorn, K. A., & Wiley, B. J. (2018). Single-Crystal Electrochemistry Reveals Why Metal Nanowires Grow. *Journal of the American Chemical Society*, *140*(44), 14740–14746. DOI: 10.1021/jacs.8b08053 PMID: 30351013

Krajewski, M., Chen, C.-H., Huang, Z.-T., & Lin, J.-Y. (2022). $Li_4Ti_5O_{12}$ Coated by Biomass-Derived Carbon Quantum Dots as Anode Material with Enhanced Electrochemical Performance for Lithium-Ion Batteries. *Energies*, *15*(20), 7715. DOI: 10.3390/en15207715

Kumar, P. S., Bae, J., Roh, J. W., Min, Y., & Lee, S. (2025). Enhanced voltage and capacitance in flexible supercapacitors using electrospun nanofiber electrolytes and $CuNi_2O_3$@N-Doped omnichannel carbon electrodes. *Nano Convergence*, *12*(1), 21. DOI: 10.1186/s40580-025-00485-2 PMID: 40301281

Kumar, R., Kumar, V. B., & Gedanken, A. (2020). Sonochemical synthesis of carbon dots, mechanism, effect of parameters, and catalytic, energy, biomedical and tissue engineering applications. *Ultrasonics Sonochemistry*, *64*, 105009. DOI: 10.1016/j.ultsonch.2020.105009 PMID: 32106066

Kumar, V. B., Tang, J., Lee, K. J., Pol, V. G., & Gedanken, A. (2016). In situ sonochemical synthesis of luminescent Sn@C-dots and a hybrid Sn@C-dots@Sn anode for lithium-ion batteries. *RSC Advances*, *6*(70), 66256–66265. DOI: 10.1039/C6RA09926B

Lai, C.-W., Hsiao, Y.-H., Peng, Y.-K., & Chou, P.-T. (2012). Facile synthesis of highly emissive carbon dots from pyrolysis of glycerol; gram scale production of carbon dots/$mSiO_2$ for cell imaging and drug release. *Journal of Materials Chemistry*, *22*(29), 14403. DOI: 10.1039/c2jm32206d

Li, H., Liu, R., Liu, Y., Huang, H., Yu, H., Ming, H., Lian, S., Lee, S.-T., & Kang, Z. (2012). Carbon quantum dots/Cu_2O composites with protruding nanostructures and their highly efficient (near) infrared photocatalytic behavior. *Journal of Materials Chemistry*, *22*(34), 17470. DOI: 10.1039/c2jm32827e

Li, H., Zhang, Y., Wang, L., Tian, J., & Sun, X. (2011). Nucleic acid detection using carbon nanoparticles as a fluorescent sensing platform. *Chemical Communications (Cambridge)*, *47*(3), 961–963. DOI: 10.1039/C0CC04326E PMID: 21079843

Li, M., Ma, C., Wang, G., Zhang, X., Dong, X., & Ma, H. (2021). Controlling the up-conversion photoluminescence property of carbon quantum dots (CQDs) by modifying its surface functional groups for enhanced photocatalytic performance of CQDs/$BiVO_4$ under a broad-spectrum irradiation. *Research on Chemical Intermediates*, *47*(8), 3469–3485. DOI: 10.1007/s11164-021-04459-x

Li, Z., Liu, F., Chen, S., Zhai, F., Li, Y., Feng, Y., & Feng, W. (2021). Single Li ion conducting solid-state polymer electrolytes based on carbon quantum dots for Li-metal batteries. *Nano Energy, 82*, 105698. DOI: 10.1016/j.nanoen.2020.105698

Liao, B., Long, P., He, B., Yi, S., Ou, B., Shen, S., & Chen, J. (2013). Reversible fluorescence modulation of spiropyran-functionalized carbon nanoparticles. *Journal of Materials Chemistry. C, Materials for Optical and Electronic Devices, 1*(23), 3716. DOI: 10.1039/c3tc00906h

Liu, H., Ye, T., & Mao, C. (2007). Fluorescent Carbon Nanoparticles Derived from Candle Soot. *Angewandte Chemie International Edition, 46*(34), 6473–6475. DOI: 10.1002/anie.200701271 PMID: 17645271

Liu, S., Ao, Z., Zou, Y., Li, H., Chen, N., Wan, Y., & Lao, Y. (2023). Enhanced ionic conductivity of PEO-based solid electrolyte through synergistic dissociation effect. *Materials Letters, 351*, 134981. DOI: 10.1016/j.matlet.2023.134981

Liu, S., Cao, X., Zhang, Y., Wang, K., Su, Q., Chen, J., He, Q., Liang, S., Cao, G., & Pan, A. (2020). Carbon quantum dot modified $Na_3V_2(PO_4)_2F_3$ as a high-performance cathode material for sodium-ion batteries. J Mater. *Journal of Materials Chemistry. A, Materials for Energy and Sustainability, 8*(36), 18872–18879. DOI: 10.1039/D0TA04307A

Liu, Y., Cai, D., Zheng, F., Qin, Z., Li, Y., Li, W., Li, A., Zhao, Y., & Zhang, J. (2024). A carbon quantum dot-decorated g-C_3N_4 composite as a sulfur hosting material for lithium–sulfur batteries. *Dalton Transactions (Cambridge, England), 53*(16), 7035–7043. DOI: 10.1039/D4DT00511B PMID: 38563460

Lu, J., Yang, J., Wang, J., Lim, A., Wang, S., & Loh, K. P. (2009). One-Pot Synthesis of Fluorescent Carbon Nanoribbons, Nanoparticles, and Graphene by the Exfoliation of Graphite in Ionic Liquids. *ACS Nano, 3*(8), 2367–2375. DOI: 10.1021/nn900546b PMID: 19702326

Lu, X., Li, S., Li, Y., Wu, F., Wu, C., & Bai, Y. (2024). From Lab to Application: Challenges and Opportunities in Achieving Fast Charging with Polyanionic Cathodes for Sodium-Ion Batteries. *Advanced Materials, 36*(36), 2407359. Advance online publication. DOI: 10.1002/adma.202407359 PMID: 38936413

Luo, P. G., Sahu, S., Yang, S.-T., Sonkar, S. K., Wang, J., Wang, H., LeCroy, G. E., Cao, L., & Sun, Y.-P. (2013). Carbon "quantum" dots for optical bioimaging. *Journal of Materials Chemistry. B, 1*(16), 2116. DOI: 10.1039/c3tb00018d PMID: 32260843

Magesh, V., Sundramoorthy, A. K., & Ganapathy, D. (2022). Recent Advances on Synthesis and Potential Applications of Carbon Quantum Dots. *Frontiers in Materials*, *9*, 906838. Advance online publication. DOI: 10.3389/fmats.2022.906838

Mahmood, A., Bai, Z., Wang, T., Lei, Y., Wang, S., Sun, B., Khan, H., Khan, K., Sun, K., & Wang, G. (2025). Enabling high-performance multivalent metal-ion batteries: Current advances and future prospects. *Chemical Society Reviews*, *54*(5), 2369–2435. DOI: 10.1039/D4CS00929K PMID: 39887968

Murakami, Y., Chiashi, S., Miyauchi, Y., Hu, M., Ogura, M., Okubo, T., & Maruyama, S. (2004). Growth of vertically aligned single-walled carbon nanotube films on quartz substrates and their optical anisotropy. *Chemical Physics Letters*, *385*(3-4), 298–303. DOI: 10.1016/j.cplett.2003.12.095

Muthukutty, B., Sathish Kumar, P., Lee, D., & Lee, S. (2024). Multichannel Carbon Nanofibers: Pioneering the Future of Energy Storage. *ACS Nano*, *18*(40), 27287–27316. DOI: 10.1021/acsnano.4c11146 PMID: 39324479

Ozyurt, D., Al Kobaisi, M., Hocking, R. K., & Fox, B. (2023). Properties, synthesis, and applications of carbon dots: A review. *Carbon Trends*, *12*, 100276. DOI: 10.1016/j.cartre.2023.100276

Pallavolu, M. R., Prabhu, S., Nallapureddy, R. R., Kumar, A. S., Banerjee, A. N., & Joo, S. W. (2023). Bio-derived graphitic carbon quantum dot encapsulated S- and N-doped graphene sheets with unusual battery-type behavior for high-performance supercapacitor. *Carbon*, *202*, 93–102. DOI: 10.1016/j.carbon.2022.10.077

Pan, Z., Brett, D. J. L., He, G., & Parkin, I. P. (2022). Progress and Perspectives of Organosulfur for Lithium–Sulfur Batteries. *Advanced Energy Materials*, *12*(8), 2103483. Advance online publication. DOI: 10.1002/aenm.202103483

Peng, J., Gao, W., Gupta, B. K., Liu, Z., Romero-Aburto, R., Ge, L., Song, L., Alemany, L. B., Zhan, X., Gao, G., Vithayathil, S. A., Kaipparettu, B. A., Marti, A. A., Hayashi, T., Zhu, J.-J., & Ajayan, P. M. (2012). Graphene Quantum Dots Derived from Carbon Fibers. *Nano Letters*, *12*(2), 844–849. DOI: 10.1021/nl2038979 PMID: 22216895

Phogat, P., Dey, S., & Wan, M. (2025). Powering the sustainable future: A review of emerging battery technologies and their environmental impact. *RSC Sustainability*, *3*(8), 3266–3306. DOI: 10.1039/D5SU00127G

Pholauyphon, W., Bulakhe, R. N., Manyam, J., In, I., & Paoprasert, P. (2022). High-performance supercapacitors using carbon dots/titanium dioxide composite electrodes and carbon dot-added sulfuric acid electrolyte. *Journal of Electroanalytical Chemistry (Lausanne, Switzerland), 910*, 116177. DOI: 10.1016/j.jelechem.2022.116177

Prabakaran, P., Arumugam, G., Ramu, P., Selvaraj, M., Assiri, M. A., Rokhum, S. L., Periyasamy, S., Arjunan, S., & Rajendran, R. (2023). Preparation and characterization of carbon quantum dots grafted Co_3O_4 nanocomposite for supercapacitors application. *Surfaces and Interfaces, 40*, 103153. DOI: 10.1016/j.surfin.2023.103153

Pradhan, A. K., Halder, S., & Chakraborty, C. (2025). Pt-nanoparticles on ZnO/carbon quantum dots: A trifunctional nanocomposite with superior electrocatalytic activity boosting direct methanol fuel cells and zinc–air batteries. J Mater. *Journal of Materials Chemistry. A, Materials for Energy and Sustainability, 13*(1), 243–256. DOI: 10.1039/D4TA05630B

Prasath, A., Athika, M., Duraisamy, E., Sharma, A. S., & Elumalai, P. (2018). Carbon-Quantum-Dot-Derived Nanostructured MnO_2 and Its Symmetrical Supercapacitor Performances. *ChemistrySelect, 3*(30), 8713–8723. DOI: 10.1002/slct.201801950

Qi, L.-Y., Zhang, Y.-W., Zuo, Z.-C., Xin, Y.-L., Yang, C.-K., Wu, B., Zhang, X.-X., & Zhou, H.-H. (2016). *In situ* quantization of ferroferric oxide embedded in 3D microcarbon for ultrahigh performance sodium-ion batteries. J Mater. *Journal of Materials Chemistry. A, Materials for Energy and Sustainability, 4*(22), 8822–8829. DOI: 10.1039/C6TA01836J

Quan, Y., Wang, G., Lu, L., Wang, Z., Xu, H., Liu, S., & Wang, D. (2020). High-performance pseudo-capacitor energy storage device based on a hollow-structured copper sulfide nanoflower and carbon quantum dot nanocomposite. *Electrochimica Acta, 353*, 136606. DOI: 10.1016/j.electacta.2020.136606

Rohith, R., Thejas Prasannakumar, A., V., M., Thomas, M., R.Mohan, R., J.Varma, S., 2024. An insight into the electrochemical performance of nanostructured V_2O_5 in aqueous neutral electrolytes and fabrication of 2V, high energy density, symmetric supercapacitor. Electrochim Acta 503, 144911. DOI: 10.1016/j.electacta.2024.144911

RasulKhan, B. B., Ponnaiah, S. K., Periakaruppan, P., Venkatachalam, G., & Balasubramanian, J.RasulKhan. (2020). A new CQDs/f-MWCNTs/GO nanocomposite electrode for arsenic (10^{-12} M) quantification in bore-well water and industrial effluents. *New Journal of Chemistry, 44*(42), 18149–18156. DOI: 10.1039/D0NJ04252H

Raza, W., Ali, F., Raza, N., Luo, Y., Kim, K.-H., Yang, J., Kumar, S., Mehmood, A., & Kwon, E. E. (2018). Recent advancements in supercapacitor technology. *Nano Energy, 52*, 441–473. DOI: 10.1016/j.nanoen.2018.08.013

Rüther, T., Bhatt, A. I., Best, A. S., Harris, K. R., & Hollenkamp, A. F. (2020). Electrolytes for Lithium (Sodium) Batteries Based on Ionic Liquids: Highlighting the Key Role Played by the Anion. *Batteries & Supercaps*, *3*(9), 793–827. DOI: 10.1002/batt.202000022

Sahu, S., Behera, B., Maiti, T. K., & Mohapatra, S. (2012). Simple one-step synthesis of highly luminescent carbon dots from orange juice: Application as excellent bio-imaging agents. *Chemical Communications*, *48*(70), 8835. DOI: 10.1039/c2cc33796g PMID: 22836910

Sead, F. F., Makasana, J., Saraswat, S. K., Rekha, M. M., Kundlas, M., Saini, S., Joshi, K. K., & Noorizadeh, H. (2025). Electrochemical behavior of carbon quantum dots as electrolyte additives for enhanced battery and supercapacitor performance. *Materials Technology*, *40*(1), 2500524. Advance online publication. DOI: 10.1080/10667857.2025.2500524

Shaker, M., Ng, S., Sadeghi Ghazvini, A. A., Javanmardi, S., Gaho, M. A., Jin, Z., & Ge, Q. (2024). Carbon/graphene quantum dots as electrolyte additives for batteries and supercapacitors: A review. *Journal of Energy Storage*, *85*, 111040. DOI: 10.1016/j.est.2024.111040

Sharma, S. K., Sharma, G., Gaur, A., Arya, A., Mirsafi, F. S., Abolhassani, R., Rubahn, H.-G., Yu, J.-S., & Mishra, Y. K. (2022). Progress in electrode and electrolyte materials: Path to all-solid-state Li-ion batteries. *Energy Advances*, *1*(8), 457–510. DOI: 10.1039/D2YA00043A

Shi, F., Xing, B., Zeng, H., Guo, H., Qu, X., Huang, G., Cao, Y., Li, P., Feng, L., & Zhang, C. (2024). Facile synthesis of ultrathin carbon nanosheets through NaCl-KCl templates coupled with ice-induced assembly strategy from carbon quantum dots as lithium-ion batteries anodes. *Journal of Alloys and Compounds*, *976*, 173325. DOI: 10.1016/j.jallcom.2023.173325

Shi, F., Xing, B., Zeng, H., Guo, H., Qu, X., Huang, G., Cao, Y., Li, P., & Zhang, C. (2023). Ice template induced assembly strategy for preparation of 3D porous carbon frameworks from low-cost carbon quantum dots for high-performance lithium-ion batteries. *Journal of Energy Storage*, *70*, 107982. DOI: 10.1016/j.est.2023.107982

Shin, D.-Y., Sung, K.-W., & Ahn, H.-J. (2021). Fluorine-doped carbon quantum dot interfacial layer on stockade-like etched copper foil for boosting Li-ion storage. *Chemical Engineering Journal*, *413*, 127563. DOI: 10.1016/j.cej.2020.127563

Sun, Y.-P., Zhou, B., Lin, Y., Wang, W., Fernando, K. A. S., Pathak, P., Meziani, M. J., Harruff, B. A., Wang, X., Wang, H., Luo, P. G., Yang, H., Kose, M. E., Chen, B., Veca, L. M., & Xie, S.-Y. (2006). Quantum-Sized Carbon Dots for Bright and Colorful Photoluminescence. *Journal of the American Chemical Society*, *128*(24), 7756–7757. DOI: 10.1021/ja062677d PMID: 16771487

Sung, K.-W., Ko, K.-Y., & Ahn, H.-J. (2023). Tailored sulfur and nitrogen co-doped carbon quantum dot interfacial layer on copper foil for highly stable and ultrafast lithium-ion capacitors. *Journal of Energy Storage*, *72*, 108797. DOI: 10.1016/j.est.2023.108797

Talat, M., Awasthi, K., Singh, V. K., & Srivastava, O. N. (2020). Investigations on Exotic Forms of Carbon: Nanotubes, Graphene, Fullerene, and Quantum Dots. In *Functionalized Nanomaterials I* (pp. 125–134). CRC Press., DOI: 10.1201/9781351021623-8

Wang, M., Fang, J., Hu, L., Lai, Y., & Liu, Z. (2017). Defects-rich graphene/carbon quantum dot composites as highly efficient electrocatalysts for aqueous zinc/air batteries. *International Journal of Hydrogen Energy*, *42*(33), 21305–21310. DOI: 10.1016/j.ijhydene.2017.07.045

Wang, T., Liu, P., Chen, X., Guo, Y., He, C., Guo, J., Liu, W., Gao, S., Lv, Y., & Liu, K. (2025). Enhancing aqueous zinc-ion energy storage performance with ion-mediating carbon quantum dots-modified separators regulating zinc deposition behavior. *Journal of Colloid and Interface Science*, *678*, 301–310. DOI: 10.1016/j.jcis.2024.08.175 PMID: 39208758

Wang, X., Feng, Y., Dong, P., & Huang, J. (2019). A Mini Review on Carbon Quantum Dots: Preparation, Properties, and Electrocatalytic Application. *Frontiers in Chemistry*, *7*, 671. Advance online publication. DOI: 10.3389/fchem.2019.00671 PMID: 31637234

Wang, X. W., Zheng, D. L., Yang, P. Z., Wang, X. E., Zhu, Q. Q., Ma, P. F., & Sun, L. Y. (2017). Preparation and electrochemical properties of NiO-Co_3O_4 composite as electrode materials for supercapacitors. *Chemical Physics Letters*, *667*, 260–266. DOI: 10.1016/j.cplett.2016.10.079

Wang, Y., & Hu, A. (2014). Carbon quantum dots: Synthesis, properties and applications. *Journal of Materials Chemistry. C, Materials for Optical and Electronic Devices*, *2*(34), 6921. DOI: 10.1039/C4TC00988F

Wang, Z., Che, H., Lu, W., Chao, Y., Wang, L., Liang, B., Liu, J., Xu, Q., & Cui, X. (2023). Application of Inorganic Quantum Dots in Advanced Lithium–Sulfur Batteries. *Advancement of Science*, *10*(19), 2301355. Advance online publication. DOI: 10.1002/advs.202301355 PMID: 37088862

Xie, W., Zhang, Q., Song, S., Cheng, X., Yang, Y., Wang, L., Ouyang, X., Xie, S., & Huang, J. (2023). High S-doped amorphous carbon/carbon quantum dots coupled micro-frame for highly efficient potassium storage. *Journal of Colloid and Interface Science*, *652*, 1522–1532. DOI: 10.1016/j.jcis.2023.08.123 PMID: 37660609

Xu, J., Zhou, P., Yuan, L., Liu, X., Ma, J., & Zhang, C. (2021). Dual lignin valorization enabled by carbon quantum dots and lithium-sulfur cathode. *Industrial Crops and Products*, *170*, 113801. DOI: 10.1016/j.indcrop.2021.113801

Xu, L., Tu, H., Zhu, F., Xiang, Y., Luo, Z., Fang, S., Deng, W., Zou, G., Hou, H., & Ji, X. (2022). Carbon dots for ultrastable solid-state batteries. *SmartMat*, *3*(2), 286–297. DOI: 10.1002/smm2.1097

Yang, H.-L., Bai, L.-F., Geng, Z.-R., Chen, H., Xu, L.-T., Xie, Y.-C., Wang, D.-J., Gu, H.-W., & Wang, X.-M. (2023). Carbon quantum dots: Preparation, optical properties, and biomedical applications. *Materials Today. Advances*, *18*, 100376. DOI: 10.1016/j.mtadv.2023.100376

Yang, R., Zhang, Y., Chen, X., Song, L., Hu, Y., & Song, Y. (2024). Locally boosted Li_2S nucleation on VO_2 by loading carbon quantum dots for soft-packaged lithium–sulfur pouch cells. *Chemical Communications*, *60*(68), 9078–9081. DOI: 10.1039/D4CC02263G PMID: 39105356

Yang, S., Ye, X., Shen, N., Sun, T., Yu, F., Liu, H., Guo, C., & Li, J. (2025). Carbon Dots Enabling High-Performances Sodium-Ion Batteries. *Chemistry (Weinheim an der Bergstrasse, Germany)*, *31*(2), e202402794. Advance online publication. DOI: 10.1002/chem.202402794 PMID: 39467772

Yao, J.-J., Liu, C., Li, J.-Y., Hu, Z.-L., Zhou, R.-Y., Guo, C.-C., Liu, X.-R., Yang, F.-F., & Zhu, Y.-R. (2023). 3D nitrogen and boron dual-doped carbon quantum dots/reduced graphene oxide aerogel for advanced aqueous and flexible quasi-solid-state zinc-ion hybrid capacitors. *Rare Metals*, *42*(7), 2307–2323. DOI: 10.1007/s12598-023-02265-5

Yu, L., Hsieh, C.-T., Keffer, D. J., Chen, H., Goenaga, G. A., Dai, S., Zawodzinski, T. A., & Harper, D. P. (2021). Hierarchical Lignin-Based Carbon Matrix and Carbon Dot Composite Electrodes for High-Performance Supercapacitors. *ACS Omega*, *6*(11), 7851–7861. DOI: 10.1021/acsomega.1c00448 PMID: 33778297

Yu, W.-J., He, W., Wang, C., Liu, F., Zhu, L., Tian, Q., Tong, H., & Guo, X. (2022). Confinement of TiO_2 quantum dots in graphene nanoribbons for high-performance lithium and sodium ion batteries. *Journal of Alloys and Compounds*, *898*, 162856. DOI: 10.1016/j.jallcom.2021.162856

Zhai, X., Zhang, P., Liu, C., Bai, T., Li, W., Dai, L., & Liu, W. (2012). Highly luminescent carbon nanodots by microwave-assisted pyrolysis. *Chemical Communications*, *48*(64), 7955. DOI: 10.1039/c2cc33869f PMID: 22763501

Zhang, F., Zhang, T., Yang, X., Zhang, L., Leng, K., Huang, Y., & Chen, Y. (2013). A high-performance supercapacitor-battery hybrid energy storage device based on graphene-enhanced electrode materials with ultrahigh energy density. *Energy & Environmental Science*, *6*(5), 1623. DOI: 10.1039/c3ee40509e

Zhang, J., Wei, S., Wang, H., Liu, H., Zhang, Y., Liu, S., Wang, Z., & Lu, X. (2021). Carbon Quantum Dots Promote Coupled Valence Engineering of V_2O_5 Nanobelts for High-Performance Aqueous Zinc-Ion Batteries. *ChemSusChem*, *14*(9), 2076–2083. DOI: 10.1002/cssc.202100223 PMID: 33751841

Zhang, R., Wang, Y., Jia, M., Xu, J., & Pan, E. (2018). One-pot hydrothermal synthesis of ZnS quantum dots/graphene hybrids as a dual anode for sodium ion and lithium ion batteries. *Applied Surface Science*, *437*, 375–383. DOI: 10.1016/j.apsusc.2017.12.110

Zhao, W., Dong, H., Xing, Z., Zhou, L., Chen, S., Liu, H., Dou, S., Zhao, Z., Xia, H., Chou, S., & Chen, M. (2024). Advances and Challenges of Porous Structure on Solid–Liquid Interfaces in Polyanionic Sodium-Ion Batteries. *Advanced Energy Materials*, *14*(42), 2402720. Advance online publication. DOI: 10.1002/aenm.202402720

Zhou, J., Booker, C., Li, R., Zhou, X., Sham, T.-K., Sun, X., & Ding, Z. (2007). An Electrochemical Avenue to Blue Luminescent Nanocrystals from Multiwalled Carbon Nanotubes (MWCNTs). *Journal of the American Chemical Society*, *129*(4), 744–745. DOI: 10.1021/ja0669070 PMID: 17243794

Zhu, H., Wang, X., Li, Y., Wang, Z., Yang, F., & Yang, X. (2009). Microwave synthesis of fluorescent carbon nanoparticles with electrochemiluminescence properties. *Chemical Communications*, *5118*(34), 5118. Advance online publication. DOI: 10.1039/b907612c PMID: 20448965

Zhu, Y., Li, J., Yun, X., Chen, X., Yao, J., & Lu, Z. (2021). Design and Fabrication of Advanced Cathode and Anode Materials for Hybrid Supercapacitors Based on Graphitic Carbon Quantum Dot-Decorated Reduced Graphene Oxide Composite Aerogels. *ACS Applied Energy Materials*, *4*(1), 714–729. DOI: 10.1021/acsaem.0c02593

Zuo, W., Li, R., Zhou, C., Li, Y., Xia, J., & Liu, J. (2017). Battery-Supercapacitor Hybrid Devices: Recent Progress and Future Prospects. *Advancement of Science*, *4*(7), 1600539. Advance online publication. DOI: 10.1002/advs.201600539 PMID: 28725528

Chapter 8
Heteroatom-Engineered Carbon Quantum Dots for High-Performance Energy Storage:
From Quantum-Level Design to Scalable Device Architectures

Matbiangthew Shadap
https://orcid.org/0000-0002-9159-6668
Karunya Institute of Technology and Sciences, India

Subin Stephen
https://orcid.org/0009-0003-6411-7814
Karunya Institute of Technology and Sciences, India

Sakunthala Ayyasamy
https://orcid.org/0000-0002-9945-6272
Karunya Institute of Technology and Sciences, India

ABSTRACT

This chapter reviews heteroatom-engineered carbon quantum dots (CQDs) for energy storage, bridging the gap between high-energy batteries and high-power supercapacitors. We examine how N, S, P, and B doping modifies electronic structure, surface properties, and crystallinity of CQDs. The study clarifies dopant roles in creating redox sites, enhancing conductivity, and optimizing charge storage mechanisms. We survey scalable synthesis methods, emphasizing biomass precursors and microwave techniques for industrial translation. Advanced electrode architectures combining

DOI: 10.4018/979-8-3373-4247-4.ch008

CQDs with graphene, MXenes, and polymers achieve capacitances greater than 1800 F/g with exceptional stability. Performance analysis covers CQD-based supercapacitors and Li/Na/Zn-ion batteries, linking material design to device metrics. Techno-economic and life cycle assessments evaluate commercial viability. Key challenges include uniform co-doping and electrode densification. The chapter concludes with perspectives on AI-driven materials discovery for next-generation energy storage.

1. INTRODUCTION

The ongoing global energy transition and emergence of large-scale high-power electronics have led to an extraordinary need for energy storage (EES) systems across a wide range of fields and applications (Abbas et al., 2019). The need for advanced electrical energy storage solutions ranges from large-capacity grid energy storage systems for renewable sources to high-performance batteries for electric vehicle propulsion and portable EES devices for wearable electronics (Permatasari et al., 2021). Nevertheless, the most common EES devices still exhibit a common trade-off relationship in their performance, as demonstrated in the Ragone plot (Khan, 2024; Permatasari et al., 2021). On the one hand, batteries such as commercialized lithium-ion batteries (LIBs) with high energy density (*Wh/kg*) but limited power density (*W/kg*) can operate for a long time but usually take minutes or hours to reach full charge/discharge capacity and usually suffer from poor cycle stability, often only lasting a few thousand cycles before severe degradation (Khan, 2024). In contrast, electrochemical capacitors (supercapacitors) exhibit a very high power density and can be fully charged/discharged in seconds, which is accompanied by ultra-stable cycling stability, often capable of more than 105 cycles (Abdel Maksoud et al., 2021). However, their energy density is significantly lower, which hinders their deployment as the sole power source in many applications (Libber et al., 2024). Therefore, research into achieving hybrid systems and new materials that could inherit the high energy/power density from batteries/supercapacitors, respectively, has received significant attention from both academia and industry (L. Liu et al., 2025).

Carbon-based materials are among the most promising and mature EES candidates to date because they are low-cost, earth-abundant, highly electrically conductive, and chemically stable (Rasal et al., 2021). While 1*D* carbon nanotubes (CNTs) and 2*D* graphene have dominated the research landscape for some time, a recent trend has emerged in the use of 0*D* carbon quantum dots (CQDs) (Kannan et al., 2024; Guan et al., 2023). CQDs are quasi-spherical, graphenic nanoclusters that are typically less than 10*nm* in diameter (Dua et al., 2023; Permatasari et al., 2021). As such, they exhibit unique properties due to their 0*D* nature, which are not inherently or readily accessible in higher-dimensional carbon allotropes (Guan et al., 2023). The most prominent of these is the quantum confinement effect (QCE), which results in a tunable, size-dependent bandgap and corresponding unique optoelectronic properties (Permatasari et al., 2021; T. Yuan et al., 2023). From an electrochemical perspective, the transition from a higher-dimensional carbon material to 0*D* CQDs is essentially an effort to increase the volumetric fraction of the most active sites involved in the charge storage process to the maximum possible extent (Dua et al., 2023; Zhai et al., 2021). Although bulk activated carbon materials capitalize on their internal surface area features and graphene maintains a superior surface-to-volume ratio compared to bulk carbon, CQDs demonstrate an edge site fraction that exceeds their total mass by several orders of magnitude (Z. Wang et al., 2019). It is well known that electrochemical reactions and ion adsorption primarily occur at the edges and surface defects of these materials (Tian et al., 2021; Z. Wang et al., 2019). Thus, the use of CQDs is more than just the adoption of a new carbon allotrope; it is the adoption of the utilization of quantum-scale material phenomena as well as a means to pack the highest possible concentration of active sites to improve device performance at the macroscopic scale (Dua et al., 2023; Zhai et al., 2021).

Although pristine CQDs are a good starting point, there are many approaches to improve their intrinsic electrochemical performance. Heteroatom engineering, the intentional addition of non-carbon elements to the carbon skeleton, is a critical accelerator for revealing the hidden potential of CQDs (Jana & Pal, 2017; Permatasari et al., 2021). For instance, doping CQDs with *N*, *S*, *P*, or *B* heteroatoms can dramatically change their physicochemical properties (Kou et al., 2020). It can fine-tune the electronic structure to create charge imbalances, introduce a high density of defects that act as active sites, improve surface wettability for better electrolyte penetration, and even enlarge the interlayer spacing of the graphenic core for faster ion diffusion (Abbas et al., 2019). These synergistically enhance both non-Faradaic charge storage via the EDLC and faradaic charge storage through pseudocapacitive redox reactions (Huang et al., 2017). A series of recent studies have emphasized the significant performance enhancements that can be achieved, such as a three-fold improvement in energy density and long-term stability over 10,000 cycles (Deng et al., 2020; Goswami et al., 2023; Pallavolu et al., 2023). Surprisingly, the current literature is fragmented. Many reviews cover either one application (SC or one type of battery), one dopant, or lab-scale synthesis methods, but often lack crucial information on scaling, sustainability, or other platforms (Kou et al., 2020). This gap in scientific knowledge is not merely an academic inconvenience. This represents a bottleneck for the field's industrial translation, as without a unifying body of knowledge, commercialization efforts will have to rely on guesswork. By compiling a multidisciplinary, transverse, and application-agnostic synthesis of the current state of knowledge, this chapter aims to fill this critical gap, offering a valuable roadmap for the field to extract mechanistic understanding, identify design rules, and guide translation efforts towards commercializing breakthroughs in heteroatom-engineered CQDs for energy storage. The structure and flow of this book chapter is shown in Figure 1.

Figure 1. Structure and flow of the book chapter.

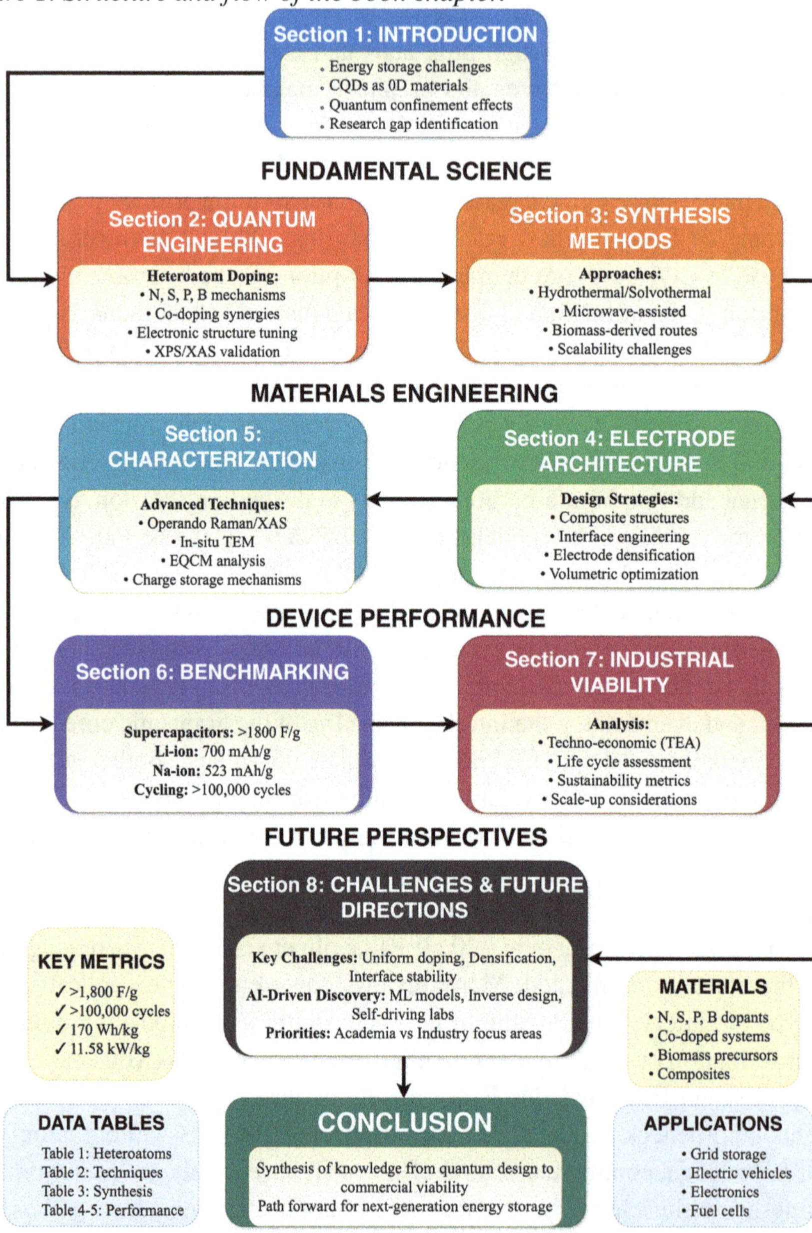

2. QUANTUM-LEVEL-ENGINEERING: THE ROLE OF HETEROATOMS

2.1. Theoretical Foundation and Mechanistic Insights

The significant improvements in the electrochemical performance by doping CQDs with heteroatoms are largely attributed to their intrinsic modification of the electronic and atomic structures (Priya et al., 2024; Pillar-Little, 2018). Theoretical investigations of such structural alterations, mostly using Density Functional Theory (DFT) and Molecular Dynamics (MD) simulations, have been conducted at the quantum level and are further supported by experimental results (Arooj et al., 2022; J. Feng et al., 2018).

a. **Nitrogen (*N*) doping:** Nitrogen is the most studied dopant, the effect of which is rationalized by the higher electronegativity of *N* relative to *C*. The substitution of a lattice carbon atom with a nitrogen atom results in a net positive charge on the neighbouring carbon atoms. The resulting charge-polarized centres provide strong sites for the electrophilic adsorption of electrolyte ions and are active sites for faradaic redox reactions, thus contributing to the pseudo-capacitance (Permatasari et al., 2021; Shi et al., 2017; Wiggins-Camacho & Stevenson, 2009). The bonding configuration of nitrogen atoms is crucial. XPS reveals three main configurations: *pyridinic-N* (*N* on the edge of a graphene plane, bonded to two C atoms), *pyrrolic-N* (*N* in a five-membered ring), and *graphitic-N* or *quaternary-N* (*N* substituting for a *C* atom in the basal plane). The various configurations differentially affect the electronic density of states (DOS) near the Fermi level (Y.-F. Lu et al., 2013; Nguyen et al., 2024; Shi et al., 2017; Wiggins-Camacho & Stevenson, 2009). Pyridinic and pyrrolic nitrogen are primarily responsible for the creation of active sites for pseudo capacitance and enhancement of electrochemical performance, while graphitic nitrogen increases the *n-type* carrier concentration, thus increasing the electrical conductivity (Y.-F. Lu et al., 2013; S. Wu et al., 2023). This observation highlights that the type of doping, not merely the overall doping concentration, is an important design variable for tuning the electrochemical response of the material.
b. **Sulphur (*S*) doping:** Sulphur has a larger atomic radius and lower electronegativity than carbon; therefore, when it is doped into carbon materials, it causes significant structural perturbations (Kadian et al., 2019). *S* doping generates lattice defects and expands the interlayer spacing between the graphenic layers of the CQD core (Kadian et al., 2019). This expanded lattice structure is particularly advantageous for battery applications, as it creates wider channels for the diffusion and intercalation of larger ions such as Na^+ and K^+, and as such,

S doping has been the subject of numerous studies on sodium and potassium batteries (Xie et al., 2023). From an electronic perspective, sulphur doping (usually a stable thiophenic *C-S-C* configuration) alters the spin density and charge distribution of the carbon lattice, creating additional active sites for redox reactions (Hemmati et al., 2023). Sulphur can also offer emissive trap states for the relaxation of photoexcited electrons, as demonstrated in studies where this approach led to improved absorbance and photoluminescence intensity in the long-wavelength regime (Kwon et al., 2013). The turn-on voltage of *S*-doped CQD field-effect transistors decreases with increasing sulphur concentration, which is in agreement with the observed optical changes (Kwon et al., 2013).

c. **Phosphorus (*P*) Doping:** Phosphorus doping introduces n-type behaviour by donating electrons to the carbon matrix and raising the Fermi level (Yashwanth et al., 2020). The modification of the band structure due to phosphorus doping can lead to increased electrical conductivity of the material (Barman et al., 2014). The *P-C* and *P-O* bonds formed on the CQD surface establish active sites that enable pseudocapacitive charge storage. Furthermore, phosphorus doping can enhance the surface wettability and aqueous electrolyte affinity, which, in turn, reduces the interfacial resistance and enables faster ion transport (S. Yang et al., 2024; Y. S. Yun et al., 2015). Phosphorus doping not only increases photocatalytic activity but also changes electronic properties. Some P-GQDs exhibit negligible fluorescence signals, which have been ascribed to P atoms creating defects in the GQD that act as large fluorescent quenching centers (Y. Wang et al., 2019; Yashwanth et al., 2020).

d. **Boron (*B*) doping:** Unlike *N* and *P*, boron is an electron-deficient *p-type* dopant that, when doped into the carbon lattice, gives rise to electron holes that can be effective as both charge carriers and active sites for desired electrochemical reactions (Boroznin, 2022; H. Hao et al., 2024). DFT simulations of *B*-doped CQDs suggest that this type of doping can increase the electrical conductivity by narrowing the energy gap of the material (Abdelsalam et al., 2025; Jana et al., 2017; Sarath Kumar et al., 2024). For example, triangular *B*-doped CQDs were predicted to have a smaller energy gap than non-doped CQDs (Sarath Kumar et al., 2024). This behaviour can lead to more favourable charge transport and higher electrochemical activity, thus leading to the potential application of boron-doped CQDs in biosensors and solar cells.

2.2. The Synergistic Advantage of Co-Doping

Heteroatom co-doping is a sophisticated technique that can be applied to CQDs. In co-doping, more than one dopant element is introduced, leading to a synergy that can optimize a wider range of material properties than that achievable through mono-element doping. In principle, this can be used to simultaneously improve several of CQDs' intrinsic properties of CQDs, broadening the scope of potential applications (Du & Guo, 2016; M. Wu et al., 2023).

a. **Nitrogen and Sulphur (*N,S*) Co-doping:** This combination has been shown to be particularly effective because of the different, yet mutually complementary benefits that the two dopants can provide (M. Wu et al., 2023; Xu et al., 2015). As mentioned earlier, nitrogen has a higher electronegativity than carbon; thus, this dopant is largely responsible for the increased number of electrochemically active sites within the CQD lattice to host pseudocapacitive faradaic redox reactions. Simultaneously, because sulphur has a larger atomic radius than carbon, it can introduce more "perturbation" to the lattice, resulting in the formation of lattice defects and an increase in the interlayer spacing between the graphenic layers of the CQD core (Xu et al., 2015). The enlarged interlayer spacing and voids in the lattice are ideal for electrochemical applications because they provide more room for the easier and rapid diffusion and intercalation of larger ions to provide faster ion transport kinetics.
b. **Nitrogen and Sulphur (*N,S*) Co-doping:** The synergetic effects of *N* and *S* co-doping give rise to CQD materials with a high number of active sites for charge storage as well as high kinetic pathways for ions to reach the active sites. Consequently, many CQDs doped with these two heteroatoms exhibit exceptional electrochemical performance, PLQYs, and stability (Somaraj et al., 2022; M. Wu et al., 2023; Xu et al., 2015). For example, it has been shown that *N,S* co-doped CQDs exhibit great stability, excitation independent emission properties, and PLQYs that can be enhanced from 11.2% to 65.1% just by the addition of a sulphur dopant (M. Wu et al., 2023).
c. **Nitrogen and Phosphorus (*N,P*) Co-doping:** This co-doping is relatively useful for aqueous-based applications, where the *N,P* co-doped CQDs are found to endow better aqueous dispersity, stability, and electrolyte tolerance (Song et al., 2019; Yashwanth et al., 2020). In this case, N doping, apart from improving the electrical conductivity and offering pseudocapacitive sites, P doping has been reported to enhance surface wettability and aqueous electrolyte affinity (S. Liu et al., 2019; Song et al., 2019). The presence of *P-C* and *P-O* groups on the CQD surface can endow the material with more active sites for pseudocapacitive charge storage and increase the surface wettability by several orders of magnitude (S. Liu et al., 2019). The incorporation of *N*- and *P*-co-doped CQDs into a composite has been shown to increase visible-light absorption, decrease the optical bandgap, and enhance photocatalytic activity and stability in water treatment processes (Song et al., 2019). The better dispersibility and stability

of the *N*- and *P*-co-doped CQDs also make them a good choice for bioimaging and sensing applications in biological media. (Song et al., 2019).

d. **Boron and Nitrogen (*B*,*N*) Co-doping:** This combination has emerged as one of the most effective *p-n* co-doping methods, offering a strategy to re-engineer the electronic structure of carbon materials (H. Liu et al., 2021). The coexistence of electron-donating nitrogen and electron-accepting boron atoms can induce strong internal polarization within the carbon matrix (H. Liu et al., 2021). This results in a high density of active *N-B* bonds with strong adsorption capability and catalytic activity, as well as a dramatic narrowing of the energy gap of the material (H. Liu et al., 2021; Zhang et al., 2022). For example, theoretical calculations on quantum-dot-based carbon nanotubes have demonstrated that the energy gap can be reduced from $2.49eV$ to as low as $0.4eV$, corresponding to a few orders of magnitude enhancement in electrical conductivity when *B*,*N* co-doping was applied (Abdelsalam et al., 2025). In carbon nanomaterials, *B*,*N* co-doping has been shown to reduce the energy gap, leading to a profound increase in electrical conductivity and improved charge-transfer capability (H. Liu et al., 2021). This unique electronic structure tuning ability has made *B*,*N* co-doped CQDs an effective metal-free electrocatalyst for the oxygen reduction reaction (ORR) and a promising candidate for electrochemical energy storage devices, including *Zn*-air batteries and vanadium redox flow batteries. In addition, *B*,*N* co-doped CQDs have been reported to have excellent stability in aqueous solutions with high ionic strength and good biocompatibility, making them promising candidates for metal ion sensing and bioimaging applications (Y. Guo et al., 2017; H. Liu et al., 2021).

Investigations of multi-heteroatom systems, such as *N*, *S*, *P*, and *B*, are another exciting area in CQD engineering (L. Hu et al., 2016; S. Lu et al., 2024). Interactions between different dopants can lead to an even higher density and variety of active sites, although controlling the synthesis to realize specific configurations is further complicated (Barman et al., 2014; L. Hu et al., 2016). The parameter space with multiple dopant types, concentrations, and arrangements is too large to be covered by brute-force experimental approaches. For this reason, this field is a particularly natural candidate for high-throughput screening via computational means and Artificial Intelligence (AI)-enabled materials discovery pipelines, which we will discuss in Section 8. The selection of a heteroatom is one of the most important design parameters, as it ultimately determines the resulting electrochemical performance of the material. Table 1 summarizes the major physicochemical properties of common single and co-dopants and their resulting effects on CQD-based materials.

Table 1. Comparative Summary of Key Physicochemical Properties and Performance Effects of Heteroatoms in CQDs

Heteroatom	Electro-negativity	Atomic Radius (pm)	Bonding Configuration	Primary Mechanism	Performance Enhancement	Typical Doping Level (%)	Ref.
Nitrogen (N)	3.04	65	*Pyridinic-N, Pyrrolic-N, Graphitic-N*	Lewis basic sites, electron donation	Improved ion adsorption, conductivity	5-15	(Ilnicka et al., 2021; T. Liu et al., 2017)
Sulfur (S)	2.58	100	*Thiophenic-S, Sulfide-S*	Lattice distortion, electron-rich regions	Expanded interlayer spacing, charge polarization	3-10	(Ghosh et al., 2025; Uppugalla et al., 2023)
Phosphorus (P)	2.19	100	*Phosphate-P, Phosphonate-P*	Large atomic size, enhanced conductivity	Enhanced rate capability, structural stability	2-8	(Rhili et al., 2023; Xu et al., 2022)
Boron (B)	2.04	85	BC_3, BC_2O	Electron deficiency, p-type doping	Increased capacitance, hole conduction	1-5	(Koroglu et al., 2025)
Oxygen (O)	3.44	60	*C-O, C=O, O-C=O*	Surface functionalization, wettability	Better electrolyte interaction, stability	10-25	(Q. Liu et al., 2023; Sousa et al., 2021)
N,S co-doping	3.04/2.58	65/100	Synergistic *N-S* interactions	Complementary electronic effects	Superior energy density, cycling life	N: 8-12, S: 4-8	(Arora et al., 2024; Uppugalla et al., 2023)
N,P co-doping	3.04/2.19	65/100	Enhanced electron density	Enhanced pseudo-capacitance	High power density, rate performance	N: 6-10, P: 3-6	(Rhili et al., 2023; Xu et al., 2022)
B,N co-doping	2.04/3.04	85/65	Balanced charge distribution	Improved cycling stability	Excellent stability, high capacitance	B: 2-4, N: 6-10	(Abdelsalam et al., 2025)
N,P,O tri-doping	3.04/2.19/3.44	65/100/60	Multiple active sites	Multiple charge storage mechanisms	Balanced energy-power performance	N: 5-8, P: 2-4, O: 15-20	(Xu et al., 2022)
B,S co-doping	2.04/2.58	85/100	Enhanced polarization	Enhanced ion accessibility	High energy density, good stability	B: 2-3, S: 4-6	(Suman et al., 2025)

2.3. Spectroscopic and Electrochemical Validation

These theoretical doping predictions can be confirmed through various characterization techniques, both modelling and experimental . X-ray Photoelectron Spectroscopy (XPS) is typically used to verify successful doping and surface modifications (Susi et al., 2015). High-resolution XPS scans can differentiate between elements, and further deconvolution can accurately assign and quantify specific bond configurations (such as *pyridinic-N* vs. *graphitic-N* by their characteristic binding energies) (Hemmati et al., 2023; Susi et al., 2015). This allows for confirmation that the synthesis method can create the desired chemical environment of the dopant of interest. This is important for compositional characterization and designing the material chemistry to have the desired dopants. Other techniques, such as X-ray Absorption Spectroscopy (XAS), including X-ray absorption near-edge structure (XANES), can provide further information about the local coordination environment and oxidation state of the dopant atoms within the carbon matrix and are also directly related to the properties and reactivity of the material (Cui et al., 2018; Sainio et al., 2021). The spectroscopic fingerprints of specific bonding environments and electronic changes can then be directly correlated to the electrochemical performance of the material. By linking these to the macroscopic functional behaviour of the material, a clear relationship between the atomic-scale structure and the observed electrochemical activity can be established. For instance, the presence of distinct redox peaks in the CV curves or a significant increase in the integrated area of the CV loop for a doped CQD electrode provides direct functional evidence that the heteroatoms participate in faradaic reactions and contribute to the overall charge storage capacity (Kannan et al., 2024).

The validation of the structure and function of the material is based on cross-checking multiple analysis methods, each of which does not provide a full picture but instead adds another piece of the puzzle. Although only some of the most common approaches, such as XPS and CV, have been used to illustrate a certain point in this chapter, in reality, many more techniques are applied to gain a holistic understanding of the physicochemical and electrochemical properties of the material. Table 2 summarizes the major characterization techniques applied for the characterization of heteroatom-doped CQDs and serves as a useful reference with a more detailed description of the type of information, ranging from nanoscale morphology, atomic structure, and optical and electrochemical properties, provided by the specific methods, their advantages, and limitations in this area of research.

Table 2. Characterization techniques for carbon quantum dots.

Characterization Technique	Information Obtained	Typical Range/ Values	Key Parameters for CQDs	Advantages	Limitations
Transmission Electron Microscopy (TEM)	Size, morphology, crystal structure	2-10 nm diameter	Particle size distribution, aggregation	Direct visualization, high resolution	Sample preparation artifacts
High-Resolution TEM (HRTEM)	Atomic-scale structure, lattice spacing	0.2-0.4 nm lattice spacing	Crystallinity, defect density	Atomic-level structural information	Limited field of view, beam damage
X-ray Photoelectron Spectroscopy (XPS)	Surface composition, bonding states	C1s: 284.8 eV, N1s: 398-402 eV	Dopant concentration, binding states	Quantitative elemental analysis	Surface sensitive, vacuum required
Fourier Transform Infrared (FTIR)	Functional groups, chemical bonds	1000-4000 cm^{-1}	Surface functionalization degree	Non-destructive, functional group ID	Limited to specific functional groups
Raman Spectroscopy	Graphitization degree, defects	D-band: ~1350 cm^{-1}, G-band: ~1580 cm^{-1}	ID/IG ratio (defect level)	Quick, non-destructive analysis	Laser heating effects possible
X-ray Diffraction (XRD)	Crystal structure, phase purity	2θ: 20-30° (broad peak)	Crystallite size, structural disorder	Bulk crystalline structure	Poor sensitivity to amorphous phases
UV-Vis Absorption Spectroscopy	Optical bandgap, electronic transitions	200-800 nm absorption	Bandgap energy, optical properties	Simple, quantitative optical properties	Limited structural information
Photoluminescence (PL) Spectroscopy	Quantum yield, emission properties	400-700 nm emission	Quantum yield, fluorescence lifetime	Sensitive to quantum effects	Environmental sensitivity
Nuclear Magnetic Resonance (NMR)	Molecular structure, carbon framework	^{13}C NMR: 100-200 ppm	Carbon environment, heteroatom bonding	Detailed molecular structure	Expensive, time-consuming
Atomic Force Microscopy (AFM)	Surface topography, mechanical properties	0.5-2 nm height	Surface roughness, layer thickness	Surface analysis, 3D imaging	Limited to surface analysis
Electrochemical Impedance Spectroscopy (EIS)	Charge transfer resistance, ion diffusion	R_{ct}: 0.1-10 Ω	Electrochemical activity, conductivity	In-situ electrochemical analysis	Requires conductive samples
Cyclic Voltammetry (CV)	Redox behaviour, capacitive properties	Scan rate: 5-100 mV/s	Capacitive vs. faradaic contribution	Real-time electrochemical behaviour	Limited potential window

continued on following page

Table 2. Continued

Characterization Technique	Information Obtained	Typical Range/ Values	Key Parameters for CQDs	Advantages	Limitations
Brunauer-Emmett-Teller (BET) Analysis	Surface area, pore structure	50-800 m^2/g	Accessible surface area, porosity	Quantitative surface area	Sample degassing required
Thermogravimetric Analysis (TGA)	Thermal stability, decomposition	Decomposition: 200-600°C	Thermal decomposition temperature	Thermal stability assessment	Destructive analysis

3. PRINCIPLES AND PRACTICES FOR SCALABLE AND SUSTAINABLE SYNTHESIS

3.1. Bottom-Up Synthesis Methodologies

The translation of heteroatom-doped CQDs from laboratory curiosities to commercially viable materials is highly dependent on the development of scalable, cost-effective, and sustainable synthesis methods for them. Bottom-up approaches, which build CQDs from molecular precursors, are generally preferred over top-down methods because of the greater control over the size, composition, and surface chemistry that they offer. A critical assessment of the most popular methods is presented below, highlighting the key trade-offs between synthesis speed, cost, energy efficiency, and product quality. The following synthesis routes are the most representative, industrially relevant, and widely adopted approaches in heteroatom-doped CQD research:

1. **Hydrothermal/Solvothermal Synthesis:** This is the most reported method in the literature owing to its operational simplicity and compatibility with water as an environment-friendly solvent (Cai et al., 2025). Carbon precursors and dopant sources are placed in an autoclave and heated to 150–250 at a constant temperature for several hours (Panigrahi, 2025). The main disadvantages of this approach are the long reaction time (generally 6–24*h*) and high energy consumption, which make this synthesis incompatible with large-scale and continuous manufacturing and are responsible for the high cost of operations (Baragau et al., n.d.).
2. **Microwave-Assisted Synthesis:** One of the most improved routes in terms of efficiency is the microwave-assisted method. The reaction time can be significantly reduced from hours to minutes using microwave irradiation as the heating

method (Cai et al., 2025). Owing to the volumetric and uniform heating nature of microwaves, more homogeneous nucleation and growth occur, resulting in CQDs with a narrower particle size distribution (Huang & Ren, 2025). The rapid synthesis, high energy efficiency, and feasibility of scaling up to continuous-flow reactors make microwave-assisted synthesis an exceptionally promising route for industrial-scale production (Cai et al., 2025).

3. **Pyrolysis and Chemical Oxidation:** Solid-phase methods include thermal decomposition (pyrolysis) or chemical oxidation of solid precursors. A significant advantage of these methods is that they can be performed without the use of solvents. This can be beneficial because costly and time-consuming downstream purification and drying steps are unnecessary (Cai et al., 2025). The disadvantage of pyrolysis is that it typically requires very high temperatures (>350), which can make the method energy-intensive (Huang & Ren, 2025). However, for some precursors, these methods can be adapted to continuous flow methods for large-scale ($>1g$) synthesis with high yield.

As can be seen, the above methods are the most commonly reported. However, the field of CQD synthesis is rich and varied, with a broad spectrum of techniques developed for specific precursors and targeted material properties. The main features, pros, and cons of all the main routes are summarized in Table 3 for easier access by users. The diversity of CQD synthesis methods is driven by the need to tailor the properties of the resulting nanomaterials for specific applications. As will be shown below, CQDs can be synthesized either as general-purpose nanoparticles with basic optical and electronic properties or with a high degree of functionalization to create materials for specific applications. Figure 2 shows the two main approach in preparing carbon quantum dots.

Figure 2. Approaches for Carbon Quantum Dot Synthesis.

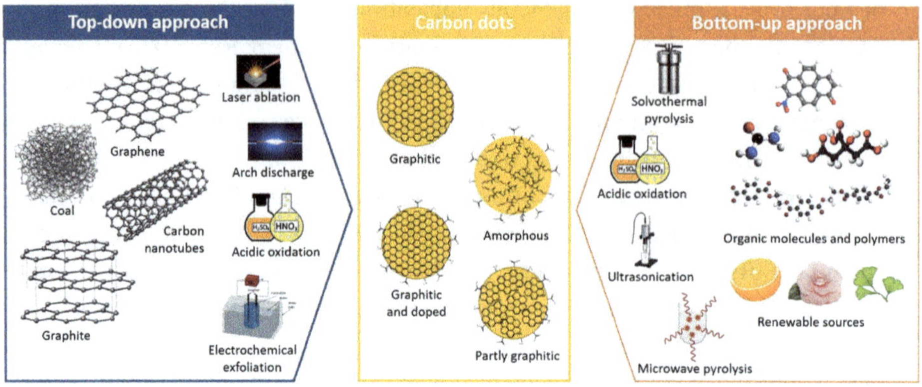

Table 3. Synthesis Methods for Heteroatom-Doped Carbon Quantum Dots

Synthesis Method	Temperature Range (°C)	Reaction Time	Product Yield (%)	Quantum Yield (%)	Key Advantages	Main Limitations	Typical Applications	Ref.
Hydrothermal	160-200	2-24 hours	20-40	10-40	Simple, scalable, uniform size	Long reaction time, energy intensive	General purpose CQDs	(Sousa et al., 2021; Y. Wang & Hu, 2014)
Microwave-assisted	120-180	5-30 minutes	15-30	20-65	Fast synthesis, high QY	Limited batch size, equipment cost	High-QY fluorescent CQDs	(T. Liu et al., 2017)
Thermal pyrolysis	400-800	2-6 hours	10-25	5-25	High doping levels, good crystallinity	High temperature, energy consumption	Heavily doped CQDs	(Elugoke et al., 2024; Y. Wang & Hu, 2014)
Electrochemical synthesis	Room temperature	2-12 hours	5-20	15-35	Mild conditions, controllable	Low yield, long reaction time	Functionalized CQDs	(Rocco et al., 2023)
Biomass-derived hydrothermal	160-180	6-12 hours	30-77	15-60	Green, sustainable, high yield	Variable precursor composition	Green energy storage CQDs	(Wareing et al., 2021)
Solvothermal	180-220	12-24 hours	25-45	20-50	Uniform size distribution	Complex purification, long time	Uniform size CQDs	(Sousa et al., 2021; Xu et al., 2022)
Sol-gel pyrolysis	300-600	2-4 hours	15-35	10-30	Controlled properties, versatile	Multi-step process, complexity	Specialty applications	(Ozyurt et al., 2023)

continued on following page

Table 3. Continued

Synthesis Method	Temperature Range (°C)	Reaction Time	Product Yield (%)	Quantum Yield (%)	Key Advantages	Main Limitations	Typical Applications	Ref.
Chemical oxidation	Room temperature	6-24 hours	10-30	5-20	Simple equipment, mild conditions	Harsh chemicals, safety concerns	Simple doped CQDs	(Al-Gaashani et al., 2019)
Alkaline-peroxide treatment	80-120	2-4 hours	60-77	25-45	Highest yield, minimal waste	Requires optimization for each precursor	High-yield biomass CQDs	(Fernandes et al., 2021; Jing et al., 2019)
One-pot synthesis	150-200	4-12 hours	25-50	20-55	Versatile, multiple dopants	Difficult to control uniformity	Multi-functional CQDs	(Suzuki et al., 2021)

3.2. The Green Chemistry Imperative- Biomass-Derived CQDs

A major driving force for sustainable practices is the use of waste biomass as the main carbon source for CQDs (Cai et al., 2025). This is consistent with the circular economy approach, which valorises feedstocks that are low-cost, abundant, and renewable, and, most importantly, otherwise discarded as waste products. A very broad range of biomass sources can be used, including agricultural residues (e.g., rice husk, sugarcane bagasse, etc.), food wastes (e.g., fruit peels, etc.), and biopolymers (cellulose, lignin, etc.) (Shah, 2024). Another, even more compelling, benefit of many biomass precursors is their high content of naturally occurring heteroatoms. For example, plant-based precursors often have high nitrogen and sulphur contents, and other sources (phytic acid) have a significant phosphorus content (Cai et al., 2025). The inherent heteroatoms can be covalently incorporated into the CQD lattice during hydrothermal carbonization/pyrolysis in a "self-doping" manner (Bao et al., 2018). This can obviate the need for synthetic, separate, and often expensive dopant precursors, further simplifying the synthesis and increasing the green factor. At the same time, an important caveat to be considered is the inevitable trade-off between the greenness of the precursor and the process in general. As noted above, the inherent heterogeneity/impurity of waste biomass can require harsh, energy-intensive, and chemically aggressive purification. In other words, it is not as straightforward as "biomass equals green," and the entire process chain must be considered in the end in a more holistic manner, for example, through life cycle assessment, as discussed in Section 7.

3.3. Controlling Dopant Incorporation- Uniformity Challenge

A major challenge in the heteroatom doping of CQDs is achieving a uniform and controllable doping process, particularly in co-doped systems (Kou et al., 2020). This means that the dopant concentration and configuration are the same from particle to particle and throughout a single particle. The doping process is considered non-uniform when the dopant distribution is uneven. The problem with non-uniform doping is that the electrochemical performance is erratic and difficult to predict, making the material unreliable and unsuitable for commercial devices (Sohal et al., 2021). The performance of a device is only as good as its weakest point. If there are regions in a device that are less well doped, those spots effectively have a higher electrical resistance or electrochemical instability. This could lead to localized "hot spots" of degradation that propagate through the electrode and ultimately result in a much shorter device lifetime (Porada et al., 2015).

Many strategies are being investigated to meet this challenge. One method involves the use of single-source precursors, which are molecules that already contain both carbon and the desired heteroatoms in their molecular structure, guaranteeing intimate and uniform mixing at the molecular level (Yin et al., 2025). Another method is post-synthesis treatment, where the as-prepared CQDs are annealed in a reactive atmosphere (such as ammonia for N-doping), which can drive dopant incorporation into the CQD lattice (Herold et al., 2025). Template-assisted synthesis can also impart the formation of CQDs with a more ordered structure and more controlled dopant locations (Rasal et al., 2021). These challenges are compounded in co-doped systems, where the reaction kinetics and incorporation efficiencies of the different heteroatoms can be vastly different, making it difficult to achieve a desired stoichiometric ratio and uniform distribution of all dopants simultaneously (Herold et al., 2025). Addressing this challenge is a major frontier for future synthesis.

3.4. Post-Synthesis Purification and Characterization

The raw CQD solutions in the post-synthesis process are always accompanied by unreacted precursors, molecular fluorophores, and amorphous carbon byproducts. These impurities are detrimental to the electrochemical performance and are removed through a series of purification steps before use. The most common approaches include dialysis, column chromatography, and centrifugation (Huang & Ren, 2025). However, these laboratory-scale techniques are generally time-consuming, expensive, and difficult to scale, which has long been a serious throughput bottleneck in practical production processes (Huang & Ren, 2025). In addition, a set of standard and uniform characterization methods is necessary for quality control and to relate the synthesis process to the material characteristics. HRTEM was used to determine the size, morphology, and crystallinity of the synthesized CQDs (Chakrabarti et al., 2025). Raman spectroscopy measures the degree of graphitization and defect density in carbon materials by calculating the intensity ratio of the D-band (disordered carbon) to the G-band (graphitic carbon) (Sohal et al., 2021). FTIR spectroscopy is another effective method for identifying the various functional groups on the CQD surface (Sohal et al., 2021). As mentioned above, XPS can further confirm the elemental composition and, more importantly, the chemical bonding states of the introduced heteroatoms (Sohal et al., 2021).

4. ELECTRODE ARCHITECTURE AND INTERFACIAL ENGINEERING

4.1. Beyond Bare CQDs- The Role of Composite Structures

Although individual heteroatom-doped CQDs exhibit remarkable intrinsic properties, further optimization of the structural design is crucial to achieve high-performance macroscopic electrodes based on their extraordinary single-particle performance (Kou et al., 2020). As-prepared bare CQDs usually readily agglomerate into large primary particles via strong van der Waals forces after electrode processing, which not only results in a dramatic decrease in the accessible surface area but also blocks ion transport channels (Y. Hu et al., 2018). Therefore, CQDs are usually combined with other structures to form composites, as described below, instead of simply adding CQDs as an active component. In the composite electrode structure, CQDs are not only the active materials but also serve as multifunctional building units, that is, conductive spacers and morphology regulators.

a) **Graphene and MXene Hybrids:** Graphene and MXenes are ideal 2D conductive scaffolds for supporting CQDs (Das et al., 2022). They offer a high surface area and in-plane conductivity for an efficient electron transport network across the electrode (Anil Kumar et al., 2023). By anchoring CQDs to the surface of 2D sheets, the sheets prevent the aggregation of CQDs, while the CQDs serve as nanoscale spacers to separate adjacent graphene or MXene layers, thus maintaining the interlayer channels that permit rapid ion diffusion of electrolyte ions into the electrode (D. Yang et al., 2025). MXenes are a family of 2D transition metal carbides and nitrides that are particularly promising supports for CQDs because of their metallic conductivity (typically superior to graphene), hydrophilic surfaces that are highly wettable, and pseudocapacitive nature (Nahirniak et al., 2023). Therefore, the hybrid architecture provides a synergistic combination of the high conductivity of the 2D scaffold and the high number of active sites of the 0D CQDs. For example, graphene quantum dot-three-dimensional graphene (GQD-3DG) composites exhibited a high specific capacitance of $268F/g$, which is over 90% higher than that of pure 3DG electrodes (Q. Chen et al., 2014).

b) **Conductive Polymer Composites:** Conducting polymers, such as polyaniline (PANI) and polypyrrole (PPy), can exhibit high theoretical pseudo-capacitance but are limited in practice by low electrical conductivity and large volume swelling and shrinking during charge-discharge cycles, which causes poor mechanical stability and rapid performance decay (Das et al., 2022). This can be remedied by incorporating CQDs into a polymer-host matrix. CQDs form an interconnected conductive network within the polymer, thereby increasing charge transport (Y. Hu et al., 2018). In addition, the strong carbon network of CQDs can provide mechanical support to the polymer, offsetting its volumetric expansion and contraction and significantly improving its cycling stability (Sead et al., 2025). This resulted in a strong synergy between the faradaic storage of the polymer and the stable conductive framework of the CQDs. For instance, graphene quantum dot-doped polyaniline composites exhibited an excellent specific capacitance of ~$1044F/g$ at $1A/g$ and moderate cyclic stability with 80.1% retention after 3000 cycles (Mondal et al., 2015).

c) **Transition-Metal Oxide/Sulfide (TMO/TMS) Composites:** TMOs and TMSs, such as cobalt oxide (Co3O4), nickel sulphide (NiS), and manganese dioxide (MnO_2), have extremely high theoretical specific capacities owing to their rich redox chemistry (Shanmugasundaram et al., 2024). However, they are intrinsically poor conductors and experience drastic mechanical degradation owing to the large volume changes accompanying the insertion/extraction of ions (Z. Tu et al., 2016). A simple yet effective strategy to address these issues is to disperse TMO/TMS nanoparticles on a conductive CQD matrix. The CQD network can ensure fast electron transport to the insulating TMO/TMS particles, while the porous and flexible carbon matrix can offer void space to buffer volume expansion and suppress the pulverization of the electrode (Shanmugasundaram et al., 2024). For example, a Co_3O_4–CQD composite exhibited a specific capacitance of $1209F/g$, which was significantly higher than that of either pristine Co_3O_4 or CQDs, demonstrating the strong synergistic effects at the nanocomposite interface (Shanmugasundaram et al., 2024).

4.2. Engineering the Electrode-Electrolyte Interface

The interface between the electrode and electrolyte in an electrochemical device often plays an essential role in the stability and performance of the device (E. Wang et al., 2022). The rich surface functional groups on CQDs, including hydroxyl ($-OH$) and carboxyl ($-COOH$) groups, enable effective interfacial engineering (E. Wang et al., 2022). The presence of these polar surface functional groups renders the electrode surface hydrophilic, significantly enhancing the wettability of the electrode by aqueous and polar organic electrolytes. This improves the intimate contact between the electrode and electrolyte and reduces the interfacial resistance, thus enabling fast transport of ions from the electrolyte to the active sites of the electrode (Sead et al., 2025). The interface is even more important in battery applications, where reactive metal anodes such as lithium, sodium, and zinc tend to react with electrolytes to form a passivation layer called the solid electrolyte interphase (SEI) on the anode surface after the first cycle (E. Wang et al., 2022). CQD-based electrodes can be engineered to form thinner, more uniform, and more chemically stable SEI (Sead et al., 2025). A stable SEI is critical for batteries because it prevents the continuous decomposition of the electrolyte while allowing the transport of ions through the SEI layer (E. Wang et al., 2022). In addition, by promoting more uniform ion deposition, CQDs may help suppress the growth of metallic dendrites, that is, needle-like structures that pierce the separator and cause short circuits, resulting in battery failure (E. Wang et al., 2022). Therefore, the enhancement of interfacial stability may be a major factor contributing to the enhanced safety and long-term cycling performance of CQD-based batteries (Brown & Li, 2024).

4.3. The Volumetric Challenge – Electrode Densification Strategies

One crucial aspect that is underappreciated in the literature is the comparison of gravimetric (e.g., *F/g*) vs. volumetric (e.g., F/cm^3) performance. Although many new nanomaterials, including CQD-based aerogels, exhibit impressive gravimetric performance, they are characterized by extremely low packing densities (Ma et al., 2021). In space-constrained applications, which many energy storage applications are ultimately intended for (e.g., portable electronics and electric vehicles), volumetric energy and power density metrics are the most important. A material that achieves record-high gravimetric capacitance is not of much commercial interest if it is too "fluffy" that the resulting device has poor volumetric performance and is unacceptably bulky. Electrode densification is not only a nice incremental step for the field but also an essential enabling step. Several advanced strategies for increasing the packing density of carbon-based electrodes without degrading their electrochemical performance have been developed. These approaches seek to develop compact electrodes that retain good ion transport pathways and a high active surface area for charge storage (E. Wang et al., 2022).

a) **Physical Compression:** The most intuitive method is mechanical pressing or hot-pressing to physically compress the electrode and reduce the void volume between carbon particles to form a denser electrode structure (Z. Guo et al., 2025). Physical compression densifies the electrode material, which can improve its adhesion, electrical conductivity, and mechanical robustness. In particular, hot-pressing can lead to enhanced contact between the electrode components, reducing the interfacial resistance and improving the electrochemical performance. However, the conditions of physical compression, such as pressure and temperature, need to be optimized. Excessive force can deform the active material and affect the pore structure, which can also influence the performance (Z. Guo et al., 2025).

b) **Electrostatic Densification:** A more advanced approach is to utilize electrostatic attraction to direct the organization of particles in an electrode into a more compact structure (C. Wu et al., 2022). For example, the assembly of negatively charged CQDs or graphene quantum dots with positively charged binders, such as chitosan, has demonstrated a much higher packing density than similar materials without directed assembly (Y. Jiang et al., 2021). Chitosan is a positively charged polysaccharide that, in addition to acting as a binder to fill the voids between particles, helps suppress the electrostatic repulsion of the particles, leading to a more uniform and dense electrode structure (Y. Jiang et al., 2021). This method can thus lead to the formation of highly dense electrodes at the molecular level, helping to improve volumetric performance

c) **Template-Assisted Assembly:** This process employs templates to direct the growth or assembly of carbon materials into dense, ordered, and porous architectures (Poolakkandy & Menamparambath, 2020). Template-based methods provide high control over the morphology and pore structure of carbon electrode materials, leading to the synthesis of 3D architectures and hierarchical porous structures that exhibit interconnected porosity for ion transport while simultaneously achieving denser packing than conventional preparation methods. This results in a better volumetric energy/power density, which is necessary for real-world applications with spatial limitations (Pan et al., 2022; S. Wu & Zhu, 2016).

The major obstacle for any densification method is the associated trade-off with electrolyte accessibility. Excessive compaction of an electrode can collapse its porous structure and obstruct ion transport pathways. The resulting increase in diffusion resistance leads to a drastic reduction in rate capability and power density (Z. Guo et al., 2025). Therefore, an ideal electrode architecture is maximally dense and has a hierarchical pore structure with macropores that act as ion-buffering reservoirs and transport highways, and meso-/micropores that provide a high surface area to enable charge storage (Quarti et al., 2023). This balance is a key topic in current electrode engineering research.

5. PROBING CHARGE STORAGE VIA ADVANCED CHARACTERIZATION

5.1. De-Convoluting Capacitive Contributions

The improved performance of the heteroatom-doped CQD electrodes can be attributed to the synergistic effect of multiple charge storage processes. Unraveling the contribution of each mechanism is key to gaining a comprehensive understanding, which is often achieved by employing advanced electrochemical characterizations (Kou et al., 2020).

a. **Electric Double-Layer Capacitance (EDLC):** This is a non-Faradaic process based on the physical adsorption of electrolyte ions at the interface of the electrode-electrolyte, which forms an electric double layer (H. Li, 2024). The stored charges are directly proportional to the accessible surface area, which is improved by the high surface area and good electrical conductivity of CQD-based electrode materials (Abbas et al., 2019). To achieve high-performance EDLC devices, carbon electrode materials with a high specific surface area, short ion diffusion length, and excellent electrical conductivity are required (H. Li, 2024). For example, *N*-doped carbon nanosheets have been reported to show remarkable capacitive performance owing to their unique structure and attained high capacitance value (X. Hao et al., 2016). The reported specific capacitance for EDLC supercapacitors is in the range of 40-300*F/g*, which is directly associated with the accessible surface area (Dubey & Guruviah, 2019).
b. **Pseudo-capacitance:** Pseudo-capacitance is the mechanism that involves the "fast and reversible faradaic (redox) reactions" taking place at or near the electrode surface material (H. Li, 2024). The pseudo-capacitance in heteroatom-doped carbonaceous materials (including CQDs) arises from heteroatoms, which play a vital role in energy storage (Huang et al., 2017). The active sites generated by nitrogen, sulphur, and phosphorus dopants and surface functional groups (e.g., quinones) can be easily oxidized and reduced by transferring charge and storing energy chemically, which occurs at a rapid rate (Abbas et al., 2019). For instance, *P*-doping can result in high capacitance due to the pseudo-capacitance of *P=O* bonds (Park et al., 2018). Moreover, there have been reports of dual-doped materials (e.g., *N*, *B*-dual-doped CQDs) with more synergistic effects of the dopants, resulting in enhanced electrochemical properties based on pseudocapacitive and carbon-based materials (Yao et al., 2023). This battery-like mechanism enables these materials to have much higher energy densities than typical carbon-based supercapacitors.

c. **Intercalation Kinetics:** This mechanism is important in the case of battery systems and is the insertion (intercalation) and removal (de-intercalation) of ions (Li^+, Na^+, Zn^+) into the host lattice of the electrode active material. In the case of CQD-based electrodes, this can occur either within the expanded interlayer spacing of the CQD's graphenic core or within a host material (e.g., a TMO) in a composite form (S. Jin et al., 2022). As heteroatom doping can expand the d-spacing of the carbon layers with dopants of larger atomic radii (for example, *S* and *P*), this can lead to preferred channels for intercalation and hence improved kinetics of this mechanism (Z. Li et al., 2017). The larger channels would allow better diffusion of the larger Na^+ ions within the host, thus improving the kinetics of intercalation. Computational studies of *Li*, *Na*, and *K* ion kinetics in CQD-based electrodes revealed the expansion of CQD interlayer spacing by intercalated atoms, which is critical for ion transport (S. Jin et al., 2022). Overall, the current mechanistic insights into this behaviour enable new design principles for carbon anodes based on reversibly binding defects and dilated turbostratic domains.

5.2. Real-Time Insights From Operando and In-Situ Techniques

While ex-situ characterization (looking at the material before and after cycling) can be revealing, it provides only static snapshots and can overlook important dynamic processes and metastable intermediate states, which are only present under device-operating conditions. To gain a mechanistic understanding, operando and in situ techniques that probe the material as it sits in an operating electrochemical cell are required (Y. Yang et al., 2017).

a. **Operando Raman Spectroscopy:** Operando Raman spectroscopy is a powerful characterization technique used to monitor the vibrational modes of the electrode material as a function of the applied potential. In the case of carbon-based materials, the technique takes advantage of the high sensitivity of specific Raman bands to local environmental changes, and thus provides insight into the charge storage mechanism. In general, the D-band (centred around $1350cm^{-1}$) is mainly attributed to defects/disorder in carbon structure (Weaving et al., 2020). On the other hand, the G-band (centred around $1580cm^{-1}$) is related to the in-plane vibrational modes of the graphitic sp^2 carbon and is sensitive to the electronic and structural properties of graphitic lattice (Weaving et al., 2020). The insertion of ions into the carbon lattice from ion intercalation is associated with a large lattice strain, which is reflected by a large shift in the G-band position (Chacón-Torres et al., 2013; Weaving et al., 2020). The spectroscopic signature indicates ion intercalation within the storage mechanism and distinguishes true intercalation from less impactful surface adsorption via G-band shifts (Xue et al., 2022). The analysis of G-band position changes through electrochemical cycling helps to determine the voltage profile area where surface adsorption or true intercalation charge storage mechanisms control the electrochemical process.

b. **In situ/operando X-ray Absorption Spectroscopy (XAS):** In situ and operando X-ray Absorption Spectroscopy (XAS) is a powerful element-specific probe that offers valuable insights into the local atomic and electronic structures of materials under electrochemical reaction conditions (Timoshenko & Cuenya, 2020). XAS includes two complementary techniques: X-ray Absorption Near Edge Structure (XANES) and Extended X-ray Absorption Fine Structure (EXAFS) (Timoshenko & Cuenya, 2020). By tuning the X-ray energy to the absorption edge of a particular dopant element (e.g., *N*, *S*, or *P*), changes in its oxidation state and coordination environment can be monitored in real time during the charging and discharging processes (S. Lu et al., 2023). This provides unambiguous evidence of the direct participation of the dopant in redox reactions and its role as a pseudocapacitive active site in the electrode material (Lopez-Astacio et al., 2024).

c. **In situ transmission electron microscopy (TEM):** In situ Transmission Electron Microscopy (TEM) provides unique insights into the morphological and structural changes of electrode materials at the nanoscale during electrochemical cycling (Liao & Zheng, 2016). The introduction of liquid cells for TEM, which can be operated in realistic liquid electrolyte environments (unlike open-cell configurations, which typically use solid lithium oxide or ionic liquids), is a notable advancement in this area (Liao & Zheng, 2016). Liquid cells allow the direct observation of dynamic processes, such as the expansion and contraction of electrode materials during ion insertion and extraction, the formation and growth of the SEI layer in real time, and the dynamics of dendrite formation on the anode (Y. Yuan et al., 2023). This provides direct and invaluable insights into the mechanisms of performance degradation and failures (Chakrabarti et al., 2025).

d. **Electrochemical quartz crystal microbalance (EQCM):** EQCM is an extremely sensitive gravimetric technique that can measure the amount of mass gained or lost on the electrode surface with nanogram-level precision during operation (Niu et al., 2023). This is possible because of its capacity to detect extremely small changes in the resonance frequency of the quartz crystal when it is coated with the electrode material, and the variations in this frequency are directly associated with alterations in the mass on the electrode surface (Buttry, 2021). EQCM has been used as an important tool to directly monitor the flux of ions and solvent molecules in and out of porous electrode materials and to understand the interplay between mass transport and charge storage (Buttry, 2021). EQCM can be used to quantitatively study complex electrochemical processes, such as ion desolvation, where large solvent shells are shed as the ions penetrate small pores in the electrode material, and to detect parasitic reactions that result in mass gain (Levi et al., 2010).

The effectiveness of these advanced characterization techniques is multiplied substantially when the results from each method are combined. With a multimodal operando setup, spectroscopic, electronic, and gravimetric data can be cross-referenced in real time, leading to a much more comprehensive understanding of the charge storage mechanism than any isolated measurement can reveal (Levi et al., 2010). A G-band shift in operando Raman spectroscopy indicative of intercalation can, for example, be connected to the simultaneous oxidation state change of the dopant as measured by XAS, which, in turn, can be cross-checked with a gravimetric measurement of the influxed ion amount using EQCM. In this way, different mechanisms of charge storage are not only observed to take place but are found to be cooperatively linked; that is, the electronic changes introduced by the dopant allow for structural changes in ion intercalation, which together greatly enhance performance. This is especially important to consider, as it has been shown that static ex-situ characterizations can be deceiving because many structural or electronic transformations can be reversible and only exist under an applied potential (Weaving et al., 2020). This changes the entire philosophy of material design. It is no longer sufficient to engineer a material with a desirable end structure; one must also consider dynamic stability and efficiency.

6. DEVICE-LEVEL PERFORMANCE BENCHMARKING

A major concern in this field is that reports on the progress of EES materials often lack standardization. Performance metrics are often reported under varying conditions, making comparisons between one work and another (so-called "apples-to-apples" comparison) potentially erroneous and misleading. For example, it is not meaningful to report a high energy density without the corresponding power density or a high specific capacitance at an unspecified current density. In this context, to provide a fair comparison between recently reported heteroatom-doped CQD-based devices, this chapter adopts consistent data normalization to benchmark the performance of SCs and batteries (Kannan et al., 2024). The performance metrics of SCs are presented in terms of specific capacitance (*F/g*) or areal capacitance (mF/cm^2), and for batteries are specific capacity (*mAh/g*) at various *C*-rates, rate capability and long-term cycling stability (% capacity retention over a certain number of cycles).

6.1. High-Performance Supercapacitors

The judicious design of heteroatom-doped CQDs has enabled remarkable achievements in this field to date, with recent reports steadily raising the bar in terms of energy and power density in supercapacitor devices. Nitrogen doping is by far the most common tactic, as detailed in Table 4, with *N*-doped CQD composites frequently exceeding 300*F/g* in specific capacitance and often delivering excellent cycling performance. Co-doping, particularly *N* co-doping with sulphur or phosphorus, typically produces the best overall performance (Permatasari et al., 2021). A combined *N*-doping strategy, which can supply redox active sites, and *S/P*-doping, which can improve conductivity and ion transport, has led to numerous devices that couple high energy density and excellent rate capability (X. Feng et al., 2021). For example, *B,S*-CQDs were reported to have an energy density of 48.0*Wh/kg*, and NAPC-650-4, a nitrogen-doped porous carbon, preserved 87.64% of its capacitance after 10,000 cycles (X. Feng et al., 2021). Another example is shown in Figure 3, which represents the electrochemical performance of a high-performance CD-intercalated MXene film electrode (D. Yang et al., 2025).

Table 4. Electrochemical Performance of Heteroatom-Doped CQDs in Supercapacitor Applications

Electrode Material	Dopant(s)	Specific Capacitance (F/g @ A/g)	Energy Density (Wh/kg)	Power Density (W/kg)	Cycling Stability (% Retention @ Cycles)	Reference
N-CQDs/Co_3O_4 Nanocomposite	N	1867 @ 1	36.9	480	96% @ 500	(Naushad et al., 2021)
Co_3O_4-CQD Composite	None (CQD composite)	1209 @ 1	13.88	684.65	100% @ 3,000 and 96% @ 5000	(Shanmugasundaram et al., 2024)
B/N-Codoped Porous Carbon	B, N	962 @ 1	108.2	450	>120% @ 5,000	(Pal et al., 2025)
NS-CQDs/$Ti_3C_2T_X$-MXene	N, S	725.7 @ 1	170.34	1290.98	98.3% @ 10,000	(Ghosh et al., 2025)
NAPC-650-4	N	353 @ 1	7.22	125	87.65% @ 10,000	(X. Feng et al., 2021)
Amine-enriched GQDs	N	595 @ 1	21.8	250	90% @ 10,000	(Permatasari et al., 2021)
B,S-CQDs	B, S	390 @ 0.1	48.0	524	98% @ 3,000	(Suman & Bhawna, 2025)

continued on following page

Table 4. Continued

Electrode Material	Dopant(s)	Specific Capacitance (F/g @ A/g)	Energy Density (Wh/kg)	Power Density (W/kg)	Cycling Stability (% Retention @ Cycles)	Reference
N, S-PCD	N, S	300.2 @ 0.2	106.7	160	97.1% @ 100,000	(Wei et al., 2023)
N/S-KAC	N, S	253 @ 0.5	26.92	566	93.91% @ 5,000	(W. Lu et al., 2022)
N-doped HPCs	N	103.7 mAh/g @ 0.1	74.07	11580	98.84% @ 10000	(Wei et al., 2023)
N, P, O tri-doped C nanocage	N, P, O	310 @ 0.5	43	137,900	Not Specified	(Wei et al., 2023)
$g\text{-}C_3N_4$ QDs/Graphene Hydrogel	N	243.2 @ 0.2	22.5	250	89.5% @ 15,000	(Ansari, 2022)
N/P-doped Graphene	N, P	210.2 @ N/A	94.6	4,500	82% @ 15,000	(Wei et al., 2023)
S-doped CQDs	S	55.7 @ 1	15.2	700	94.5% @ 10,000	(Suman & Bhawna, 2025)
CDs-Intercalated MXene Film	None (CQD composite)	396.4 @ 1	21.2 (Wh/L)*	12,000 (W/L)*	100% @ 100,000	(D. Yang et al., 2025)
N, S-doped graphitic CQDs with sepia melanin	N, S	180 mF/cm^2 @ 10 mV/s (areal)	11 $\mu Wh/cm^2$ (areal)	102mW/cm^2	~92% @ 10000	(Gouda et al., 2021)

Figure 3. Electrochemical performance of the CDs-MF and p-MF electrodes.

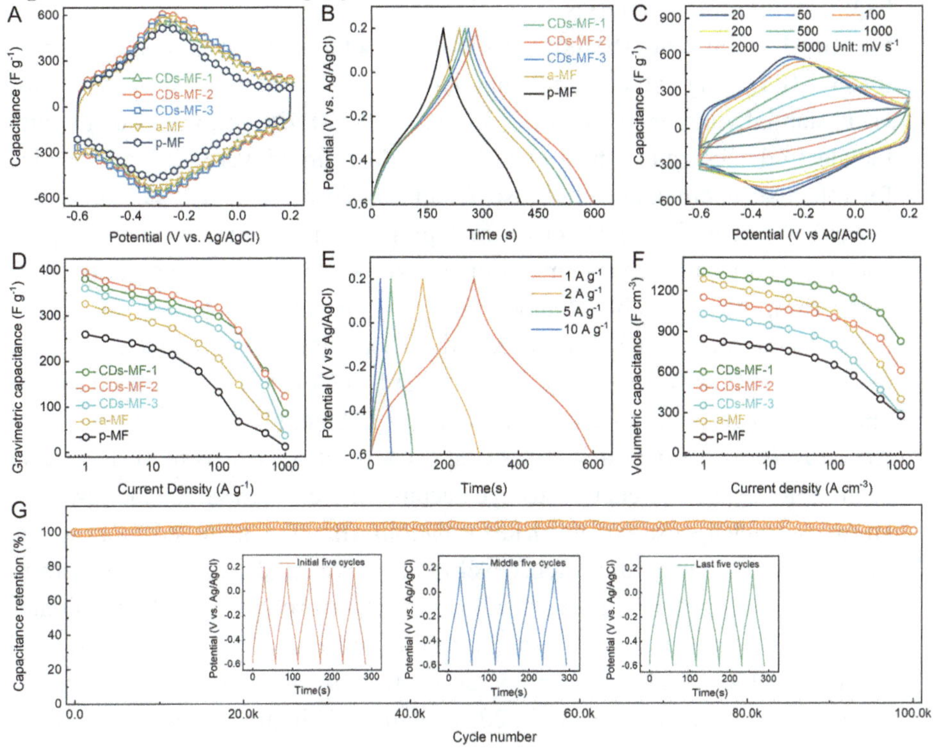

Reprinted with permission from (D. Yang et al., 2025), CC BY 4.0.

6.2. Next-Generation Batteries

Heteroatom-doped CQDs have been successfully applied as anode or cathode materials in battery systems and as conductive additives to enhance the performance of other active materials. The functions of CQDs in batteries include increasing the conductivity, buffering large volume changes of the host structure during ion intercalation, and providing more active sites for charge storage. Table 5 presents the performance characteristics of CQD-based electrodes in different battery systems.

a. **Lithium-Ion Batteries (LIBs):** As the current market leader, efforts are being made to improve the capacity and cycle life of LIBs. In the anode, CQD composites with high-capacity hosts, such as silicon or metal oxides, have the potential to form stable anodes with resistance to large strains induced during lithiation/delithiation (Gurushankar et al., 2025; Bie et al., 2023). Doping with heteroatoms can also improve the conductivity and lithiation of carbon matrices

(X. Jiang et al., 2024). Table 5 presents the performance of heteroatom-doped CQD in lithium-ion batteries.

b. **Sodium-Ion Batteries (SIBs):** SIBs have the potential to be a low-cost alternative to LIBs owing to the natural abundance of sodium. The ionic radius of Na^+ is substantially larger (0.102*nm*) than that of Li^+ (0.076*nm*), which challenges the electrode design as a host material with large interlayer spacing is needed for effective intercalation and also results in slower kinetics (H. Liu et al., 2016; Y. Yu et al., 2020). Heteroatom doping is beneficial in this case. Sulphur and phosphorus are two common dopants known to enlarge the carbon lattice and form more hospitable channels for Na^+ transport (Z. Li et al., 2017). As demonstrated in Table 5, *N,B* co-doped carbon nanosheets yielded a capacity of 307.1*mAh/g* with outstanding stability for 8,000 cycles, and *S*-doped carbon nanofibers reached a capacity of 460*mAh/g* (Y. Yu et al., 2020).

c. **Zinc-Ion Batteries/Capacitors (ZIBs/ZICs):** Aqueous ZIBs hold significant promise for grid-scale energy storage because of their high safety and low cost. However, challenges such as dendrite formation on the zinc anode and poor cathode material performance limit their development (Wei et al., 2023). Heteroatom-doped carbon materials have emerged as top-performing cathode candidates for ZICs. The high surface area and tuneable active sites of heteroatom-doped carbons enable efficient adsorption/desorption of Zn^{2+} ions and involvement in redox reactions (X. Feng et al., 2021). +-doped hierarchical porous carbons have achieved a capacity of 103.7*mAh/g* and an impressive power density of 11.58*kW/kg* (Wei et al., 2023).

d. **Magnesium-ion batteries (MIBs):** The high volumetric capacity and natural abundance of Mg make MIBs a promising "post-lithium" battery material. However, the divalent character of Mg^{2+} results in strong electrostatic interactions with the host matrix, which gives rise to sluggish diffusion kinetics and poor cycling stability. Heteroatom-doped CQDs provide a smart strategy for preparing composites with a high specific surface area and abundant active sites. As demonstrated in Table 5, a composite of *N*-doped CQDs with reduced graphene oxide (*rGO*) exhibited a reversible capacity of 259.1*mAh/g* at 0.1*A/g* (Qin et al., 2023). *N*-doping not only promotes the conductivity of the material but also produces defect sites for faster transport and accommodation of divalent Mg^2 ions, resulting in enhanced electrochemical performance.

While heteroatom-doped CQDs have most often been applied to the chemistries detailed above, they have also been applied to other key energy technologies. Heteroatom-doped CQDs have been used in aqueous batteries and fuel cells. Although aqueous batteries are often considered to have lower voltage and energy densities than their air-free counterparts, the safety, high conductivity, and low cost of alkaline aqueous batteries make them strong candidates for future large-scale energy storage technologies. The electrode-electrolyte interface of these devices is strongly affected by heteroatom doping (X. Feng et al., 2021). The enhanced wettability of this material and the abundance of active sites for fast redox chemistry have been observed to lead to a high specific capacitance and good long-term cycling stability. This material can also be used as a metal-free catalyst. Platinum-based catalysts are often used in fuel cells; however, the expensive and unsustainable nature of this material often becomes a limiting factor for these devices. *N*-doped GQDs show great promise as metal-free catalysts for fuel cells and can also induce an efficient near four-electron transfer pathway for the oxygen reduction reaction (ORR) (X. Li et al., 2022).

Table 5. CQD-Based Electrode Performance in Various Battery Systems

Electrode Material	Battery System	Heteroatom/ Dopant	Specific Capacity (mAh g^{-1})	Rate Capability	Cycling Performance	Reference
N-doped CQDs/ reduced graphene oxide/porous Fe2O3 composite aerogel	Alkaline aqueous	N	274.1 at 1 A/g	Capacity retention: 72.1% (5 A/g), 58.9% (50 A/g), 46.2% (100 A/g)19	80.4% retention over 5000 cycles at 3 A/ get al., 2019	(X. Yun et al., 2019)
VSe_2 composite with CQD	Potassium-ion	CQD	362 @ 100 mA/g	255 mAh/g @ 1 A/g	76.7% @ 200 cycles	(Y. Feng et al., 2022)
Boron and nitrogen composite with nano carbons	Sodium-ion	B,N	307.1 @ 0.1 A/g	157 mAh/g @ 4 A/g	Stable over 8,000 cycles	(Y. Yu et al., 2020)
Sulphur and nitrogen composite with nanosheets	Sodium-ion	S, N	523.7 @ 0.2 A/g	Not specified	Stable over 3,000 cycles @ 5 A/g	(Y. Yu et al., 2020)
GQD-coated VO2	Sodium-ion	GQD	306 @ 100 mA/g	110 mAh/g @ 18 A/g	>88% @ 1,500 cycles (for NIB)	(Chakrabarti et al., 2025)
Boron and nitrogen co-doped carbon quantum dots (BN-CQDs)	Lithium-ion	B, N	130.4 at 3 A/ gAhn, 2022	Ultrafast charge/ discharge with 88% capacity retention at high rates	Stable over 1000 cycles	(K. Kim & Ahn, 2022)
N,F co-doped carbon quantum dots coated on Ni foam current collector (NF-CQD/ Ni)	Lithium-air	N, F	Approx. 5122 at 100 mA/g	High electrochemical activity allowing high current densities	Stable over 42 cycles	(Ma et al., 2023)
Nitrogen-doped graphene quantum dots (N-GQD)	Lithium-ion	N	740 @ 50 mA/g	Maintains high capacity even at elevated current densities	90% retention over 500 cycles @ 200 mA/g	(Saroja et al., 2020)
Nitrogen-doped graphene quantum dots (N-GQD)	Sodium-ion	N	215 @ 50 mA/g	Maintains high capacity even at elevated current densities	66% retention over 500 cycles @ 200 mA/g	(Saroja et al., 2020)

continued on following page

Table 5. Continued

Electrode Material	Battery System	Heteroatom/ Dopant	Specific Capacity (mAh g^{-1})	Rate Capability	Cycling Performance	Reference
Boron-doped graphene quantum dots (N-GQD)	Lithium-ion	B	1097 @ 50 mA/g	Maintains high capacity even at elevated current densities	95.7% retention over 500 cycles @ 200 mA/g	(Saroja et al., 2020)
Boron-doped graphene quantum dots (N-GQD)	Sodium-ion	B	310 @ 50 mA/g	Maintains high capacity even at elevated current densities	53% retention over 500 cycles @ 200 mA/g	(Saroja et al., 2020)
N,S doped CQD encapsulated within graphene sheets	Supercapattery	N, S	21.7 @ 1A/g	Higher rates not specified	94.5% retention over 10000 cycles	(Pallavolu et al., 2023)
Boron, nitrogen dual doped carbon	Zinc-ion	B,N	315.3 F/g @ 0.05 A/g	241.1 F/g @ 40 A/g	98.4% @ 20,000 cycles	(Wei et al., 2023)
Nitrogen and sulphur co-doped carbon dots (N,S-CDs) as electrolyte additive	Sodium metal	N, S	Not applicable (electrolyte additive)	Enhances uniform sodium plating and suppresses dendrite formation	Coulombic efficiency ~99% after 250 cycles; >1200 h stable cycling	(H. Tu et al., 2024)
Nitrogen doped carbon-coated NiS1.97 quantum dots (NiS1.97 QDs@NC) Nitrogen doped Carbon Coating on NiS1.97 QD)	Magnesium-ion	N	259.1 @ 0.1 A/g; 96.0 @ 1.0 A/g	Good rate performance with stable capacity at high current densities23	Demonstrates good cycling stability with large capacity retention at elevated current densities3	(Qin et al., 2023)
CuS quantum dots encapsulated in porous carbon nanorods (CuS-QD@C nanorods)	Magnesium-ion	CQD	323.7 mAh/g @ 100 mA/g after 100 cycles	111.8 mAh/g @ 1.0 A/g	98.5 mAh/g after 1000 cycles at 1.0 A/g	(Fei et al., 2023)
Magnesium copper phosphate nanoparticles doped with graphene quantum dots (GQDs-MgCuPO4)	Supercapattery (with Mg-ion involvement)	Graphene Quantum Dots (GQDs)	1188 C/g @ 1.0 A/g	Not Specified	93% retention after 10,000 cycles	(Albalwi et al., 2024)
N,S co-doped CQDs sandwiched with graphene (N,S-CQDs/rGO)	Supercapacitor / Fuel Cell	N, S	Dual performance as electrode material for supercapacitors and fuel cell catalysts with specific capacitance 156 F/g	High energy and power density of 21.7 Wh/kg and 78.12 W/kg at scan rate 1 mV/s.	Robust cyclic stability	(Samantara et al., 2015)
Heteroatom-doped porous carbon nanosheets (N-doped)	Supercapacitor (general)	N	243 F/g @ 0.5 A/g (capacitance)	High rate capability (179 F/g at 20 A/g)2016	Excellent cycle stability.	(X. Hao et al., 2016)

continued on following page

Table 5. Continued

Electrode Material	Battery System	Heteroatom/ Dopant	Specific Capacity (mAh g^{-1})	Rate Capability	Cycling Performance	Reference
Ultra-small SnS quantum dots anchored onto Nitrogen-enriched carbon (SnS QDs@ NC)	Sodium-ion	N	281 F/g (initial) @ 100 mA/g	Excellent cycling stability at high current density	75% capacity retention after 500 cycles at 1000 mA/g	(Veerasubramani et al., 2020)
Nitrogen-doped carbon quantum dot (N-CQD) in CoS2/ mesoporous carbon composites	Lithium-ion	N	392 mAh/g @ 2.0 A/g	Not Specified	77% capacity retention after 100 cycles at 1000 mA/g	(Lee et al., 2024)

7. BRIDGING LAB-SCALE SUCCESS TO INDUSTRIAL VIABILITY

7.1. Techno-Economic Analysis (TEA)

While academic research efforts are typically centred on maximizing electrochemical performance, metrics such as cost, scalability, and environmental impact become more relevant when transitioning towards industrial applications (Argüelles-Cruz et al., 2023; U.S. Department of Energy, Industrial Efficiency & Decarbonization Office, 2023). Techno-Economic Analysis (TEA) is a valuable tool for determining the commercial viability of heteroatom-doped CQD technologies (Argüelles-Cruz et al., 2023). TEA can be used to identify primary cost drivers, help prioritize research efforts, and de-risk technical concerns associated with scale-up through process design.

a. **Precursor Costs:** The price of the starting material, also known as the feedstock, is another important cost factor. Syntheses dependent on costly high-purity chemical precursors are unlikely to find a market where established technologies exist because the material price is too high (Zhu et al., 2020). The use of waste biomass provides a significant economic benefit by converting a material that would otherwise need to be disposed of into a starting material (Zhu et al., 2020). Agricultural residues, food waste, and industrial byproducts can sometimes be acquired at close to zero or even negative costs, resulting in immediate savings in material costs and in accordance with the concepts of sustainability and green chemistry (Chávez-García et al., 2024). Biomass waste can be converted to value-added CQDs using a green synthesis approach for CQD fabrication owing to its low cost, sustainability, environmental benignity, and potential for commercialization (Zhu et al., 2020).

b. **Synthesis and Processing Costs:** The energy required for the synthesis process is a significant operational cost in CQD production. As previously mentioned, conventional hydrothermal methods are energy-intensive because of their prolonged reaction times and elevated temperatures (Q. Li & Jin, 2025). Hydrothermal synthesis offers several advantages, including high reactivity, facile control, and the ability to utilize readily available precursors. However, the energy consumption associated with its application in CQD production on a larger scale must be considered (Q. Li & Jin, 2025). In contrast, rapid synthesis methods, such as microwave-assisted heating, can drastically reduce the energy costs per unit of product owing to their high efficiency and rapidity (Huang & Ren, 2025). Microwave synthesis can be performed in a matter of minutes, making it suitable for large-scale production and achieving high quantum yields. Moreover, the capital costs of equipment should be considered; techniques that operate at ambient pressure and use straightforward reactors are more economically desirable than those that necessitate high-pressure autoclaves or complex vacuum systems (Saari et al., 2023).

c. **Downstream processing and purification costs:** Another often overlooked cost is associated with downstream processing (recovery and purification steps). Elaborate multi-step purification processes are not only energy- and chemical-intensive but also result in lower yields of the final product, which significantly contributes to the overall production cost (Jing et al., 2019). Therefore, synthesis routes that intrinsically afford high-purity CQDs or can be easily and efficiently scaled up with relatively simple purification methods are economically advantageous (X. Wang et al., 2019). Minimalist approaches that require fewer purification steps are essential for the cost-effectiveness and scalability of industrial CQD production.

7.2. Life Cycle Assessment (LCA) and Sustainability

In addition to economic concerns, the environmental sustainability of a technology is increasingly important for the social license to operate and the long-term prospects of a specific technology. The increasing demand for "greener" and more sustainable technologies has created a demand for systematic and objective information on the environmental impacts of various technological systems and their associated products. The concept of Life Cycle Assessment (LCA) can be employed to systematically evaluate the environmental aspects of resource use and environmental releases "from cradle to gate" (when the product leaves the factory) or "cradle to grave" (through disposal or recycling) (Upreti et al., 2015). LCA has matured and developed over the years into a common form of energy analysis to encompass environmental burden analysis and is now taking a more encompassing turn towards Life Cycle Sustainability Analysis (LCSA) that integrates not only environmental but also economic and social aspects (Guinée et al., 2010; Upreti et al., 2015).

a. **Environmental impact of synthesis and materials:** The carbon quantum dot (CQD) LCA has revealed that the "green" nature of biomass precursors is often overplayed, and the terminology needs to be qualified and standardized (Lesch et al., n.d.). Although using waste biomass is an environmentally friendly option that avoids the need to manufacture and use virgin chemical precursors, the energy required to process and manipulate the material, including transport, drying, and hydrothermal treatment, can dominate the overall carbon footprint (Lesch et al., n.d.). In one study, a high-yield synthesis process was found to be unsustainable because of its extremely high electricity input, most of which originated from fossil fuel sources (Fernandes et al., 2022). The goal is a low-impact process for the entire system: using local waste, which eliminates transport energy, low-energy synthesis processes, and green solvents (preferably water) or solvent-free techniques.
b. **Circular Economy Principles:** In the context of EES devices, the end-of-life stage is an often-overlooked but important aspect of overall sustainability. Designing CQD-based electrodes with the end-of-life phase in mind is crucial for meeting circular economy principles. This involves selecting appropriate binders and substrates that can be easily separated from the active material, as well as developing methods to recover and regenerate CQDs or the materials from which they are synthesized, once the device reaches the end of its useful life (Glogic et al., 2022). The use of earth-abundant carbon and common heteroatoms, such as those in tin disulfide, compared to rarer and potentially toxic heavy metals that are prevalent in other battery chemistries, provides a significant built-in advantage for CQD-based materials in terms of long-term sustainability and resource security (Matek et al., 2019). A large motivation for the move towards earth-abundant elements is to reduce the environmental and supply chain impacts of organic additives/by-products and scarce critical raw materials. TEA and LCA can be used to close the loop on early stage materials research by adding a much-needed reality check that prevents the scientific community from doggedly pursuing high-performance materials with little path to economic or environmentally viable implementation.

Figure 4. Infographic on nanomaterials and nano safety.

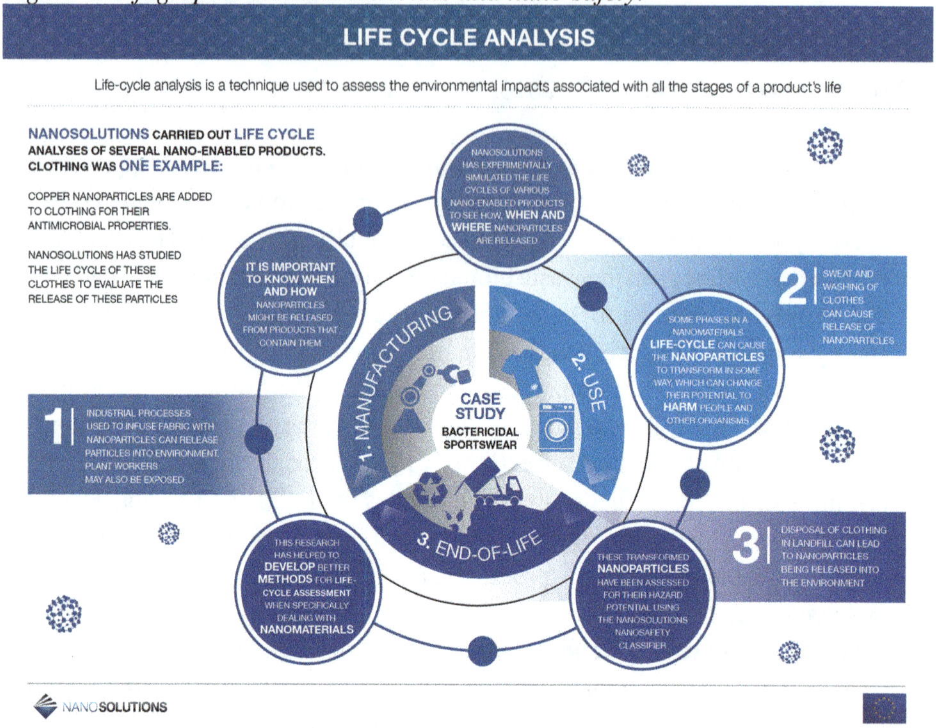

Reproduced from Wikimedia Commons, licensed under Creative Commons Attribution-Share Alike 4.0 International (CC BY-SA 4.0). © Wikimedia Commons contributors. Reproduced as is (Wikimedia Commons contributors, 2020).

By quantifying emissions, resource use, and energy use, LCA, when combined with TEA, can help identify the main cost drivers, set priorities for research and development efforts, and reduce technical risks for scale-up. Ultimately, the comprehensive results support decision-makers in making more informed choices between alternative options by considering all major environmental impacts (Fernandes et al., 2021; Saari et al., 2023). Figure 4 illustrates the major steps in the Life Cycle Assessment (LCA) of a nano-enabled product. The analysis considers the full life cycle from raw material extraction/manufacturing ("cradle") to product use and end-of-life scenarios, such as recycling or disposal (Wikimedia Commons Contributors, 2020).

8. CHALLENGES AND EMERGING DIRECTIONS IN RESEARCH

8.1. Overcoming Key Hurdles

Even though significant advances have been made towards addressing the grand challenges that prevent heteroatom-doped CQDs from having a broad-spread practical application (Rasal et al., 2021), there are some issues related to synthesis and processing, performance optimization, stability, and applications that are yet to be resolved. These grand challenges are strongly coupled and need to be addressed together in a holistic materials design and device engineering strategy that encompasses advanced synthesis, innovative densification, and robust interface engineering to ultimately overcome the major hurdles towards heteroatom-doped CQD electrode implementation.

a. **Uniform Co-doping:** As previously mentioned in several sections of this chapter, one of the most important challenges in synthesis remains the precise and homogeneous control of the distribution of different dopant types and their bonding configurations (Im et al., 2023). Owing to the stochastic nature of most bottom-up synthetic approaches, it is difficult to precisely control the location and stoichiometry of dopants, resulting in material heterogeneity and inconsistent performance (Im et al., 2023). This fundamental randomness is considered to be the key hindrance to reaching the theoretical limits predicted for optimally co-doped structures. This non-uniformity of dopants directly affects the electronic and optical properties, leading to less reproducible and consistent CQDs for high-level applications (John et al., 2021).
b. **Electrode Densification:** The large gap between gravimetric and volumetric performance remains a key challenge for practical device integration in energy storage technologies. The development of approaches to increase the packing density of such highly porous nanomaterials without severely hampering the ion transport pathways is highly desirable (Z. Guo et al., 2025). This issue is related to both doping and stability, as densification creates mechanical stress that is further amplified by inhomogeneities in the material and can damage the fragile electrode-electrolyte interface. For example, achieving a high packing density with a high number of multimodal pores and good volumetric performance is difficult, even with careful synthesis approaches (H. Jin et al., 2019).
c. **Long-Term Interfacial Stability:** The long-term stability of an EES device is ultimately limited by the stability of its interfaces (Zhai et al., 2021). Heteroatom doping is known to create more active surfaces for enhanced performance in interfacial reactions. However, these sites can also be more active towards the electrolyte itself, leading to parasitic side reactions and gradual loss of activity

over thousands of cycles (Y. Yu et al., 2020). Therefore, it is critical to ensure that the engineered electrode-electrolyte interface is chemically and mechanically robust for the entire lifetime of the device, including demanding applications such as grid storage or electrical vehicles (EVs), which require decades of continuous service (X. Feng et al., 2021).

d. **Theoretical understanding and computational modelling:** In contrast to advances in synthesis, there is still a lack of theoretical understanding and accurate computational modelling of heteroatom-doped CQDs (J. Yu et al., 2021). The highly irregular and diverse micro/nanostructures of CQDs, as well as the complex ways in which dopants can be incorporated into the carbon lattice, make it difficult to accurately predict their structure-property relationships. Computational methods, including Density Functional Theory (DFT), often face challenges in accurately simulating dopant-induced defects, electronic states, and excitonic effects in these materials (Min et al., 2024). This hinders the rational design of CQDs with tailored properties and the prediction of optimal doping strategies for CQDs. Additionally, multi-scale modelling approaches that can connect atomic-level doping effects to the macroscopic performance of CQD-based devices are required, as this remains an unresolved issue.

8.2. The Role of AI-Driven Materials Discovery

The vast chemical-combinatorial space of the material design library for heteroatom-doped CQDs (including combinations of dopants, dopant concentration and configurations, carbon precursors, synthesis conditions, and composite architectures) renders traditional, intuition-based, trial-and-error experimentation very inefficient. In this regard, Artificial Intelligence (AI) and Machine Learning (ML) will likely prompt a change in paradigm, allowing for orders of magnitude in efficiency and speed for discoveries and applications to be made in this area (C. Chen et al., 2020; Lv et al., 2021).

Accelerated Screening and Synthesis Optimization: ML models trained on existing experimental and computational datasets can predict the properties of unsynthesized CQD materials (Duman & Jalilov, 2024). These models can be used to screen thousands of potential dopant combinations in silico and identify the most promising candidates for experimental validation, dramatically accelerating the discovery process and enabling the exploration of vast chemical spaces (Y. Li et al., 2025). Additionally, AI algorithms can be employed to optimize the complex multivariable parameter space of synthesis (e.g., temperature, time, pressure, and precursor ratios) to achieve specific target properties, such as maximizing the quantum yield or ensuring uniform doping (Duman & Jalilov, 2024). For example, ML-based techniques have been used to guide the synthesis of carbon dots with improved fluorescent quantum yields by optimizing hydrothermal parameters, and the most important features have been identified, including the precursor mass and alkaline catalyst volume (Han et al., 2020). In this example, the ML-based technique was used not only for property prediction but also to guide and accelerate the carbon dot synthesis and improve the process-related properties (Han et al., 2020). The careful choice and preprocessing of large datasets with knowledge from the domain of materials science are key factors in the performance of ML-based material design approaches (Maleki et al., 2022).

Inverse Design: Inverse design, in which the researcher specifies the desired device performance and the model optimizes the material and structure to meet that end goal, is the long-term goal of AI incorporation in this field (Ganti et al., 2025). Thus, the model must effectively work backward to determine the necessary material composition, atomic structure, and synthesis protocol required to meet the desired performance targets (Pollice et al., 2021). In this way, the general philosophy of "form follows function" in a completely different context can be leveraged to identify tailor-made materials with the desired property or function, starting first and foremost with what the desired property is. Generative models, such as deep neural networks and generative adversarial networks, are commonly used for inverse design and can learn from known data to generate new materials with desired properties (C. Chen & Gu, 2020; B. Kim et al., 2020). Many of these advanced ML techniques use backpropagation or other methods to calculate an analytical gradient of the objective function to optimize rapidly and avoid being stuck in a local minimum in the design space. In addition, the active learning strategies outlined in the previous section can be combined with inverse design to reduce the quantity of training data required to identify high-performing candidates by at least an order of magnitude.

Data-driven materials intelligence ecosystems: The ability to acquire and store "big data" has outstripped the ability to extract information from it, leading to data-driven discovery being the so-called "fourth paradigm" of science in the field of materials science (Agrawal & Choudhary, 2016). Machine learning and AI are fuelling the expansion of materials intelligence ecosystems by making it possible to achieve dramatic efficiencies and faster discoveries (Batra et al., 2020). The data-driven approach is further accelerated by targeted initiatives, such as the Materials Genome Initiative (MGI), with the goal of achieving orders-of-magnitude improvements in time-to-insight and accelerating cost-effective material discovery (Agrawal & Choudhary, 2016). Proof-of-principle successes in data-driven material discovery and design efforts have demonstrated the efficacy of such approaches, while highlighting potential pitfalls and open challenges (Pollice et al., 2021). The growing adoption and implementation of large-scale data-driven approaches are expected to enable leaps in the progress of material discovery and design campaigns. The systematic collection and analysis of data and their feedback into ML algorithms form an ongoing feedback loop that can be used to refine material design strategies and gain insights into the processing-structure-property-performance relationships (Agrawal & Choudhary, 2016; Pollice et al., 2021).

Integration with Automation and high-throughput techniques: This integration also extends to automated experimental platforms (self-driving laboratories) to further accelerate discovery and optimization processes (MacLeod et al., 2020). Self-driving laboratories autonomously design, perform, and learn from materials science experiments in a completely self-driven iterative loop and have the potential to dramatically shorten the multi-decade timelines for discovering and optimizing materials, with the promise of reducing the time-to-discover-and-optimize commercially viable materials to months (MacLeod et al., 2020). Machine learning also finds great application in high-throughput virtual screening for rapidly evaluating a vast number of material candidates based on computational predictions (Nandy et al., 2021; Pollice et al., 2021). The integration of AI-driven design and discovery with automated experimentation and high-throughput screening is a powerful approach for efficient material innovation, opening new avenues for exploring wide and complex chemical spaces that were previously inaccessible (Nandy et al., 2021). It also allows for the rapid identification of structure-property relationships, thereby facilitating the accelerated discovery of materials with desired properties (Nandy et al., 2021).

8.3. Directions for Future Research

Guided by the insights and analysis in this chapter, it is possible to provide a straightforward outline of what may be expected and needed for future research, with academia and industry having different but not mutually exclusive priorities.

1. For Academia:
   a. Persistent efforts in advanced operando/in-situ characterizations, with the end goal of pushing the fundamental understanding of the charge storage mechanisms and degradation at the atomic level, are required.
   b. The incorporation of new doping species and the design of more complex multi-heteroatom co-doping systems continue the trend of finding synergistic enhancements.
   c. The generation of high-quality and standardized experimental/computational datasets for open-source databases for training the next generation of materials discovery AI/ML models.
2. For Industry:
   a. The development of continuous-flow, low-energy, and low-cost synthesis protocols, ideally built around waste biomass precursors, is essential.

b. Standardization and quality control of material synthesis, purification, and characterization methods are required to produce consistent and reliable products.
c. Greater effort is required towards electrode densification, slurry formulations, and full cell assembly, with a particular focus on the translation from material performance to manufacturable high-volumetric-performance devices.

CONCLUSION

In summary, this chapter provides a detailed overview of the role of heteroatom doping in carbon quantum dots (CQDs) for next-generation energy storage applications. Starting from a quantum-mechanical understanding of the impact of heteroatoms on the electronic structure and formation of electrochemically active sites, the discussion has spanned the entire development pathway from scalable synthesis approaches, especially from biomass precursors, to integration into functional composite systems and device-level performance. The importance of optimizing material properties at the nanoscale, as well as upscaling to the micro- and macroscale, has been highlighted, with recent achievements in specific capacitance and cycling stability pointing to the remarkable potential of heteroatom-doped CQDs. Operando characterization has provided deep insights into the charge storage mechanisms, which involve a combination of EDLC, pseudocapacitance, and intercalation processes. Performance benchmarking has demonstrated the versatility of these materials for supercapacitors and various battery chemistries, offering a platform to address the energy-power dichotomy. The challenges of achieving a uniform distribution of heteroatoms, closing the gap in volumetric performance through electrode densification, and establishing standardized synthesis and characterization protocols have been recognized as critical for commercialization. The incorporation of techno-economic analysis and life cycle assessment, along with the emerging field of AI-driven materials discovery, marks the maturation of this field and sets the stage for accelerated development. Looking ahead, as the electrification of the grid, transportation, and portable electronics demands solutions for grid-scale renewable energy storage, electric vehicles, and advanced energy storage for consumer electronics, heteroatom-doped CQDs present a compelling proposition for these applications. They promise to enable a shift towards energy storage devices that are not only high-performing and sustainable but also economically competitive. This chapter presents a current review that suggests significant exponential growth in the field over the next decade. The key to unlocking this potential is to ensure a balance between scientific exploration and practical scalability, and between pushing the limits of performance and ensuring industrial relevance.

REFERENCES

Abbas, Q., Raza, R., Shabbir, I., & Olabi, A. G. (2019). Heteroatom doped high porosity carbon nanomaterials as electrodes for energy storage in electrochemical capacitors: A review. *Journal of Science. Advanced Materials and Devices*, *4*(3), 341–352. DOI: 10.1016/j.jsamd.2019.07.007

Abdel Maksoud, M. I. A., Fahim, R. A., Shalan, A. E., Abd Elkodous, M., Olojede, S. O., Osman, A. I., Farrell, C., Al-Muhtaseb, A. H., Awed, A. S., Ashour, A. H., & Rooney, D. W. (2021). Advanced materials and technologies for supercapacitors used in energy conversion and storage: A review. *Environmental Chemistry Letters*, *19*(1), 375–439. DOI: 10.1007/s10311-020-01075-w

Abdelsalam, H., Sakr, M. A. S., Teleb, N. H., Abd-Elkader, O. H., & Zhang, Q. (2025). Quantum-dot-based carbon nanotubes: Design, doping, and lithium storage for high-capacity energy applications. *Langmuir*, *41*(27), 17803–17814. DOI: 10.1021/acs.langmuir.5c01572 PMID: 40615346

Agrawal, A., & Choudhary, A. (2016). Perspective: Materials informatics and big data: Realization of the "fourth paradigm" of science in materials science. *APL Materials*, *4*(5), 053208. Advance online publication. DOI: 10.1063/1.4946894

Al-Gaashani, R., Najjar, A., Zakaria, Y., Mansour, S., & Atieh, M. A. (2019). Xps and structural studies of high quality graphene oxide and reduced graphene oxide prepared by different chemical oxidation methods. *Ceramics International*, *45*(11), 14439–14448. DOI: 10.1016/j.ceramint.2019.04.165

Albalwi, S., Al-Ojeery, A., Hassan, H., Iqbal, M. W., Afzal, A., & Aly, K. (2024). Incorporation of graphene quantum dots into magnesium copper phosphate (GQDs-MgCuPO4) for supercapattery and electrocatalysis applications. *Physica Scripta*. https://iopscience.iop.org/article/10.1088/1402-4896/ad9117

Anil Kumar, Y., Koyyada, G., Ramachandran, T., Kim, J. H., Sajid, S., Moniruzzaman, M., Alzahmi, S., & Obaidat, I. M. (2023). Carbon Materials as a Conductive Skeleton for Supercapacitor Electrode Applications: A Review. *Nanomaterials (Basel, Switzerland)*, *13*(6), 1049. DOI: 10.3390/nano13061049 PMID: 36985942

Ansari, S. A. (2022). Graphene Quantum Dots: Novel Properties and Their Applications for Energy Storage Devices. *Nanomaterials (Basel, Switzerland)*, *12*(21), 3814. DOI: 10.3390/nano12213814 PMID: 36364590

Argüelles-Cruz, A., Shah, K. B., Logeshwaran, J., & Shrestha, A. (2023). Genetic algorithm based techno-economic optimization of an isolated hybrid energy system. *ICTACT Journal on Microelectronics*, *8*(4), 1447–1450. DOI: 10.21917/ijme.2023.0249

Arooj, M., Parambath, J. B. M., Ali, N., Khan, A., Malik, S., Bilal, M., & Mohamed, A. A. (2022). Experimental and theoretical review on covalent coupling and elemental doping of carbon nanomaterials for environmental photocatalysis. *Critical Reviews in Solid State and Material Sciences*, *48*(2), 215–256. DOI: 10.1080/10408436.2022.2049697

Arora, G., Sabran, N. S., Ng, C. Y., Low, F. W., & Jun, H. K. (2024). Applications of carbon quantum dots in electrochemical energy storage devices. *Heliyon*, *10*(15), e35543. DOI: 10.1016/j.heliyon.2024.e35543 PMID: 39166025

Bao, R., Chen, Z., Zhao, Z., Sun, X., Zhang, J., Hou, L., & Yuan, C. (2018). Green and facile synthesis of nitrogen and phosphorus co-doped carbon quantum dots towards fluorescent ink and sensing applications. *Nanomaterials (Basel, Switzerland)*, *8*(6), 386. DOI: 10.3390/nano8060386 PMID: 29857502

Baragau, I.-A., Power, N. P., Morgan, D. J., Lobo, R. A., Titirici, M.-M., Middelkoop, V., Diaz, A., & Dunn, S. (n.d.). *An Efficient Continuous Hydrothermal Flow Synthesis of Carbon Quantum Dots from a Targeted Biomass Precursor for On-Off Metal Ions Nano-Sensing*.

Barman, M. K., Jana, B., Bhattacharyya, S., & Patra, A. (2014). Photophysical properties of doped carbon dots (n, p, and b) and their influence on electron/hole transfer in carbon dots–nickel (ii) phthalocyanine conjugates. *The Journal of Physical Chemistry. C, Nanomaterials and Interfaces*, *118*(34), 20034–20041. DOI: 10.1021/jp507080c

Batra, R., Song, L., & Ramprasad, R. (2020). Emerging materials intelligence ecosystems propelled by machine learning. *Nature Reviews. Materials*, *6*(8), 655–678. DOI: 10.1038/s41578-020-00255-y

Bie, X., Dong, Y., Xiong, M., Wang, B., Chen, Z., Zhang, Q., Liu, Y., & Huang, R. (2023). Nitrogen-Doped Carbon Matrix to Optimize Cycling Stability of Lithium Ion Battery Anode from SiOx Materials. *Inorganics*, *12*(1), 9. DOI: 10.3390/inorganics12010009

Boroznin, S. V. (2022). Carbon nanostructures containing boron impurity atoms: Synthesis, physicochemical properties and potential applications. *Modern Electronic Materials*, *8*(1), 23–42. DOI: 10.3897/j.moem.8.1.84317

Brown, M., & Li, D. (2024). Interfacial engineering for high performance carbon-based perovskite solar cells. *Frontiers in Energy Research*, *12*, 1463024. DOI: 10.3389/fenrg.2024.1463024

Buttry, D. A. (2021). Applications of the quartz crystal microbalance to electrochemistry. *Electroanalytical Chemistry*, •••, 1–85. DOI: 10.1201/9781003209003-1

Cai, D., Zhong, X., Xu, L., Xiong, Y., Deng, W., Zou, G., Hou, H., & Ji, X. (2025). Biomass-derived carbon dots: Synthesis, modification and application in batteries. *Chemical Science (Cambridge)*, *16*(12), 4937–4970. DOI: 10.1039/D4SC08659G PMID: 40046072

Chacón-Torres, J. C., Wirtz, L., & Pichler, T. (2013). Manifestation of charged and strained graphene layers in the raman response of graphite intercalation compounds. *ACS Nano*, *7*(10), 9249–9259. DOI: 10.1021/nn403885k PMID: 24025089

Chakrabarti, A., Majumdar, P., Alessandri, E., & Roemer, C. (2025). Recent advancement of quantum dot-based nanocomposites as electrode materials for secondary batteries. *Energies*, *18*(3), 630. DOI: 10.3390/en18030630

Chávez-García, D., Guzman, M., Sanchez, V., & Cadena-Nava, R. D. (2024). Green synthesis of biomass-derived carbon quantum dots for photocatalytic degradation of methylene blue. *Beilstein Journal of Nanotechnology*, *15*, 755–766. DOI: 10.3762/bjnano.15.63 PMID: 38952416

Chen, C., & Gu, G. X. (2020). Generative deep neural networks for inverse materials design using backpropagation and active learning. *Advancement of Science*, *7*(5), 1902607. Advance online publication. DOI: 10.1002/advs.201902607 PMID: 32154072

Chen, C., Zuo, Y., Ye, W., Li, X., Deng, Z., & Ong, S. P. (2020). A critical review of machine learning of energy materials. *Advanced Energy Materials*, *10*(8), 1903242. Advance online publication. DOI: 10.1002/aenm.201903242

Chen, Q., Hu, Y., Hu, C., Cheng, H., Zhang, Z., Shao, H., & Qu, L. (2014). Graphene quantum dots–three-dimensional graphene composites for high-performance supercapacitors. *Physical Chemistry Chemical Physics*, *16*(36), 19307–19313. DOI: 10.1039/C4CP02761B PMID: 25100222

Cui, H., Guo, Y., Guo, L., Wang, L., Zhou, Z., & Peng, Z. (2018). Heteroatom-doped carbon materials and their composites as electrocatalysts for co2 reduction. *Journal of Materials Chemistry. A, Materials for Energy and Sustainability*, *6*(39), 18782–18793. DOI: 10.1039/C8TA07430E

Das, H. T., Barai, P., Dutta, S., Das, N., Das, P., Roy, M., Alauddin, M., & Barai, H. R. (2022). Polymer Composites with Quantum Dots as Potential Electrode Materials for Supercapacitors Application: A Review. *Polymers*, *14*(5), 1053. DOI: 10.3390/polym14051053 PMID: 35267876

Deng, Y., Ji, Y., Chen, F., Ren, F., & Tan, S. (2020). Superior performance of flexible solid-state supercapacitors enabled by ultrafine graphene quantum dot-decorated porous carbon spheres. *New Journal of Chemistry*, *44*(32), 13591–13597. DOI: 10.1039/D0NJ03163A

Domingo-Tafalla, B., Martínez-Ferrero, E., Franco, F., & Palomares-Gil, E. (2022). Applications of Carbon Dots for the Photocatalytic and Electrocatalytic Reduction of CO2. *Molecules (Basel, Switzerland)*, *27*(3), 1081. DOI: 10.3390/molecules27031081 PMID: 35164346

Du, Y., & Guo, S. (2016). Chemically doped fluorescent carbon and graphene quantum dots for bioimaging, sensor, catalytic and photoelectronic applications. *Nanoscale*, *8*(5), 2532–2543. DOI: 10.1039/C5NR07579C PMID: 26757977

Dua, S., Kumar, P., Pani, B., Kaur, A., Khanna, M. K., & Bhatt, G. (2023). Stability of carbon quantum dots: A critical review. *RSC Advances*, *13*(20), 13845–13861. DOI: 10.1039/D2RA07180K PMID: 37181523

Dubey, R., & Guruviah, V. (2019). Review of carbon-based electrode materials for supercapacitor energy storage. *Ionics*, *25*(4), 1419–1445. DOI: 10.1007/s11581-019-02874-0

Duman, A. N., & Jalilov, A. S. (2024). Machine learning for carbon dot synthesis and applications. *Materials Advances*, *5*(18), 7097–7112. DOI: 10.1039/D4MA00505H

Elugoke, S. E., Uwaya, G. E., Quadri, T. W., & Ebenso, E. E. (2024). Carbon Quantum Dots: Basics, Properties, and Fundamentals. In Berdimurodov, E., Verma, D. K., & Guo, L. (Eds.), *ACS Symposium Series* (Vol. 1465, pp. 3–42). American Chemical Society. DOI: 10.1021/bk-2024-1465.ch001

Fei, Y., Man, Y., Sun, J., Du, Y., Chen, B., Bao, J., & Zhou, X. (2023). Implanting cus quantum dots into carbon nanorods for efficient magnesium-ion batteries. *Small*, *19*(34), 2301954. Advance online publication. DOI: 10.1002/smll.202301954 PMID: 37086143

Feng, J., Dong, H., Pang, B., Shao, F., Zhang, C., Yu, L., & Dong, L. (2018). Theoretical study on the optical and electronic properties of graphene quantum dots doped with heteroatoms. *Physical Chemistry Chemical Physics*, *20*(22), 15244–15252. DOI: 10.1039/C8CP01403E PMID: 29789854

Feng, X., Bai, Y., Liu, M., Li, Y., Yang, H., Wang, X., & Wu, C. (2021). Untangling the respective effects of heteroatom-doped carbon materials in batteries, supercapacitors and the orr to design high performance materials. *Energy &Amp. Energy & Environmental Science*, *14*(4), 2036–2089. DOI: 10.1039/D1EE00166C

Feng, Y., Xu, C., Wu, K., Guo, Y., Wu, S., Guo, Z., He, M., & Xue, M. (2022). Carbon quantum dot-coated vse2 nanosheets as anodes for high-performance potassium-ion batteries. *ACS Applied Nano Materials*, *5*(9), 13136–13148. DOI: 10.1021/acsanm.2c02907

Fernandes, S., Esteves Da Silva, J. C. G., & Pinto Da Silva, L. (2022). Life Cycle Assessment-Based Comparative Study between High-Yield and "Standard" Bottom-Up Procedures for the Fabrication of Carbon Dots. *Materials (Basel)*, *15*(10), 3446. DOI: 10.3390/ma15103446 PMID: 35629474

Fernandes, S., Silva, J. C. G. E., & Silva, L. P. (2021). Comparative life cycle assessment of high-yield synthesis routes for carbon dots. *NanoImpact*, *23*, 100332. DOI: 10.1016/j.impact.2021.100332 PMID: 35559833

Ganti, S. V. S., Wölfel, L., & Kuenneth, C. (2025). Ai-driven discovery of high performance polymer electrodes for next-generation batteries. *Journal of Polymer Science*, *63*(21), 4631–4639. Advance online publication. DOI: 10.1002/pol.20250198

Ghosh, S., Bera, S., Kapuria, A., Debnath, A., Das, P., Su, Y., & Saha, S. K. (2025). Nitrogen and sulfur co-doped carbon quantum dot-decorated ti3c2tx-mxenes as electrode materials for high-performance symmetric supercapacitors. *Nanoscale*, *17*(26), 15731–15748. Advance online publication. DOI: 10.1039/D4NR05037A PMID: 40531092

Glogic, E., Kamali, A. K., Keppetipola, N. M., Alonge, B., Kumara, G., Sonnemann, G., Toupance, T., & Cojocaru, L. (2022). Life cycle assessment of supercapacitor electrodes based on activated carbon from coconut shells. *ACS Sustainable Chemistry &Amp. ACS Sustainable Chemistry & Engineering*, *10*(46), 15025–15034. DOI: 10.1021/acssuschemeng.2c03239

Goswami, M., Mandal, S., & Pillai, V. K. (2023). Effect of hetero-atom doping on the electrocatalytic properties of graphene quantum dots for oxygen reduction reaction. *Scientific Reports*, *13*(1), 5182. DOI: 10.1038/s41598-023-31854-8 PMID: 36997535

Gouda, A., Manioudakis, J., Naccache, R., Soavi, F., & Santato, C. (2021). 3d network of sepia melanin and n- and, s-doped graphitic carbon quantum dots for sustainable electrochemical capacitors. *Advanced Sustainable Systems*, *5*(10), 2100152. Advance online publication. DOI: 10.1002/adsu.202100152

Guan, X., Li, Z., Geng, X., Lei, Z., Karakoti, A., Wu, T., Kumar, P., Yi, J., & Vinu, A. (2023). Emerging Trends of Carbon-Based Quantum Dots: Nanoarchitectonics and Applications. *Small, 19*(17), 2207181. DOI: 10.1002/smll.202207181 PMID: 36693792

Guinée, J. B., Heijungs, R., Huppes, G., Zamagni, A., Masoni, P., Buonamici, R., Ekvall, T., & Rydberg, T. (2010). Life cycle assessment: Past, present, and future. *Environmental Science &Amp. Technology (Elmsford, N.Y.), 45*(1), 90–96. DOI: 10.1021/es101316v PMID: 20812726

Guo, Y., Chen, Y., Cao, F., Wang, L., Wang, Z., & Leng, Y. (2017). Hydrothermal synthesis of nitrogen and boron doped carbon quantum dots with yellow-green emission for sensing Cr(vi), anti-counterfeiting and cell imaging. *RSC Advances, 7*(76), 48386–48393. DOI: 10.1039/C7RA09785A

Guo, Z., Tian, Z., Liu, Y., Huang, Y., Li, W., Han, X., He, S., Mao, H. K., Zhang, C., Duan, G., & Jiang, S. (2025). A densification-activation strategy toward hierarchical towards self-supporting thick carbon electrodes for high-power supercapacitors. *Chemical Communications, 61*(35), 6474–6477. DOI: 10.1039/D5CC01049G PMID: 40176733

Gurushankar, K., Karuppaiya, C., Tari, V., & Kannan, K. (2025). Exploring MXenes and MBenes in Metal Sulfur and Selenium Battery Technologies. In *Novel Energy Storage and Conversion Technologies for Two-Dimensional MXenes and MBenes* (pp. 105–120). IGI Global Scientific Publishing. DOI: 10.4018/979-8-3693-9546-2.ch005

Han, Y., Tang, B., Wang, L., Bao, H., Lu, Y., Guan, C., Zhang, L., Le, M., Liu, Z., & Wu, M. (2020). Machine-learning-driven synthesis of carbon dots with enhanced quantum yields. *ACS Nano, 14*(11), 14761–14768. DOI: 10.1021/acsnano.0c01899 PMID: 32960048

Hao, H., Wang, M., Cao, Y., He, J., Yang, Y., Zhao, C., & Yan, L. (2024). Boron-doped engineering for carbon quantum dots-based memristors with controllable memristance stability. *Small Methods, 8*(9), 2301454. Advance online publication. DOI: 10.1002/smtd.202301454 PMID: 38204209

Hao, X., Wang, J., Ding, B., Shen, L., Xu, Y., Wang, Y., Chang, Z., Dou, H., Lu, X., & Zhang, X. (2016). Heteroatom-doped porous carbon nanosheets: General preparation and enhanced capacitive properties. *Chemistry (Weinheim an der Bergstrasse, Germany), 22*(46), 16668–16674. DOI: 10.1002/chem.201602922 PMID: 27704674

Hemmati, A., Emadi, H., & Nabavi, S. R. (2023). Green Synthesis of Sulfur- and Nitrogen-Doped Carbon Quantum Dots for Determination of L-DOPA Using Fluorescence Spectroscopy and a Smartphone-Based Fluorimeter. *ACS Omega*, *8*(23), 20987–20999. DOI: 10.1021/acsomega.3c01795 PMID: 37332813

Herold, F., De Oliveira, D., Baade, G., Friedland, J., Güttel, R., Claeys, M., & Rønning, M. (2025). Is Carbon Heteroatom Doping the Key to Active and Stable Carbon Supported Cobalt Fischer–Tropsch Catalysts? *ACS Catalysis*, *15*(8), 6673–6689. DOI: 10.1021/acscatal.4c08092 PMID: 40270880

Hu, L., Hou, J., Ma, Y., Li, H., & Zhai, T. (2016). Multi-heteroatom self-doped porous carbon derived from swim bladders for large capacitance supercapacitors. *Journal of Materials Chemistry. A, Materials for Energy and Sustainability*, *4*(39), 15006–15014. DOI: 10.1039/C6TA06337C

Hu, Y., Chen, W., Lei, T., Zhou, B., Jiao, Y., Yan, Y., Du, X., Huang, J., Wu, C., Wang, X., Wang, Y., Chen, B., Xu, J., Wang, C., & Xiong, J. (2018). Carbon quantum dots–modified interfacial interactions and ion conductivity for enhanced high current density performance in lithium–sulfur batteries. *Advanced Energy Materials*, *9*(7), 1802955. Advance online publication. DOI: 10.1002/aenm.201802955

Huang, Z., Liu, T., Song, Y., Li, Y., & Liu, X. (2017). Balancing the electrical double layer capacitance and pseudocapacitance of hetero-atom doped carbon. *Nanoscale*, *9*(35), 13119–13127. DOI: 10.1039/C7NR04234E PMID: 28849857

Huang, Z., & Ren, L. (2025). Large Scale Synthesis of Carbon Dots and Their Applications: A Review. *Molecules (Basel, Switzerland)*, *30*(4), 774. DOI: 10.3390/molecules30040774 PMID: 40005085

Ilnicka, A., Skorupska, M., Szkoda, M., Zarach, Z., Kamedulski, P., Zielinski, W., & Lukaszewicz, J. P. (2021). Combined effect of nitrogen-doped functional groups and porosity of porous carbons on electrochemical performance of supercapacitors. *Scientific Reports*, *11*(1), 18387. DOI: 10.1038/s41598-021-97932-x PMID: 34526635

Im, M. J., Kim, J. I., Hyeong, S., Moon, B. J., & Bae, S. (2023). From pristine to heteroatom-doped graphene quantum dots: An essential review and prospects for future research. *Small*, *19*(47), 2304497. Advance online publication. DOI: 10.1002/smll.202304497 PMID: 37496316

Jana, J., Ganguly, M., Chandrakumar, K. R. S., Rao, G. M., & Pal, T. (2017). Boron precursor-dependent evolution of differently emitting carbon dots. *Langmuir*, *33*(2), 573–584. DOI: 10.1021/acs.langmuir.6b04100 PMID: 28024393

Jana, J., & Pal, T. (2017). An account of doping in carbon dots for varied applications. *Natural Resources &Amp. Engineering*, *2*(1), 5–12. DOI: 10.1080/23802693.2017.1421506

Jiang, X., Xin, Y., He, B., Zhang, F., & Tian, H. (2024). Effect of Heteroatom Doping on Electrochemical Properties of Olivine LiFePO4 Cathodes for High-Performance Lithium-Ion Batteries. *Materials (Basel)*, *17*(6), 1299. DOI: 10.3390/ma17061299 PMID: 38541453

Jiang, Y., Li, J., Jiang, Z., Shi, M., Sheng, R., Liu, Z., Zhang, S., Cao, Y., Wei, T., & Fan, Z. (2021). Large-surface-area activated carbon with high density by electrostatic densification for supercapacitor electrodes. *Carbon*, *175*, 281–288. DOI: 10.1016/j.carbon.2021.01.016

Jin, H., Feng, X., Li, J., Li, M., Xia, Y., Yuan, Y., Yang, C., Dai, B., Lin, Z., Wang, J., Lü, J., & Wang, S. (2019). Heteroatom-doped porous carbon materials with unprecedented high volumetric capacitive performance. *Angewandte Chemie International Edition*, *58*(8), 2397–2401. DOI: 10.1002/anie.201813686 PMID: 30620788

Jin, S., Allam, O., Lee, K., Lim, J., Lee, M. J., Loh, S. H., Jang, S. S., & Lee, S. W. (2022). Carbon quantum dot modified reduced graphene oxide framework for improved alkali metal ion storage performance. *Small*, *18*(35), 2202898. Advance online publication. DOI: 10.1002/smll.202202898 PMID: 35927029

Jing, S., Zhao, Y., Sun, R., Zhong, L., & Peng, X. (2019). Facile and high-yield synthesis of carbon quantum dots from biomass-derived carbons at mild condition. *ACS Sustainable Chemistry &Amp. ACS Sustainable Chemistry & Engineering*, *7*(8), 7833–7843. DOI: 10.1021/acssuschemeng.9b00027

John, V. L., Nair, Y., & Vinod, T. P. (2021). Doping and surface modification of carbon quantum dots for enhanced functionalities and related applications. *Particle &Amp. Particle & Particle Systems Characterization*, *38*(11), 2100170. Advance online publication. DOI: 10.1002/ppsc.202100170

Kadian, S., Manik, G., Kalkal, A., Singh, M., & Chauhan, R. P. (2019). Effect of sulfur doping on fluorescence and quantum yield of graphene quantum dots: An experimental and theoretical investigation. *Nanotechnology*, *30*(43), 435704. DOI: 10.1088/1361-6528/ab3566 PMID: 31342919

Kannan, K., Chanda, D., Gautam, J., Behera, A., Meshesha, M. M., Jang, S. G., & Yang, B. (2023). Hydrothermally synthesized mixed metal oxide nanocomposites for electrochemical water splitting and photocatalytic hydrogen production. *International Journal of Hydrogen Energy*, *48*(93), 36412–36426. DOI: 10.1016/j.ijhydene.2023.06.042

Kannan, K., Chanda, D., Gautam, J., Meshesha, M. M., Jang, S. G., & Yang, B. L. (2023). Facial synthesis of pp heterojunction composites: Evaluation of their electrochemical properties with photovoltaics-electrolyzer water splitting using two-electrode system. *International Journal of Hydrogen Energy*, *48*(37), 13814–13826. DOI: 10.1016/j.ijhydene.2022.12.235

Kannan, K., Chinnaiah, K., Gurushankar, K., Krishnamoorthi, R., Chen, Y. S., Murphin Kumar, P. S., & Li, Y. Y. (2024). Investigation of the electrochemical behavior of CuO-NiO-Co3O4 nanocomposites for enhanced supercapacitor applications. *Materials (Basel)*, *17*(16), 3976. DOI: 10.3390/ma17163976 PMID: 39203154

Kannan, K., Gautam, J., Chanda, D., Meshesha, M. M., Jang, S. G., & Yang, B. L. (2023). Two-dimensional MAX supported copper oxide/nickel Oxide/MAX as an efficient and novel photocatalyst for hydrogen evolution. *International Journal of Hydrogen Energy*, *48*(20), 7273–7283. DOI: 10.1016/j.ijhydene.2022.10.131

Kannan, K., Radhika, D., Gnanasangeetha, D., Krishna, L. S., & Gurushankar, K. (2021). Y3+ and Sm3+ co-doped mixed metal oxide nanocomposite: Structural, electrochemical, photocatalytic, and antibacterial properties. *Applied Surface Science Advances*, *4*, 100085. DOI: 10.1016/j.apsadv.2021.100085

Kannan, K., Radhika, D., Kasai, R. D., Gnanasangeetha, D., Palani, G., Gurushankar, K., Koutavarapu, R., Lee, D.-Y., & Shim, J. (2022). Facile fabrication of novel ceria-based nanocomposite (CYO-CSO) via co-precipitation: Electrochemical, photocatalytic and antibacterial performances. *Journal of Molecular Structure*, *1256*, 132519. DOI: 10.1016/j.molstruc.2022.132519

Kannan, K., & Tari, V. (Eds.). (2024). *Novel energy storage and conversion technologies for two-dimensional MXenes and MBenes*. https://doi.org/ ISBN: 9798369395462, eISBN: 9798369395486DOI: 10.4018/979-8-3693-9546-2

Khan, M. (2024). Innovations in battery technology: Enabling the revolution in electric vehicles and energy storage. *British Journal of Multidisciplinary and Advanced Studies*, *5*(1), 23–41. DOI: 10.37745/bjmas.2022.0414

Kim, B., Lee, S., & Kim, J. (2020). Inverse design of porous materials using artificial neural networks. *Science Advances*, *6*(1), eaax9324. Advance online publication. DOI: 10.1126/sciadv.aax9324 PMID: 31922005

Kim, K., & Ahn, H. (2022). Surface functional group-tailored b and n co-doped carbon quantum dot anode for lithium-ion batteries. *International Journal of Energy Research*, *46*(6), 8367–8375. DOI: 10.1002/er.7738

Koroglu, D., Bingol, H., & Uralcan, B. (2025). Enhanced flexible supercapacitors with boron-doped graphene electrodes and carbon quantum dot gel electrolytes. *RSC Advances*, *15*(7), 5011–5019. DOI: 10.1039/D4RA06990K PMID: 39957823

Kou, X., Jiang, S., Park, S., & Meng, L. (2020). A review: Recent advances in preparations and applications of heteroatom-doped carbon quantum dots. *Dalton Transactions (Cambridge, England)*, *49*(21), 6915–6938. DOI: 10.1039/D0DT01004A PMID: 32400806

Kwon, W., Lim, J., Lee, J., Park, T., & Rhee, S. (2013). Sulfur-incorporated carbon quantum dots with a strong long-wavelength absorption band. *Journal of Materials Chemistry. C, Materials for Optical and Electronic Devices*, *1*(10), 2002. DOI: 10.1039/c3tc00683b

Lee, H. R., Kim, Y. S., Lee, S. Y., Son, U. H., Lee, S., & Joh, H. (2024). Bifunctional effects of nitrogen-doped carbon quantum dots on cos2/mesoporous carbon composites for high-performance lithium-ion batteries. *Applied Surface Science*, *664*, 160228. DOI: 10.1016/j.apsusc.2024.160228

Lesch, R., Visser, E. D., Seroka, N. S., & Khotseng, L. (n.d.). *Biomass-Derived Carbon-Based Nanomaterials: Current Research.* Trends, and Challenges.

Levi, M. D., Levy, N., Sigalov, S., Salitra, G., Aurbach, D., & Maier, J. (2010). Electrochemical quartz crystal microbalance (eqcm) studies of ions and solvents insertion into highly porous activated carbons. *Journal of the American Chemical Society*, *132*(38), 13220–13222. DOI: 10.1021/ja104391g PMID: 20828121

Li, H. (2024). A review of edlc and pseudocapacitance with synergistic integration of carbon-based and metal oxide materials for enhanced electrochemical energy storage. *Applied and Computational Engineering*, *61*(1), 231–244. DOI: 10.54254/2755-2721/61/20240968

Li, Q., & Jin, Q. (2025). Holistic comparison of hydrothermal carbonization versus slow pyrolysis for biochar production: Techno-economic analysis and life cycle assessment. *ACS Sustainable Chemistry & Engineering*, *13*(31), 12382–12395. DOI: 10.1021/acssuschemeng.5c02249

Li, X., Liu, G., Zheng, H., Sun, K., Wan, L., Cao, J., Asif, S., Cao, Y., Si, W., Wang, F., & Bokhari, A. (2022). Recent advances on heteroatom-doped porous carbon—Based electrocatalysts for oxygen reduction reaction. *Energies*, *16*(1), 128. DOI: 10.3390/en16010128

Li, Y., Wang, S., Lv, Z., Wang, Z., Zhao, Y., Xie, Y., Xu, Y., Qian, L., Yang, Y., Zhao, Z., & Zhang, J. (2025). Transforming the synthesis of carbon nanotubes with machine learning models and automation. *Matter*, *8*(1), 101913. DOI: 10.1016/j.matt.2024.11.007

Li, Z., Bommier, C., Chong, Z. S., Jian, Z., Surta, T. W., Wang, X., Xing, Z., Neuefeind, J. C., Stickle, W. F., Dolgos, M., Greaney, P. A., & Ji, X. (2017). Mechanism of Na-Ion Storage in Hard Carbon Anodes Revealed by Heteroatom Doping. *Advanced Energy Materials*, *7*(18), 1602894. DOI: 10.1002/aenm.201602894

Liao, H., & Zheng, H. (2016). Liquid cell transmission electron microscopy. *Annual Review of Physical Chemistry*, *67*(1), 719–747. DOI: 10.1146/annurev-physchem-040215-112501 PMID: 27215823

Libber, M., Gariya, N., & Kumar, M. (2024). A comprehensive analysis of supercapacitors with current limitations and emerging trends in research. *Journal of Solid State Electrochemistry*, *29*(2), 513–527. DOI: 10.1007/s10008-024-06107-x

Liu, H., Jia, M., Zhu, Q., Cao, B., Chen, R., Wang, Y., Wu, F., & Xu, B. (2016). 3d-0d graphene-fe3o4 quantum dot hybrids as high-performance anode materials for sodium-ion batteries. *ACS Applied Materials &Amp. Interfaces*, *8*(40), 26878–26885. DOI: 10.1021/acsami.6b09496 PMID: 27642811

Liu, H., Liu, Z., Zhang, J., Zhi, L., & Wu, M. (2021). Boron and nitrogen co-doped carbon dots for boosting electrocatalytic oxygen reduction. *New Carbon Materials*, *36*(3), 585–593. DOI: 10.1016/S1872-5805(21)60043-4

Liu, L., Zhang, X., Liu, Y., & Gong, X. (2025). Electrochemical energy storage devices—Batteries, supercapacitors, and battery–supercapacitor hybrid devices. *ACS Applied Electronic Materials*, *7*(6), 2233–2270. DOI: 10.1021/acsaelm.5c00069

Liu, Q., Ge, K., Wu, X., Zhu, Z., Zhu, Y., & Bi, H. (2023). Preparation of Carbon Dots@r-GO Nanocomposite with an Enhanced Pseudo-Capacitance. *Molecules (Basel, Switzerland)*, *28*(2), 541. DOI: 10.3390/molecules28020541 PMID: 36677598

Liu, S., Gao, D., Li, J., Hui, K. S., Yin, Y., Hui, K. N., & Jun, S. C. (2019). Phosphorus dual-site driven cos2@s, n co-doped porous carbon nanosheets for flexible quasi-solid-state supercapacitors. *Journal of Materials Chemistry. A, Materials for Energy and Sustainability*, *7*(46), 26618–26630. DOI: 10.1039/C9TA09646A

Liu, T., Cui, Z., Zhou, J., Wang, Y., & Zou, Z. (2017). Synthesis of Pyridinic-Rich N, S Co-doped Carbon Quantum Dots as Effective Enzyme Mimics. *Nanoscale Research Letters*, *12*(1), 375. DOI: 10.1186/s11671-017-2149-y PMID: 28549380

Lopez-Astacio, H. J. H., Vargas-Pérez, B. L., Valle-Perez, A. D., Pollock, C. J., & Cunci, L. (2024). Open-source electrochemical cell for in situ x-ray absorption spectroscopy in transmission and fluorescence modes. *Journal of Synchrotron Radiation*, *31*(2), 322–327. DOI: 10.1107/S1600577524000122 PMID: 38306299

Lu, S., Cheng, C., Shi, Y., Wu, Y., Zhang, Z., & Zhang, B. (2023). Unveiling the structural transformation and activity origin of heteroatom-doped carbons for hydrogen evolution. *Proceedings of the National Academy of Sciences of the United States of America*, *120*(20), e2300549120. Advance online publication. DOI: 10.1073/pnas.2300549120 PMID: 37155878

Lu, S., Xiao, Q., Yang, W., Wang, X., Guo, T., Xie, Q., & Ruan, Y. (2024). Multi-heteroatom-doped porous carbon with high surface adsorption energy of potassium derived from biomass waste for high-performance supercapacitors. *International Journal of Biological Macromolecules*, *258*, 128794. DOI: 10.1016/j.ijbiomac.2023.128794 PMID: 38110166

Lu, W., Hao, L., & Wang, Y. (2022). Highly active N, S Co-Doped UltramicroCodopedarbon for High-Performance Supercapacitor Electrodes. *Micromachines*, *13*(6), 905. DOI: 10.3390/mi13060905 PMID: 35744519

Lu, Y.-F., Lo, S.-T., Lin, J.-C., Zhang, W., Lu, J.-Y., Liu, F.-H., Tseng, C.-M., Lee, Y.-H., Liang, C.-T., & Li, L.-J. (2013). Nitrogen-Doped Graphene Sheets Grown by Chemical Vapor Deposition: Synthesis and Influence of Nitrogen Impurities on Carrier Transport. *ACS Nano*, *7*(8), 6522–6532. DOI: 10.1021/nn402102y PMID: 23879622

Lv, C., Zhou, X., Zhong, L., Yan, C., Srinivasan, M., Seh, Z. W., Liu, C., Pan, H., Li, S., Wen, Y., & Yan, Q. (2021). Machine learning: An advanced platform for materials development and state prediction in lithium-ion batteries. *Advanced Materials*, *34*(25), 2101474. Advance online publication. DOI: 10.1002/adma.202101474 PMID: 34490683

Ma, Y., Chen, D., Fang, Z., Zheng, Y., Li, W., Xu, S., Lu, X., Shao, G., Liu, Q., & Yang, W. (2021). High energy density and extremely stable supercapacitors based on carbon aerogels with 100% capacitance retention up to 65,000 cycles. *Proceedings of the National Academy of Sciences of the United States of America*, *118*(21), e2105610118. DOI: 10.1073/pnas.2105610118 PMID: 34011610

Ma, Y., Sung, K.-W., & Ahn, H.-J. (2023). N- and F-Co-Doped Carbon QuaCodopeds Coated on a Ni Foam Substrate as Current Collector for Highly Stable Li-Air Batteries. *International Journal of Energy Research*, *2023*, 1–11. DOI: 10.1155/2023/5310171

MacLeod, B. P., Parlane, F. G. L., Morrissey, T. D., Häse, F., Roch, L. M., Dettelbach, K. E., Moreira, R., Yunker, L. P. E., Rooney, M. B., Deeth, J. R., Lai, V., Ng, G. J., Situ, H., Zhang, R., Elliott, M. S., Haley, T. H., Dvořák, D., Aspuru-Guzik, A., Hein, J. E., & Berlinguette, C. P. (2020). Self-driving laboratory for accelerated discovery of thin-film materials. *Science Advances*, *6*(20), eaaz8867. Advance online publication. DOI: 10.1126/sciadv.aaz8867 PMID: 32426501

Maleki, R., Asadnia, M., & Razmjou, A. (2022). Artificial Intelligence-Based Material Discovery for Clean Energy Future. *Advanced Intelligent Systems*, *4*(10), 2200073. DOI: 10.1002/aisy.202200073

Matek, B., Dorn, J., & Divita, F. (2019). *Life cycle assessment: C4V lithium-ion battery cells for electric vehicles. Final report* (Nos. 19–44). New York State Energy Research and Development Authority (NYSERDA). https://www.nyserda.ny.gov/-/media/Project/Nyserda/Files/Publications/Research/Transportation/19-44-Lifecycle-Assessment-of-CV4-Li-ion-Battery-Cells-for-Electric-Vehicles.pdf

Min, Y., Qin, B., Si, C., Sun, X., & Li, B. (2024). Electrochemical reactions catalyzed by carbon dots from computational investigations: Functional groups, dopants, and defects. *Journal of Materials Chemistry. A, Materials for Energy and Sustainability*, *12*(5), 2520–2560. DOI: 10.1039/D3TA06361E

Mondal, S., Rana, U., & Malik, S. (2015). Graphene quantum dot-doped polyaniline nanofiber as high performance supercapacitor electrode materials. *Chemical Communications*, *51*(62), 12365–12368. DOI: 10.1039/C5CC03981A PMID: 26095974

Nahirniak, S., Ray, A., & Saruhan, B. (2023). Challenges and Future Prospects of the MXene-Based Materials for Energy Storage Applications. *Batteries*, *9*(2), 126. DOI: 10.3390/batteries9020126

Nandy, A., Duan, C., Taylor, M. G., Liu, F., Steeves, A. H., & Kulik, H. J. (2021). Computational discovery of transition-metal complexes: From high-throughput screening to machine learning. *Chemical Reviews*, *121*(16), 9927–10000. DOI: 10.1021/acs.chemrev.1c00347 PMID: 34260198

Naushad, M., Ahamad, T., Ubaidullah, M., Ahmed, J., Ghafar, A. A., Al-Sheetan, K. M., & Arunachalam, P. (2021). Nitrogen-doped carbon quantum dots (N-CQDs)/Co3O4 nanocomposite for high performance supercapacitor. *Journal of King Saud University. Science*, *33*(1), 101252. DOI: 10.1016/j.jksus.2020.101252

Nguyen, K. G., Huš, M., Baragau, I., Bowen, J., Heil, T., Nicolaev, A., Abramiuc, L. E., Sapelkin, A., Sajjad, M. T., & Kellici, S. (2024). Engineering Nitrogen-Doped Carbon Quantum Dots: Tailoring Optical and Chemical Properties through Selection of Nitrogen Precursors. *Small*, *20*(24), 2310587. DOI: 10.1002/smll.202310587 PMID: 38546418

Niu, L., Yang, L., Yang, J., Chen, M., Zeng, L., Duan, P., Wu, T., Pameté, E., Presser, V., & Feng, G. (2023). Understanding the charging of supercapacitors by electrochemical quartz crystal microbalance. *Industrial Chemistry & Materials*, *1*(2), 175–187. DOI: 10.1039/D2IM00038E

Ozyurt, D., Kobaisi, M. A., Hocking, R. K., & Fox, B. (2023). Properties, synthesis, and applications of carbon dots: A review. *Carbon Trends*, *12*, 100276. DOI: 10.1016/j.cartre.2023.100276

Pal, M., Pal, P., & Nandi, M. (2025). B/n-codoped porous carbco-dopedmetal-free and binder-free electrode material for supercapacitors. *Chemistry (Weinheim an der Bergstrasse, Germany)*, *31*(34), e202500800. Advance online publication. DOI: 10.1002/chem.202500800 PMID: 40351145

Pallavolu, M. R., Prabhu, S., Nallapureddy, R. R., Kumar, A. S., Banerjee, A. N., & Joo, S. W. (2023). Bio-derived graphitic carbon quantum dot encapsulated s- and n-doped graphene sheets with unusual battery-type behavior for high-performance supercapacitor. *Carbon*, *202*, 93–102. DOI: 10.1016/j.carbon.2022.10.077

Pan, Z., Yang, J., Kong, J., Loh, X. J., Wang, J., & Liu, Z. (2022). "Porous and Yet Dense" Electrodes for High-Volumetric-Performance Electrochemical Capacitors: Principles, Advances, and Challenges. *Advancement of Science*, *9*(4), 2103953. DOI: 10.1002/advs.202103953 PMID: 34796698

Panigrahi, S. (2025). *Carbon Quantum Dots Prepared by Hydrothermal Method for Drug Delivery, Photocatalysis, and Bioimaging Applications*. Chemistry. DOI: 10.26434/chemrxiv-2025-w4117

Park, S. K., Lee, H., Choi, M. S., Suh, D. H., Nakhanivej, P., & Park, H. S. (2018). Straightforward and controllable synthesis of heteroatom-doped carbon dots and nanoporous carbons for surface-confined energy and chemical storage. *Energy Storage Materials*, *12*, 331–340. DOI: 10.1016/j.ensm.2017.10.008

Permatasari, F. A., Irham, M. A., Bisri, S. Z., & Iskandar, F. (2021). Carbon-Based Quantum Dots for Supercapacitors: Recent Advances and Future Challenges. *Nanomaterials (Basel, Switzerland)*, *11*(1), 91. DOI: 10.3390/nano11010091 PMID: 33401630

Pillar-Little, T. J.Jr. (2018). Carbon Quantum Dots: Bridging the Gap Between Chemical Structure and Material Properties [University of Kentucky Libraries; PDF]. DOI: 10.13023/ETD.2018.108

Pollice, R., Gomes, G. P., Aldeghi, M., Hickman, R. J., Krenn, M., Lavigne, C., Lindner-D'Addario, M., Nigam, A., Ser, C. T., Yao, Z., & Aspuru-Guzik, A. (2021). Data-driven strategies for accelerated materials design. *Accounts of Chemical Research*, *54*(4), 849–860. DOI: 10.1021/acs.accounts.0c00785 PMID: 33528245

Poolakkandy, R. R., & Menamparambath, M. M. (2020). Soft-template-assisted synthesis: A promising approach for the fabrication of transition metal oxides. *Nanoscale Advances*, *2*(11), 5015–5045. DOI: 10.1039/D0NA00599A PMID: 36132034

Porada, S., Schipper, F., Aslan, M., Antonietti, M., Presser, V., & Fellinger, T. (2015). Capacitive deionization using biomass-based microporous salt-templated heteroatom-doped carbons. *ChemSusChem*, *8*(11), 1867–1874. DOI: 10.1002/cssc.201500166 PMID: 25970654

Priya, A. S., Premanand, R., Ragupathi, I., Bhaviripudi, V. R., Aepuru, R., Kannan, K., & Shanmugaraj, K. (2024). Comprehensive review of hydrogel synthesis, characterization, and emerging applications. *Journal of Composites Science*, *8*(11), 457. DOI: 10.3390/jcs8110457

Qin, N., Tong, B., Ling, X., Shi, J., Wei, W., & Mi, L. (2023). Fabrication of nitrogen-doped carbon-coated nis1.97 quantum dots for advanced magnesium-ion batteries. *Langmuir*, *39*(45), 16111–16117. DOI: 10.1021/acs.langmuir.3c02250 PMID: 37924327

Quarti, M., Bayer, A., & Bessler, W. G. (2023). Trade-off between energy density and fast-charge capability of lithium-ion batteries: A model-based design study of cells with thick electrodes. *Electrochemical Science Advances*, *3*(1), e2100161. DOI: 10.1002/elsa.202100161

Rasal, A. S., Yadav, S., Yadav, A., Kashale, A. A., Manjunatha, S. T., Altaee, A., & Chang, J. (2021). Carbon quantum dots for energy applications: A review. *ACS Applied Nano Materials*, *4*(7), 6515–6541. DOI: 10.1021/acsanm.1c01372

Rhili, K., Chergui, S., Abergo-Martinez, J. C., El Douhaibi, A. S., & Siaj, M. (2023). Phosphorus and nitrogen co-doped reduced graphene oxide as superior electrode nanomaterials for supercapacitors. *Materials Advances*, *4*(21), 5263–5272. DOI: 10.1039/D3MA00553D

Rocco, D., Moldoveanu, V. G., Feroci, M., Bortolami, M., & Vetica, F. (2023). Electrochemical Synthesis of Carbon Quantum Dots. *ChemElectroChem*, *10*(3), e202201104. DOI: 10.1002/celc.202201104 PMID: 37502311

Saari, J., Martinez, C. M., Kuparinen, K., Lipiäinen, S., Hamaguchi, M., Cardoso, M., Vakkilainen, E., & Haider, M. (2023). Techno-economic analysis of hydrothermal carbonization of pulp mill biosludge. *Tappi Journal*, *22*(3), 206–216. DOI: 10.32964/TJ22.3.206

Sainio, S., Wester, N., Aarva, A., Titus, C. J., Nordlund, D., Kauppinen, E. I., Leppänen, E., Palomäki, T., Koehne, J. E., Pitkänen, O., Kordas, K., Kim, M., Lipsanen, H., Mozetič, M., Caro, M. A., Meyyappan, M., Koskinen, J., & Laurila, T. (2021). Trends in Carbon, Oxygen, and Nitrogen Core in the X-ray Absorption Spectroscopy of Carbon Nanomaterials: A Guide for the Perplexed. *The Journal of Physical Chemistry. C, Nanomaterials and Interfaces*, *125*(1), 973–988. DOI: 10.1021/acs.jpcc.0c08597

Samantara, A. K., Sahu, S. C., Ghosh, A., & Jena, B. K. (2015). Sandwiched graphene with nitrogen, sulphur co-doped cqds: An efficient metal-free material for energy storage and conversion applications. *Journal of Materials Chemistry. A, Materials for Energy and Sustainability*, *3*(33), 16961–16970. DOI: 10.1039/C5TA03376D

Sarath Kumar, C. B., Reji, R. P., Sivalingam, Y., Kawazoe, Y., & Surya, V. J. (2024). Carbon and boron nitride quantum dots as optical sensor probes for selective detection of toxic metals in drinking water: A quantum chemical prediction through structure- and morphology-dependent electronic and optical properties. *RSC Advances*, *14*(38), 28182–28200. DOI: 10.1039/D4RA04843A PMID: 39234523

Saroja, A. P. V. K., & Garapati, M. S. (2020). Facile synthesis of heteroatom doped and undoped graphene quantum dots as active materials for reversible lithium and sodium ions storage. *Applied Surface Science*, *504*, 144430. DOI: 10.1016/j.apsusc.2019.144430

Sead, F. F., Makasana, J., Saraswat, S. K., Rekha, M., Kundlas, M., Saini, S., Joshi, K. K., & Noorizadeh, H. (2025). Electrochemical behavior of carbon quantum dots as electrolyte additives for enhanced battery and supercapacitor performance. *Materials Technology*, *40*(1), 2500524. Advance online publication. DOI: 10.1080/10667857.2025.2500524

Shah, S. S. (2024). Biomass-derived carbon materials for advanced metal-ion hybrid supercapacitors: A step towards more sustainable energy. *Batteries*, *10*(5), 168. DOI: 10.3390/batteries10050168

Shanmugasundaram, E., Vellaisamy, K., Ganesan, V., Narayanan, V., Saleh, N., & Thambusamy, S. (2024). Dual Applications of Cobalt-Oxide-Grafted Carbon Quantum Dot Nanocomposite for Two Electrode Asymmetric Supercapacitors and Photocatalytic Behavior. *ACS Omega*, *9*(12), 14101–14117. DOI: 10.1021/acsomega.3c09594 PMID: 38559980

Shi, R., Li, Z., Yu, H., Shang, L., Zhou, C., Waterhouse, G. I. N., Wu, L., & Zhang, T. (2017). Effect of nitrogen doping level on the performance of n-doped carbon quantum dot/tio2 composites for photocatalytic hydrogen evolution. *ChemSusChem*, *10*(22), 4650–4656. DOI: 10.1002/cssc.201700943 PMID: 28671326

Sohal, N., Maity, B., & Basu, S. (2021). Recent advances in heteroatom-doped graphene quantum dots for sensing applications. *RSC Advances*, *11*(41), 25586–25615. DOI: 10.1039/D1RA04248C PMID: 35478909

Somaraj, G., Mathew, S., Abraham, T., Ambady, K. G., Mohan, C., & Mathew, B. (2022). Nitrogen and sulfur co-doped carbon quantum dots for sensing applications: A review. *ChemistrySelect*, *7*(19), e202200473. Advance online publication. DOI: 10.1002/slct.202200473

Song, S., Wu, K., Wu, H., Guo, J., & Zhang, L. (2019). Multi-shelled zno decorated with nitrogen and phosphorus co-doped carbon quantum dots: Synthesis and enhanced photodegradation activity of methylene blue in aqueous solutions. *RSC Advances*, *9*(13), 7362–7374. DOI: 10.1039/C9RA00168A PMID: 35519954

Sousa, H. B. A., Martins, C. S. M., & Prior, J. A. V. (2021). You Don't Learn That in School: An Updated Practical Guide to Carbon Quantum Dots. *Nanomaterials (Basel, Switzerland)*, *11*(3), 611. DOI: 10.3390/nano11030611 PMID: 33804394

Suman, R., Rani, G., Bhawna, , Kumar, H., Devi, J., & Siddharth, . (2025). B, S doped zero-dimensional carbon quantum dots from bio-waste for high density supercapacitors. *Carbon Letters*. Advance online publication. DOI: 10.1007/s42823-025-00945-z

Suman, R., Rani, G., Bhawna, , Kumar, H., Devi, J., & Siddharth, . (2025, July 10). G., Bhawna, Kumar, H., Devi, J., & Siddharth. (2025). B, S doped zero-dimensional carbon quantum dots from bio-waste for high density supercapacitors. *Carbon Letters*. Advance online publication. DOI: 10.1007/s42823-025-00945-z

Susi, T., Pichler, T., & Ayala, P. (2015). X-ray photoelectron spectroscopy of graphitic carbon nanomaterials doped with heteroatoms. *Beilstein Journal of Nanotechnology*, *6*, 177–192. DOI: 10.3762/bjnano.6.17 PMID: 25671162

Suzuki, K., Miyamura, H., & Balachandran, J. (2021). One-pot hydrothermal synthesis of carbon dots-immobilized hydrozincite for ZnO-based nanocomposite lighting applications. *Journal of Asian Ceramic Societies*, *9*(4), 1473–1480. DOI: 10.1080/21870764.2021.1995939

Tian, L., Li, Z., Wang, P., Zhai, X., Wang, X., & Li, T. (2021). Carbon quantum dots for advanced electrocatalysis. *Journal of Energy Chemistry*, *55*, 279–294. DOI: 10.1016/j.jechem.2020.06.057

Timoshenko, J., & Cuenya, B. R. (2020). In situ/operando electrocatalyst characterization by x-ray absorption spectroscopy. *Chemical Reviews*, *121*(2), 882–961. DOI: 10.1021/acs.chemrev.0c00396 PMID: 32986414

Tu, H., Zhang, Y., Wu, J., Li, Y., Liu, H., Deng, W., Zou, G., Hou, H., & Ji, X. (2024). Revealing electrochemical process of functional carbon dots stabilized sodium metal anode: Co-deposition and strengthened sei films. *Advanced Functional Materials*, *35*(3), 2413488. Advance online publication. DOI: 10.1002/adfm.202413488

Tu, Z., Yang, G., Song, H., & Wang, C. (2016). Amorphous zno quantum dot/mesoporous carbon bubble composites for a high-performance lithium-ion battery anode. *ACS Applied Materials &Amp. Interfaces*, *9*(1), 439–446. DOI: 10.1021/acsami.6b13113 PMID: 27966898

Uppugalla, S., Pothu, R., Boddula, R., Desai, M. A., & Al-Qahtani, N. (2023). Nitrogen and sulfur co-doped activated carbon nanosheets for high-performance coin cell supercapacitor device with outstanding cycle stability. *Emergent Materials*, *6*(4), 1167–1176. DOI: 10.1007/s42247-023-00503-1

Upreti, G., Dhingra, R., Naidu, S., Atuahene, I., & Sawhney, R. (2015). Life cycle assessment of nanomaterials. *Green Processes for Nanotechnology*, 393–408. DOI: 10.1007/978-3-319-15461-9_14

U.S. Department of Energy. Industrial Efficiency & Decarbonization Office. (2023). *Techno-economic, energy, & carbon heuristic tool for early-stage technologies (TECHTEST) tool.* https://www.energy.gov/eere/iedo/techno-economic-energy-carbon-heuristic-tool-early-stage-technologies-techtest-tool

Veerakumari, P., Rao, A. V., Devi, K. A., Rao, P. V., Raziya, S., Shanmugan, S., & Kannan, K. (2025). A Review on Tailoring Metal–Organic Frameworks for Electrochemical Applications: Advances in Synthesis and Structure for Energy Storage. *iScience*.

Veerasubramani, G. K., Park, M., Choi, J., & Kim, D. (2020). Ultrasmall sns quantum dots anchored onto nitrogen-enriched carbon nanospheres as an advanced anode material for sodium-ion batteries. *ACS Applied Materials &Amp. Interfaces*, *12*(6), 7114–7124. DOI: 10.1021/acsami.9b18997 PMID: 31944653

Wang, E., Wan, J., Guo, Y., Zhang, Q., He, W., Zhang, C., Chen, W., Yan, H., Xue, D., Fang, T., Wang, F., Wen, R., Xin, S., Yin, Y., & Guo, Y. (2022). Mitigating electron leakage of solid electrolyte interface for stable sodium-ion batteries. *Angewandte Chemie International Edition*, *62*(4), e202216354. Advance online publication. DOI: 10.1002/anie.202216354 PMID: 36440597

Wang, X., Feng, Y., Dong, P., & Huang, J. (2019). A mini review on carbon quantum dots: Preparation, properties, and electrocatalytic application. *Frontiers in Chemistry*, *7*, 671. Advance online publication. DOI: 10.3389/fchem.2019.00671 PMID: 31637234

Wang, Y., & Hu, A. (2014). Carbon quantum dots: Synthesis, properties and applications. *Journal of Materials Chemistry. C, Materials for Optical and Electronic Devices*, *2*(34), 6921. DOI: 10.1039/C4TC00988F

Wang, Y., Jing, Y., Wang, L., Kong, W., Wang, S., Wang, Z., Li, Y., & Lu, Q. (2019). Photoluminescence of graphene quantum dots doped with different elements. *Kexue Tongbao*, *64*(4), 411–418. DOI: 10.1360/N972018-00986

Wang, Z., Xie, Y., Lei, Z., Lu, Y., Wei, G., Liu, S., Xu, C., Zhang, Z., Wang, X., Rao, L., & Chen, J. (2019). Quantitative analysis of surface sites on carbon dots and their interaction with metal ions by a potentiometric titration method. *Analytical Chemistry*, *91*(15), 9690–9697. DOI: 10.1021/acs.analchem.9b01225 PMID: 31245999

Wareing, T., Gentile, P., & Phan, A. N. (2021). Biomass-based carbon dots: Current development and future perspectives. *ACS Nano*, *15*(10), 15471–15501. DOI: 10.1021/acsnano.1c03886 PMID: 34559522

Weaving, J. S., Lim, A., Millichamp, J., Neville, T. P., Ledwoch, D., Kendrick, E., McMillan, P. F., Shearing, P. R., Howard, C. A., & Brett, D. J. L. (2020). Elucidating the sodiation mechanism in hard carbon by operando raman spectroscopy. *ACS Applied Energy Materials*, *3*(8), 7474–7484. DOI: 10.1021/acsaem.0c00867

Wei, F., Zeng, Y., Guo, Y., Li, J., Zhu, S., Gao, S., Zhang, H., & He, X. (2023). Recent progress on the heteroatom-doped carbon cathode for zinc ion hybrid capacitors. *Chemical Engineering Journal*, *468*, 143576. DOI: 10.1016/j.cej.2023.143576

Wiggins-Camacho, J. D., & Stevenson, K. J. (2009). Effect of nitrogen concentration on capacitance, density of states, electronic conductivity, and morphology of n-doped carbon nanotube electrodes. *The Journal of Physical Chemistry. C, Nanomaterials and Interfaces*, *113*(44), 19082–19090. DOI: 10.1021/jp907160v

Wikimedia Commons contributors. (2020). *Infographic – Lifecycle analysis of engineered nanomaterials (ENM)* [Graphic]. https://commons.wikimedia.org/wiki/File:Infographic_%E2%80%93_Lifecycle_analysis_of_engineered_nanomaterials_(ENM).jpg

Wu, C., Yin, Z., Wang, W., Liu, Y., Yu, J., & Liu, Y. (2022). Phase behavior and aggregate transition based on co-assembly of negatively charged carbon dots and a ph-responsive tertiary amine cationic surfactant. *Langmuir*, *38*(45), 13771–13781. DOI: 10.1021/acs.langmuir.2c01895 PMID: 36318637

Wu, M., Li, J., Wu, Y., Gong, X., & Wu, M. (2023). Design of a synthetic strategy to achieve enhanced fluorescent carbon dots with sulfur and nitrogen codoping and its muco-dopingional applications. *Small*, *19*(42), 2302764. Advance online publication. DOI: 10.1002/smll.202302764 PMID: 37330653

Wu, S., Wang, X., Zhu, Y., He, P., Yu, X., Qin, F., Yao, Y., & Ren, L. (2023). Different types of nitrogen-doped cqds loaded by cop used as oer electrocatalysts. *Inorganic Chemistry Communications*, *153*, 110872. DOI: 10.1016/j.inoche.2023.110872

Wu, S., & Zhu, Y. (2016). Highly densified carbon electrode materials towards practical supercapacitor devices. *Science China Materials*, *60*(1), 25–38. DOI: 10.1007/s40843-016-5109-4

Xie, W., Zhang, Q., Song, S., Cheng, X., Yang, Y., Wang, L., Ouyang, X., Xie, S., & Huang, J. (2023). High s-doped amorphous carbon/carbon quantum dots coupled micro-frame for highly efficient potassium storage. *Journal of Colloid and Interface Science*, *652*, 1522–1532. DOI: 10.1016/j.jcis.2023.08.123 PMID: 37660609

Xu, Q., Liu, Y., Gao, C., Wei, J., Zhou, H., Chen, Y., Dong, C., Sreenivasan, S. T., Li, N., & Xia, Z. (2015). Synthesis, mechanistic investigation, and application of photoluminescent sulfur and nitrogen co-doped carbon dots. *Journal of Materials Chemistry. C, Materials for Optical and Electronic Devices*, *3*(38), 9885–9893. DOI: 10.1039/C5TC01912E

Xu, Q., Niu, Y., Li, J., Yang, Z., Gao, J., Ding, L., Ni, H., Zhu, P., Liu, Y., Tang, Y., Lv, Z.-P., Peng, B., Hu, T. S., Zhou, H., & Xu, C. (2022). Recent progress of quantum dots for energy storage applications. *Carbon Neutrality*, *1*(1), 13. DOI: 10.1007/s43979-022-00002-y

Xue, L., Li, Y., Hu, A., Zhou, M., Chen, W., Lei, T., Yan, Y., Huang, J., Yang, C., Wang, X., Hu, Y., & Xiong, J. (2022). In situ/operando raman techniques in lithium–sulfur batteries. *Small Structures*, *3*(3), 2100170. Advance online publication. DOI: 10.1002/sstr.202100170

Yang, D., Cui, H., Soomro, R. A., Zhang, P., & Xu, B. (2025). Flexible, highly conductive, and microscopically disordered MXene film electrode constructed via in-situ carbon dots intercalation toward high-performatowardspercapacitors. *Energy Materials*, *5*(1). Advance online publication. DOI: 10.20517/energymater.2024.45

Yang, S., Xu, Z., Wang, S., Sun, J., Zhao, D., Cao, B., & Wang, X. (2024). Hydrophilic and nanocrystalline carbon quantum dots enable highly reversible zinc-ion batteries. *Green Chemistry*, *26*(12), 7293–7301. DOI: 10.1039/D4GC01434K

Yang, Y., Liu, X., Dai, Z., Yuan, F., Bando, Y., Golberg, D., & Wang, X. (2017). In situ electrochemistry of rechargeable battery materials: Status report and perspectives. *Advanced Materials*, *29*(31), 1606922. Advance online publication. DOI: 10.1002/adma.201606922 PMID: 28627135

Yao, J., Liu, C., Li, J., Hu, Z., Zhou, R., Guo, C., Liu, X., Yang, F., & Zhu, Y. (2023). 3d nitrogen and boron dual-doped carbon quantum dots/reduced graphene oxide aerogel for advanced aqueous and flexible quasi-solid-state zinc-ion hybrid capacitors. *Rare Metals*, *42*(7), 2307–2323. DOI: 10.1007/s12598-023-02265-5

Yashwanth, H., Rondiya, S. R., Dzade, N. Y., Dhole, S., Phase, D. M., & Hareesh, K. (2020). Enhanced photocatalytic activity of n, p, co-doped carbon quantum dots: An insight from experimental and computational approach. *Vacuum*, *180*, 109589. DOI: 10.1016/j.vacuum.2020.109589

Yin, T., Guo, Y., Huang, X., Yang, X., Qin, L., Ning, T., Tan, L., Li, L., & Zou, K. (2025). Heteroatom Doping Strategy of Advanced Carbon for Alkali Metal-Ion Capacitors. *Batteries*, *11*(2), 69. DOI: 10.3390/batteries11020069

Yu, J., Yong, X., Tang, Z., Yang, B., & Lu, S. (2021). Theoretical understanding of structure–property relationships in luminescence of carbon dots. *The Journal of Physical Chemistry Letters*, *12*(32), 7671–7687. DOI: 10.1021/acs.jpclett.1c01856 PMID: 34351771

Yu, Y., Chen, Z., Yu, H., Zhang, X., Liu, T., Xia, M., Zheng, R., Shui, M., & Shu, J. (2020). Heteroatom-doped carbon-based materials for lithium and sodium ion batteries. *Energy Storage Materials*, *32*, 65–90. DOI: 10.1016/j.ensm.2020.07.027

Yuan, T., Yuan, F., Sui, L., Zhang, Y., Li, Y., Li, X., Tan, Z., & Fan, L. (2023). Carbon quantum dots with near-unity quantum yield bandgap emission for electroluminescent light-emitting diodes. *Angewandte Chemie*, *135*(20), e202218568. Advance online publication. DOI: 10.1002/ange.202218568 PMID: 36924197

Yuan, Y., Pu, S. D., Gao, X., & Robertson, A. W. (2023). The application of in situ liquid cell tem in advanced battery research. *Energy Materials*. Advance online publication. DOI: 10.20517/energymater.2023.14

Yun, X., Li, J., Chen, X., Chen, H., Li, X., Xiang, K., Chen, W., Liao, H., & Zhu, Y. (2019). Porous fe2o3 modified by nitrogen-doped carbon quantum dots/reduced graphene oxide composite aerogel as a high-capacity and high-rate anode material for alkaline aqueous batteries. *ACS Applied Materials &Amp. Interfaces*, *11*(40), 36970–36984. DOI: 10.1021/acsami.9b12827 PMID: 31487152

Yun, Y. S., Kim, D. H., Hong, S. J., Park, M., Park, Y., Kim, B. H., Jin, H., & Kang, K. (2015). Microporous carbon nanosheets with redox-active heteroatoms for pseudocapacitive charge storage. *Nanoscale*, *7*(37), 15051–15058. DOI: 10.1039/C5NR04231C PMID: 26315977

Zhai, Y., Zhang, B., Shi, R., Zhang, S., Liu, Y., Wang, B., Zhang, K., Waterhouse, G. I. N., Zhang, T., & Lu, S. (2021). Carbon dots as new building blocks for electrochemical energy storage and electrocatalysis. *Advanced Energy Materials*, *12*(6), 2103426. Advance online publication. DOI: 10.1002/aenm.202103426

Zhang, J., Geng, W., Shi, L., Yang, C. C., Zhang, X., Geng, Y., Hirani, R. A. K., Xu, X., Wei, J., Jing, Y., Zhang, S., Zhang, H., Wang, S., & Sun, H. (2022). One–pot synthesis of boron and nitrogen co–doped nanocarboco-dopedfficient catalytic reduction of nitrophenols. *Chemical Engineering Journal*, *439*, 135733. DOI: 10.1016/j.cej.2022.135733

Zhu, L., Shen, D., Wu, C., & Gu, S. (2020). State-of-the-art on the preparation, modification, and application of biomass-derived carbon quantum dots. *Industrial &Amp. Industrial & Engineering Chemistry Research*, *59*(51), 22017–22039. DOI: 10.1021/acs.iecr.0c04760

Chapter 9
Carbon Quantum Dots With Tailored Heterogeneous Atoms for Electrochemical Sensors

Sreejitha Raj
Department of Mechanical Engineering, Amrita Vishwa Vidyapeetham, Amritapuri, India

Lakshmi Krishnaa Suresh
Department of Mechanical Engineering, Amrita Vishwa Vidyapeetham, Amritapuri, India

Akhila Raman
Department of Mechanical Engineering, Amrita Vishwa Vidyapeetham, Amritapuri, India

Saritha Appukuttan
Department of Mechanical Engineering, Amrita Vishwa Vidyapeetham, Amritapuri, India

Gopika Venu
Department of Mechanical Engineering, Amrita Vishwa Vidyapeetham, Amritapuri, India

A. Santhy
https://orcid.org/0000-0002-7748-7975
Department of Mechanical Engineering, Amrita Vishwa Vidyapeetham, Amritapuri, India

ABSTRACT

Carbon quantum dots (CQDs) have emerged as an exciting class of carbon nanomaterials due to their small size, high surface area, and excellent electrochemical activity. However, pristine CQDs often face challenges such as limited selectivity, moderate sensitivity, and insufficient stability under complex operating conditions. To overcome these limitations, tailoring CQDs with heterogeneous atoms has gained significant attention. The introduction of heteroatoms into the carbon lattice not only modulates the electronic structure but also enriches the surface

DOI: 10.4018/979-8-3373-4247-4.ch009

chemistry, creating abundant active sites for redox reactions. Such engineered CQDs have shown remarkable potential in electrochemical sensing. This chapter presents a comprehensive overview of the synthesis approaches, characterization techniques, and application landscape of heterogeneous atom-tailored CQDs for electrochemical sensors. Finally, perspectives on emerging research opportunities are provided, aiming to inspire the development of next-generation electrochemical sensing platforms driven by engineered CQDs.

1. INTRODUCTION

Carbon Quantum Dots (CQDs) are ultrasmall, zero-dimensional, quasi-spherical carbon-based nanoparticles, typically less than 10 nm in size. Its remarkable physicochemical properties include high photoluminescence, low toxicity, good biocompatibility, and chemical inertness. These cluster properties uniquely position them for major roles in bioimaging, sensing, catalysis, optoelectronics, and therapeutic delivery etc. In many aspects, CQDs can take the place of other well-known quantum dots, like perovskite-based, Si-based, etc., due to their excellent compatibility and economical preparation techniques (Lim et al., 2015). CQDs are typically derived from carbon-rich precursors spanning both natural and synthetic materials via adopting either the bottom-up or top-down strategy (M. Liu et al., 2016)(Fawaz et al., 2023) (Emam, 2024). When the bulk carbon precursors, including graphite or CNTs, are systematically broken down into nanoscale domains via prominent techniques like laser ablation, arc-discharge, ball-milling, and plasma treatment. Meanwhile, the bottom-up approach assembles CQDs from small natural molecular precursors such as fruit peels, glucose, citric acid, chitosan, etc., which are subjected to controlled temperature and pressure processes like hydrothermal, solvothermal, microwave, and ultrasonic-assisted treatments (Singh, Inderbir, Riya Arora, Hardik Dhiman, 2018) (Eliboev et al., 2025). Thus, depending on the method of preparation and precursors, the size, shape, composition, and surface functionalities may directly affect its diverse physicochemical properties and applications. The strong photoluminescence effect with low toxicity enables CQDs to be used as a fluorescent for cellular and tissue imaging targeted for drug delivery and therapeutic monitoring. Also, the color-emitting property of CQDs makes them ideal for optoelectronics devices like light-emitting diodes (LEDs) and display devices. Additionally, CQDs have revolutionized the field of electrochemical sensors, providing unmatched sensitivity, selectivity, and operational stability through their exceptional electron transfer capabilities and rich surface functionalities. This enables precise detection of environmental toxins, heavy metals, organic pollutants, and biologically relevant molecules (Arora et al., 2024; Kannan K. & Tari V., 2024). Modifying such an electrochemical sensor that converts chemical information of the analyte into an electrical signal via redox reaction at the electrode interface is necessary for its advancement (Rao & Prakash, 2025). Also, combustion-based flame technique can be used even for single atom as well as dual atom doping of carbon quantum dot where the dopant concentration can be adjusted too (Han et al., 2015).

Generally, electrochemical sensors are fabricated using a variety of materials designed to optimize their chemical stability, including carbon-based materials, metals and metal oxides, conductive polymers, quantum dots, and metal organic frameworks (Rafiq et al., 2024). Moreover, CQDs, with their excellent electron transfer kinetics, large reactive surfaces, and versatile surface functionality, interact smoothly with the analytes and boost their performance. The conjugated π-electron system with sp^2 hybridization provides efficient electron transport with less resistance at the electrode interface. However, despite their numerous advantages, CQDs do possess certain limitations when employed alone in electrochemical sensing that reduce their usefulness in specified applications (Molaei, 2020). For instance, the partial crystallinity to a largely amorphous-like carbon structure of CQDs can influence the electronic properties, particularly conductivity. Sometimes, the surface defect and non-uniformity in size during synthesis can hinder the selectivity in sensing applications. To overcome these inherent limitations associated with pristine CQDs, the strategy involving tailoring or doping with heterogeneous atoms has been employed in recent years (Kou et al., 2020) (Elugoke et al., 2024). The integration of non-carbon atoms like non-metallic nitrogen, sulphur, boron, phosphorus, oxygen, zinc, cobalt, etc. onto the CQDs surface. Since these atoms differ chemically and electrically from carbon, they enable the modification of the electronic structure, surface chemistry, and physicochemical properties (Sohal et al., 2021). Silicon is considered as a biocompatible element which offers a promising way for the doping of carbon quantum dots (Qian et al., 2014). This can be achieved through multiple synthesis routes, and with control over the dopant precursor, concentration, and distribution, fine-tuned CQDs specifically for electrochemical applications can be created. Therefore, exploration of heterogenous atom-tailored CQDs can be of paramount importance as they have a vital potential use in electrochemical sensors.

2. SYNTHESIS METHODS OF HETEROGENEOUS ATOM-TAILORED CQDS

The synthesis of heterogeneous atom-tailored CQDs has primarily been achieved through several bottom-up approaches, including hydrothermal, microwave-assisted, pyrolysis, solvothermal, and electrochemical routes. Hydrothermal synthesis is versatile for multi-doping. Microwave-assisted or pyrolysis techniques deliver high yields and fast synthesis, while solvothermal techniques enable controlled Nitrogen or phosphorus incorporation.

2.1. Hydrothermal Synthesis

The hydrothermal approach involves the thermal treatment of molecular carbon and heteroatom precursors within sealed autoclaves. This method is extensively employed for the doping of single, dual, and multiple heteroatoms, as it allows for the introduction of dopants into the reaction mixture, facilitating their incorporation during the carbonisation process. Standard hydrothermal syntheses of CQDs are conducted at temperatures ranging from 150 to 220 °C over extended periods. These processes utilise various carbon sources, including sugars, amino acids, or small molecules, in conjunction with heteroatom-containing acids or amines to introduce dopants into the resulting materials.

Hydrothermal and solvothermal methods are among the most widely employed techniques for synthesising heteroatom-doped CQDs, owing to their simplicity, scalability, and ability to incorporate various dopants. These bottom-up approaches typically involve the reaction of carbon-rich precursors in aqueous or organic solvents under elevated temperature and pressure in sealed Teflon-lined autoclaves. By carefully selecting the precursors and reaction conditions, efficient heteroatom doping can be achieved to tune the structural and optical properties of CQDs. For instance, Dey et al. synthesized nitrogen-doped CQDs (N-CQDs) using glucose and urea (1.0 g each) under hydrothermal conditions at 180 °C for 10 hours, resulting in CQDs with a nitrogen content of 18–22% and excellent fluorescence characteristics (Dey et al., 2014a). Similarly, Aparicio et al. employed kraft lignin as a carbon source and ethylenediamine (EDA) as a nitrogen dopant, optimizing the hydrothermal synthesis via response surface methodology. Their work demonstrated a significant increase in nitrogen content and quantum yield, underscoring the effectiveness of lignin waste as a sustainable precursor (Aparício et al., 2025). Green synthesis approaches have also gained traction, as seen in the work of Alhabardi et al., who utilized banana (Musa paradisiaca) biomass in a one-step hydrothermal carbonization process at 300 °C for 2 hours, leveraging the intrinsic nitrogen in the biomass for doping (Alhabardi et al., 2025). Furthermore, heteroatom co-doping has been explored to further enhance performance. One such example includes phosphorus and nitrogen co-doped CQDs synthesized using small molecular precursors under hydrothermal conditions at 200 °C for 6 hours, enabling tunable optical and electronic characteristics ideal for advanced sensing and bioimaging applications (Yadav & Lahariya, 2025a). Overall, hydrothermal synthesis provides a versatile and environmentally benign platform for producing high-quality, heteroatom-tailored CQDs with customizable functionalities. These methods not only reduce synthetic complexity and environmental impact but also allow for fine-tuning of CQD composition, making them highly promising for applications in sensing, bioimaging, and optoelectronics.

2.2. Microwave-Assisted Synthesis

Microwave-assisted synthesis has emerged as a highly efficient and rapid method for producing heteroatom-doped CQDs, offering advantages such as uniform temperature distribution, reduced reaction time, and improved energy efficiency compared to conventional thermal techniques. This method enables precise control over the doping process and structural properties of CQDs, making it ideal for tailoring their photophysical behaviours. A notable example is the rapid synthesis of nitrogen-doped CQDs from glucose and urea (1.0 g each) under microwave irradiation at 700 W for just 5 minutes, as demonstrated by Dey et al., achieving high nitrogen content and strong fluorescence, while significantly reducing synthesis time compared to hydrothermal routes (Dey et al., 2014b). Another study by Serpone et al. utilized a microwave-assisted solvothermal approach to prepare PEG-passivated graphene quantum dots (GQDs) using graphene oxide and polyethylene glycol (PEG-6000) at 200 °C for 30 minutes, resulting in high quantum yield due to effective surface passivation by oxygen-containing groups from PEG (Hagiwara et al., 2021). Yadav and Lahariya further expanded this method by co-doping boron and nitrogen into CQDs using boric acid and melamine as precursors, achieving excellent fluorescence sensing properties through a 15-minute microwave-assisted reaction at 180 °C (Yadav & Lahariya, 2025b). Similarly, De et al. synthesized nitrogen-enriched CQDs via microwave irradiation of citric acid and ethylenediamine at 800 W for 3 minutes, producing nanodots with enhanced reactive oxygen species (ROS) detection capabilities (De et al., 2023). Collectively, these studies illustrate the versatility and speed of microwave-assisted synthesis in producing high-quality, heteroatom-tailored CQDs suitable for applications in sensing, imaging, and optoelectronics.

2.3. Thermal Pyrolysis and Carbonization

Thermal pyrolysis and carbonization are robust techniques for synthesizing heteroatom-doped CQDs, involving the high-temperature decomposition of organic precursors in controlled atmospheres. These methods enable precise control over the degree of carbonization, particle size, and heteroatom incorporation by tuning parameters such as temperature, heating rate, and atmosphere composition. This makes them particularly suitable for fabricating CQDs with well-defined compositions and functionalities. Several studies summarized in recent articles have identified the thermal pyrolysis of citric acid combined with amino-containing compounds, such as histidine or other amino acids, as a commonly employed bottom-up method for the synthesis of N-CQDs (Gulati et al., 2023; Q. Liu et al., 2025). This approach leverages the dual functionality of amino acids, which serve both as nitrogen dopants and carbon precursors during pyrolysis. According to these articles, thermal decomposition of such precursor systems results in CQDs with significant nitrogen incorporation, which is known to influence surface chemistry, electronic structure, and optical behaviour, particularly enhancing photoluminescence tunability. Collectively, thermal pyrolysis and carbonization provide a versatile and scalable route for producing heteroatom-doped CQDs with tunable properties suitable for advanced applications in catalysis, sensing, and bioimaging.

2.4. Electrochemical Synthesis

Electrochemical synthesis has gained considerable attention as a versatile and controllable technique for producing heteroatom-doped CQDs. This method utilizes electrochemical reactions under applied potential and current to drive the carbonization and doping processes, offering real-time control over reaction parameters and in-situ heteroatom incorporation. One of the notable examples is the electrochemical synthesis of N-CQDs using mono-ethanolamine and citric acid as precursors. Fakheri et al. demonstrated that by modulating potential and current density, high-quality N-CQDs could be obtained with consistent doping profiles, while also comparing the method favourably against traditional one-pot techniques (Fakheri et al., 2023). Expanding this approach, they also synthesized nitrogen and fluorine co-doped CQDs by adding ammonium fluoride into the precursor mixture, showing that simultaneous N,F co-doping can be precisely achieved through electrochemical control. Similarly, the incorporation of sulphur alongside nitrogen was realized by introducing thiourea into the system, highlighting the adaptability of this method for generating N,S co-doped CQDs under the same electrochemical conditions. Beyond nitrogen-based doping, electrochemical routes have also been employed to introduce boron, using glucose and boric acid as precursors. This enabled fine-tuning of boron incorporation levels, which directly influenced the optical properties of the resulting CQDs. Collectively, these examples underline the electrochemical method's strength in enabling flexible, multi-heteroatom doping under mild conditions, with enhanced control over composition and functional properties. This makes it particularly valuable for designing CQDs tailored for specific applications in catalysis, biosensing, and optoelectronics.

2.5. Chemical Oxidation and Exfoliation Methods

Chemical oxidation and exfoliation techniques represent powerful top-down approaches for the synthesis of heteroatom-doped CQDs, primarily relying on strong oxidizing agents to fragment bulk carbon materials and facilitate heteroatom incorporation. These methods offer high efficiency in breaking down graphitic or amorphous carbon structures while simultaneously introducing oxygenated or heteroatom-containing functional groups. Similarly, coal-derived N,S co-doped CQDs were produced using an ultrasonic-assisted wet oxidation method, in which heteroatoms inherently present in the coal structure were preserved during the oxidation process. Boruah et al. highlighted this as an eco-friendly strategy for converting low-cost and abundant coal sources into functional CQDs (Boruah et al., 2023). Additional heteroatom doping can be achieved through post-oxidation treatment; for example, carbon fibres were oxidized using HNO_3 and subsequently treated with ammonia solution, facilitating nitrogen doping through surface amination. These oxidation-based methods are especially useful for incorporating heteroatoms such as oxygen, nitrogen, and sulphur, and are compatible with scalable production, making them attractive for industrial applications in catalysis, sensing, and energy conversion. Table 1 summarizes the advantages, limitations and applications of different synthesis methods of doping CQDs with heterogeneous atoms

Table 1. Advantages, limitations, and applications of CQDs doped with heterogeneous atoms synthesized by different methods

Sl. No.	Synthesis Method	Advantageous	Limitations	Applications
1	Hydrothermal Method	• Simple and Scalable. • Tuning of CQDs by adjusting the temperature, precursor and reaction time. • Versatile for various carbon sources (Biomass carbon). • Integration of multiple heterogeneous atoms, and functional groups.	• Long reaction time • High temperature and Pressure (Autoclave)	• Sensing • Bioimaging • Optoelectronics • Electrocatalysis
2	Microwave-Assisted	• Rapid synthesis • High yields • Operates in a uniform temperature • Integrate multiple heterogeneous atoms.	• Limited scalability (Microwave reactor size) • Control over particle size.	• Fertiliser synthesis • Photoluminescence • Sensing
3	Thermal Pyrolysis and Carbonization	• Simple procedure. • Applicable from bulk to large-scale synthesis. • Tunable particle size for advanced applications. • Effective for high-temperature stable heterogeneous atoms.	• Low control over functionalization and surface groups. • Ineffective for low-temperature stable heteroatoms. •	• Opto-Electronic • Bioimaging • Photoluminescent • Sensing
4	Electrochemical Method	• Selective incorporation of various heteroatoms. • Short reaction time.	• Limited surface functionalities. • Some solvents may be hazardous.	• Sensing • Catalysis • Bioimaging • Optoelectronics
5	Chemical Oxidation and Exfoliation Method	• Highly efficient and tunable method. • Addition of Multiple heterogeneous atoms. • Easier conversion of bulk carbon to smaller sizes.	• Need for strong exfoliation agents. • Chance of unwanted byproduct formation.	• Energy Conversion, • Nanoparticle synthesis • Sensing

3. CHARACTERIZATION AND PROPERTIES OF HETEROGENEOUS ATOM-TAILORED CQDS

CQDs are one of the most emerging categories of carbon-based nanomaterials that have attracted considerable interest due to their unique combination of nanoscale dimensions, versatile surface chemistry, and tunable electronic and optical properties (Khan et al., 2022). Typically, CQDs are less than 10 nanometers in diameter, and they exhibit quantum confinement and abundant edge sites that give rise to unusual behaviors, which are not observed in bulk carbon. Over recent years, researchers have advanced the concept of tailoring CQDs through heterogeneous atom incorporation, where foreign elements such as nitrogen, sulfur, phosphorus, boron, and other transition metals are introduced into the carbon framework (D. L. Zhao & Chung, 2018). This atomistic modification has proven to be an efficient means of altering the electronic band structure, surface charge density, defect states, and overall reactivity of the material, thereby unveiling new opportunities for applications across catalysis, energy storage, sensing, optoelectronics and bioimaging (F. Zhao et al., 2023).

Carbon Nanodots are almost spherical in structure exhibiting either amorphous or crystalline structure with a strong photoluminescence. The combination of polymer and carbon characteristics is seen in carbonized polymer dots, offering flexible surface functionalities (Mindivan & Göktaş, 2023). The introduction of foreign atoms into crystal lattice in heteroatom-doped CQDs will enhance the optical, electronic and catalytic performance for energy applications and sensors (Mindivan & Göktaş, 2023; Pillar-Little et al., 2018).

3.1. Characterization of CQDs

The characterization of heterogeneous atom-tailored CQDs is performed with a combination of techniques to unravel their structure and property relationships. There are different characterization techniques for unraveling different properties of CQDs as shown in **Fig 1**. Among them, XPS and Raman spectroscopy are considered to be the most reliable characterisation technique for the confirmation of heteroatom doping by revealing dopant bonding environments and structural disorder, whereas HRTEM and XRD provide corresponding morphological and crystallinity information. High-resolution transmission electron microscopy (HRTEM) is commonly employed to determine the internal surface morphology and lattice fringes that indicate degree of graphitic ordering, while atomic force microscopy provides thickness information confirming the zero-dimensional nature of these dots (H. Singh et al., 2020a). Singh et al. (H. Singh et al., 2020b) characterized the CQDs morphology by HRTEM and the chemical analysis was analyzed from EDS Spectra. Similarly, Rahul et al. (Das & Mohapatra, 2017a) synthesized and characterized CQDs doped with three hetero atoms like boron, sulfur and nitrogen. The photoluminescence of the doped CQDs was analyzed using Horiba FluoroMax-4 fluorimeter. Also, the microstructural analysis as well as the elemental analysis were carried out by HRTEM (FEI Technai G2 30ST) which was operated at 300 kV. X-ray diffraction reveals their amorphous-to-semicrystalline character with a broad diffraction peak.

Smrithi et al. (Das & Mohapatra, 2017b) has done the structural and morphological characterizations of N-CQDs using Bruker D8 advance diffractometer for XRD analysis using a Cu K alpha radiation with a wavelength of 0.15406 nm. Photoluminescence was done using the instrument Hitachi F 2700 spectrofluorometer where slit width was 5 nm at a rate of 1500 nm min^{-1} at 700 V. Raman spectroscopy was frequently used to evaluate the degree of wall graphitization and intensity ratio of sp^2 hybridized and sp^3 hybridized carbon atoms (P. Zhao et al., 2022a). Surface chemistry plays a defining role in these systems, and therefore X-ray photoelectron spectroscopy is indispensable for identifying the bonding environment of both carbon and dopant atoms, allowing researchers to verify the successful introduction of heteroatoms such as pyridinic or graphitic nitrogen, thiophene sulfur, or phosphate groups (H. Singh et al., 2020b; P. Zhao et al., 2022b). Vaibhav Naik et al. (Zheng et al., 2013) synthesized and characterised nitrogen doped carbon dots using different characterisation techniques. They employed UV-VIS spectroscopy, which showed a peak around 278 nm and showing the presence of π-π^* transition of the aromatic sp^2 carbon of the CQDs.

Figure 1. Schematic representation of major characterization techniques used for heteroatom-tailored CQDs

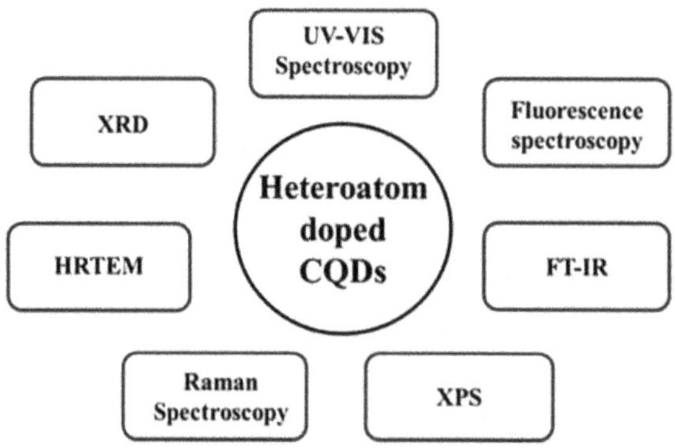

Fourier transform infrared spectroscopy offers further insights into the functional groups that decorate the surface, ranging from hydroxyl and carboxyl to amine and sulfonate moieties, all of which dictate solubility and chemical activity. In terms of optical characterization, ultraviolet–visible spectroscopy tracks absorption edge shifts that occur upon doping, while photoluminescence and time-resolved PL measurements shed light on emission behavior and charge recombination dynamics (P. Zhao et al., 2022b). Electrochemical methods such as cyclic voltammetry (CV) and electrochemical impedance spectroscopy (EIS) provide valuable information on charge transfer kinetics, which are particularly relevant for catalysis and electrochemical devices (Mindivan & Göktaş, 2023; P. Zhao et al., 2022b; Gurushankar et al., 2025; Kannan et al., 2023).

3.2. Electrical Properties of CQDs

The electrical properties of CQDs are greatly influenced by inclusion of heteroatoms. Pristine CQDs generally exhibit wide bandgaps in the range of two to three electron volts, with limited conductivity arising from small sp^2 domains embedded within an amorphous carbon matrix (Guan et al., 2023). The introduction of nitrogen narrows the bandgap by introducing electron-donating sites, thereby enhancing n-type conductivity, while sulfur or phosphorus doping can act as electron acceptors and introduce localized states that facilitate charge separation (P. Zhao et al., 2022b). Co-doping with multiple elements often generates synergistic effects that balance donor and acceptor sites, yielding superior charge transport compared to single-atom doping (P. Zhao et al., 2022b). Hafez et al. (Hafez et al., 2025a) synthesized CQDs doped with multiple hetero atom like sulfur, silicon and phosphorus which showed outstanding bifunctional electrocatalytic performance for applications like oxygen and hydrogen evolution reactions especially in the alkaline solution. The adsorption of silicon atoms helped in improved electrocatalytic performance which was calculated by the DFT calculations. Similar study by Cisem et al. (Hafez et al., 2025b) showed that the boron-doped CQDs can convert ultraviolet into blue light when used as an additive in perovskite solar cells based on $MAPbI_3$. The results revealed that optimum addition of the doped CQDs passivated the grain boundaries and act as down-converters and thereby increasing the carrier lifetime and photocurrent. Ghosh et al. (Ghosh et al., 2025) revealed that decorating the conductive MXene sheets with heteroatoms like N,S-co-doped CQDs can increase accessible active sites as well as electronic coupling, distributing higher area/specific capacitance and enhanced symmetry in supercapacitors. The synthesised hybrid materials showed high C_{sp}, high energy density and excellent electrochemical stability showing substantial potential in the field of storage device applications.

3.3. Mechanical Properties of CQDs

Although direct mechanical testing of individual CQDs is challenging due to their small size, their effect can be evaluated when incorporated into composite matrices. Heteroatom-tailored CQDs have been shown to reinforce polymer films and hydrogels by enhancing tensile strength, modulus, and flexibility (El-Shamy, 2019a). This improvement arises not only from the intrinsic rigidity of graphitic domains but also from the strong interfacial bonding promoted by dopant-induced functional groups that interact favorably with the host matrix. For instance, nitrogen-containing surface groups form hydrogen bonds with polymers, while sulfur and phosphorus moieties enhance crosslinking potential (El-Shamy, 2019b; Latif et al., 2023). Jo et al. (Jo et al., 2022) prepared CQDs doped with phosphorus atom formed hydrogen-bond in the form of a graft with a-WO_3, bearing structural deformation during repetitive cycling. The functional groups and the P-CQDs help in improving the Li ion wettability as well as electrical conductivity. This in turn enhances the electrochemical activity and kinetics of the devices which maintained flexibility and stable electrocapacitive response. Yu et al. (Yu et al., 2022a) synthesized carbon dots doped with nitrogen and phosphorus from m-phenylenediamine and phosphoric acid via solvothermal method which was then incorporated into Poly vinyl alcohol (PVA). The study proved that the N, P-CQDs dispersion in PVA produced strong interfacial bonding which simultaneously improves flame resistance and mechanical performance of the films compared to neat PVA.

3.4. Physio-Chemical Properties of CQDs

The physicochemical properties of heterogeneous atom-tailored CQDs are equally important. Their surface is naturally abundant in oxygen-containing groups, namely, hydroxyl groups and carboxyl groups, which confer hydrophilicity and enable colloidal stability in aqueous environments. Zhi et al. (Yu et al., 2022b) shown that phosphorus doping in polymeric CQDs was capable of increasing resistance to UV-B photobleaching compared with undoped counterparts, attributed to P-induced crosslinking/energy-dissipation pathways. Singh et al. (V. Singh et al., 2019) studied about nitrogen-doped CQDs which were manufactured from arginine/citric acid which exhibited strong colloidal stability in physiological media with negligible cytotoxicity both in vitro/in vivo.

3.5. Optical Properties of CQDs

The optical properties of CQDs are one of the most promising features of CQDs, that has gained their attention worldwide. Pure CQDs have strong absorption in the ultraviolet and blue regions, with photoluminescence that is often excitation-dependent and size-tunable due to quantum confinement (Kalpana et al., 2023). On the other hand, doping changes electronic transitions and causes defect states, producing red-shifted emissions that extend into the visible and near-infrared regions. Nitrogen and sulfur co-doping in particular yields broad emission spectra and facilitates deeper tissue imaging in biomedical applications (Lakhera et al., 2024a). The quantum yield of CQDs, which is typically moderate, can be substantially improved through both surface passivation and dopant inclusion, sometimes reaching values like semiconductor quantum dots but without associated toxicity (Lakhera et al., 2024b). Pal et al. (Pal et al., 2019) produced boron-doped CQDs by microwave synthesis which resulted in high photoluminescence quantum yield of 25.8%, while comparing with undoped was about 16.9% and 11.4% for phosphorus-doped counterparts. The improved emission as well as response behavior permitted effective ratiometric sensing of pH in cancer cells, demonstrating how the boron atom doping lifts the optical performance by producing new emission centers. Similarly, Rohan et al. (Ambade et al., 2023) produced CQDs which is co-doped with both nitrogen as well as sulfur atoms which is then combined into a TiO_2–rGO matrix on a flexible fabric. This composite was capable of degrading 98% of methylene blue while compared to 67% with N-only CQD composites under the visible light. Therefore, Co-doping allowed enhanced bandgap, morphology, and a Z-scheme electron transfer pathway which suggestively improved the photocatalytic activity of the CQDs. Another study by Phang et al. (Phang et al., 2022) reported about boron-doped CQD and graphitic carbon nitride hybrid which was made-up to produce a homojunction photocatalyst. The results revealed that the optimum composite was found to degrade 96.8% of Rhodamine B in 4 hours under visible light, representing a reaction rate constant of about 3.6× higher than g-C_3N_4 composite alone. Thus, boron doping remarkably enhanced charge separation and extended light absorption range.

The hetero atom-tailored CQDs have potential applications in bioelectronics, smart textiles, and solar energy systems and it is expected to drive future technological breakthroughs. In summary, heterogeneous atom-tailored CQDs possess a unique combination of nanoscale engineering and functionality of materials. They have high electrical conductivity, mechanical strength, chemical reactivity, optical tenability, which can be modified by the introduction of heteroatoms. In coming future, with progress in synthesis and characterization techniques, CQDs have the potential to have next generation applications in the field of environmental technologies, electronics, sensing and energy storage.

4. APPLICATION OF HETEROGENEOUS ATOM-TAILORED CQDS IN ELECTROCHEMICAL SENSORS

Electrochemical sensors are analytical devices used to detect various types of analytes in a specific and selective way by converting electrochemical reactions into electrically measurable signals (Murali et al., 2024). Heterogeneous atom-doped carbon quantum dots, especially N-CQDs, were extensively utilized in electrochemical sensing applications for the detection of heavy metals, determination of biomolecules and biomarkers, environmental monitoring, food safety analysis, healthcare applications, etc. Upon introducing the heterogeneous atoms, many surface and edge defects are created, resulting in vacancies and tailored functional groups that can serve as an active site for interacting with targeted ions or molecules. These surface defects increase both the conductivity and reactivity of CQDs, facilitating much faster intermolecular electron transfer. However, the extensive results from Raman spectroscopy revealed the degree of these defects in the literature reports.

This section of the chapter briefly describes the applications of heterogeneous atom-doped CQDs in electrochemical sensors.

4.1. Electrochemical Sensors for Biomolecules and Biomarkers

In 2024, Saithip Pakapongpan et al. developed an electrochemical sensor for the determination of amyloid-β42 (Aβ42), an Alzheimer's disease biomarker using N-CQDs-graphene nanohybrid (NCG). (Pakapongpan et al., 2024). Molecular imprinting technology was adopted for the sensing process and polypyrolle was used for the imprinting process. The molecular imprinted polypyrolle (MIP) and NCG nanohybrid was modified on screen printed electrode (SPE). Nitrogen doped carbon dots were synthesized via hydrothermal method and graphene powder was added to the nitrogen doped carbon dots solution and followed by sonication to get NCG nanohybrid. The synergistic effect of the NCG nanohybrid enhances the electrochemical sensing characteristics of MIP sensor for the detection of Aβ42 using ferri-ferrocyanide as the redox system. The MIP- NCG sensor exhibited linear response in picomolar levels from 5 to 70 pg/mL with a limit of detection (LOD) of 1 pg/mL. Hence, the developed sensing device would be a good analytical tool for the diagnosis of Alzheimer's disease in future. The results showed that the combination of NCD-G and molecularly imprinted polymer (MIP) improved the sensitivity as well as the selectivity of the sensor.

The applicability N-CQDs for electrochemical sensing was explained in the work done by M. Ganesan et al. in 2021 (Ganesan et al., 2021). The sensor utilized N-CQDs modified with iron oxide and multiwalled carbon nanotube (MWCNT) for the electrochemical determination of 5-fluorouracil, uric acid and xanthine for clinical applications. N-CQDs and iron oxide composite were synthesized using a hydrothermal method for improved efficiency in electrochemical sensing. The modified electrode showed good sensitivity with detection limits of 0.106, 0.092, and 0.019 μM for uric acid, xanthine, and 5-fluorouracil, respectively. The combination of N-CQDs with iron oxide nano particles enhanced charge transfer rates in electrochemical applications. The modified electrodes were characterized using CV and EIS to assess the electrochemical behaviour of the electrodes. Differential pulse voltammetry (DPV) was also used for the electrochemical analysis of the analytes. The sensor showed good electrochemical performances in spiked urine samples, indicating its appropriateness for practical utility of this sensor in health care monitoring.

Another study in 2019 introduces N-CQD and SnO_2 composite as an efficient material with enhanced properties for voltammetric sensing of riboflavin (Muthusankar, Rajkumar, et al., 2019). The nitrogen doping in carbon quantum dots helped to increase the electrochemical efficiency of the sensors towards the analytical detection of riboflavin. The modified electrode exhibited a fast electrochemical response with high sensitivity and a wide linear range of 0.05-306 µM. The N-CQD/SnO_2 nanocomposite is prepared using a simple ultrasonic method and applied for sensitive riboflavin detection in powder and riboflavin tablets. Nevertheless, real-world deployment would benefit from broader real-matrix validation and efficient interference studies against structurally related vitamins and common excipients. The significance of N-CQDs in the determination of epinephrine was demonstrated in another work by molecular imprinting technique (Yola & Atar, 2019). Molecularly imprinted polymers are highlighted as effective and inexpensive recognition elements for sensors and the study focused on developing a graphitic carbon nitride/N-CQDs composite for selective epinephrine detection in urine samples. The sensor revealed high selectivity towards epinephrine compared to other interfering species, thereby signifying its practical application in clinical aspects. Also, the sensor showed high stability over 60 days, specifying the feasibility for long-term use in medical diagnostics.

In another work, nitrogen and phosphorus doped carbon quantum dots (N,P-CQD) from banana flower bract extract via hydrothermal synthesis for the selective detection of catecholamine, dopamine (Padmapriya et al., 2023). The N,P-CQD sensor showed a wide linear detection range from 6.0 µM to 0.1 mM and the N,P-CQD successfully detects dopamine with a LOD as low as 630 pM in real-time samples. Another example for dual heterogeneous-atom doped carbon quantum dots was for electrochemical sensing was the monitoring of anticancer drug, Imatinib mesylate using a hybrid nanocomposite of N,S-doped carbon dots/carbon nanotube and poly(amidoamine) dendrimer (Saleh Mohammadnia et al., 2023a) and the schematic representation of the synthesis is given in **Fig 2.** The figure highlights how controlled heteroatom incorporation during the synthetic process improves electron-transfer pathways and active-site density, which leads to enhanced electrochemical performance. Imatinib mesylate was electrochemically quantified in blood-serum samples and the work utilized electrochemical techniques, such as CV and DPV, for analyzing imatinib mesylate. The presence of nitrogen and sulphur in carbon dots greatly improved the electrochemical performance of the sensor towards imatinib mesylate detection. The N,S-doped carbon dots-based hybrid sensor demonstrates high selectivity and sensitivity for Imatinib mesylate detection in clinical settings.

Figure 2. Schematic illustration of the synthesis process of nitrogen and sulfur co-doped CQDs/CNT hybrids using poly(amidoamine) dendrimer templates

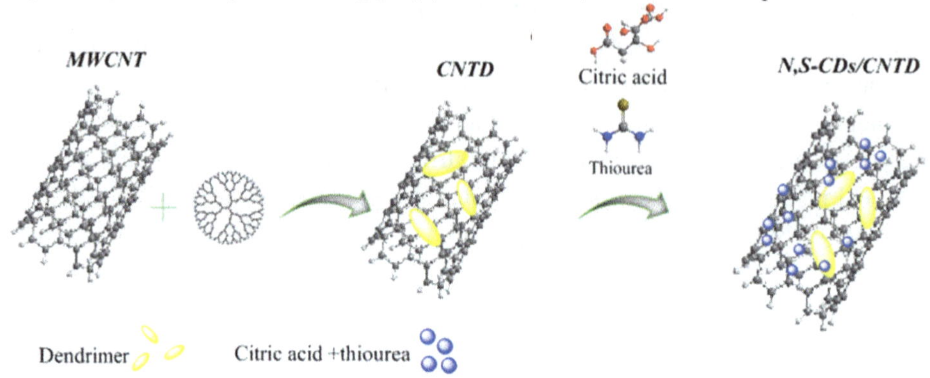

(Saleh Mohammadnia et al., 2023b)

N,S-doped carbon dots and MXene nanocomposite was used for the fabrication of a dopamine sensor by electrochemical techniques (Hou et al., 2024). The combination of N, S-CQDs and MXene enhanced the interlayer spacing and specific surface area of MXene, making it a fascinating material for sensing applications. Also, the presence of N, S-doped carbon dots increased the stability, and electron transfer rates of MXene nanosheets thereby facilitating a better quantification of dopamine by electrochemical technique. The study emphasizes the potential application of the heterogenous- atom tailored carbon dots in real sample analysis for early disease diagnosis.

4.2. Electrochemical Sensors for Organic Environmental Pollutants and Heavy Metals

Heterogeneous- atom doped carbon quantum dots were also applied for the environmental monitoring via electrochemical sensing methods. For example, recently an electrochemical sensing strategy was developed for the detection of the insecticide carbendazim in environmental samples (Velmurugan et al., 2024). The study presented the design of N-CQDs encapsulated in a UiO-66 zirconium metal-organic framework (MOF) for carbendazim detection. The N-CQDs@UiO -66- zirconium MOF exhibited a linear range of 0.02-126 μM and a LOD of 5.8 nM. This work highlighted the necessity for sensitive detection methods for carbendazim in vegetables and environmental samples to protect human beings from severe environmental health hazards. The study emphases on selective analysis of carbendazim and hence a systematic interference testing with other pesticides/organic contaminants is done to enhance their practical utility. Catechol and hydroquinone are organic compounds which shows toxic effects to environment and the detection of these compounds from environmental samples are very crucial (Jyothi et al., 2024). The effect of nitrogen doping in the sensing performance of CQDs towards catechol and hydroquinone was described by Cheng Rong and co-workers in 2025 (Rong et al., 2025). Nitrogen doping and malic acid as precursors facilitated the improved electrochemical performance of the sensor with a linear detection range of 1-500 μM for hydroquinone and 1-200 μM for catechol. Also, the electrochemical sensing of the analytes has positively influenced by the increased electroactive surface area due to synergistic effect of N-doping and introduction polar functional groups on CQDs. The research pinpoints the potential of N-CQDs in environmental monitoring, showcasing their practical applications.

Another research study investigated the synthesis and characterization of N-CQDs-αFe_2O_3 - poly aniline nanocomposites, for the electrochemical sensing of Cd^{2+} in water samples (Abinaya et al., 2025). The study focused mainly on aqueous solutions, and data on interference from co-existing metal ions (e.g., Pb^{2+}, Zn^{2+}, Cu^{2+}) and the electrode was found to be a selective tool for detecting cadmium ions in the presence of other heavy metals. This selectivity proved the synergistic effect of nanocomposite containing N-CQDs, αFe_2O_3 and poly aniline. Hence, from the literature reports, the heterogenous atom-doped CQDs plays a significant role in the electrochemical detection of environmental pollutants including toxic organic compounds, heavy metals etc.

4.3. Electrochemical Sensors for Pharmaceutical Analysis and Other Applications

The applications of N-CQDs in electrochemical sensing was well explained in the work done by Ganesan Muthusankar and Co- researchers (Muthusankar et al., 2020). The work demonstrated the fabrication of an electrochemical sensor for simultaneous detection of anticancer drug Flutamide (FLU) and antibiotic Nitrofurantoin (NF) using CV and DPV techniques. Real time application was done in human urine samples with known concentrations of FLU and NF by standard addition method. The introduction N-CQDs enhances charge transfer in hybrid nanocomposites thereby making this sensor an excellent device in clinical analysis in future. Furthermore, while proof of urine analyses was found to be successful, endorsement on larger and more diverse biological mediums would be crucial to create clinical reliability and strength for real-world use. In another work, sulfur-doped carbon quantum dots (S-CQDs) decorated with iron oxide nanoparticles were employed for the electrocatalytic oxidation of the olanzapine, an antipsychotic drug (Muthusankar, Sangili, et al., 2018). The nanocomposite was synthesized by hydrothermal method and was successfully applied to determine olanzapine in urine samples, accounting its practical utility. The sensor achieved a LOD of 0.006 μM with the synergetic effects of iron oxide and S-CQDs in the composite material. Another study highlights the use of a N-CQDs modified copper oxide on a glassy carbon electrode (GCE) for the detection of the analyte aspirin, in food samples (Muthusankar, Sasikumar, et al., 2018). The N-CQD/Cu2O/GCE sensor displayed excellent repeatability and stability, thereby improving its reliability for practical use in food safety analysis. CV, DPV and EIS were used for the electrochemical characterization, and the results showed the effectiveness of this sensor for the analytical quantification of aspirin in a wide linear range of 1-907 mM and LOD of 0.002 mM.

Ramadhass Keerthika Devi et al. reported significance of nitrogen doping in carbon quantum dots for the electrochemical sensing applications (Devi et al., 2022). The work focused on the electrochemical detection of sulfadiazine, an antibiotic drug, using N-CQDs and Mn_3O_4 composite. Sulpha-based antibiotics, such as sulfadiazine, are widely used in veterinary medicine and the unwarranted use of sulfadiazine leads to environmental pollution and chronic toxicity. The device manifested a recovery efficiency of around 98% for sulfadiazine in urine samples, indicating its practical feasibility.

Also, a comparative study of other existing literature reports for the electrochemical sensors based on heterogenous- atom tailored carbon quantum dots are given in **Table 2**.

Table 2. Literature Comparison of Heterogeneous Atom-Tailored CQDs-based Electrochemical Sensors for Enhanced Sensing Applications

Modification	Analyte	Real samples	Linear range	Reference
N,S-CQD- Bi_2O_3 composite	Aniline	Tap and river water	5–35 nM	(John & Mathew, 2023)
F,N-CDs modified with laccase	Catechol	Tap and lake water	12–450 μM	(L. Liu et al., 2019)
N-CQDs	Tetracycline	Tap and river water	5-30 nM	(John et al., 2022)
N,S-CQDs	Tert-butylhydroquinone	Coconut oil	0.05-10 μM	(Karthikeyan et al., 2019)
N-CQDs -GO hybrid	Cd and Pb ions	Tap and lake water	11.24–11241 μg/L for Cd^{2+} and 20.72–10360 μg/L for Pb^{2+}	(Li et al., 2018)
WP6-N-CQDs	Trinitrotoluene	Tap and lake water	0.001–1 μM and 1–20 μM	(Ran et al., 2018)

continued on following page

Table 2. Continued

Modification	Analyte	Real samples	Linear range	Reference
N,S-CQDs	Tartrazine	Soft drinks and candies	5–35 nM	(John et al., 2023)
N-CQD/HP-Cu_2O/ MWCNT/GCE	Caffeic acid	Red wine samples	0.05–43 µM	(Muthusankar, Sethupathi, et al., 2019)
F-CQDs	Ketoconazole	pharmaceutical formulations and tap water samples	0.037 to 11.13 µM	(Dos Santos et al., 2025)

In conclusion, heterogenous atom-tailored carbon quantum dots have emerged as promising candidates for electrochemical sensing applications owing to their nanoscale dimensions and versatile surface chemistry. Their successful integration into sensing platforms has enabled the detection of various chemical and biological targets with high selectivity and sensitivity. These findings suggest that heterogenous atom-tailored carbon quantum dots -based electrodes finds a greater potential for future developments in analytical and diagnostic technologies for healthcare, clinical and environmental sectors.

5. CHALLENGES AND LIMITATIONS

Although heterogeneous atom-tailored CQDs show remarkable potential in advancing electrochemical sensing, several critical challenges limit their translation into practical technologies. The foremost concern lies in the reproducibility of synthesis. Small changes in precursor composition, dopant concentration, or reaction environment can lead to substantial variations in particle size, crystallinity, and surface chemistry. These inconsistencies often result in unreliable sensor performance and hinder the ability to establish standardized fabrication protocols (Kar et al., 2024). Furthermore, controlling the precise incorporation of single or multiple heteroatoms without generating unwanted structural defects remains a considerable obstacle. Excessive doping can disturb the carbon lattice, while insufficient doping may fail to provide the intended electronic modifications (Deshsorn et al., 2023).

Another major limitation relates to the stability of CQDs during prolonged operation. Tailored CQDs are susceptible to aggregation, photobleaching, and chemical degradation when exposed to light, oxygen, or complex aqueous media. These stability issues reduce the reproducibility of electrochemical responses and shorten sensor lifetimes, making them less suitable for continuous or field-based monitoring. Although surface passivation and hybridization strategies can partially improve stability, they may simultaneously compromise conductivity or introduce new processing complexities (Tavan et al., 2025). A deeper challenge arises from the incomplete understanding of structure–property relationships. While experimental evidence clearly demonstrates improvements in conductivity, charge transfer, and selectivity with heteroatom doping, the exact mechanisms governing these enhancements remain insufficiently explored (Poh & Pumera, 2015). Without clear insight into how nitrogen, sulphur, phosphorus, boron, or multi-heteroatom doping influences electron density, catalytic activity, or molecular recognition, tailoring CQDs for specific analytes often relies on empirical approaches rather than rational design principles.

Device-level integration presents additional difficulties. Ensuring strong adhesion of doped CQDs to electrode surfaces, achieving homogeneous dispersion, and maintaining compatibility with flexible or miniaturized platforms are still unresolved issues. The fabrication processes sometimes require binders or stabilizers, which may block active sites or interfere with electron transfer. Moreover, scaling up the synthesis and electrode modification procedures remains costly and time-consuming, raising concerns about economic viability for widespread applications (Guo et al., 2021). Finally, most reported studies focus on controlled laboratory conditions, whereas real-world environments involve complex matrices with numerous interfering species. In such scenarios, issues of selectivity, durability, and reproducibility become more pronounced (Deivayanai et al., 2025). Therefore, while heteroatom-tailored CQDs represent an exciting direction, significant progress in reproducible synthesis, stability improvement, mechanistic understanding, and device engineering is essential for their reliable use in next-generation electrochemical sensors.

However, several challenges persist in the synthesis and application of heteroatom-tailored CQDs, advanced technologies are being developed to address these. One way is the incorporation of machine-learning (ML) methods and data-driven optimization in synthesis. These applications have helped optimise, synthesise, evaluate, and utilise CQDs for various research and development purposes. The approach of ML not only paves the way for development in CQDs but also expands the understanding of close relations between precursor composition and synthesis methods with CDs properties. Considering relationships between precursor composition, synthesis temperature, reaction time, and product quality, ML algorithms can predict conditions where uniform dopant integration is attained and can guide the reproducibility of certain optical or electrochemical properties. Such predictive modelling can vividly decrease experimental trial-and-error and enable specific tuning of heteroatom configuration for targeted functionalities.

6. FUTURE PERSPECTIVES AND RESEARCH OPPORTUNITIES

The progress in heterogeneous atom-tailored CQDs has already demonstrated their value in electrochemical sensing, yet much broader opportunities lie ahead. One of the most pressing directions is the refinement of synthesis techniques that can provide precise, reproducible, and scalable doping strategies (Wang et al., 2025). Emerging methods such as hydrothermal, microwave-assisted, and electrochemical routes offer potential, but they need to be optimized for consistent heteroatom incorporation and environmentally sustainable production. Exploring systematic single, dual, and multi-heteroatom doping will help establish clearer design rules, enabling researchers to move from trial-based fabrication to purposeful engineering of CQDs. Equally important is the need for a deeper understanding of the structure–property relationships that govern sensing performance. At present, the exact role of different dopants in altering electron density, catalytic activity, or molecular affinity remains only partly understood. Coupling experimental studies with computational modeling and advanced in situ spectroscopic analysis will be essential for unraveling these mechanisms. This knowledge will accelerate the rational design of CQDs with targeted functionalities for selective detection of metals, biomolecules, or environmental toxins. Improving stability presents another rich research opportunity (Sethulekshmi et al., 2025). Strategies such as surface passivation, hybridization with robust nanostructures, or encapsulation in protective matrices may extend sensor lifetimes without compromising sensitivity (Balakrishnan et al., 2023). In parallel, the development of multifunctional CQDs that can perform dual or multi-analyte detection could broaden their utility in healthcare, food safety, and environmental monitoring. At the device level, future efforts should focus on integrating CQDs into practical sensing platforms through scalable methods such as inkjet printing, spray deposition, or flexible substrate embedding. Interdisciplinary collaborations between chemists, engineers, and industry partners will be crucial in translating laboratory findings into portable, affordable, and user-friendly sensor devices (Manzoor et al., 2023).

Finally, real-world testing must become a research priority. Most existing studies report performance under controlled laboratory conditions, whereas field environments present far more complex challenges. With progress in these areas, heterogeneous atom-tailored CQDs are poised to evolve from a promising nanomaterial to a cornerstone of next-generation electrochemical sensing technologies.

7. CONCLUSION

Heterogeneous atom-tailored CQDs represent an exciting frontier in electrochemical sensing, combining the unique advantages of nanoscale carbon materials with the versatility offered by heteroatom engineering. Their tunable electronic structure, abundant active sites, and adaptable surface chemistry enable improved conductivity, enhanced sensitivity, and greater selectivity compared to pristine CQDs or traditional nanomaterials. Recent advances in single, dual, and multi-heteroatom doping strategies highlight how precise modifications at the atomic level can translate into significant performance gains across a range of sensing applications, from environmental monitoring to healthcare diagnostics. Despite these achievements, challenges related to reproducible synthesis, long-term stability, scalable device integration, and real-world applicability remain unresolved. Addressing these issues will require deeper insights into structure–property relationships, improved synthetic control, and stronger connections between material design and device engineering. Importantly, practical validation under complex sample conditions must become central to future research efforts to ensure reliability beyond laboratory settings. Looking ahead, the convergence of advanced synthetic methods, computational modelling, and interdisciplinary collaborations offers an opportunity to push CQD-based sensors toward practical deployment. Future advancement in heteroatom-tailored CQDs will strongly depend on interdisciplinary collaborations that bridge materials science, chemistry, and device engineering. Integration of these doped CQDs with flexible substrates and microelectronic platforms could lead to their application in wearable and portable biosensors. Such cooperative approaches would expand the scope of CQD-based sensing from laboratory research toward practical, real-world technologies. By overcoming current barriers, heterogeneous atom-tailored CQDs can move beyond proof-of-concept studies to define a new generation of robust, cost-effective, and sustainable electrochemical sensing platforms capable of addressing pressing societal needs.

REFERENCES

Abinaya, S., Girija, S., Baby, A., Banu, A. A., & Jose, S. P. (2025). Nitrogen doped carbon quantum dots@ α-Fe2O3/PANI nanocomposite based electrochemical sensor for cadmium ion detection. *Inorganic Chemistry Communications*, *179*, 114846. DOI: 10.1016/j.inoche.2025.114846

Alhabardi, S. A., Almutairi, J. A., Al-Hamoud, G. A., Zaki Zakaraya, Z., Hani, U., Ali Ahmed Asiri, Z., Al Freahat, I., Alsafadi, W., Abu Dayyih, W., Khan, S. L., & Begh, M. (2025). Musa Paradisiaca derived intrinsically heteroatom doped carbon dots as antioxidant and controlled drug release behavior. *PLoS One*, *20*(8), e0329116. DOI: 10.1371/journal.pone.0329116 PMID: 40811709

Ambade, R. B., Ali, M., Lee, K. H., Jeong, W., Jeong, S. H., & Han, T. H. (2023). Nitrogen and Sulfur Co-Doped Carbon Quantum Dot-Engineered TiO2 Graphene on Carbon Fabric for Photocatalysis Applications. *ACS Applied Nano Materials*, *6*(17), 15782–15794. DOI: 10.1021/acsanm.3c02661

Aparício, R. R., Henning, E., Arantes, M. S. T., Jesus, M. B., Hansel, F. A., Dias, I. A., Roman, L. S., Schiavon, M. A., de Muniz, G. I. B., Magalhães, W. L. E., & de Cademartori, P. H. G. (2025). Techno-economic analysis and optimization of ethylenediamine-doped carbon dots production from kraft lignin via hydrothermal process. *International Journal of Biological Macromolecules*, *323*, 147132. DOI: 10.1016/j.ijbiomac.2025.147132 PMID: 40865817

Arora, G., Sabran, N. S., Ng, C. Y., Low, F. W., & Jun, H. K. (2024). Applications of carbon quantum dots in electrochemical energy storage devices. *Heliyon*, *10*(15), e35543. https://doi.org/https://doi.org/10.1016/j.heliyon.2024.e35543. DOI: 10.1016/j.heliyon.2024.e35543 PMID: 39166025

Balakrishnan, T., Ang, W. L., & Mahmoudi, E. (2023). Enhancing photoluminescence stability of blue emissive nitrogen doped graphene quantum dots by encapsulating inside zeolitic imidazole framework-8. *Journal of Solid State Chemistry*, *328*, 124343. https://doi.org/https://doi.org/10.1016/j.jssc.2023.124343. DOI: 10.1016/j.jssc.2023.124343

Boruah, A., Bora, S., Thakur, A., Dutta, H. S., & Saikia, B. K. (2023). Solid-State Phosphors from Coal-Derived Carbon Quantum Dots. *ACS Omega*, *8*(28), 25410–25423. DOI: 10.1021/acsomega.3c02884 PMID: 37483255

Das, R. K., & Mohapatra, S. (2017a). Highly luminescent, heteroatom-doped carbon quantum dots for ultrasensitive sensing of glucosamine and targeted imaging of liver cancer cells. *Journal of Materials Chemistry. B*, *5*(11), 2190–2197. DOI: 10.1039/C6TB03141B PMID: 32263692

Das, R. K., & Mohapatra, S. (2017b). Highly luminescent, heteroatom-doped carbon quantum dots for ultrasensitive sensing of glucosamine and targeted imaging of liver cancer cells. *Journal of Materials Chemistry. B*, *5*(11), 2190–2197. DOI: 10.1039/C6TB03141B PMID: 32263692

De, R., Jo, K. W., Lee, B. H., Some, S., & Kim, K.-T. (2023). Microwave-assisted rapid synthesis of nitrogen-enriched amphibious carbon quantum dots for sensitive detection of ROS and multiple other applications. *Journal of Materials Chemistry. B*, *11*(26), 6024–6043. DOI: 10.1039/D3TB00614J PMID: 37272382

Deivayanai, V. C., Karishma, S., Swaminaathan, P., Saravanan, A., Vickram, A. S., Anbarasu, K., & Yaashikaa, P. R. (2025). Nanohybrid approaches for precise analytical screening of trace contaminants in complex environmental and biological matrices. *Trends in Analytical Chemistry*, *191*, 118358. https://doi.org/https://doi.org/10.1016/j.trac.2025.118358. DOI: 10.1016/j.trac.2025.118358

Deshsorn, K., Payakkachon, K., Chaisrithong, T., Jitapunkul, K., Lawtrakul, L., & Iamprasertkun, P. (2023). Unlocking the Full Potential of Heteroatom-Doped Graphene-Based Supercapacitors through Stacking Models and SHAP-Guided Optimization. *Journal of Chemical Information and Modeling*, *63*(16), 5077–5088. DOI: 10.1021/acs.jcim.3c00670 PMID: 37635637

Devi, R. K., Ganesan, M., Chen, T.-W., Chen, S.-M., Al-onazi, W. A., Al-Mohaimeed, A. M., Elshikh, M. S., & Yu, Y.-Y. (2022). 3D-nanocubes of N-doped carbon quantum dots adorned manganese oxide: A functional electrocatalyst for the sensitive detection of sulfadiazine. *Colloids and Surfaces. A, Physicochemical and Engineering Aspects*, *648*, 129141. DOI: 10.1016/j.colsurfa.2022.129141

Dey, S., Chithaiah, P., Belawadi, S., Biswas, K., & Rao, C. N. R. (2014a). New methods of synthesis and varied properties of carbon quantum dots with high nitrogen content. *Journal of Materials Research*, *29*(3), 383–391. DOI: 10.1557/jmr.2013.295

Dey, S., Chithaiah, P., Belawadi, S., Biswas, K., & Rao, C. N. R. (2014b). New methods of synthesis and varied properties of carbon quantum dots with high nitrogen content. *Journal of Materials Research*, *29*(3), 383–391. DOI: 10.1557/jmr.2013.295

Dos Santos, C. M., Silva, F. W. L., de Souza, O. P. L., Canevari, T. C., Santelli, R. E., Saez, V. M., & Cincotto, F. H. (2025). Sensitive electrochemical sensor based on hybrid material F-doped carbon quantum dots (Cdots) and NiO nanoparticles: detection of ketoconazole in pharmaceutical formulations and tap water samples: CM dos Santos et al. *Analytical and Bioanalytical Chemistry*, 1–13.

El-Shamy, A. G. (2019a). Novel conducting PVA/Carbon quantum dots (CQDs) nanocomposite for high anti-electromagnetic wave performance. *Journal of Alloys and Compounds*, *810*, 151940. https://doi.org/https://doi.org/10.1016/j.jallcom.2019.151940. DOI: 10.1016/j.jallcom.2019.151940

El-Shamy, A. G. (2019b). Novel conducting PVA/Carbon quantum dots (CQDs) nanocomposite for high anti-electromagnetic wave performance. *Journal of Alloys and Compounds*, *810*, 151940. https://doi.org/https://doi.org/10.1016/j.jallcom.2019.151940. DOI: 10.1016/j.jallcom.2019.151940

Eliboev, I., Ishankulov, A., Berdimurodov, E., Chulpanov, K., Nazarov, M., Jamshid, B., Toshpulotov, B., Tukhtaeva, R., Demir, M., Rashidova, K., Jalilov, F., & Polvonov, K. (2025). Advancing analytical chemistry with carbon quantum dots: A comprehensive review. *Analytical Methods*, *17*(13), 2627–2649. DOI: 10.1039/D4AY02237H PMID: 40104848

Elugoke, S. E., Uwaya, G. E., Quadri, T. W., & Ebenso, E. E. (2024). Carbon Quantum Dots: Basics, Properties, and Fundamentals. In *Carbon Dots: Recent Developments and Future Perspectives* (Vol. 1465, pp. 1–3). American Chemical Society. https://doi.org/doi:10.1021/bk-2024-1465.ch001

Emam, H. E. (2024). Carbon quantum dots derived from polysaccharides: Chemistry and potential applications. *Carbohydrate Polymers*, *324*, 121503. https://doi.org/https://doi.org/10.1016/j.carbpol.2023.121503. DOI: 10.1016/j.carbpol.2023.121503 PMID: 37985091

Fakheri, M., Fatemi, S., & Rahimi Kakolaki, R. (2023). Comparative study of one-pot and facile methods to synthesize codoped <scp>CQDs</scp> with low band gap and photovoltaic properties. *Canadian Journal of Chemical Engineering*, *101*(8), 4480–4492. DOI: 10.1002/cjce.24765

Fawaz, W., Hasian, J., & Alghoraibi, I. (2023). Synthesis and physicochemical characterization of carbon quantum dots produced from folic acid. *Scientific Reports*, *13*(1), 18641. DOI: 10.1038/s41598-023-46084-1 PMID: 37903841

Ganesan, M., Ramadhass, K. D., Chuang, H.-C., & Gopalakrishnan, G. (2021). Synthesis of nitrogen-doped carbon quantum dots@ Fe2O3/multiwall carbon nanotubes ternary nanocomposite for the simultaneous electrochemical detection of 5-fluorouracil, uric acid, and xanthine. *Journal of Molecular Liquids*, *331*, 115768. DOI: 10.1016/j.molliq.2021.115768

Ghosh, S., Bera, S., Kapuria, A., Debnath, A., Das, P., Su, Y.-K., & Saha, S. K. (2025). Nitrogen and sulfur co-doped carbon quantum dot-decorated Ti3C2Tx-MXenes as electrode materials for high-performance symmetric supercapacitors. *Nanoscale*, *17*(26), 15731–15748. DOI: 10.1039/D4NR05037A PMID: 40531092

Guan, X., Li, Z., Geng, X., Lei, Z., Karakoti, A., Wu, T., Kumar, P., Yi, J., & Vinu, A. (2023). Emerging Trends of Carbon-Based Quantum Dots: Nanoarchitectonics and Applications. *Small*, *19*(17), 2207181. https://doi.org/https://doi.org/10.1002/smll.202207181. DOI: 10.1002/smll.202207181 PMID: 36693792

Gulati, S., Baul, A., Amar, A., Wadhwa, R., Kumar, S., & Varma, R. S. (2023). Eco-Friendly and Sustainable Pathways to Photoluminescent Carbon Quantum Dots (CQDs). *Nanomaterials (Basel, Switzerland)*, *13*(3), 554. DOI: 10.3390/nano13030554 PMID: 36770515

Guo, R., Li, L., Wang, B., Xiang, Y., Zou, G., Zhu, Y., Hou, H., & Ji, X. (2021). Functionalized carbon dots for advanced batteries. *Energy Storage Materials*, *37*, 8–39. https://doi.org/https://doi.org/10.1016/j.ensm.2021.01.020. DOI: 10.1016/j.ensm.2021.01.020

Gurushankar, K., Karuppaiya, C., Tari, V., & Kannan, K. (2025). Exploring MXenes and MBenes in Metal Sulfur and Selenium Battery Technologies. In *Novel Energy Storage and Conversion Technologies for Two-Dimensional MXenes and MBenes* (pp. 105–120). IGI Global Scientific Publishing. DOI: 10.4018/979-8-3693-9546-2.ch005

Hafez, N. S., Ebeid, E.-Z. M., Khalifa, A., Azim, S. A., El-Shaer, A., Sakr, M. A. S., & Hegazy, M. B. Z. (2025a). Modulating carbon quantum dots using multi-heteroatom doping as a bifunctional electrocatalyst for efficient oxygen and hydrogen evolution reactions in alkaline electrolytes. *Journal of Materials Chemistry. A, Materials for Energy and Sustainability*, *13*(29), 23454–23462. DOI: 10.1039/D5TA03457D

Hafez, N. S., Ebeid, E.-Z. M., Khalifa, A., Azim, S. A., El-Shaer, A., Sakr, M. A. S., & Hegazy, M. B. Z. (2025b). Modulating carbon quantum dots using multi-heteroatom doping as a bifunctional electrocatalyst for efficient oxygen and hydrogen evolution reactions in alkaline electrolytes. *Journal of Materials Chemistry. A, Materials for Energy and Sustainability*, *13*(29), 23454–23462. DOI: 10.1039/D5TA03457D

Hagiwara, K., Horikoshi, S., & Serpone, N. (2021). Luminescent monodispersed carbon quantum dots by a microwave solvothermal method toward bioimaging applications. *Journal of Photochemistry and Photobiology A Chemistry*, *415*, 113310. DOI: 10.1016/j.jphotochem.2021.113310

Han, Y., Tang, D., Yang, Y., Li, C., Kong, W., Huang, H., Liu, Y., & Kang, Z. (2015). Non-metal single/dual doped carbon quantum dots: A general flame synthetic method and electro-catalytic properties. *Nanoscale*, *7*(14), 5955–5962. Advance online publication. DOI: 10.1039/C4NR07116F PMID: 25771786

Hou, Y., Yan, F., Bai, X., Wei, X., Chen, L., & Fu, Y. (2024). Designing advanced 0D/2D heterojunctions of N, S-co-doped carbon dots/Ti3C2Tx MXene nanosheets in electrochemical sensor applications. *Ceramics International*, *50*(7), 10735–10745. DOI: 10.1016/j.ceramint.2023.12.389

Jo, M.-H., Kim, K.-H., & Ahn, H.-J. (2022). P-doped carbon quantum dot graft-functionalized amorphous WO3 for stable and flexible electrochromic energy-storage devices. *Chemical Engineering Journal*, *445*, 136826. https://doi.org/https://doi.org/10.1016/j.cej.2022.136826. DOI: 10.1016/j.cej.2022.136826

John, B. K., John, N., Korah, B. K., Thara, C., Abraham, T., & Mathew, B. (2022). Nitrogen-doped carbon quantum dots as a highly selective fluorescent and electrochemical sensor for tetracycline. *Journal of Photochemistry and Photobiology A Chemistry*, *432*, 114060. DOI: 10.1016/j.jphotochem.2022.114060

John, B. K., & Mathew, B. (2023). Nitrogen and sulphur co-doped carbon quantum dot integrated bismuth oxide nanocomposite for photocatalytic degradation and electrochemical sensing applications. *Optical Materials*, *139*, 113819. DOI: 10.1016/j.optmat.2023.113819

John, B. K., Mathew, S., John, N., Mathew, J., & Mathew, B. (2023). Hydrothermal synthesis of N, S-doped carbon quantum dots as a dual mode sensor for azo dye tartrazine and fluorescent ink applications. *Journal of Photochemistry and Photobiology A Chemistry*, *436*, 114386. DOI: 10.1016/j.jphotochem.2022.114386

Jyothi, S. P., Vinod, D., Chandran, D., Antherjanam, S., Saraswathyamma, B., Balaraman, V., & Rajamani, R. (2024). Poly-4-amino-6-chloro-1, 3-benzene disulfonamide decorated pencil graphite electrode for the simultaneous electrochemical quantification of catechol and hydroquinone. *Journal of Applied Electrochemistry*, *54*(8), 1819–1831. DOI: 10.1007/s10800-024-02067-2

Kalpana, R., Sakthi Vignesh, N., Vinothini, K., Rajan, M., Ashokkumar, B., Brindhadevi, K., Thuy Lan Chi, N., Pugazhendhi, A., & Varalakshmi, P. (2023). Carbon quantum dots (CQD) fabricated from Exiguobacterium sp. VK2 exopolysaccharide (EPS) using hydrothermal reaction and its biodiesel applications. *Fuel, 333*, 126426. https://doi.org/https://doi.org/10.1016/j.fuel.2022.126426. DOI: 10.1016/j.fuel.2022.126426

Kannan, K., Chanda, D., Gautam, J., Behera, A., Meshesha, M. M., Jang, S. G., & Yang, B. (2023). Hydrothermally synthesized mixed metal oxide nanocomposites for electrochemical water splitting and photocatalytic hydrogen production. *International Journal of Hydrogen Energy*, *48*(93), 36412–36426. DOI: 10.1016/j.ijhydene.2023.06.042

Kannan, K., Chanda, D., Gautam, J., Meshesha, M. M., Jang, S. G., & Yang, B. L. (2023). Facial synthesis of pp heterojunction composites: Evaluation of their electrochemical properties with photovoltaics-electrolyzer water splitting using two-electrode system. *International Journal of Hydrogen Energy*, *48*(37), 13814–13826. DOI: 10.1016/j.ijhydene.2022.12.235

Kannan, K., Gautam, J., Chanda, D., Meshesha, M. M., Jang, S. G., & Yang, B. L. (2023). Two-dimensional MAX supported copper oxide/nickel Oxide/MAX as an efficient and novel photocatalyst for hydrogen evolution. *International Journal of Hydrogen Energy*, *48*(20), 7273–7283. DOI: 10.1016/j.ijhydene.2022.10.131

Kannan, K., Radhika, D., Gnanasangeetha, D., Lakkaboyana, S. K., Sadasivuni, K. K., Gurushankar, K., & Hanafiah, M. M. (2021). Photocatalytic and antimicrobial properties of microwave synthesized mixed metal oxide nanocomposite. *Inorganic Chemistry Communications*, *125*, 108429. DOI: 10.1016/j.inoche.2020.108429

Kannan, K., Sliem, M. H., Abdullah, A. M., Sadasivuni, K. K., & Kumar, B. (2020). Fabrication of ZnO-Fe-MXene based nanocomposites for efficient CO2 reduction. *Catalysts*, *10*(5), 549. DOI: 10.3390/catal10050549

Kannan, K., & Tari, V. (Eds.). (2024). Novel energy storage and conversion technologies for two-dimensional MXenes and MBenes. ISBN: 9798369395462, eISBN: 9798369395486DOI: 10.4018/979-8-3693-9546-2

Kar, D. K., v, P., Si, S., Panigrahi, H., & Mishra, S. (2024). Carbon Dots and Their Polymeric Nanocomposites: Insight into Their Synthesis, Photoluminescence Mechanisms, and Recent Trends in Sensing Applications. *ACS Omega*, *9*(10), 11050–11080. DOI: 10.1021/acsomega.3c07612 PMID: 38497004

Karthikeyan, R., Nelson, D. J., Ajith, A., & John, S. A. (2019). Hetero atoms doped carbon dots modified electrodes for the sensitive and selective determination of phenolic anti-oxidant in coconut oil. *Journal of Electroanalytical Chemistry (Lausanne, Switzerland)*, *848*, 113297. DOI: 10.1016/j.jelechem.2019.113297

Khan, M. E., Mohammad, A., & Yoon, T. (2022). State-of-the-art developments in carbon quantum dots (CQDs): Photo-catalysis, bio-imaging, and bio-sensing applications. *Chemosphere*, *302*, 134815. https://doi.org/https://doi.org/10.1016/j.chemosphere.2022.134815. DOI: 10.1016/j.chemosphere.2022.134815 PMID: 35526688

Khan, M. M. (2021). *Introduction and fundamentals of chalcogenides and chalcogenides-based nanomaterials*. Elsevier., DOI: 10.1016/B978-0-12-820498-6.00001-9

Kou, X., Jiang, S., Park, S.-J., & Meng, L.-Y. (2020). A review: Recent advances in preparations and applications of heteroatom-doped carbon quantum dots. *Dalton Transactions (Cambridge, England)*, *49*(21), 6915–6938. DOI: 10.1039/D0DT01004A PMID: 32400806

Lakhera, S., Devlal, K., Rana, M., & Dhuliya, V. (2024a). Influence of the substitution of different functional groups on the gas sensing and light harvesting efficiency of zero-dimensional coronene quantum dot: A first principle DFT study. *Spectrochimica Acta. Part A: Molecular and Biomolecular Spectroscopy*, *308*, 123737. https://doi.org/https://doi.org/10.1016/j.saa.2023.123737. DOI: 10.1016/j.saa.2023.123737 PMID: 38064960

Lakhera, S., Devlal, K., Rana, M., & Dhuliya, V. (2024b). Influence of the substitution of different functional groups on the gas sensing and light harvesting efficiency of zero-dimensional coronene quantum dot: A first principle DFT study. *Spectrochimica Acta. Part A: Molecular and Biomolecular Spectroscopy*, *308*, 123737. https://doi.org/https://doi.org/10.1016/j.saa.2023.123737. DOI: 10.1016/j.saa.2023.123737 PMID: 38064960

Latif, Z., Ali, M., Lee, E.-J., Zubair, Z., & Lee, K. H. (2023). Thermal and Mechanical Properties of Nano-Carbon-Reinforced Polymeric Nanocomposites: A Review. In *Journal of Composites Science* (Vol. 7, Issue 10). DOI: 10.3390/jcs7100441

Li, L., Liu, D., Shi, A., & You, T. (2018). Simultaneous stripping determination of cadmium and lead ions based on the N-doped carbon quantum dots-graphene oxide hybrid. *Sensors and Actuators. B, Chemical*, *255*, 1762–1770. DOI: 10.1016/j.snb.2017.08.190

Lim, C. S., Hola, K., Ambrosi, A., Zboril, R., & Pumera, M. (2015). Graphene and carbon quantum dots electrochemistry. *Electrochemistry Communications, 52*, 75–79. https://doi.org/https://doi.org/10.1016/j.elecom.2015.01.023. DOI: 10.1016/j.elecom.2015.01.023

Liu, L., Anwar, S., Ding, H., Xu, M., Yin, Q., Xiao, Y., Yang, X., Yan, M., & Bi, H. (2019). Electrochemical sensor based on F, N-doped carbon dots decorated laccase for detection of catechol. *Journal of Electroanalytical Chemistry (Lausanne, Switzerland), 840*, 84–92. DOI: 10.1016/j.jelechem.2019.03.071

Liu, M., Xu, Y., Niu, F., Gooding, J. J., & Liu, J. (2016). Carbon quantum dots directly generated from electrochemical oxidation of graphite electrodes in alkaline alcohols and the applications for specific ferric ion detection and cell imaging. *Analyst, 141*(9), 2657–2664. DOI: 10.1039/C5AN02231B PMID: 26878217

Liu, Q., Chen, H., Mi, R., Min, X., Fang, M., Wu, X., Huang, Z., & Liu, Y. (2025). Biomass-Derived Carbon Dots: Preparation, Properties, and Applications. *Nanomaterials (Basel, Switzerland), 15*(16), 1279. DOI: 10.3390/nano15161279 PMID: 40863859

Manzoor, S., Dar, A. H., Dash, K. K., Pandey, V. K., Srivastava, S., Bashir, I., & Khan, S. A. (2023). Carbon dots applications for development of sustainable technologies for food safety: A comprehensive review. *Applied Food Research, 3*(1), 100263. https://doi.org/https://doi.org/10.1016/j.afres.2023.100263. DOI: 10.1016/j.afres.2023.100263

Mindivan, F., & Göktaş, M. (2023). The green synthesis of carbon quantum dots (CQDs) and characterization of polycaprolactone (PCL/CQDs) films. *Colloids and Surfaces. A, Physicochemical and Engineering Aspects, 677*, 132446. https://doi.org/https://doi.org/10.1016/j.colsurfa.2023.132446. DOI: 10.1016/j.colsurfa.2023.132446

Molaei, M. J. (2020). Principles, mechanisms, and application of carbon quantum dots in sensors: A review. *Analytical Methods, 12*(10), 1266–1287. DOI: 10.1039/C9AY02696G

Murali, A. S., Sajeevan, B., Kumar, S. M. S., Govindasamy, M., & Saraswathyamma, B. (2024). Facile formation of surface functionalised α-Fe2O3 nanoparticles and their role in developing dual sensors towards cysteamine quantification. *Materials Research Bulletin, 174*, 112734. DOI: 10.1016/j.materresbull.2024.112734

Muthusankar, G., Devi, R. K., & Gopu, G. (2020). Nitrogen-doped carbon quantum dots embedded Co3O4 with multiwall carbon nanotubes: An efficient probe for the simultaneous determination of anticancer and antibiotic drugs. *Biosensors & Bioelectronics*, *150*, 111947. DOI: 10.1016/j.bios.2019.111947 PMID: 31818763

Muthusankar, G., Rajkumar, C., Chen, S.-M., Karkuzhali, R., Gopu, G., Sangili, A., Sengottuvelan, N., & Sankar, R. (2019). Sonochemical driven simple preparation of nitrogen-doped carbon quantum dots/SnO2 nanocomposite: A novel electrocatalyst for sensitive voltammetric determination of riboflavin. *Sensors and Actuators. B, Chemical*, *281*, 602–612. DOI: 10.1016/j.snb.2018.10.145

Muthusankar, G., Sangili, A., Chen, S.-M., Karkuzhali, R., Sethupathi, M., Gopu, G., Karthick, S., Devi, R. K., & Sengottuvelan, N. (2018). In situ assembly of sulfur-doped carbon quantum dots surrounded iron (III) oxide nanocomposite; a novel electrocatalyst for highly sensitive detection of antipsychotic drug olanzapine. *Journal of Molecular Liquids*, *268*, 471–480. DOI: 10.1016/j.molliq.2018.07.059

Muthusankar, G., Sasikumar, R., Chen, S.-M., Gopu, G., Sengottuvelan, N., & Rwei, S.-P. (2018). Electrochemical synthesis of nitrogen-doped carbon quantum dots decorated copper oxide for the sensitive and selective detection of non-steroidal anti-inflammatory drug in berries. *Journal of Colloid and Interface Science*, *523*, 191–200. DOI: 10.1016/j.jcis.2018.03.095 PMID: 29621646

Muthusankar, G., Sethupathi, M., Chen, S.-M., Devi, R. K., Vinoth, R., Gopu, G., Anandhan, N., & Sengottuvelan, N. (2019). N-doped carbon quantum dots@ hexagonal porous copper oxide decorated multiwall carbon nanotubes: A hybrid composite material for an efficient ultra-sensitive determination of caffeic acid. *Composites. Part B, Engineering*, *174*, 106973. DOI: 10.1016/j.compositesb.2019.106973

Padmapriya, A., Thiyagarajan, P., Devendiran, M., Kalaivani, R. A., & Shanmugharaj, A. M. (2023). Electrochemical sensor based on N,P–doped carbon quantum dots derived from the banana flower bract (Musa acuminata) biomass extract for selective and picomolar detection of dopamine. *Journal of Electroanalytical Chemistry (Lausanne, Switzerland)*, *943*, 117609. https://doi.org/https://doi.org/10.1016/j.jelechem.2023.117609. DOI: 10.1016/j.jelechem.2023.117609

Pakapongpan, S., Poo-arporn, Y., Ninket, S., & Poo-arporn, R. P. (2024). A disposable electrochemical sensor for amyloid-β42 protein based on molecular imprinted polymers with nitrogen doped carbon dots-graphene nanohybrid. *Microchemical Journal*, *206*, 111559. DOI: 10.1016/j.microc.2024.111559

Pal, A., Ahmad, K., Dutta, D., & Chattopadhyay, A. (2019). Boron doped carbon dots with unusually high photoluminescence quantum yield for ratiometric intracellular pH sensing. *ChemPhysChem*, *20*(8), 1018–1027. DOI: 10.1002/cphc.201900140 PMID: 30891892

Phang, S. J., Lee, J., Wong, V.-L., Tan, L.-L., & Chai, S.-P. (2022). Synergistic effects of the hybridization between boron-doped carbon quantum dots and n/n-type g-C3N4 homojunction for boosted visible-light photocatalytic activity. *Environmental Science and Pollution Research International*, *29*(27), 41272–41292. DOI: 10.1007/s11356-021-18253-0 PMID: 35088270

Pillar-Little, T. J., Wanninayake, N., Nease, L., Heidary, D. K., Glazer, E. C., & Kim, D. Y. (2018). Superior photodynamic effect of carbon quantum dots through both type I and type II pathways: Detailed comparison study of top-down-synthesized and bottom-up-synthesized carbon quantum dots. *Carbon*, *140*, 616–623. https://doi.org/https://doi.org/10.1016/j.carbon.2018.09.004. DOI: 10.1016/j.carbon.2018.09.004

Poh, H. L., & Pumera, M. (2015). p-Element-Doped Graphene: Heteroatoms for Electrochemical Enhancement. *ChemElectroChem*, *2*(2), 190–199. https://doi.org/https://doi.org/10.1002/celc.201402307. DOI: 10.1002/celc.201402307

Qian, Z., Shan, X., Chai, L., Ma, J., Chen, J., & Feng, H. (2014). Si-doped carbon quantum dots: A facile and general preparation strategy, bioimaging application, and multifunctional sensor. *ACS Applied Materials & Interfaces*, *6*(9), 6797–6805. Advance online publication. DOI: 10.1021/am500403n PMID: 24707855

Rafiq, K., Sadia, I., Abid, M. Z., Waleed, M. Z., Rauf, A., & Hussain, E. (2024). Scientific Insights into the Quantum Dots (QDs)-Based Electrochemical Sensors for State-of-the-Art Applications. *ACS Biomaterials Science & Engineering*, *10*(12), 7268–7313. DOI: 10.1021/acsbiomaterials.4c01256 PMID: 39499739

Ran, X., Qu, Q., Qian, X., Xie, W., Li, S., Li, L., & Yang, L. (2018). Water-soluble pillar [6] arene functionalized nitrogen-doped carbon quantum dots with excellent supramolecular recognition capability and superior electrochemical sensing performance towards TNT. *Sensors and Actuators. B, Chemical*, *257*, 362–371. DOI: 10.1016/j.snb.2017.10.185

Rao, K. K., & Prakash, B. (2025). Analyzing Functionalized Carbon Quantum Dots in Electrode Engineering for the Precise Electrochemical Detection of a Neurotransmitter: Achieving Laboratory Accuracy with Portable Sensors. *ACS Sensors*, *10*(5), 3423–3432. DOI: 10.1021/acssensors.4c03298 PMID: 40261822

Rong, C., Huang, Y., & Zheng, X. (2025). Sensitive determination of hydroquinone and catechol using an electrochemical sensor based on nitrogen-doped malic acid carbon quantum dots. *Materials Chemistry and Physics, 329*, 130077. DOI: 10.1016/j.matchemphys.2024.130077

Saleh Mohammadnia, M., Roghani-Mamaqani, H., Ghalkhani, M., & Hemmati, S. (2023a). A modified electrochemical sensor based on N, S-doped carbon dots/carbon nanotube-poly (amidoamine) dendrimer hybrids for imatinib mesylate determination. *Biosensors (Basel), 13*(5), 547. DOI: 10.3390/bios13050547 PMID: 37232908

Saleh Mohammadnia, M., Roghani-Mamaqani, H., Ghalkhani, M., & Hemmati, S. (2023b). A Modified Electrochemical Sensor Based on N,S-Doped Carbon Dots/Carbon Nanotube-Poly(Amidoamine) Dendrimer Hybrids for Imatinib Mesylate Determination. *Biosensors (Basel), 13*(5), 547. Advance online publication. DOI: 10.3390/bios13050547 PMID: 37232908

Sethulekshmi, A. S., Aparna, A., Parvathi, P., Pathak, R., Punetha, V. D., Selvaraj, M., & Saritha, A. (2025). Advances in doped carbon quantum dots: Synthesis, mechanisms, and applications in sensing technologies. *Chemical Engineering Journal, 514*, 163262. https://doi.org/https://doi.org/10.1016/j.cej.2025.163262. DOI: 10.1016/j.cej.2025.163262

Singh, H., Bamrah, A., Khatri, M., & Bhardwaj, N. (2020a). One-pot hydrothermal synthesis and characterization of carbon quantum dots (CQDs). *Materials Today: Proceedings, 28*, 1891–1894. https://doi.org/https://doi.org/10.1016/j.matpr.2020.05.297. DOI: 10.1016/j.matpr.2020.05.297

Singh, H., Bamrah, A., Khatri, M., & Bhardwaj, N. (2020b). One-pot hydrothermal synthesis and characterization of carbon quantum dots (CQDs). *Materials Today: Proceedings, 28*, 1891–1894. https://doi.org/https://doi.org/10.1016/j.matpr.2020.05.297. DOI: 10.1016/j.matpr.2020.05.297

Singh, I., Arora, R., Dhiman, H., & Pahwa, R. (2018). Carbon quantum dots: Synthesis, characterization and biomedical applications. *Turkish Journal of Pharmaceutical Sciences, 15*(2), 219–230. DOI: 10.4274/tjps.63497 PMID: 32454664

Singh, V., Kashyap, S., Yadav, U., Srivastava, A., Singh, A. V., Singh, R. K., Singh, S. K., & Saxena, P. S. (2019). Nitrogen doped carbon quantum dots demonstrate no toxicity under in vitro conditions in a cervical cell line and in vivo in Swiss albino mice. *Toxicology Research, 8*(3), 395–406. DOI: 10.1039/C8TX00260F PMID: 31160973

Sohal, N., Maity, B., & Basu, S. (2021). Recent advances in heteroatom-doped graphene quantum dots for sensing applications. *RSC Advances*, *11*(41), 25586–25615. DOI: 10.1039/D1RA04248C PMID: 35478909

Tavan, M., Yousefian, Z., Bakhtiar, Z., Rahmandoust, M., & Mirjalili, M. H. (2025). Carbon quantum dots: Multifunctional fluorescent nanomaterials for sustainable advances in biomedicine and agriculture. *Industrial Crops and Products*, *231*, 121207. https://doi.org/https://doi.org/10.1016/j.indcrop.2025.121207. DOI: 10.1016/j.indcrop.2025.121207

Velmurugan, S., Traiwatcharanon, P., Jiwanti, P. K., Cheng, S.-H., & Wongchoosuk, C. (2024). Highly selective carbendazim fungicide sensing performance of nitrogen-doped carbon quantum dots encapsulated aminated-UiO-66 zirconium metal-organic framework electrocatalyst. *Electrochimica Acta*, *479*, 143911. DOI: 10.1016/j.electacta.2024.143911

Wang, Z., Changotra, R., Dasog, M., Singh Selopal, G., Yang, J., & He, Q. S. (2025). Carbon quantum dots: Synthesis via hydrothermal processing, doping strategies, integration with photocatalysts, and their application in photocatalytic hydrogen production. *Sustainable Materials and Technologies, 44*, e01386. https://doi.org/https://doi.org/10.1016/j.susmat.2025.e01386

Yadav, R., & Lahariya, V. (2025a). Investigation of Fluorescence Sensing Capabilities in Boron-Nitrogen Co-Doped Carbon Quantum Dots Towards Fe (III) and Hg(II) Ions. *Journal of Fluorescence*. Advance online publication. DOI: 10.1007/s10895-025-04487-5 PMID: 40824367

Yadav, R., & Lahariya, V. (2025b). Investigation of Fluorescence Sensing Capabilities in Boron-Nitrogen Co-Doped Carbon Quantum Dots Towards Fe (III) and Hg(II) Ions. *Journal of Fluorescence*. Advance online publication. DOI: 10.1007/s10895-025-04487-5 PMID: 40824367

Yola, M. L., & Atar, N. (2019). Development of molecular imprinted sensor including graphitic carbon nitride/N-doped carbon dots composite for novel recognition of epinephrine. *Composites. Part B, Engineering*, *175*, 107113. DOI: 10.1016/j.compositesb.2019.107113

Yu, S., Lu, S., Tan, D., & Zhu, Y. (2022a). Nitrogen and phosphorus co-doped carbon dots for developing highly flame retardant poly (vinyl alcohol) composite films. *European Polymer Journal*, *164*, 110970. https://doi.org/https://doi.org/10.1016/j.eurpolymj.2021.110970. DOI: 10.1016/j.eurpolymj.2021.110970

Yu, S., Lu, S., Tan, D., & Zhu, Y. (2022b). Nitrogen and phosphorus co-doped carbon dots for developing highly flame retardant poly (vinyl alcohol) composite films. *European Polymer Journal*, *164*, 110970. https://doi.org/https://doi.org/10.1016/j.eurpolymj.2021.110970. DOI: 10.1016/j.eurpolymj.2021.110970

Zhao, D. L., & Chung, T.-S. (2018). Applications of carbon quantum dots (CQDs) in membrane technologies: A review. *Water Research*, *147*, 43–49. https://doi.org/https://doi.org/10.1016/j.watres.2018.09.040. DOI: 10.1016/j.watres.2018.09.040 PMID: 30296608

Zhao, F., Li, X., Zuo, M., Liang, Y., Qin, P., Wang, H., Wu, Z., Luo, L., Liu, C., & Leng, L. (2023). Preparation of photocatalysts decorated by carbon quantum dots (CQDs) and their applications: A review. *Journal of Environmental Chemical Engineering*, *11*(2), 109487. https://doi.org/https://doi.org/10.1016/j.jece.2023.109487. DOI: 10.1016/j.jece.2023.109487

Zhao, P., Song, X., Dong, M., Sun, H., Wu, W., Zhang, R., Sun, M., & Zhao, X. (2022a). Preparation and characterization of CQDs/SBS composites and its application performance as asphalt modifier. *Construction & Building Materials*, *320*, 126312. https://doi.org/https://doi.org/10.1016/j.conbuildmat.2022.126312. DOI: 10.1016/j.conbuildmat.2022.126312

Zhao, P., Song, X., Dong, M., Sun, H., Wu, W., Zhang, R., Sun, M., & Zhao, X. (2022b). Preparation and characterization of CQDs/SBS composites and its application performance as asphalt modifier. *Construction & Building Materials*, *320*, 126312. https://doi.org/https://doi.org/10.1016/j.conbuildmat.2022.126312. DOI: 10.1016/j.conbuildmat.2022.126312

Zheng, M., Xie, Z., Qu, D., Li, D., Du, P., Jing, X., & Sun, Z. (2013). On–Off–On Fluorescent Carbon Dot Nanosensor for Recognition of Chromium(VI) and Ascorbic Acid Based on the Inner Filter Effect. *ACS Applied Materials & Interfaces*, *5*(24), 13242–13247. DOI: 10.1021/am4042355 PMID: 24274643

Compilation of References

Abbas, Q., Raza, R., Shabbir, I., & Olabi, A. G. (2019). Heteroatom doped high porosity carbon nanomaterials as electrodes for energy storage in electrochemical capacitors: A review. *Journal of Science. Advanced Materials and Devices*, *4*(3), 341–352. DOI: 10.1016/j.jsamd.2019.07.007

Abd Rani, U., Ng, L. Y., Ng, C. Y., Mahmoudi, E., Ng, Y. S., & Mohammad, A. W. (2021). Sustainable production of nitrogen-doped carbon quantum dots for photocatalytic degradation of methylene blue and malachite green. *Journal of Water Process Engineering*, *40*, 101816. DOI: 10.1016/j.jwpe.2020.101816

Abdel Maksoud, M. I. A., Fahim, R. A., Shalan, A. E., Abd Elkodous, M., Olojede, S. O., Osman, A. I., Farrell, C., Al-Muhtaseb, A. H., Awed, A. S., Ashour, A. H., & Rooney, D. W. (2021). Advanced materials and technologies for supercapacitors used in energy conversion and storage: A review. *Environmental Chemistry Letters*, *19*(1), 375–439. DOI: 10.1007/s10311-020-01075-w

Abdelsalam, H., Sakr, M. A. S., Teleb, N. H., Abd-Elkader, O. H., & Zhang, Q. (2025). Quantum-dot-based carbon nanotubes: Design, doping, and lithium storage for high-capacity energy applications. *Langmuir*, *41*(27), 17803–17814. DOI: 10.1021/acs.langmuir.5c01572 PMID: 40615346

Abid, . (2022). Synthesis of nanomaterials using various top-down and bottom-up approaches, influencing factors, advantages, and disadvantages: A review. In *Advances in Colloid and Interface Science* (Vol. 300). Elsevier B.V., DOI: 10.1016/j.cis.2021.102597

Abinaya, S., Girija, S., Baby, A., Banu, A. A., & Jose, S. P. (2025). Nitrogen doped carbon quantum dots@ α-Fe2O3/PANI nanocomposite based electrochemical sensor for cadmium ion detection. *Inorganic Chemistry Communications*, *179*, 114846. DOI: 10.1016/j.inoche.2025.114846

Adeyinka, A. M., Esan, O. C., Ijaola, A. O., & Farayibi, P.-K. (2024). Advancements in hybrid energy storage systems for enhancing renewable energy-to-grid integration. *Sustainable Energy Research, 11*(1), 26. DOI: 10.1186/s40807-024-00120-4

Aghamali, , Khosravi, M., Hamishehkar, H., Modirshahla, N., & Behnajady, M. A. (2018). Synthesis and characterization of high efficient photoluminescent sunlight driven photocatalyst of N-Carbon Quantum Dots. *Journal of Luminescence, 201*, 265–274. DOI: 10.1016/j.jlumin.2018.04.061

Agrawal, A., & Choudhary, A. (2016). Perspective: Materials informatics and big data: Realization of the "fourth paradigm" of science in materials science. *APL Materials, 4*(5), 053208. Advance online publication. DOI: 10.1063/1.4946894

Ahmadian, E., Janas, D., Eftekhari, A., & Zare, N. (2022). Application of carbon nanotubes in sensing/monitoring of pancreas and liver cancer. *Chemosphere, 302*, 134826. Advance online publication. DOI: 10.1016/j.chemosphere.2022.134826 PMID: 35525455

Ahmed, H. B., El-Shahat, M., Allayeh, A. K., & Emam, H. E. (2023). Maillard reaction for nucleation of polymer quantum dots from chitosan-glucose conjugate: Antagonistic for cancer and viral diseases. *International Journal of Biological Macromolecules, 224*, 858–870. DOI: 10.1016/j.ijbiomac.2022.10.172 PMID: 36306909

Al Ragib, A., Al Amin, A., Alanazi, Y. M., Kormoker, T., Uddin, M., Siddique, M. A. B., & Barai, H. R. (2023). Multifunctional carbon dots in nanomaterial surface modification: A descriptive review. *Carbon Research, 2*(1), 37. DOI: 10.1007/s44246-023-00069-x

Alam, A.-M., Park, B.-Y., Ghouri, Z. K., Park, M., & Kim, H.-Y. (2015). Synthesis of carbon quantum dots from cabbage with down- and up-conversion photoluminescence properties: Excellent imaging agent for biomedical applications. *Green Chemistry, 17*(7), 3791–3797. DOI: 10.1039/C5GC00686D

Alaqel, , Alqahtani, A. S., Alharbi, A., Althobaiti, Y. S., Bamaga, A. K., Algarni, M. A., Almrasy, A. A., & Almalki, A. H. (2023). Spectrofluorometric quantitative analysis of aripiprazole based on quenching of natural derived carbon quantum dots in spiked human plasma. *Scientific Reports, 13*(1), 21048. Advance online publication. DOI: 10.1038/s41598-023-47392-2 PMID: 38030673

Albalwi, S., Al-Ojeery, A., Hassan, H., Iqbal, M. W., Afzal, A., & Aly, K. (2024). Incorporation of graphene quantum dots into magnesium copper phosphate (GQDs-MgCuPO4) for supercapattery and electrocatalysis applications. *Physica Scripta*. https://iopscience.iop.org/article/10.1088/1402-4896/ad9117

Alfei, S., Reggio, C., & Zuccari, G. (2025). Carbon Nanotubes as Excellent Adjuvants for Anticancer Therapeutics and Cancer Diagnosis: A Plethora of Laboratory Studies Versus Few Clinical Trials. *Cells, 14*(14), 1052. DOI: 10.3390/cells14141052 PMID: 40710305

Al-Gaashani, R., Najjar, A., Zakaria, Y., Mansour, S., & Atieh, M. A. (2019). Xps and structural studies of high quality graphene oxide and reduced graphene oxide prepared by different chemical oxidation methods. *Ceramics International, 45*(11), 14439–14448. DOI: 10.1016/j.ceramint.2019.04.165

Alhabardi, S. A., Almutairi, J. A., Al-Hamoud, G. A., Zaki Zakaraya, Z., Hani, U., Ali Ahmed Asiri, Z., Al Freahat, I., Alsafadi, W., Abu Dayyih, W., Khan, S. L., & Begh, M. (2025). Musa Paradisiaca derived intrinsically heteroatom doped carbon dots as antioxidant and controlled drug release behavior. *PLoS One, 20*(8), e0329116. DOI: 10.1371/journal.pone.0329116 PMID: 40811709

Ali, A., Rahimian Koloor, S. S., Alshehri, A. H., & Arockiarajan, A. (2023). Carbon nanotube characteristics and enhancement effects on the mechanical features of polymer-based materials and structures – A review. *Journal of Materials Research and Technology, 24*, 6495–6521. DOI: 10.1016/j.jmrt.2023.04.072

Ali, , Riaz, R., Anjum, A. S., Sun, K. C., Li, H., Ahn, S. J., Jeong, S. H., & Ko, M. J. (2021). Microwave-assisted ultrafast in-situ growth of N-doped carbon quantum dots on multiwalled carbon nanotubes as an efficient electrocatalyst for photovoltaics. *Journal of Colloid and Interface Science, 586*, 349–361. DOI: 10.1016/j.jcis.2020.10.098 PMID: 33187666

Alkahtani, M. E., Elbadawi, M., Chapman, C. A., Green, R. A., Gaisford, S., Orlu, M., & Basit, A. W. (2024). Electroactive polymers for on-demand drug release. *Advanced Healthcare Materials, 13*(3), 2301759. DOI: 10.1002/adhm.202301759 PMID: 37861058

Alkattan, N. (2014). *Magnetic Carbon Nanotubes as a Theranostic Platform for Drug Delivery and Magnetic Resonance Imaging*.

Alotaibi, B. S., Buabeid, M., Ibrahim, N. A., Kharaba, Z. J., Ijaz, M., Noreen, S., & Murtaza, G. (2021). Potential of nanocarrier-based drug delivery systems for brain targeting: A current review of literature. *International Journal of Nanomedicine, 16*, 7517–7533. DOI: 10.2147/IJN.S333657 PMID: 34795481

Alsuraifi, A., Mouzan, M. M., Algzaare, A., Aqeel, Z., Al-Essa, U. I. R., Mohammed, N. A., & Ayad, A. (2025). Quantum Dots: Catalysts for a New Era of Precision Medicine and Biomedical Innovation. *Journal of Fluorescence*, •••, 1–64. DOI: 10.1007/s10895-025-04452-2 PMID: 41051600

Amali, A. J., Hoshino, H., Wu, C., Ando, M., & Xu, Q. (2014). From Metal–Organic Framework to Intrinsically Fluorescent Carbon Nanodots. *Chemistry (Weinheim an der Bergstrasse, Germany)*, *20*(27), 8279–8282. DOI: 10.1002/chem.201402982 PMID: 24889063

Ambade, R. B., Ali, M., Lee, K. H., Jeong, W., Jeong, S. H., & Han, T. H. (2023). Nitrogen and Sulfur Co-Doped Carbon Quantum Dot-Engineered TiO2 Graphene on Carbon Fabric for Photocatalysis Applications. *ACS Applied Nano Materials*, *6*(17), 15782–15794. DOI: 10.1021/acsanm.3c02661

Anil Kumar, Y., Koyyada, G., Ramachandran, T., Kim, J. H., Sajid, S., Moniruzzaman, M., Alzahmi, S., & Obaidat, I. M. (2023). Carbon Materials as a Conductive Skeleton for Supercapacitor Electrode Applications: A Review. *Nanomaterials (Basel, Switzerland)*, *13*(6), 1049. DOI: 10.3390/nano13061049 PMID: 36985942

Anjali Devi, J., Anulekshmi, A., Salini, S., Aparna, R., & George, S. (2017). Boronic acid functionalized nitrogen doped carbon dots for fluorescent turn-on detection of dopamine. *Microchimica Acta*, *184*(10), 4081–4090. DOI: 10.1007/s00604-017-2433-7

Ansari, S. A. (2022). Graphene Quantum Dots: Novel Properties and Their Applications for Energy Storage Devices. *Nanomaterials (Basel, Switzerland)*, *12*(21), 3814. DOI: 10.3390/nano12213814 PMID: 36364590

Aparício, R. R., Henning, E., Arantes, M. S. T., Jesus, M. B., Hansel, F. A., Dias, I. A., Roman, L. S., Schiavon, M. A., de Muniz, G. I. B., Magalhães, W. L. E., & de Cademartori, P. H. G. (2025). Techno-economic analysis and optimization of ethylenediamine-doped carbon dots production from kraft lignin via hydrothermal process. *International Journal of Biological Macromolecules*, *323*, 147132. DOI: 10.1016/j.ijbiomac.2025.147132 PMID: 40865817

Argüelles-Cruz, A., Shah, K. B., Logeshwaran, J., & Shrestha, A. (2023). Genetic algorithm based techno-economic optimization of an isolated hybrid energy system. *ICTACT Journal on Microelectronics*, *8*(4), 1447–1450. DOI: 10.21917/ijme.2023.0249

Arkan, , Barati, A., Rahmanpanah, M., Hosseinzadeh, L., Moradi, S., & Hajialyani, M. (2018). Green synthesis of carbon dots derived from walnut oil and an investigation of their cytotoxic and apoptogenic activities toward cancer cells. *Advanced Pharmaceutical Bulletin*, *8*(1), 149–155. DOI: 10.15171/apb.2018.018 PMID: 29670850

Arooj, M., Parambath, J. B. M., Ali, N., Khan, A., Malik, S., Bilal, M., & Mohamed, A. A. (2022). Experimental and theoretical review on covalent coupling and elemental doping of carbon nanomaterials for environmental photocatalysis. *Critical Reviews in Solid State and Material Sciences*, *48*(2), 215–256. DOI: 10.1080/10408436.2022.2049697

Arora, G., Sabran, N. S., Ng, C. Y., Low, F. W., & Jun, H. K. (2024). Applications of carbon quantum dots in electrochemical energy storage devices. *Heliyon*, *10*(15), e35543. DOI: 10.1016/j.heliyon.2024.e35543 PMID: 39166025

Arsalani, N., Nezhad-Mokhtari, P., & Jabbari, E. (2019). Microwave-assisted and one-step synthesis of PEG passivated fluorescent carbon dots from gelatin as an efficient nanocarrier for methotrexate delivery. *Artificial Cells, Nanomedicine, and Biotechnology*, *47*(1), 540–547. DOI: 10.1080/21691401.2018.1562460 PMID: 30829085

Arshada, F., & Sk, M. P. (2019). Aggregation-induced red shift in N, S-doped chiral carbon dot emissions for moisture sensing. *New Journal of Chemistry*, *43*(33), 13240–13248. DOI: 10.1039/C9NJ03009C

Arumugham, , Alagumuthu, M., Amimodu, R. G., Munusamy, S., & Iyer, S. K. (2020). A sustainable synthesis of green carbon quantum dot (CQD) from Catharanthus roseus (white flowering plant) leaves and investigation of its dual fluorescence responsive behavior in multi-ion detection and biological applications. *Sustainable Materials and Technologies*, *23*, e00138. Advance online publication. DOI: 10.1016/j.susmat.2019.e00138

Atchudan, R., Immanuel Edison, T. N. J., Shanmugam, M., Perumal, S., Somanathan, T., & Lee, Y. R. (2021). Sustainable synthesis of carbon quantum dots from banana peel waste using hydrothermal process for in vivo bioimaging. *Physica E, Low-Dimensional Systems and Nanostructures*, *126*, 114417. DOI: 10.1016/j.physe.2020.114417

Ayaz, F., Alas, M. O., & Genc, R. (2020). Differential immunomodulatory effect of carbon dots influenced by the type of surface passivation agent. *Inflammation*, *43*(2), 777–783. DOI: 10.1007/s10753-019-01165-0 PMID: 31873835

Baghel, P., Sakhiya, A. K., & Kaushal, P. (2022). Ultrafast growth of carbon nanotubes using microwave irradiation: Characterization and its potential applications. *Heliyon*, *8*(10), e10943. DOI: 10.1016/j.heliyon.2022.e10943 PMID: 36276756

Baker, , & Baker, G. A. (2010). Luminescent Carbon Nanodots: Emergent Nanolights. *Angewandte Chemie International Edition*, *49*(38), 6726–6744. DOI: 10.1002/anie.200906623 PMID: 20687055

Balakrishnan, T., Ang, W. L., & Mahmoudi, E. (2023). Enhancing photoluminescence stability of blue emissive nitrogen doped graphene quantum dots by encapsulating inside zeolitic imidazole framework-8. *Journal of Solid State Chemistry*, *328*, 124343. https://doi.org/https://doi.org/10.1016/j.jssc.2023.124343. DOI: 10.1016/j.jssc.2023.124343

Balogun, M.-S., Luo, Y., Lyu, F., Wang, F., Yang, H., Li, H., Liang, C., Huang, M., Huang, Y., & Tong, Y. (2016). Carbon Quantum Dot Surface-Engineered VO_2 Interwoven Nanowires: A Flexible Cathode Material for Lithium and Sodium Ion Batteries. *ACS Applied Materials & Interfaces*, *8*(15), 9733–9744. DOI: 10.1021/acsami.6b01305 PMID: 27028048

Bao, H., Liu, Y., Li, H., Qi, W., & Sun, K. (2023). Luminescence of carbon quantum dots and their application in biochemistry. *Heliyon*, *9*(10), e20317. DOI: 10.1016/j.heliyon.2023.e20317 PMID: 37790961

Bao, R., Chen, Z., Zhao, Z., Sun, X., Zhang, J., Hou, L., & Yuan, C. (2018). Green and facile synthesis of nitrogen and phosphorus co-doped carbon quantum dots towards fluorescent ink and sensing applications. *Nanomaterials (Basel, Switzerland)*, *8*(6), 386. DOI: 10.3390/nano8060386 PMID: 29857502

Bao, W., Ma, H., Wang, N., & He, Z. (2019). pH-sensitive carbon quantum dots–doxorubicin nanoparticles for tumor cellular targeted drug delivery. *Polymers for Advanced Technologies*, *30*(11), 2664–2673. DOI: 10.1002/pat.4696

Bao, Z., Lu, L., Wang, J., Zhu, H., Dang, X., Chen, Y., Yin, X., Song, Y., & Wen, X. (2025). Application of quantum dots in cancer diagnosis and treatment: Advances and perspectives. *Nano Research*, *18*(1), 10–26599. DOI: 10.1007/s10895-025-04598-z PMID: 41139377

Baragau, I.-A., Power, N. P., Morgan, D. J., Lobo, R. A., Titirici, M.-M., Middelkoop, V., Diaz, A., & Dunn, S. (n.d.). *An Efficient Continuous Hydrothermal Flow Synthesis of Carbon Quantum Dots from a Targeted Biomass Precursor for On-Off Metal Ions Nano-Sensing*.

Baragau, , Power, N. P., Morgan, D. J., Heil, T., Lobo, R. A., Roberts, C. S., Titirici, M.-M., Dunn, S., & Kellici, S. (2020). Continuous hydrothermal flow synthesis of blue-luminescent, excitation-independent nitrogen-doped carbon quantum dots as nanosensors. *Journal of Materials Chemistry. A, Materials for Energy and Sustainability*, *8*(6), 3270–3279. DOI: 10.1039/C9TA11781D

Barik, R., & Ingole, P. P. (2020). Challenges and prospects of metal sulfide materials for supercapacitors. *Current Opinion in Electrochemistry*, *21*, 327–334. DOI: 10.1016/j.coelec.2020.03.022

Barman, M. K., Jana, B., Bhattacharyya, S., & Patra, A. (2014). Photophysical properties of doped carbon dots (n, p, and b) and their influence on electron/hole transfer in carbon dots–nickel (ii) phthalocyanine conjugates. *The Journal of Physical Chemistry. C, Nanomaterials and Interfaces*, *118*(34), 20034–20041. DOI: 10.1021/jp507080c

Başoğlu, , Ocak, Ü., & Gümrükçüoğlu, A. (2020). Synthesis of Microwave-Assisted Fluorescence Carbon Quantum Dots Using Roasted–Chickpeas and its Applications for Sensitive and Selective Detection of Fe^{3+} Ions. *Journal of Fluorescence*, *30*(3), 515–526. DOI: 10.1007/s10895-019-02428-7 PMID: 32152829

Basso, L., Cazzanelli, M., Orlandi, M., & Miotello, A. (2020). Nanodiamonds: Synthesis and application in sensing, catalysis, and the possible connection with some processes occurring in space. *Applied Sciences (Basel, Switzerland)*, *10*(12), 1–28. DOI: 10.3390/app10124094

Batra, R., Song, L., & Ramprasad, R. (2020). Emerging materials intelligence ecosystems propelled by machine learning. *Nature Reviews. Materials*, *6*(8), 655–678. DOI: 10.1038/s41578-020-00255-y

Becker, W., Bergmann, A., Biscotti, G. L., & Rueck, A. (2004). Advanced time-correlated single photon counting techniques for spectroscopy and imaging in biomedical systems. *Commercial and Biomedical Applications of Ultrafast Lasers IV*, *5340*, 104. DOI: 10.1117/12.529143

Bian, J., Huang, C., Wang, L., Hung, T., Daoud, W. A., & Zhang, R. (2014). Carbon Dot Loading and TiO_2 Nanorod Length Dependence of Photoelectrochemical Properties in Carbon Dot/TiO_2 Nanorod Array Nanocomposites. *ACS Applied Materials & Interfaces*, *6*(7), 4883–4890. DOI: 10.1021/am4059183 PMID: 24601482

Bie, X., Dong, Y., Xiong, M., Wang, B., Chen, Z., Zhang, Q., Liu, Y., & Huang, R. (2023). Nitrogen-Doped Carbon Matrix to Optimize Cycling Stability of Lithium Ion Battery Anode from SiOx Materials. *Inorganics*, *12*(1), 9. DOI: 10.3390/inorganics12010009

Bolshakova, O., Zherebyatieva, O., & Sarantseva, S. V. (2025). Fullerenes in vivo. Toxicity and protective effects. *Nanotoxicology*, *19*(3), 233–258. DOI: 10.1080/17435390.2025.2471273 PMID: 40015266

Bonferoni, M. C., Rassu, G., Gavini, E., Sorrenti, M., Catenacci, L., Torre, M. L., Perteghella, S., Ansaloni, L., Maestri, M., & Giunchedi, P. (2021). Electrochemotherapy of Deep-Seated Tumors: State of Art and Perspectives as Possible "EPR Effect Enhancer" to Improve Cancer Nanomedicine Efficacy. *Cancers 2021, Vol. 13, Page 4437*, *13*(17), 4437. DOI: 10.3390/cancers13174437

Boroznin, S. V. (2022). Carbon nanostructures containing boron impurity atoms: Synthesis, physicochemical properties and potential applications. *Modern Electronic Materials*, *8*(1), 23–42. DOI: 10.3897/j.moem.8.1.84317

Boruah, A., Bora, S., Thakur, A., Dutta, H. S., & Saikia, B. K. (2023). Solid-State Phosphors from Coal-Derived Carbon Quantum Dots. *ACS Omega*, *8*(28), 25410–25423. DOI: 10.1021/acsomega.3c02884 PMID: 37483255

Bottini, M., Balasubramanian, C., Dawson, M. I., Bergamaschi, A., Bellucci, S., & Mustelin, T. (2006). Isolation and characterization of fluorescent nanoparticles from pristine and oxidized electric arc-produced single-walled carbon nanotubes. *The Journal of Physical Chemistry B*, *110*(2), 831–836. DOI: 10.1021/jp055503b PMID: 16471611

Bourlinos, A. B., Stassinopoulos, A., Anglos, D., Zboril, R., Karakassides, M., & Giannelis, E. P. (2008). Surface Functionalized Carbogenic Quantum Dots. *Small*, *4*(4), 455–458. DOI: 10.1002/smll.200700578 PMID: 18350555

Bourlinos, A. B., Trivizas, G., Karakassides, M. A., Baikousi, M., Kouloumpis, A., Gournis, D., Bakandritsos, A., Hola, K., Kozak, O., Zboril, R., Papagiannouli, I., Aloukos, P., & Couris, S. (2015). Green and simple route toward boron doped carbon dots with significantly enhanced non-linear optical properties. *Carbon*, *83*, 173–179. DOI: 10.1016/j.carbon.2014.11.032

Bouzas-Ramos, D., Canga, J. C., Mayo, J. C., Sainz, R. M., Encinar, J. R., & Costa-Fernandez, J. M. (2019). Carbon quantum dots co-doped with nitrogen and lanthanides for multimodal imaging. *Advanced Functional Materials*, *29*(38), 1903884. DOI: 10.1002/adfm.201903884

Brachi., (2020). Synthesis of fluorescent carbon quantum dots (CQDs) through the mild thermal treatment of agro-industrial residues assisted by γ-alumina. *Biomass Conversion and Biorefinery*, *10*(4), 1301–1312. DOI: 10.1007/s13399-019-00503-4

Brown, M., & Li, D. (2024). Interfacial engineering for high performance carbon-based perovskite solar cells. *Frontiers in Energy Research*, *12*, 1463024. DOI: 10.3389/fenrg.2024.1463024

Buttry, D. A. (2021). Applications of the quartz crystal microbalance to electrochemistry. *Electroanalytical Chemistry*, •••, 1–85. DOI: 10.1201/9781003209003-1

Cai, D., Zhong, X., Xu, L., Xiong, Y., Deng, W., Zou, G., Hou, H., & Ji, X. (2025). Biomass-derived carbon dots: Synthesis, modification and application in batteries. *Chemical Science (Cambridge)*, *16*(12), 4937–4970. DOI: 10.1039/D4SC08659G PMID: 40046072

Cai, R., Xiao, L., Liu, M., Du, F., & Wang, Z. (2021). Recent advances in functional carbon quantum dots for antitumour. *International Journal of Nanomedicine*, *16*, 7195–7229. DOI: 10.2147/IJN.S334012 PMID: 34720582

Cai, S., Chang, G., Hu, J., Wu, J., Luo, Y., Zou, G., Hou, H., & Ji, X. (2023). N, S-Doped Carbon Dots as Additive for Suppression of Zinc Dendrites in Alkaline Electrolyte †. *Chinese Journal of Chemistry*, *41*(14), 1697–1704. DOI: 10.1002/cjoc.202200799

Caņado, L. G., Takai, K., Enoki, T., Endo, M., Kim, Y. A., Mizusaki, H., Jorio, A., Coelho, L. N., Magalhães-Paniago, R., & Pimenta, M. A. (2006). General equation for the determination of the crystallite size la of nanographite by Raman spectroscopy. *Applied Physics Letters*, *88*(16), 163106. Advance online publication. DOI: 10.1063/1.2196057

Cánchig, M. B., López, F., Morales-Navarro, Z., Debut, A., Vizuete, K., Terencio, T., Caetano, M., & Saucedo-Vázquez, J. P. (2025). Enhanced selectivity of carbon quantum dots for metal ion detection through surface modification by heteroatom doping: A study on optical properties and theoretical approach. *Carbon Trends*, *18*, 100445. DOI: 10.1016/j.cartre.2024.100445

Cao, K., Tian, Z., Zhang, X., Wang, Y., & Zhu, Q. (2023). Green preparation of graphene oxide nanosheets as adsorbent. *Scientific Reports*, *13*(1), 1–11. DOI: 10.1038/s41598-023-36595-2 PMID: 37291198

Çeşme, M., Eskalen, H., & Başkaya, S. K. (2022). Fluorescent carbon dots from vegetable and fruit wastes and their applications. In Ray, R. C. (Ed.), *Fruits and vegetable wastes*. Springer., DOI: 10.1007/978-981-16-9527-8_15

Chacón-Torres, J. C., Wirtz, L., & Pichler, T. (2013). Manifestation of charged and strained graphene layers in the raman response of graphite intercalation compounds. *ACS Nano*, *7*(10), 9249–9259. DOI: 10.1021/nn403885k PMID: 24025089

Chahal., ... (2021b). Green synthesis of carbon dots and their applications. In *RSC Advances* (Vol. 11, Issue 41, pp. 25354–25363). Royal Society of Chemistry. DOI: 10.1039/D1RA04718C

Chai, , He, H., Fan, H., Kang, X., & Song, X. (2019). A hydrothermal-carbonization process for simultaneously production of sugars, graphene quantum dots, and porous carbon from sugarcane bagasse. *Bioresource Technology*, *282*, 142–147. DOI: 10.1016/j.biortech.2019.02.126 PMID: 30852334

Chakrabarti, A., Majumdar, P., Alessandri, E., & Roemer, C. (2025). Recent advancement of quantum dot-based nanocomposites as electrode materials for secondary batteries. *Energies*, *18*(3), 630. DOI: 10.3390/en18030630

Chao-Mujica, F. J., Garcia-Hernández, L., Camacho-López, S., Camacho-López, M., Camacho-López, M. A., Reyes Contreras, D., Pérez-Rodríguez, A., Peña-Caravaca, J. P., Páez-Rodríguez, A., Darias-Gonzalez, J. G., Hernandez-Tabares, L., Arias de Fuentes, O., Prokhorov, E., Torres-Figueredo, N., Reguera, E., & Desdin-García, L. F. (2021). Carbon quantum dots by submerged arc discharge in water: Synthesis, characterization, and mechanism of formation. *Journal of Applied Physics*, *129*(16), 163301. Advance online publication. DOI: 10.1063/5.0040322

Chávez-García, D., Guzman, M., Sanchez, V., & Cadena-Nava, R. D. (2024). Green synthesis of biomass-derived carbon quantum dots for photocatalytic degradation of methylene blue. *Beilstein Journal of Nanotechnology*, *15*, 755–766. DOI: 10.3762/bjnano.15.63 PMID: 38952416

Chawre, Y., Satnami, M. L., Kujur, A. B., Ghosh, K. K., Nagwanshi, R., Karbhal, I., Pervez, S., & Deb, M. K. (2023). Förster Resonance Energy Transfer between Multicolor Emissive N-Doped Carbon Quantum Dots and Gold Nanorods for the Detection of H2O2, Glucose, Glutathione, and Acetylcholinesterase. *ACS Applied Nano Materials*, *6*(9), 8046–8058. DOI: 10.1021/acsanm.3c01518

Chen, B., Yang, Q., Yang, Y., Chen, J., Yan, B., Gu, Y., Fu, R., & Chen, S. (2025). PANI doped with carbon quantum dots for the asymmetric high-performance solid-state electrochromic supercapacitor. *Journal of Power Sources*, *633*, 236407. DOI: 10.1016/j.jpowsour.2025.236407

Chen, C., & Gu, G. X. (2020). Generative deep neural networks for inverse materials design using backpropagation and active learning. *Advancement of Science*, *7*(5), 1902607. Advance online publication. DOI: 10.1002/advs.201902607 PMID: 32154072

Chen, C., Zhao, D., Hu, T., Sun, J., & Yang, X. (2017). Highly fluorescent nitrogen and sulfur co-doped graphene quantum dots for an inner filter effect-based cyanide sensor. *Sensors and Actuators. B, Chemical*, *241*, 779–788. DOI: 10.1016/j.snb.2016.11.010

Chen, C., Zuo, Y., Ye, W., Li, X., Deng, Z., & Ong, S. P. (2020). A critical review of machine learning of energy materials. *Advanced Energy Materials*, *10*(8), 1903242. Advance online publication. DOI: 10.1002/aenm.201903242

Chen, G., Wu, S., Hui, L., Zhao, Y., Ye, J., Tan, Z., Zeng, W., Tao, Z., Yang, L., & Zhu, Y. (2016). Assembling carbon quantum dots to a layered carbon for high-density supercapacitor electrodes. *Scientific Reports*, *6*(1), 19028. DOI: 10.1038/srep19028 PMID: 26754463

Cheng, H., Chen, S., Li, M., Lu, Y., Chen, H., Fang, X., Qiu, H., Wang, W., Jiang, C., & Zheng, Y. (2023). Ultra-stable dual-color phosphorescence Carbon-Dot@Silica material for advanced anti-counterfeiting. *Dyes and Pigments*, *208*, 110827. DOI: 10.1016/j.dyepig.2022.110827

Chen, Q., Hu, Y., Hu, C., Cheng, H., Zhang, Z., Shao, H., & Qu, L. (2014). Graphene quantum dots–three-dimensional graphene composites for high-performance supercapacitors. *Physical Chemistry Chemical Physics*, *16*(36), 19307–19313. DOI: 10.1039/C4CP02761B PMID: 25100222

Chen, S., Chen, Y., Xu, H., Lyu, M., Zhang, X., Han, Z., Liu, H., Yao, Y., Xu, C., Sheng, J., Xu, Y., Gao, L., Gao, N., Zhang, Z., Peng, L. M., & Li, Y. (2023). Single-walled carbon nanotubes synthesized by laser ablation from coal for field-effect transistors. *Materials Horizons*, *10*(11), 5185–5191. DOI: 10.1039/D3MH01053H PMID: 37724683

Chen, X., Gong, F., Cao, Z., Zou, W., & Gu, T. (2018). Highly cysteine-selective fluorescent nanoprobes based on ultrabright and directly synthesized carbon quantum dots. *Analytical and Bioanalytical Chemistry*, *410*(12), 2961–2970. DOI: 10.1007/s00216-018-0980-3 PMID: 29523940

Chen, Y., Lam, J. W., Kwok, R. T., Liu, B., & Tang, B. Z. (2019). Aggregation-induced emission: Fundamental understanding and future developments. *Materials Horizons*, *6*(3), 428–433. DOI: 10.1039/C8MH01331D

Cho, S., Kim, H., Song, D., Jung, J., Park, S., Jo, H., Seo, S., Han, C., Park, S., Kwon, W., & Han, H. (2024). Insights into glucose-derived carbon dot synthesis via Maillard reaction: From reaction mechanism to biomedical applications. *Scientific Reports*, *14*(1), 31325. DOI: 10.1038/s41598-024-82767-z PMID: 39733004

Chu, K.-W., Lee, S. L., Chang, C.-J., & Liu, L. (2019). Recent progress of carbon dot precursors and photocatalysis applications. *Polymers*, *11*(4), 689. DOI: 10.3390/polym11040689 PMID: 30995724

Chung, S., Revia, R. A., & Zhang, M. (2021). Graphene Quantum Dots and Their Applications in Bioimaging, Biosensing, and Therapy. *Advanced Materials*, *33*(22), 1904362. Advance online publication. DOI: 10.1002/adma.201904362 PMID: 31833101

Cortes, , Falomir, E., Lancis, J., & Mínguez-Vega, G. (2024). Pulsed laser fragmentation synthesis of carbon quantum dots (CQDs) as fluorescent probes in non-enzymatic glucose detection. *Applied Surface Science*, *665*, 160326. Advance online publication. DOI: 10.1016/j.apsusc.2024.160326

Cruz, R. A. T., Soriano, A. N., de Yro, P. A. N., Quiachon, G. M. O., Emolaga, C. S., Ysulat, M. L. M., Bigol, U. G., & Basilia, B. A. (2019). Functionalized Carbon-based Quantum Dots: Optical Characterization and Potential Application as Bio-fluorophore. *IOP Conference Series. Materials Science and Engineering*, *559*(1), 012003. DOI: 10.1088/1757-899X/559/1/012003

Cui, H., Guo, Y., Guo, L., Wang, L., Zhou, Z., & Peng, Z. (2018). Heteroatom-doped carbon materials and their composites as electrocatalysts for co2 reduction. *Journal of Materials Chemistry. A, Materials for Energy and Sustainability*, *6*(39), 18782–18793. DOI: 10.1039/C8TA07430E

Cui, L., Ren, X., Wang, J., & Sun, M. (2020). Synthesis of homogeneous carbon quantum dots by ultrafast dual-beam pulsed laser ablation for bioimaging. *Materials Today Nano*, *12*, 100091. Advance online publication. DOI: 10.1016/j.mtnano.2020.100091

D'souza, A. A., & Shegokar, R. (2016). Polyethylene glycol (PEG): A versatile polymer for pharmaceutical applications. *Expert Opinion on Drug Delivery*, *13*(9), 1257–1275. DOI: 10.1080/17425247.2016.1182485 PMID: 27116988

Dager, A., Uchida, T., Maekawa, T., & Kumar, D. S. (2019). Synthesis and characterization of mono-disperse carbon quantum dots from fennel seeds: Photoluminescence analysis using machine learning. *Scientific Reports*, *9*(1), 14004. DOI: 10.1038/s41598-019-50397-5 PMID: 31570739

Dager, , Baliyan, A., Kurosu, S., Maekawa, T., & Tachibana, M. (2020). Ultrafast synthesis of carbon quantum dots from fenugreek seeds using microwave plasma enhanced decomposition: Application of C-QDs to grow fluorescent protein crystals. *Scientific Reports*, *10*(1), 12333. Advance online publication. DOI: 10.1038/s41598-020-69264-9 PMID: 32704038

Dai, B., Wu, C., Lu, Y., Deng, D., & Xu, S. (2017). Synthesis and formation mechanism of s-doped carbon dots from low-molecule-weight organics. *Journal of Luminescence*, *183*, 1–6. DOI: 10.1016/j.jlumin.2017.04.054

Dang, D. K., Sundaram, C., Ngo, Y. L. T., Chung, J. S., Kim, E. J., & Hur, S. H. (2018). One-pot solid-state synthesis of highly fluorescent N and S co-doped carbon dots and their use as a fluorescent probe for Ag^+ detection in aqueous solution. *Sensors and Actuators. B, Chemical*, *255*, 3284–3291. DOI: 10.1016/j.snb.2017.09.155

Das, H. T., Barai, P., Dutta, S., Das, N., Das, P., Roy, M., Alauddin, M., & Barai, H. R. (2022). Polymer Composites with Quantum Dots as Potential Electrode Materials for Supercapacitors Application: A Review. *Polymers, 14*(5), 1053. DOI: 10.3390/polym14051053 PMID: 35267876

Das, , Mondal, S., & Ghosh, D. (2024). Carbon quantum dots in bioimaging and biomedicines. *Frontiers in Bioengineering and Biotechnology, 11*, 1333752. DOI: 10.3389/fbioe.2023.1333752 PMID: 38318419

Das, R. K., & Mohapatra, S. (2017). Highly luminescent, heteroatom-doped carbon quantum dots for ultrasensitive sensing of glucosamine and targeted imaging of liver cancer cells. *Journal of Materials Chemistry. B, 5*(11), 2190–2197. DOI: 10.1039/C6TB03141B PMID: 32263692

Das, S. S., Bharadwaj, P., Bilal, M., Barani, M., Rahdar, A., Taboada, P., & Kyzas, G. Z. (2020). Stimuli-responsive polymeric nanocarriers for drug delivery, imaging, and theragnosis. *Polymers, 12*(6), 1397. DOI: 10.3390/polym12061397 PMID: 32580366

De Yro, P. A. N., Quaichon, G. M. O., Cruz, R. A. T., Emolaga, C. S., Que, M. C. O., Magdaluyo, E. R., & Basilia, B. A. (2019). *Hydrothermal synthesis of carbon quantum dots from biowaste for bio-imaging.*, DOI: 10.1063/1.5094310

De, B., Kumar, M., Mandal, B. B., & Karak, N. (2015). An in situ prepared photoluminescent transparent biocompatible hyperbranched epoxy/carbon dot nanocomposite. *RSC Advances, 5*(91), 74692–74704. DOI: 10.1039/C5RA12131K

Deivayanai, V. C., Karishma, S., Swaminaathan, P., Saravanan, A., Vickram, A. S., Anbarasu, K., & Yaashikaa, P. R. (2025). Nanohybrid approaches for precise analytical screening of trace contaminants in complex environmental and biological matrices. *Trends in Analytical Chemistry, 191*, 118358. https://doi.org/https://doi.org/10.1016/j.trac.2025.118358. DOI: 10.1016/j.trac.2025.118358

Demchenko, A. P. (Ed.). (2010). *Advanced Fluorescence Reporters in Chemistry and Biology II. 9.* DOI: 10.1007/978-3-642-04701-5

Deng, X., Shao, Z., & Zhao, Y. (2021). Solutions to the Drawbacks of Photothermal and Photodynamic Cancer Therapy. *Advancement of Science, 8*(3), 2002504. DOI: 10.1002/advs.202002504 PMID: 33552860

Deng, Y., Ji, Y., Chen, F., Ren, F., & Tan, S. (2020). Superior performance of flexible solid-state supercapacitors enabled by ultrafine graphene quantum dot-decorated porous carbon spheres. *New Journal of Chemistry, 44*(32), 13591–13597. DOI: 10.1039/D0NJ03163A

De, R., Jo, K. W., Lee, B. H., Some, S., & Kim, K.-T. (2023). Microwave-assisted rapid synthesis of nitrogen-enriched amphibious carbon quantum dots for sensitive detection of ROS and multiple other applications. *Journal of Materials Chemistry. B*, *11*(26), 6024–6043. DOI: 10.1039/D3TB00614J PMID: 37272382

Deshsorn, K., Payakkachon, K., Chaisrithong, T., Jitapunkul, K., Lawtrakul, L., & Iamprasertkun, P. (2023). Unlocking the Full Potential of Heteroatom-Doped Graphene-Based Supercapacitors through Stacking Models and SHAP-Guided Optimization. *Journal of Chemical Information and Modeling*, *63*(16), 5077–5088. DOI: 10.1021/acs.jcim.3c00670 PMID: 37635637

Devi, R. K., Ganesan, M., Chen, T.-W., Chen, S.-M., Al-onazi, W. A., Al-Mohaimeed, A. M., Elshikh, M. S., & Yu, Y.-Y. (2022). 3D-nanocubes of N-doped carbon quantum dots adorned manganese oxide: A functional electrocatalyst for the sensitive detection of sulfadiazine. *Colloids and Surfaces. A, Physicochemical and Engineering Aspects*, *648*, 129141. DOI: 10.1016/j.colsurfa.2022.129141

Dey, S., Chithaiah, P., Belawadi, S., Biswas, K., & Rao, C. N. R. (2014a). New methods of synthesis and varied properties of carbon quantum dots with high nitrogen content. *Journal of Materials Research*, *29*(3), 383–391. DOI: 10.1557/jmr.2013.295

Dhandapani, E., Maadeswaran, P., Raj, R. M., Raj, V., Kandiah, K., & Duraisamy, N. (2023). A potential forecast of carbon quantum dots (CQDs) as an ultrasensitive and selective fluorescence probe for Hg (II) ions sensing. *Materials Science and Engineering B*, *287*, 116098. DOI: 10.1016/j.mseb.2022.116098

Dhariwal, J., Rao, G. K., & Vaya, D. (2024). Recent advancements towards the green synthesis of carbon quantum dots as an innovative and eco-friendly solution for metal ion sensing and monitoring. *RSC Sustainability*, *2*(1), 11–36. DOI: 10.1039/D3SU00375B

Dimos, K. (2016). Carbon quantum dots: Surface passivation and functionalization. *Current Organic Chemistry*, *20*(6), 682–695. DOI: 10.2174/138527281966150730220948

Ding, C., Tong, L., Feng, J., & Fu, J. (2016). Recent advances in stimuli-responsive treatment. *Molecules (Basel, Switzerland)*, *21*(12), 1715. DOI: 10.3390/molecules21121715 PMID: 27999414

Ding, C., Zhu, A., & Tian, Y. (2014). Functional surface engineering of C-dots for fluorescent biosensing and in vivo bioimaging. *Accounts of Chemical Research*, *47*(1), 20–30. DOI: 10.1021/ar400023s PMID: 23911118

Ding, H., Li, X.-H., Chen, X.-B., Wei, J.-S., Li, X.-B., & Xiong, H.-M. (2020). Surface states of carbon dots and their influences on luminescence. *Journal of Applied Physics*, *127*(23), 231101. DOI: 10.1063/1.5143819

Ding, H., Yu, S.-B., Wei, J.-S., & Xiong, H.-M. (2016). Full-Color Light-Emitting Carbon Dots with a Surface-State-Controlled Luminescence Mechanism. *ACS Nano*, *10*(1), 484–491. DOI: 10.1021/acsnano.5b05406 PMID: 26646584

Ding, L., Agrawal, P., Singh, S. K., Chhonker, Y. S., Sun, J., & Murry, D. J. (2024). Polymer-based drug delivery systems for cancer therapeutics. *Polymers*, *16*(6), 843. DOI: 10.3390/polym16060843 PMID: 38543448

Ding, R., Li, Z., Xiong, Y., Wu, W., Yang, Q., & Hou, X. (2023). Electrochemical (bio) sensors for the detection of organophosphorus pesticides based on nanomaterial-modified electrodes: A review. *Critical Reviews in Analytical Chemistry*, *53*(8), 1766–1791. DOI: 10.1080/10408347.2022.2041391 PMID: 35235478

Domingo-Tafalla, B., Martínez-Ferrero, E., Franco, F., & Palomares-Gil, E. (2022). Applications of Carbon Dots for the Photocatalytic and Electrocatalytic Reduction of CO2. *Molecules (Basel, Switzerland)*, *27*(3), 1081. DOI: 10.3390/molecules27031081 PMID: 35164346

Dong, X., Awak, M. A., Tomlinson, N., Tang, Y., Sun, Y. P., & Yang, L. (2017). Antibacterial effects of carbon dots in combination with other antimicrobial reagents. *PLoS One*, *12*(9), e0185324. DOI: 10.1371/journal.pone.0185324 PMID: 28934346

Dong, X., Liang, W., Meziani, M. J., Sun, Y. P., & Yang, L. (2020). Carbon dots as potent antimicrobial agents. *Theranostics*, *10*(2), 671–686. DOI: 10.7150/thno.39863 PMID: 31903144

Dong, Y., Cai, J., You, X., & Chi, Y. (2015). Sensing applications of luminescent carbon based dots. *Analyst*, *140*(22), 7468–7486. DOI: 10.1039/C5AN01487E PMID: 26396996

Dong, Y., Pang, H., Yang, H., Guo, C., Shao, J., Chi, Y., Li, C., & Yu, T. (2013). Carbon-based dots co-doped with nitrogen and sulfur for high quantum yield and excitation-independent emission. *Angewandte Chemie International Edition*, *52*(30), 7800–7804. DOI: 10.1002/anie.201301114 PMID: 23761198

Dos Santos, C. M., Silva, F. W. L., de Souza, O. P. L., Canevari, T. C., Santelli, R. E., Saez, V. M., & Cincotto, F. H. (2025). Sensitive electrochemical sensor based on hybrid material F-doped carbon quantum dots (Cdots) and NiO nanoparticles: detection of ketoconazole in pharmaceutical formulations and tap water samples: CM dos Santos et al. *Analytical and Bioanalytical Chemistry*, 1–13.

Dresselhaus, M. S., Jorio, A., Hofmann, M., Dresselhaus, G., & Saito, R. (2010). Perspectives on Carbon Nanotubes and Graphene Raman Spectroscopy. *Nano Letters*, *10*(3), 751–758. DOI: 10.1021/nl904286r PMID: 20085345

Dristant, U., Mukherjee, K., Saha, S., & Maity, D. (2023). An Overview of Polymeric Nanoparticles-Based Drug Delivery System in Cancer Treatment. *Technology in Cancer Research & Treatment*, *22*, 15330338231152084. DOI: 10.1177/15330338231152083 PMID: 36718541

Dua., ... (2023). Stability of carbon quantum dots: a critical review. In *RSC Advances* (Vol. 13, Issue 20, pp. 13845–13861). Royal Society of Chemistry. DOI: 10.1039/D2RA07180K

Dubey, R., & Guruviah, V. (2019). Review of carbon-based electrode materials for supercapacitor energy storage. *Ionics*, *25*(4), 1419–1445. DOI: 10.1007/s11581-019-02874-0

Duman, A. N., & Jalilov, A. S. (2024). Machine learning for carbon dot synthesis and applications. *Materials Advances*, *5*(18), 7097–7112. DOI: 10.1039/D4MA00505H

Duong, , Minh Hoa, N., Toan, L. D., Chuyen, P. T., Hien, N. T. T., Hong Minh, P., Thanh Binh, N., & Anh Thi, L. (2025). Plasma-assisted green fabrication of fluorescent carbon quantum dot–silver nanocomposites. *Journal of Dispersion Science and Technology*, *46*(9), 1356–1364. Advance online publication. DOI: 10.1080/01932691.2024.2440422

Du, Y., & Guo, S. (2016). Chemically doped fluorescent carbon and graphene quantum dots for bioimaging, sensor, catalytic and photoelectronic applications. *Nanoscale*, *8*(5), 2532–2543. DOI: 10.1039/C5NR07579C PMID: 26757977

Egerton, R. F. (2005). Physical principles of electron microscopy: An introduction to TEM, SEM, and AEM. *Physical Principles of Electron Microscopy: An Introduction to TEM, SEM, and AEM*, 1–202. https://doi.org/DOI: 10.1007/B136495/COVER

Elabed, S., Sheirf, A., & Ali, M. (2025). Nanostructures for cancer therapeutics and diagnostics: Recent advances and future outlook. *Radiation Physics and Chemistry*, *226*, 112295. DOI: 10.1016/j.radphyschem.2024.112295

Eliboev, I., Ishankulov, A., Berdimurodov, E., Chulpanov, K., Nazarov, M., Jamshid, B., Toshpulotov, B., Tukhtaeva, R., Demir, M., Rashidova, K., Jalilov, F., & Polvonov, K. (2025). Advancing analytical chemistry with carbon quantum dots: A comprehensive review. *Analytical Methods*, *17*(13), 2627–2649. DOI: 10.1039/D4AY02237H PMID: 40104848

El-Shamy, A. G. (2019a). Novel conducting PVA/Carbon quantum dots (CQDs) nanocomposite for high anti-electromagnetic wave performance. *Journal of Alloys and Compounds*, *810*, 151940. https://doi.org/https://doi.org/10.1016/j.jallcom.2019.151940. DOI: 10.1016/j.jallcom.2019.151940

Elugoke, S. E., Uwaya, G. E., Quadri, T. W., & Ebenso, E. E. (2024). Carbon Quantum Dots: Basics, Properties, and Fundamentals. In *Carbon Dots: Recent Developments and Future Perspectives* (Vol. 1465, pp. 1–3). American Chemical Society. https://doi.org/doi:10.1021/bk-2024-1465.ch001

Elugoke, S. E., Uwaya, G. E., Quadri, T. W., & Ebenso, E. E. (2024). Carbon Quantum Dots: Basics, Properties, and Fundamentals. In Berdimurodov, E., Verma, D. K., & Guo, L. (Eds.), *ACS Symposium Series* (Vol. 1465, pp. 3–42). American Chemical Society. DOI: 10.1021/bk-2024-1465.ch001

Emam, H. E. (2024). Carbon quantum dots derived from polysaccharides: Chemistry and potential applications. *Carbohydrate Polymers*, *324*, 121503. https://doi.org/https://doi.org/10.1016/j.carbpol.2023.121503. DOI: 10.1016/j.carbpol.2023.121503 PMID: 37985091

Escudero., ... (2021). Molecular bottom-up approaches for the synthesis of inorganic and hybrid nanostructures. In *Inorganics* (Vol. 9, Issue 7). DOI: 10.3390/inorganics9070058

Facure, M. H. M., Schneider, R., Mercante, L. A., & Correa, D. S. (2022). Rational hydrothermal synthesis of graphene quantum dots with optimized luminescent properties for sensing applications. *Materials Today. Chemistry*, *23*, 100755. DOI: 10.1016/j.mtchem.2021.100755

Fakheri, M., Fatemi, S., & Rahimi Kakolaki, R. (2023). Comparative study of one-pot and facile methods to synthesize codoped <scp>CQDs</scp> with low band gap and photovoltaic properties. *Canadian Journal of Chemical Engineering*, *101*(8), 4480–4492. DOI: 10.1002/cjce.24765

Fawaz, W., Hasian, J., & Alghoraibi, I. (2023). Synthesis and physicochemical characterization of carbon quantum dots produced from folic acid. *Scientific Reports*, *13*(1), 18641. DOI: 10.1038/s41598-023-46084-1 PMID: 37903841

Fei, Y., Man, Y., Sun, J., Du, Y., Chen, B., Bao, J., & Zhou, X. (2023). Implanting cus quantum dots into carbon nanorods for efficient magnesium-ion batteries. *Small*, *19*(34), 2301954. Advance online publication. DOI: 10.1002/smll.202301954 PMID: 37086143

Feng, H., & Qian, Z. (2018). Functional carbon quantum dots: A versatile platform for chemosensing and biosensing. *Chemical Record (New York, N.Y.), 18*(5), 491–505. DOI: 10.1002/tcr.201700055 PMID: 29171708

Feng, J., Dong, H., Pang, B., Shao, F., Zhang, C., Yu, L., & Dong, L. (2018). Theoretical study on the optical and electronic properties of graphene quantum dots doped with heteroatoms. *Physical Chemistry Chemical Physics, 20*(22), 15244–15252. DOI: 10.1039/C8CP01403E PMID: 29789854

Feng, X., Bai, Y., Liu, M., Li, Y., Yang, H., Wang, X., & Wu, C. (2021). Untangling the respective effects of heteroatom-doped carbon materials in batteries, supercapacitors and the orr to design high performance materials. *Energy &Amp. Energy & Environmental Science, 14*(4), 2036–2089. DOI: 10.1039/D1EE00166C

Feng, Y., Xu, C., Wu, K., Guo, Y., Wu, S., Guo, Z., He, M., & Xue, M. (2022). Carbon quantum dot-coated vse2 nanosheets as anodes for high-performance potassium-ion batteries. *ACS Applied Nano Materials, 5*(9), 13136–13148. DOI: 10.1021/acsanm.2c02907

Fernandes, N. B., Shenoy, R. U. K., Kajampady, M. K., DCruz, C. E. M., Shirodkar, R. K., Kumar, L., & Verma, R. (2022). Fullerenes for the treatment of cancer: An emerging tool. *Environmental Science and Pollution Research International, 29*(39), 58607–58627. DOI: 10.1007/s11356-022-21449-7 PMID: 35790637

Fernandes, S., Esteves Da Silva, J. C. G., & Pinto Da Silva, L. (2022). Life Cycle Assessment-Based Comparative Study between High-Yield and "Standard" Bottom-Up Procedures for the Fabrication of Carbon Dots. *Materials (Basel), 15*(10), 3446. DOI: 10.3390/ma15103446 PMID: 35629474

Fernandes, S., Silva, J. C. G. E., & Silva, L. P. (2021). Comparative life cycle assessment of high-yield synthesis routes for carbon dots. *NanoImpact, 23*, 100332. DOI: 10.1016/j.impact.2021.100332 PMID: 35559833

Ferrari, C. (2007). Raman spectroscopy of graphene and graphite: Disorder, electron–phonon coupling, doping and nonadiabatic effects. *Solid State Communications, 143*(1-2), 47–57. DOI: 10.1016/j.ssc.2007.03.052

Fu, C., Qin, X., Zhang, J., Zhang, T., Song, Y., Yang, J., Wu, G., Luo, D., Jiang, N., & Bikker, F. J. (2023). In vitro and in vivo toxicological evaluation of carbon quantum dots originating from Spinacia oleracea. *Heliyon, 9*(2), e13422. DOI: 10.1016/j.heliyon.2023.e13422 PMID: 36820041

Fu, L., Zheng, Y., Li, X., Liu, X., Te Lin, C., & Karimi-Maleh, H. (2023). Strategies and Applications of Graphene and Its Derivatives-Based Electrochemical Sensors in Cancer Diagnosis. *Molecules (Basel, Switzerland)*, *28*(18), 6719. Advance online publication. DOI: 10.3390/molecules28186719 PMID: 37764496

Fu, Q., Li, N., Lu, K., Dong, Z., & Yang, Y. (2024). The emerging development of nitrogen and sulfur co-doped carbon dots: Synthesis methods, influencing factors and applications. *Materials Today. Chemistry*, *37*, 102032. DOI: 10.1016/j.mtchem.2024.102032

Fu, X., Sheng, L., Yu, Y., Ma, M., Cai, Z., & Huang, X. (2018). Rapid and universal detection of ovalbumin based on N, O, P-co-doped carbon dots-fluorescence resonance energy transfer technology. *Sensors and Actuators. B, Chemical*, *269*, 278–287. DOI: 10.1016/j.snb.2018.04.134

Ganesan, M., Ramadhass, K. D., Chuang, H.-C., & Gopalakrishnan, G. (2021). Synthesis of nitrogen-doped carbon quantum dots@ Fe2O3/multiwall carbon nanotubes ternary nanocomposite for the simultaneous electrochemical detection of 5-fluorouracil, uric acid, and xanthine. *Journal of Molecular Liquids*, *331*, 115768. DOI: 10.1016/j.molliq.2021.115768

Ganti, S. V. S., Wölfel, L., & Kuenneth, C. (2025). Ai-driven discovery of high performance polymer electrodes for next-generation batteries. *Journal of Polymer Science*, *63*(21), 4631–4639. Advance online publication. DOI: 10.1002/pol.20250198

Gao, H., Liu, Z., Song, L., Guo, W., Gao, W., Ci, L., Rao, A., Quan, W., Vajtai, R., & Ajayan, P. M. (2012). Synthesis of S-doped graphene by liquid precursor. *Nanotechnology*, *23*(27), 275605. DOI: 10.1088/0957-4484/23/27/275605 PMID: 22710561

Gao, S., Xu, B., Sun, J., & Zhang, Z. (2024). Nanotechnological advances in cancer: Therapy a comprehensive review of carbon nanotube applications. *Frontiers in Bioengineering and Biotechnology*, *12*, 1351787. DOI: 10.3389/fbioe.2024.1351787 PMID: 38562672

Gao, , Yin, Q., Meng, X., He, X., & Xin, Z. (2022). Excellent behaviors of highly dispersed Ni-based catalyst in CO methanation synthesized by in-situ hydrothermal method with carbon quantum dots assisted. *Fuel*, *310*, 121813. Advance online publication. DOI: 10.1016/j.fuel.2021.121813

Gayen, B., Palchoudhury, S., & Chowdhury, J. (2019). Carbon dots: A mystic star in the world of nanoscience. *Journal of Nanomaterials*, *3451307*, 1–19. Advance online publication. DOI: 10.1155/2019/3451307

Ge, G., Li, L., Wang, D., Chen, M., Zeng, Z., Xiong, W., & Guo, C. (2021). Carbon dots: Synthesis, properties and biomedical applications. *Journal of Materials Chemistry. B*, *9*(33), 6553–6575. DOI: 10.1039/D1TB01077H PMID: 34328147

Ghosh, R. (2022). *Carbon Nanotubes: Methods, Achievements, Challenges, and Future Directions*. DOI: 10.36227/techrxiv.19502455.v2

Ghosh, S., Bera, S., Kapuria, A., Debnath, A., Das, P., Su, Y., & Saha, S. K. (2025). Nitrogen and sulfur co-doped carbon quantum dot-decorated ti3c2tx-mxenes as electrode materials for high-performance symmetric supercapacitors. *Nanoscale*, *17*(26), 15731–15748. Advance online publication. DOI: 10.1039/D4NR05037A PMID: 40531092

Ghosh, T., Das, T. K., Das, P., Banerji, P., & Das, N. C. (2022). Current scenario and recent advancement of doped carbon dots: A short review scientocracy update (2013–2022). *Carbon Letters*, *32*(4), 953–977. DOI: 10.1007/s42823-022-00339-5

Giordano, M. G., Seganti, G., Bartoli, M., & Tagliaferro, A. (2023). An overview on carbon quantum dots optical and chemical features. *Molecules (Basel, Switzerland)*, *28*(6), 2772. DOI: 10.3390/molecules28062772 PMID: 36985743

Gismondi, A., Reina, G., Orlanducci, S., Mizzoni, F., Gay, S., Terranova, M. L., & Canini, A. (2015). Nanodiamonds coupled with plant bioactive metabolites: A nanotech approach for cancer therapy. *Biomaterials*, *38*, 22–35. DOI: 10.1016/j.biomaterials.2014.10.057 PMID: 25457980

Glogic, E., Kamali, A. K., Keppetipola, N. M., Alonge, B., Kumara, G., Sonnemann, G., Toupance, T., & Cojocaru, L. (2022). Life cycle assessment of supercapacitor electrodes based on activated carbon from coconut shells. *ACS Sustainable Chemistry &Amp. ACS Sustainable Chemistry & Engineering*, *10*(46), 15025–15034. DOI: 10.1021/acssuschemeng.2c03239

Goenka, S., Sant, V., & Sant, S. (2014). Graphene-based nanomaterials for drug delivery and tissue engineering. *Journal of Controlled Release*, *173*(1), 75–88. DOI: 10.1016/j.jconrel.2013.10.017 PMID: 24161530

Gomes, M. F., Gomes, Y. F., Lopes-Moriyama, A., de Barros Neto, E. L., & de Souza, C. P. (2019a). Design of carbon quantum dots via hydrothermal carbonization synthesis from renewable precursors. *Biomass Conversion and Biorefinery*, *9*(4), 689–694. DOI: 10.1007/s13399-019-00387-4

Gopi, C. V. M., & Ramesh, R. (2024). Review of battery-supercapacitor hybrid energy storage systems for electric vehicles. *Results in Engineering*, *24*, 103598. DOI: 10.1016/j.rineng.2024.103598

Goswami, M., Mandal, S., & Pillai, V. K. (2023). Effect of hetero-atom doping on the electrocatalytic properties of graphene quantum dots for oxygen reduction reaction. *Scientific Reports, 13*(1), 5182. DOI: 10.1038/s41598-023-31854-8 PMID: 36997535

Gouda, A., Manioudakis, J., Naccache, R., Soavi, F., & Santato, C. (2021). 3d network of sepia melanin and n- and, s-doped graphitic carbon quantum dots for sustainable electrochemical capacitors. *Advanced Sustainable Systems, 5*(10), 2100152. Advance online publication. DOI: 10.1002/adsu.202100152

Gou, Z., Tang, K., Zeng, C., Yuan, H., Zhang, C., Huang, Y., Qu, T., Xin, Q., Zhao, Y., Zeng, G., Yang, J., Xie, P., Yang, S. T., & Tang, X. (2024). Photothermal therapy of xenografted tumor by carbon nanoparticles-Fe(II) complex. *Colloids and Surfaces. B, Biointerfaces, 240*, 113968. DOI: 10.1016/j.colsurfb.2024.113968 PMID: 38788472

Guan, G., Lu, L., Meng, W., Ye, T., Zhang, Y., & Guo, F. (2023). Nb_2CT_x MXene Anchored with Carbon Quantum Dots for Lithium-Ion Batteries. *ACS Applied Nano Materials, 6*(24), 23620–23629. DOI: 10.1021/acsanm.3c05424

Guan, M., Dong, H., Ge, J., Chen, D., Sun, L., Li, S., Wang, C., Yan, C., Wang, P., & Shu, C. (2015). Multifunctional upconversion-nanoparticles-trismethylpyridylporphyrin-fullerene nanocomposite: A near-infrared light-triggered theranostic platform for imaging-guided photodynamic therapy. *NPG Asia Materials, 7*(7), e205–e205. DOI: 10.1038/am.2015.82

Guan, X., Li, Z., Geng, X., Lei, Z., Karakoti, A., Wu, T., Kumar, P., Yi, J., & Vinu, A. (2023). Emerging trends of carbon-based quantum dots: Nanoarchitectonics and applications. *Small, 19*(17), 2207181. DOI: 10.1002/smll.202207181 PMID: 36693792

Guerrero-Gonzalez, R., Vázquez-Dávila, F., Saucedo-Flores, E., Ruelas, R., Ceballos-Sánchez, O., & Pelayo, J. E. (2023). Green approach synthesis of carbon quantum dots from agave bagasse and their use to boost seed germination and plant growth. *SN Applied Sciences, 5*(8), 204. Advance online publication. DOI: 10.1007/s42452-023-05428-2

Guilbault, G. (2020). Practical fluorescence. In *Practical Fluorescence* (2nd ed.). CRC Press., DOI: 10.1201/9781003066514

Guinée, J. B., Heijungs, R., Huppes, G., Zamagni, A., Masoni, P., Buonamici, R., Ekvall, T., & Rydberg, T. (2010). Life cycle assessment: Past, present, and future. *Environmental Science &Amp. Technology (Elmsford, N.Y.), 45*(1), 90–96. DOI: 10.1021/es101316v PMID: 20812726

Gulati, S., Baul, A., Amar, A., Wadhwa, R., Kumar, S., & Varma, R. S. (2023). Eco-Friendly and Sustainable Pathways to Photoluminescent Carbon Quantum Dots (CQDs). *Nanomaterials (Basel, Switzerland), 13*(3), 554. DOI: 10.3390/nano13030554 PMID: 36770515

Guo, J., Gao, F., Li, D., Luo, X., Sun, Y., Wang, X., Ran, Z., Wu, Q., & Li, S. (2020). Novel Strategy of Constructing Hollow Ga_2O_3@N-CQDs as a Self-Healing Anode Material for Lithium-Ion Batteries. *ACS Sustainable Chemistry & Engineering, 8*(36), 13692–13700. DOI: 10.1021/acssuschemeng.0c03756

Guo, L., Huang, Y., Su, A., Shang, T., Qu, M., Shi, W., Qiang, Y., & Marzouki, R. (2025). Boosting the performance of alkaline Al-air batteries by interfacial reconstruction via N-doped carbon quantum dots. *Journal of Energy Storage, 127*, 117136. DOI: 10.1016/j.est.2025.117136

Guo, R., Li, L., Wang, B., Xiang, Y., Zou, G., Zhu, Y., Hou, H., & Ji, X. (2021). Functionalized carbon dots for advanced batteries. *Energy Storage Materials, 37*, 8–39. https://doi.org/https://doi.org/10.1016/j.ensm.2021.01.020. DOI: 10.1016/j.ensm.2021.01.020

Guo, R., Li, T., & Shi, S. (2019). Aggregation-induced emission enhancement of carbon quantum dots and applications in light emitting devices. *Journal of Materials Chemistry. C, Materials for Optical and Electronic Devices*, *7*(17), 5148–5154. DOI: 10.1039/C9TC01138B

Guo, Y., Chen, Y., Cao, F., Wang, L., Wang, Z., & Leng, Y. (2017). Hydrothermal synthesis of nitrogen and boron doped carbon quantum dots with yellow-green emission for sensing Cr (VI), anti-counterfeiting and cell imaging. *RSC Advances, 7*(76), 48386–48393. DOI: 10.1039/C7RA09785A

Guo, Z., Tian, Z., Liu, Y., Huang, Y., Li, W., Han, X., He, S., Mao, H. K., Zhang, C., Duan, G., & Jiang, S. (2025). A densification-activation strategy toward hierarchical towards self-supporting thick carbon electrodes for high-power supercapacitors. *Chemical Communications*, *61*(35), 6474–6477. DOI: 10.1039/D5CC01049G PMID: 40176733

Gupta, D., Roy, P., Sharma, R., Kasana, R., Rathore, P., & Gupta, T. K. (2024). Recent nanotheranostic approaches in cancer research. *Clinical and Experimental Medicine*, *24*(1), 8. DOI: 10.1007/s10238-023-01262-3 PMID: 38240834

Gupta, V., Kannan, K., Tari, V., & Chandra, N. (2024). Bioengineered silver nanoparticles using Brassica oleracea sub sp. botrytis (L.) for enhanced antibacterial activity. *Physics and Chemistry of Solid State*, *25*(2), 303–310. DOI: 10.15330/pcss.25.2.303-310

Gurushankar, K., Karuppaiya, C., Tari, V., & Kannan, K. (2025). Exploring MXenes and MBenes in Metal Sulfur and Selenium Battery Technologies. In *Novel Energy Storage and Conversion Technologies for Two-Dimensional MXenes and MBenes* (pp. 105–120). IGI Global Scientific Publishing. DOI: 10.4018/979-8-3693-9546-2.ch005

Gu, Z., Li, D., Zheng, C., Kang, Y., Wöll, C., & Zhang, J. (2017). MOF-Templated Synthesis of Ultrasmall Photoluminescent Carbon-Nanodot Arrays for Optical Applications. *Angewandte Chemie International Edition*, *56*(24), 6853–6858. DOI: 10.1002/anie.201702162 PMID: 28481003

Gyulai, , Ouanzi, F., Bertóti, I., Mohai, M., Kolonits, T., Horváti, K., & Bősze, S. (2019). Chemical structure and in vitro cellular uptake of luminescent carbon quantum dots prepared by solvothermal and microwave assisted techniques. *Journal of Colloid and Interface Science*, *549*, 150–161. DOI: 10.1016/j.jcis.2019.04.058 PMID: 31029843

Hafez, N. S., Ebeid, E.-Z. M., Khalifa, A., Azim, S. A., El-Shaer, A., Sakr, M. A. S., & Hegazy, M. B. Z. (2025a). Modulating carbon quantum dots using multi-heteroatom doping as a bifunctional electrocatalyst for efficient oxygen and hydrogen evolution reactions in alkaline electrolytes. *Journal of Materials Chemistry. A, Materials for Energy and Sustainability*, *13*(29), 23454–23462. DOI: 10.1039/D5TA03457D

Hagiwara, , Horikoshi, S., & Serpone, N. (2021). Luminescent monodispersed carbon quantum dots by a microwave solvothermal method toward bioimaging applications. *Journal of Photochemistry and Photobiology A Chemistry*, *415*, 113310. Advance online publication. DOI: 10.1016/j.jphotochem.2021.113310

Hailing, Y., Xiufang, L., Lili, W., Baoqiang, L., Kaichen, H., Yongquan, H., & Hong, S. (2020). Doxorubicin-loaded fluorescent carbon dots with PEI passivation as a drug delivery system for cancer therapy. *Nanoscale*, *12*(33), 17222–17237. DOI: 10.1039/D0NR01236J PMID: 32671377

Hamad, I., Harb, A. A., & Bustanji, Y. (2024). Liposome-Based Drug Delivery Systems in Cancer Research: An Analysis of Global Landscape Efforts and Achievements. *Pharmaceutics*, *16*(3), 400. DOI: 10.3390/pharmaceutics16030400 PMID: 38543294

Han, Y., Tang, B., Wang, L., Bao, H., Lu, Y., Guan, C., Zhang, L., Le, M., Liu, Z., & Wu, M. (2020). Machine-learning-driven synthesis of carbon dots with enhanced quantum yields. *ACS Nano*, *14*(11), 14761–14768. DOI: 10.1021/acsnano.0c01899 PMID: 32960048

Han, Y., Tang, D., Yang, Y., Li, C., Kong, W., Huang, H., Liu, Y., & Kang, Z. (2015). Non-metal single/dual doped carbon quantum dots: A general flame synthetic method and electro-catalytic properties. *Nanoscale*, *7*(14), 5955–5962. Advance online publication. DOI: 10.1039/C4NR07116F PMID: 25771786

Hao, H., Wang, M., Cao, Y., He, J., Yang, Y., Zhao, C., & Yan, L. (2024). Boron-doped engineering for carbon quantum dots-based memristors with controllable memristance stability. *Small Methods*, *8*(9), 2301454. Advance online publication. DOI: 10.1002/smtd.202301454 PMID: 38204209

Hao, X., Wang, J., Ding, B., Shen, L., Xu, Y., Wang, Y., Chang, Z., Dou, H., Lu, X., & Zhang, X. (2016). Heteroatom-doped porous carbon nanosheets: General preparation and enhanced capacitive properties. *Chemistry (Weinheim an der Bergstrasse, Germany)*, *22*(46), 16668–16674. DOI: 10.1002/chem.201602922 PMID: 27704674

Haugstad, G. (2012). Atomic Force Microscopy: Understanding Basic Modes and Advanced Applications. *Atomic Force Microscopy: Understanding Basic Modes and Advanced Applications*. DOI: 10.1002/9781118360668

He, G., Shu, M., Yang, Z., Ma, Y., Huang, D., Xu, S., Wang, Y., Hu, N., Zhang, Y., & Xu, L. (2017). Microwave formation and photoluminescence mechanisms of multi-states nitrogen doped carbon dots. *Applied Surface Science*, *422*, 257–265. DOI: 10.1016/j.apsusc.2017.05.036

Hemmati, A., Emadi, H., & Nabavi, S. R. (2023). Green Synthesis of Sulfur- and Nitrogen-Doped Carbon Quantum Dots for Determination of L-DOPA Using Fluorescence Spectroscopy and a Smartphone-Based Fluorimeter. *ACS Omega*, *8*(23), 20987–20999. DOI: 10.1021/acsomega.3c01795 PMID: 37332813

Hernandez-Tabares, , Darias-Gonzalez, J. G., Chao-Mujica, F. J., Ledo-Pereda, L. M., Antuch, M., Carrillo-Barroso, E., Chong-Quero, J. E., Reguera, E., & Desdin-Garcia, L. F. (2021). Stabilization Methods in the Submerged Arc Discharge Synthesis of Carbon Nanostructures. *Journal of Nanomaterials*, *2021*, 1–12. Advance online publication. DOI: 10.1155/2021/6550809

Herold, F., De Oliveira, D., Baade, G., Friedland, J., Güttel, R., Claeys, M., & Rønning, M. (2025). Is Carbon Heteroatom Doping the Key to Active and Stable Carbon Supported Cobalt Fischer–Tropsch Catalysts? *ACS Catalysis*, *15*(8), 6673–6689. DOI: 10.1021/acscatal.4c08092 PMID: 40270880

Hettiarachchi, S. D., Graham, R. M., Mintz, K. J., Zhou, Y., Vanni, S., Peng, Z., & Leblanc, R. M. (2019). Triple conjugated carbon dots as a nano-drug delivery model for glioblastoma brain tumors. *Nanoscale*, *11*(13), 6192–6205. DOI: 10.1039/C8NR08970A PMID: 30874284

Hoan, B. T., Tam, P. D., & Pham, V.-H. (2019). Green Synthesis of Highly Luminescent Carbon Quantum Dots from Lemon Juice. *Journal of Nanotechnology*, *2019*(1), 2852816. DOI: 10.1155/2019/2852816

Hoang, V. C., Dave, K., & Gomes, V. G. (2019). Carbon quantum dot-based composites for energy storage and electrocatalysis: Mechanism, applications and future prospects. *Nano Energy*, *66*, 104093. DOI: 10.1016/j.nanoen.2019.104093

Hoa, , Toan, L. D., Tran, N., Hung, L. X., & Thi, L. A. (2025). Microplasma-synthesized Citrus-derived carbon quantum dots: Antibacterial properties and nanoprobe sensitivity. *RSC Advances*, *15*(31), 25362–25371. DOI: 10.1039/D5RA02150B PMID: 40677949

Hola, K., Bourlinos, A. B., Kozak, O., Berka, K., Siskova, K. M., Havrdova, M., Tucek, J., Safarova, K., Otyepka, M., Giannelis, E. P., & Zboril, R. (2014). Photoluminescence effects of graphitic core size and surface functional groups in carbon dots: COO^- induced red-shift emission. *Carbon*, *70*, 279–286. DOI: 10.1016/j.carbon.2014.01.008

Hong, , Deng, X., Liang, J., Li, J., Tao, Y., & Tan, K. (2019). One-step hydrothermal synthesis of down/up-conversion luminescence F-doped carbon quantum dots for label-free detection of Fe3+. *Microchemical Journal*, *151*, 104217. Advance online publication. DOI: 10.1016/j.microc.2019.104217

Hosseini, S. M., Mohammadnejad, J., Najafi-Taher, R., Zadeh, Z. B., Tanhaei, M., & Ramakrishna, S. (2023a). Multifunctional Carbon-Based Nanoparticles: Theranostic Applications in Cancer Therapy and Diagnosis. *ACS Applied Bio Materials*, *6*(4), 1323–1338. DOI: 10.1021/acsabm.2c01000 PMID: 36921253

Hou, H., Banks, C. E., Jing, M., Zhang, Y., & Ji, X. (2015). Carbon quantum dots and their derivative 3D porous carbon frameworks for sodium-ion batteries with ultralong cycle life. *Advanced Materials*, *27*(47), 7861–7866. DOI: 10.1002/adma.201503816 PMID: 26506218

Hou, Y., Lu, Q., Deng, J., Li, H., & Zhang, Y. (2015). One-pot electrochemical synthesis of functionalized fluorescent carbon dots and their selective sensing for mercury ion. *Analytica Chimica Acta*, *866*, 69–74. DOI: 10.1016/j.aca.2015.01.039 PMID: 25732694

Hou, Y., Yan, F., Bai, X., Wei, X., Chen, L., & Fu, Y. (2024). Designing advanced 0D/2D heterojunctions of N, S-co-doped carbon dots/Ti3C2Tx MXene nanosheets in electrochemical sensor applications. *Ceramics International*, *50*(7), 10735–10745. DOI: 10.1016/j.ceramint.2023.12.389

Hsu, P.-C., & Chang, H.-T. (2012). Synthesis of high-quality carbon nanodots from hydrophilic compounds: Role of functional groups. *Chemical Communications*, *48*(33), 3984. DOI: 10.1039/c2cc30188a PMID: 22422194

Huang, , Cui, Y., Liu, M., Chen, J., Wan, Q., Wen, Y., Deng, F., Zhou, N., Zhang, X., & Wei, Y. (2018). A one-step ultrasonic irradiation assisted strategy for the preparation of polymer-functionalized carbon quantum dots and their biological imaging. *Journal of Colloid and Interface Science*, *532*, 767–773. DOI: 10.1016/j.jcis.2018.07.099 PMID: 30130727

Huang, S., Yang, E., Yao, J., Liu, Y., & Xiao, Q. (2018). Red emission nitrogen, boron, sulfur co-doped carbon dots for "on-off-on" fluorescent mode detection of Ag^+ ions and l-cysteine in complex biological fluids and living cells. *Analytica Chimica Acta*, *1035*, 192–202. DOI: 10.1016/j.aca.2018.06.051 PMID: 30224139

Huang, Z., Liu, T., Song, Y., Li, Y., & Liu, X. (2017). Balancing the electrical double layer capacitance and pseudocapacitance of hetero-atom doped carbon. *Nanoscale*, *9*(35), 13119–13127. DOI: 10.1039/C7NR04234E PMID: 28849857

Huang, Z., & Ren, L. (2025). Large Scale Synthesis of Carbon Dots and Their Applications: A Review. *Molecules (Basel, Switzerland)*, *30*(4), 774. DOI: 10.3390/molecules30040774 PMID: 40005085

Huan, H., Wu, W., Feng, Q., Tang, Y., Li, S., Deng, H., Xiang, X., Zu, X., & Li, Y. (2026). Li-ion battery based on B, N-graphene prepared by ball milling. *Journal of Physics and Chemistry of Solids*, *208*, 113002. DOI: 10.1016/j.jpcs.2025.113002

Hu, C., Channa, A. I., Xia, L., Li, X., Li, Z., Wang, Z. M., & Tong, X. (2025). Colloidal InAs quantum dots: Synthesis, properties, and optoelectronic devices. *Advanced Functional Materials*, *35*(32), 2500280. DOI: 10.1002/adfm.202500280

Hu, L., Hou, J., Ma, Y., Li, H., & Zhai, T. (2016). Multi-heteroatom self-doped porous carbon derived from swim bladders for large capacitance supercapacitors. *Journal of Materials Chemistry. A, Materials for Energy and Sustainability*, *4*(39), 15006–15014. DOI: 10.1039/C6TA06337C

Hu, S. L., Niu, K. Y., Sun, J., Yang, J., Zhao, N. Q., & Du, X. W. (2009). One-step synthesis of fluorescent carbon nanoparticles by laser irradiation. *Journal of Materials Chemistry*, *19*(4), 484–488. DOI: 10.1039/B812943F

Hu, S., Liu, J., Yang, J., Wang, Y., & Cao, S. (2011). Laser synthesis and size tailor of carbon quantum dots. *Journal of Nanoparticle Research*, *13*(12), 7247–7252. DOI: 10.1007/s11051-011-0638-y

Hu, S., Tian, R., Dong, Y., Yang, J., Liu, J., & Chang, Q. (2013). Modulation and effects of surface groups on photoluminescence and photocatalytic activity of carbon dots. *Nanoscale*, *5*(22), 11665–11671. DOI: 10.1039/c3nr03893a PMID: 24101143

Hu, Y., Chen, W., Lei, T., Zhou, B., Jiao, Y., Yan, Y., Du, X., Huang, J., Wu, C., Wang, X., Wang, Y., Chen, B., Xu, J., Wang, C., & Xiong, J. (2018). Carbon quantum dots–modified interfacial interactions and ion conductivity for enhanced high current density performance in lithium–sulfur batteries. *Advanced Energy Materials*, *9*(7), 1802955. Advance online publication. DOI: 10.1002/aenm.201802955

Hu, Y., Yang, J., Jia, L., & Yu, J. S. (2015). Ethanol in aqueous hydrogen peroxide solution: Hydrothermal synthesis of highly photoluminescent carbon dots as multifunctional nanosensors. *Carbon*, *93*, 999–1007. DOI: 10.1016/j.carbon.2015.06.018

Huy, H.. (2024). A self-powered photodetector through facile processing using polyethyleneimine/carbon quantum dots for highly sensitive UVC detection. *RSC Advances*, *14*(18), 12360–12371. DOI: 10.1039/D3RA08538D PMID: 38633486

Ijaz, H., Mahmood, A., Abdel-Daim, M. M., Sarfraz, R. M., Zaman, M., Zafar, N., Alshehery, S., Salem-Bekhit, M. M., Ali, M. A., Eltayeb, L. B., & Benguerba, Y. (2023). Review on carbon nanotubes (CNTs) and their chemical and physical characteristics, with particular emphasis on potential applications in biomedicine. *Inorganic Chemistry Communications*, *155*, 111020. DOI: 10.1016/j.inoche.2023.111020

Ilnicka, A., Skorupska, M., Szkoda, M., Zarach, Z., Kamedulski, P., Zielinski, W., & Lukaszewicz, J. P. (2021). Combined effect of nitrogen-doped functional groups and porosity of porous carbons on electrochemical performance of supercapacitors. *Scientific Reports*, *11*(1), 18387. DOI: 10.1038/s41598-021-97932-x PMID: 34526635

Im, M. J., Kim, J. I., Hyeong, S., Moon, B. J., & Bae, S. (2023). From pristine to heteroatom-doped graphene quantum dots: An essential review and prospects for future research. *Small*, *19*(47), 2304497. Advance online publication. DOI: 10.1002/smll.202304497 PMID: 37496316

Inayat, A., Albalawi, K., Rehman, A. U., Saad, A. Y., Saleh, E. A. M., Alamri, M. A., El-Zahhar, A. A., Haider, A., & Abbas, S. M. (2023). Tunable synthesis of carbon quantum dots from the biomass of spent tea leaves as supercapacitor electrode. *Materials Today. Communications*, *34*, 105479. DOI: 10.1016/j.mtcomm.2023.105479

Iqbal, A., Iqbal, K., Xu, L., Li, B., Gong, D., Liu, X., Guo, Y., Liu, W., Qin, W., & Guo, H. (2018). Heterogeneous synthesis of nitrogen-doped carbon dots prepared via anhydrous citric acid and melamine for selective and sensitive turn on-off-on detection of Hg (II), glutathione and its cellular imaging. *Sensors and Actuators. B, Chemical*, *255*, 1130–1138. DOI: 10.1016/j.snb.2017.08.130

Jaldurgam, F. F., Ahmad, Z., & Touati, F. (2021). Synthesis and performance of large-scale cost-effective environment-friendly nanostructured thermoelectric materials. *Nanomaterials (Basel, Switzerland)*, *11*(5), 1091. DOI: 10.3390/nano11051091 PMID: 33922455

Jana, J., Ganguly, M., Chandrakumar, K. R. S., Rao, G. M., & Pal, T. (2017). Boron Precursor-Dependent Evolution of Differently Emitting Carbon Dots. *Langmuir*, *33*(3), 573–584. DOI: 10.1021/acs.langmuir.6b04100 PMID: 28024393

Jana, J., & Pal, T. (2017). An account of doping in carbon dots for varied applications. *Natural Resources &Amp. Engineering*, *2*(1), 5–12. DOI: 10.1080/23802693.2017.1421506

Janus, , Piątkowski, M., & Radwan-Pragłowska, J. (2019). Microwave-assisted synthesis and characterization of poly(L-lysine)-based polymer/carbon quantum dot nanomaterials for biomedical purposes. *Materials (Basel)*, *12*(23), 3825. Advance online publication. DOI: 10.3390/ma12233825 PMID: 31766363

Javed, M., Saqib, A. N. S., Ali, B., Faizan, M., Anang, D. A., Iqbal, Z., & Abbas, S. M. (2019). Carbon quantum dots from glucose oxidation as a highly competent anode material for lithium and sodium-ion batteries. *Electrochimica Acta*, *297*, 250–257. DOI: 10.1016/j.electacta.2018.11.167

Jeffery, G. A. (1957). Elements of x-ray diffraction (Cullity, BD). *Journal of Chemical Education*, *34*(4), A178. DOI: 10.1021/ed034pA178

Jiang, H., Chen, F., Lagally, M. G., & Denes, F. S. (2010). New Strategy for Synthesis and Functionalization of Carbon Nanoparticles. *Langmuir*, *26*(3), 1991–1995. DOI: 10.1021/la9022163 PMID: 20038107

Jiang, K., Sun, S., Zhang, L., Lu, Y., Wu, A., Cai, C., & Lin, H. (2015). Red, Green, and Blue Luminescence by Carbon Dots: Full-Color Emission Tuning and Multicolor Cellular Imaging. *Angewandte Chemie International Edition*, *54*(18), 5360–5363. DOI: 10.1002/anie.201501193 PMID: 25832292

Jiang, X., Xin, Y., He, B., Zhang, F., & Tian, H. (2024). Effect of Heteroatom Doping on Electrochemical Properties of Olivine LiFePO4 Cathodes for High-Performance Lithium-Ion Batteries. *Materials (Basel)*, *17*(6), 1299. DOI: 10.3390/ma17061299 PMID: 38541453

Jiang, Y., Li, J., Jiang, Z., Shi, M., Sheng, R., Liu, Z., Zhang, S., Cao, Y., Wei, T., & Fan, Z. (2021). Large-surface-area activated carbon with high density by electrostatic densification for supercapacitor electrodes. *Carbon*, *175*, 281–288. DOI: 10.1016/j.carbon.2021.01.016

Jiang, Y., Yin, C., Mo, J., Wang, X., Wang, T., Li, G., & Zhou, Q. (2023). Recent progress in carbon dots for anti-pathogen applications in oral cavity. *Frontiers in Cellular and Infection Microbiology, 13*, 1251309. DOI: 10.3389/fcimb.2023.1251309 PMID: 37780847

Jian, X., Yang, H. M., Li, J. G., Zhang, E. H., Cao, L. L., & Liang, Z. H. (2017). Flexible all-solid-state high-performance supercapacitor based on electrochemically synthesized carbon quantum dots/polypyrrole composite electrode. *Electrochimica Acta, 228*, 483–493. DOI: 10.1016/j.electacta.2017.01.082

Jing, S., Zhao, Y., Sun, R., Zhong, L., & Peng, X. (2019). Facile and high-yield synthesis of carbon quantum dots from biomass-derived carbons at mild condition. *ACS Sustainable Chemistry &Amp. ACS Sustainable Chemistry & Engineering, 7*(8), 7833–7843. DOI: 10.1021/acssuschemeng.9b00027

Jin, H., Feng, X., Li, J., Li, M., Xia, Y., Yuan, Y., Yang, C., Dai, B., Lin, Z., Wang, J., Lü, J., & Wang, S. (2019). Heteroatom-doped porous carbon materials with unprecedented high volumetric capacitive performance. *Angewandte Chemie International Edition, 58*(8), 2397–2401. DOI: 10.1002/anie.201813686 PMID: 30620788

Jin, S., Allam, O., Lee, K., Lim, J., Lee, M. J., Loh, S. H., Jang, S. S., & Lee, S. W. (2022). Carbon quantum dot modified reduced graphene oxide framework for improved alkali metal ion storage performance. *Small, 18*(35), 2202898. Advance online publication. DOI: 10.1002/smll.202202898 PMID: 35927029

John, B. K., John, N., Korah, B. K., Thara, C., Abraham, T., & Mathew, B. (2022). Nitrogen-doped carbon quantum dots as a highly selective fluorescent and electrochemical sensor for tetracycline. *Journal of Photochemistry and Photobiology A Chemistry, 432*, 114060. DOI: 10.1016/j.jphotochem.2022.114060

John, B. K., John, N., Mathew, S., Korah, B. K., Punnoose, M. S., & Mathew, B. (2022). Fluorescent carbon quantum dots as a novel solution and paper strip-based dual sensor for the selective detection of Cr (VI) ions. *Diamond and Related Materials, 126*, 109138. DOI: 10.1016/j.diamond.2022.109138

John, B. K., & Mathew, B. (2023). Nitrogen and sulphur co-doped carbon quantum dot integrated bismuth oxide nanocomposite for photocatalytic degradation and electrochemical sensing applications. *Optical Materials, 139*, 113819. DOI: 10.1016/j.optmat.2023.113819

John, B. K., Mathew, S., John, N., Mathew, J., & Mathew, B. (2023). Hydrothermal synthesis of N, S-doped carbon quantum dots as a dual mode sensor for azo dye tartrazine and fluorescent ink applications. *Journal of Photochemistry and Photobiology A Chemistry, 436*, 114386. DOI: 10.1016/j.jphotochem.2022.114386

John, V. L., Nair, Y., & Vinod, T. P. (2021). Doping and surface modification of carbon quantum dots for enhanced functionalities and related applications. *Particle &Amp. Particle & Particle Systems Characterization, 38*(11), 2100170. Advance online publication. DOI: 10.1002/ppsc.202100170

Jo, M.-H., Kim, K.-H., & Ahn, H.-J. (2022). P-doped carbon quantum dot graft-functionalized amorphous WO3 for stable and flexible electrochromic energy-storage devices. *Chemical Engineering Journal, 445*, 136826. https://doi.org/https://doi.org/10.1016/j.cej.2022.136826. DOI: 10.1016/j.cej.2022.136826

Jorio, A., Saito, R., Dresselhaus, G., & Dresselhaus, M. S. (2011). *Raman Spectroscopy in Graphene Related Systems*. Raman Spectroscopy in Graphene Related Systems., DOI: 10.1002/9783527632695

Jorns, M., & Pappas, D. (2021). A review of fluorescent carbon dots, their synthesis, physical and chemical characteristics, and applications. *Nanomaterials (Basel, Switzerland), 11*(6), 1448. DOI: 10.3390/nano11061448 PMID: 34070762

Joseph, J., & Anappara, A. A. (2016a). Microwave-assisted hydrothermal synthesis of UV-emitting carbon dots from tannic acid. *New Journal of Chemistry, 40*(9), 8110–8117. DOI: 10.1039/C6NJ02107G

Joseph, J., & Anappara, A. A. (2016b). White light emission of carbon dots by creating different emissive traps. *Journal of Luminescence, 178*, 128–133. DOI: 10.1016/j.jlumin.2016.05.051

Joseph, J., & Anappara, A. A. (2017a). Ellagic acid-functionalized fluorescent carbon dots for ultrasensitive and selective detection of mercuric ions via quenching. *Journal of Luminescence, 192*, 761–766. DOI: 10.1016/j.jlumin.2017.08.008

Joseph, J., & Anappara, A. A. (2017b). Long Life–time Room–temperature Phosphorescence of Carbon Dots in Aluminum Sulfate. *ChemistrySelect, 2*(14), 4058–4062. DOI: 10.1002/slct.201700786

Joseph, J., & Anappara, A. A. (2022). Comparative photoluminescence study of nitrogen-doped carbon dots co-doped with boron and sulphur. *Luminescence, 37*(9), 1475–1481. DOI: 10.1002/bio.4320 PMID: 35797229

Jyothi, S. P., Vinod, D., Chandran, D., Antherjanam, S., Saraswathyamma, B., Balaraman, V., & Rajamani, R. (2024). Poly-4-amino-6-chloro-1, 3-benzene disulfonamide decorated pencil graphite electrode for the simultaneous electrochemical quantification of catechol and hydroquinone. *Journal of Applied Electrochemistry, 54*(8), 1819–1831. DOI: 10.1007/s10800-024-02067-2

Kadian, S., Manik, G., Kalkal, A., Singh, M., & Chauhan, R. P. (2019). Effect of sulfur doping on fluorescence and quantum yield of graphene quantum dots: An experimental and theoretical investigation. *Nanotechnology*, *30*(43), 435704. DOI: 10.1088/1361-6528/ab3566 PMID: 31342919

Kakaei, K., Khodadoost, S., Gholipour, M., & Shouraei, N. (2021). Core-shell polyaniline functionalized carbon quantum dots for supercapacitor. *Journal of Physics and Chemistry of Solids*, *148*, 109753. DOI: 10.1016/j.jpcs.2020.109753

Kalaiyarasan, G., Joseph, J., & Kumar, P. (2020). Phosphorus-doped carbon quantum dots as fluorometric probes for iron detection. *ACS Omega*, *5*(35), 22278–22288. DOI: 10.1021/acsomega.0c02627 PMID: 32923785

Kalpana, R., Sakthi Vignesh, N., Vinothini, K., Rajan, M., Ashokkumar, B., Brindhadevi, K., Thuy Lan Chi, N., Pugazhendhi, A., & Varalakshmi, P. (2023). Carbon quantum dots (CQD) fabricated from Exiguobacterium sp. VK2 exopolysaccharide (EPS) using hydrothermal reaction and its biodiesel applications. *Fuel*, *333*, 126426. https://doi.org/https://doi.org/10.1016/j.fuel.2022.126426. DOI: 10.1016/j.fuel.2022.126426

Kamaruzaman, N. A. N., Fadzil, N. I., Alli, Y. A., Tan, T., Abdullah, J., & Abdul Rashid, S. (2025). Optimized synthesis and characterization of red-shifted nitrogen-doped carbon quantum dots for highly sensitive detection of Fe^{3+} ions in aqueous media. *Sensors and Actuators. A, Physical*, *394*, 116895. DOI: 10.1016/j.sna.2025.116895

Kandasamy, G. (2019). Recent Advancements in Doped/Co-Doped Carbon Quantum Dots for Multi-Potential Applications. *C*, *5*(2), 24.

Kannan et al., (2021). Gd3+ and Y3+ co-doped mixed metal oxide nanohybrids for photocatalytic and antibacterial applications. Nano express, 2(1), 010014.

Kannan, K., & Tari, V. (Eds.). (2024). Novel energy storage and conversion technologies for two-dimensional MXenes and MBenes. ISBN: 9798369395462, eISBN: 9798369395486DOI: 10.4018/979-8-3693-9546-2

Kannan, K., Radhika, D., Reddy, K. R., Raghu, A. V., Sadasivuni, K. K., Palani, G., & Gurushankar, K. (2021). Gd3+ and Y3+ co-doped mixed metal oxide nanohybrids for photocatalytic and antibacterial applications. Nano express, 2(1), 010014.

Kannan, K., Chanda, D., Gautam, J., Behera, A., Meshesha, M. M., Jang, S. G., & Yang, B. (2023). Hydrothermally synthesized mixed metal oxide nanocomposites for electrochemical water splitting and photocatalytic hydrogen production. *International Journal of Hydrogen Energy*, *48*(93), 36412–36426. DOI: 10.1016/j.ijhydene.2023.06.042

Kannan, K., Chanda, D., Gautam, J., Meshesha, M. M., Jang, S. G., & Yang, B. L. (2023). Facial synthesis of pp heterojunction composites: Evaluation of their electrochemical properties with photovoltaics-electrolyzer water splitting using two-electrode system. *International Journal of Hydrogen Energy*, *48*(37), 13814–13826. DOI: 10.1016/j.ijhydene.2022.12.235

Kannan, K., Chinnaiah, K., Gurushankar, K., Krishnamoorthi, R., Chen, Y. S., Murphin Kumar, P. S., & Li, Y. Y. (2024). Investigation of the electrochemical behavior of CuO-NiO-Co3O4 nanocomposites for enhanced supercapacitor applications. *Materials (Basel)*, *17*(16), 3976. DOI: 10.3390/ma17163976 PMID: 39203154

Kannan, K., Gautam, J., Chanda, D., Meshesha, M. M., Jang, S. G., & Yang, B. L. (2023). Two-dimensional MAX supported copper oxide/nickel Oxide/MAX as an efficient and novel photocatalyst for hydrogen evolution. *International Journal of Hydrogen Energy*, *48*(20), 7273–7283. DOI: 10.1016/j.ijhydene.2022.10.131

Kannan, K., Sliem, M. H., Abdullah, A. M., Sadasivuni, K. K., & Kumar, B. (2020). Fabrication of ZnO-Fe-MXene based nanocomposites for efficient CO2 reduction. *Catalysts*, *10*(5), 549. DOI: 10.3390/catal10050549

Kannan, , Radhika, D., Gnanasangeetha, D., Krishna, L. S., & Gurushankar, K. (2021). Y3+ and Sm3+ co-doped mixed metal oxide nanocomposite: Structural, electrochemical, photocatalytic, and antibacterial properties. *Applied Surface Science Advances*, *4*, 100085. DOI: 10.1016/j.apsadv.2021.100085

Kannan, , Radhika, D., Gnanasangeetha, D., Lakkaboyana, S. K., Sadasivuni, K. K., Gurushankar, K., & Hanafiah, M. M. (2021). Photocatalytic and antimicrobial properties of microwave synthesized mixed metal oxide nanocomposite. *Inorganic Chemistry Communications*, *125*, 108429. DOI: 10.1016/j.inoche.2020.108429

Kannan, , Radhika, D., Kasai, R. D., Gnanasangeetha, D., Palani, G., Gurushankar, K., Koutavarapu, R., Lee, D.-Y., & Shim, J. (2022). Facile fabrication of novel ceria-based nanocomposite (CYO-CSO) via co-precipitation: Electrochemical, photocatalytic and antibacterial performances. *Journal of Molecular Structure*, *1256*, 132519. DOI: 10.1016/j.molstruc.2022.132519

Karakoçak, B. B., Laradji, A., Primeau, T., Berezin, M. Y., Li, S., & Ravi, N. (2020). Hyaluronan-conjugated carbon quantum dots for bioimaging use. *ACS Applied Materials & Interfaces, 13*(1), 277–286. DOI: 10.1021/acsami.0c20088 PMID: 33355448

Kar, D. K., v, P., Si, S., Panigrahi, H., & Mishra, S. (2024). Carbon Dots and Their Polymeric Nanocomposites: Insight into Their Synthesis, Photoluminescence Mechanisms, and Recent Trends in Sensing Applications. *ACS Omega, 9*(10), 11050–11080. DOI: 10.1021/acsomega.3c07612 PMID: 38497004

Karthikeyan, R., Nelson, D. J., Ajith, A., & John, S. A. (2019). Hetero atoms doped carbon dots modified electrodes for the sensitive and selective determination of phenolic anti-oxidant in coconut oil. *Journal of Electroanalytical Chemistry (Lausanne, Switzerland), 848*, 113297. DOI: 10.1016/j.jelechem.2019.113297

Kashyap, B. K., Singh, V. V., Solanki, M. K., Kumar, A., Ruokolainen, J., & Kesari, K. K. (2023). Smart Nanomaterials in Cancer Theranostics: Challenges and Opportunities. *ACS Omega, 8*(16), 14290–14320. DOI: 10.1021/acsomega.2c07840 PMID: 37125102

Kaur, R., & Badea, I. (2013). Nanodiamonds as novel nanomaterials for biomedical applications: Drug delivery and imaging systems. *International Journal of Nanomedicine, 8*, 203. DOI: 10.2147/IJN.S37348 PMID: 23326195

Keenan, J. M., Zhou, Y., & Leblanc, R. M. (2019). Recent development of carbon quantum dots regarding their optical properties, photoluminescence mechanism, and core structure. *Nanoscale, 11*(11), 4634–4652. DOI: 10.1039/C8NR10059D PMID: 30834912

Keramat, A., Kadkhoda, J., Farahzadi, R., Fathi, E., & Davaran, S. (2022). The Potential of Graphene Oxide and Reduced Graphene Oxide in Diagnosis and Treatment of Cancer. *Current Medicinal Chemistry, 29*(26), 4529–4546. DOI: 10.2174/0929867329666220208092157 PMID: 35135444

Khairol Anuar, N. K., Tan, H. L., Lim, Y. P., So'aib, M. S., & Abu Bakar, N. F. (2021). A review on multifunctional carbon-dots synthesized from biomass waste: Design/fabrication, characterization and applications. *Frontiers in Energy Research, 9*, 626549. DOI: 10.3389/fenrg.2021.626549

Khajuria, A., Alajangi, H. K., Sharma, A., Kaur, H., Sharma, P., Negi, S., Kumari, L., Trivedi, M., Yadav, A. K., Kumar, R., Raghuvanshi, R. S., Kaur, I. P., Tyagi, R. K., Jaiswal, P. K., Lim, Y. B., Barnwal, R. P., & Singh, G. (2024). Theranostics: Aptamer-assisted carbon nanotubes as MRI contrast and photothermal agent for breast cancer therapy. *Discover Nano, 19*(1), 145. DOI: 10.1186/s11671-024-04095-5 PMID: 39256285

Khan, M. (2024). Innovations in battery technology: Enabling the revolution in electric vehicles and energy storage. *British Journal of Multidisciplinary and Advanced Studies*, *5*(1), 23–41. DOI: 10.37745/bjmas.2022.0414

Khan, M. E., Mohammad, A., & Yoon, T. (2022). State-of-the-art developments in carbon quantum dots (CQDs): Photo-catalysis, bio-imaging, and bio-sensing applications. *Chemosphere*, *302*, 134815. https://doi.org/https://doi.org/10.1016/j.chemosphere.2022.134815. DOI: 10.1016/j.chemosphere.2022.134815 PMID: 35526688

Khan, M. M. (2021). *Introduction and fundamentals of chalcogenides and chalcogenides-based nanomaterials*. Elsevier., DOI: 10.1016/B978-0-12-820498-6.00001-9

Khorasani, A., Shahbazi-Gahrouei, D., & Safari, A. (2023). Recent Metal Nanotheranostics for Cancer Diagnosis and Therapy: A Review. *Diagnostics (Basel)*, *13*(5), 833. DOI: 10.3390/diagnostics13050833 PMID: 36899980

Kim., ... (2017). Synthesis of nanoparticles by laser ablation: A review. In *KONA Powder and Particle Journal* (Vol. 2017, Issue 34, pp. 80–90). Hosokawa Powder Technology Foundation. DOI: 10.14356/kona.2017009

Kim, B., Lee, S., & Kim, J. (2020). Inverse design of porous materials using artificial neural networks. *Science Advances*, *6*(1), eaax9324. Advance online publication. DOI: 10.1126/sciadv.aax9324 PMID: 31922005

Kim, K., & Ahn, H. (2022). Surface functional group-tailored b and n co-doped carbon quantum dot anode for lithium-ion batteries. *International Journal of Energy Research*, *46*(6), 8367–8375. DOI: 10.1002/er.7738

Kim, M. J., Alvarez, S., Chen, Z., Fichthorn, K. A., & Wiley, B. J. (2018). Single-Crystal Electrochemistry Reveals Why Metal Nanowires Grow. *Journal of the American Chemical Society*, *140*(44), 14740–14746. DOI: 10.1021/jacs.8b08053 PMID: 30351013

Kim, T. E., Jang, H. J., Park, S. W., Wei, J., Cho, S., Park, W. I., & Jung, Y. K. (2021). Folic acid functionalized carbon dot/polypyrrole nanoparticles for specific bioimaging and photothermal therapy. *ACS Applied Bio Materials*, *4*(4), 3453–3461. DOI: 10.1021/acsabm.1c00018 PMID: 35014429

Knoblauch, R., & Geddes, C. D. (2020). Carbon nanodots in photodynamic antimicrobial therapy: A review. *Materials (Basel)*, *13*(18), 4004. DOI: 10.3390/ma13184004 PMID: 32927673

Kohli, H. K., & Parab, D. (2025). Green synthesis of carbon quantum dots and applications: An insight. *Next Materials*, *8*, 100527. DOI: 10.1016/j.nxmate.2025.100527

Ko, , Nafiujjaman, M., Cherukula, K., Lee, S. J., Hong, S.-J., Lim, H.-N., Park, C. H., Park, I.-K., Lee, Y.-K., & Kwon, I. K. (2018). Microwave-Assisted Synthesis of Biocompatible Silk Fibroin-Based Carbon Quantum Dots. *Particle & Particle Systems Characterization*, *35*(3), 1700300. Advance online publication. DOI: 10.1002/ppsc.201700300

Kong., ... (2024). Carbon Quantum Dots: Properties, Preparation, and Applications. In *Molecules* (Vol. 29, Issue 9). Multidisciplinary Digital Publishing Institute (MDPI). DOI: 10.3390/molecules29092002

Koroglu, D., Bingol, H., & Uralcan, B. (2025). Enhanced flexible supercapacitors with boron-doped graphene electrodes and carbon quantum dot gel electrolytes. *RSC Advances*, *15*(7), 5011–5019. DOI: 10.1039/D4RA06990K PMID: 39957823

Kostarelos, K., Lacerda, L., Pastorin, G., Wu, W., Wieckowski, S., Luangsivilay, J., Godefroy, S., Pantarotto, D., Briand, J. P., Muller, S., Prato, M., & Bianco, A. (2007). Cellular uptake of functionalized carbon nanotubes is independent of functional group and cell type. *Nature Nanotechnology*, *2*(2), 108–113. DOI: 10.1038/nnano.2006.209 PMID: 18654229

Koutsogiannis., ... (2020). Advances in fluorescent carbon dots for biomedical applications. In *Advances in Physics: X* (Vol. 5, Issue 1). Taylor and Francis Ltd. DOI: 10.1080/23746149.2020.1758592

Kou, X., Jiang, S., Park, S.-J., & Meng, L.-Y. (2020). A review: Recent advances in preparations and applications of heteroatom-doped carbon quantum dots. *Dalton Transactions (Cambridge, England)*, *49*(21), 6915–6938. DOI: 10.1039/D0DT01004A PMID: 32400806

Krajewski, M., Chen, C.-H., Huang, Z.-T., & Lin, J.-Y. (2022). $Li_4Ti_5O_{12}$ Coated by Biomass-Derived Carbon Quantum Dots as Anode Material with Enhanced Electrochemical Performance for Lithium-Ion Batteries. *Energies*, *15*(20), 7715. DOI: 10.3390/en15207715

Kumar et al., (2023). Technoeconomic analysis of biofuel production from agricultural residues through pyrolysis. *Green Approach to Alternative Fuel for a Sustainable Future*, 463–469. DOI: 10.1016/B978-0-12-824318-3.00001-1

Kumar, A., Kumar, D., & Saikia, M. (2023). A review on plant derived carbon quantum dots for bio-imaging. *Materials Advances*, *4*(18), 3951–3966. DOI: 10.1039/D3MA00254C

Kumar, A., Kumar, I., & Gathania, A. K. (2022). Synthesis, characterization and potential sensing application of carbon dots synthesized via the hydrothermal treatment of cow milk. *Scientific Reports*, *12*(1), 22495. DOI: 10.1038/s41598-022-26906-4 PMID: 36577768

Kumar, , Mirsky, S. K., Shaked, N. T., & Gazit, E. (2024). High Quantum Yield Amino Acid Carbon Quantum Dots with Unparalleled Refractive Index. *ACS Nano*, *18*(3), 2421–2433. DOI: 10.1021/acsnano.3c10792 PMID: 38190624

Kumar, P. S., Bae, J., Roh, J. W., Min, Y., & Lee, S. (2025). Enhanced voltage and capacitance in flexible supercapacitors using electrospun nanofiber electrolytes and $CuNi_2O_3$@N-Doped omnichannel carbon electrodes. *Nano Convergence*, *12*(1), 21. DOI: 10.1186/s40580-025-00485-2 PMID: 40301281

Kumar, P., Bhatt, G., Kaur, R., Dua, S., & Kapoor, A. (2020). Synthesis and modulation of the optical properties of carbon quantum dots using microwave radiation. *Fullerenes, Nanotubes, and Carbon Nanostructures*, *28*(9), 724–731. DOI: 10.1080/1536383X.2020.1752679

Kumar, P., Dua, S., Kaur, R., Kumar, M., & Bhatt, G. (2022). A review on advancements in carbon quantum dots and their application in photovoltaics. *RSC Advances*, *12*(8), 4714–4732. DOI: 10.1039/D1RA08452F PMID: 35425490

Kumar, P., Mahapatra, A., Kumar, S., & Pradhan, B. (2023). Photovoltaic application of carbon quantum dots. In *Carbon Quantum Dots for Sustainable Energy and Optoelectronics* (pp. 135–158). Woodhead Publishing., DOI: 10.1016/B978-0-323-90895-5.00017-5

Kumar, R., Kumar, V. B., & Gedanken, A. (2020). Sonochemical synthesis of carbon dots, mechanism, effect of parameters, and catalytic, energy, biomedical and tissue engineering applications. *Ultrasonics Sonochemistry*, *64*, 105009. DOI: 10.1016/j.ultsonch.2020.105009 PMID: 32106066

Kumar, V. B., Tang, J., Lee, K. J., Pol, V. G., & Gedanken, A. (2016). In situ sonochemical synthesis of luminescent Sn@C-dots and a hybrid Sn@C-dots@Sn anode for lithium-ion batteries. *RSC Advances*, *6*(70), 66256–66265. DOI: 10.1039/C6RA09926B

Kundu, S., Yadav, R. M., Narayanan, T. N., Shelke, M. V., Vajtai, R., Ajayan, P. M., & Pillai, V. K. (2015). Synthesis of N, F and S co-doped graphene quantum dots. *Nanoscale*, *7*(27), 11515–11519. DOI: 10.1039/C5NR02427G PMID: 26087457

Kush, P., Kumar, P., & Singh, R. (2024). Functionalized Carbon Nanostructures in Cancer Diagnosis and Therapy. *Handbook of Functionalized Carbon Nanostructures: From Synthesis Methods to Applications*, 2091–2130. DOI: 10.1007/978-3-031-32150-4_60

Kwon, W., Lim, J., Lee, J., Park, T., & Rhee, S. (2013). Sulfur-incorporated carbon quantum dots with a strong long-wavelength absorption band. *Journal of Materials Chemistry. C, Materials for Optical and Electronic Devices*, *1*(10), 2002. DOI: 10.1039/c3tc00683b

Kyesmen, P. I., Onoja, A., & Amah, A. N. (2016). Fullerenes synthesis by combined resistive heating and arc discharge techniques. *SpringerPlus*, *5*(1), 1323. DOI: 10.1186/s40064-016-2994-7 PMID: 27563518

Lai, C.-W., Hsiao, Y.-H., Peng, Y.-K., & Chou, P.-T. (2012). Facile synthesis of highly emissive carbon dots from pyrolysis of glycerol; gram scale production of carbon dots/mSiO_2 for cell imaging and drug release. *Journal of Materials Chemistry*, *22*(29), 14403. DOI: 10.1039/c2jm32206d

Lakhera, S., Devlal, K., Rana, M., & Dhuliya, V. (2024a). Influence of the substitution of different functional groups on the gas sensing and light harvesting efficiency of zero-dimensional coronene quantum dot: A first principle DFT study. *Spectrochimica Acta. Part A: Molecular and Biomolecular Spectroscopy*, *308*, 123737. https://doi .org/https://doi.org/10.1016/j.saa.2023.123737. DOI: 10.1016/j.saa.2023.123737 PMID: 38064960

Lakowicz, J. (2006). Principles of fluorescence spectroscopy. In *Principles of Fluorescence Spectroscopy*. Springer US., DOI: 10.1007/978-0-387-46312-4

Lakshmanan, A., Surendran, P., Sakthy Priya, S., Balakrishnan, K., Rameshkumar, P., Kannan, K., Mahalakshmi, K., Gayathri, V., & Vinitha, G. (2024). Investigations on structural, dielectric, and third-order nonlinear optical studies of $Ni_{0.5}Zn_{0.5}Fe_2O_4$ nanoparticles by microwave combustion method with different fuels (glycine, urea and citric acid). *Materials Letters*, *360*, 135973. DOI: 10.1016/j.matlet.2024.135973

Lan, M., Zhao, S., Zhang, Z., Yan, L., Guo, L., Niu, G., Zhang, J., Zhao, J., Zhang, H., Wang, P., Zhu, G., Lee, C.-S., & Zhang, W. (2017). Two-photon-excited near-infrared emissive carbon dots as multifunctional agents for fluorescence imaging and photothermal therapy. *Nano Research*, *10*(9), 3113–3123. DOI: 10.1007/s12274-017-1528-0

Latif, Z., Ali, M., Lee, E.-J., Zubair, Z., & Lee, K. H. (2023). Thermal and Mechanical Properties of Nano-Carbon-Reinforced Polymeric Nanocomposites: A Review. In *Journal of Composites Science* (Vol. 7, Issue 10). DOI: 10.3390/jcs7100441

Latiza, R. J. P., Olay, J., Eguico, C., Yan, R. J., & Rubi, R. V. (2025). Surface matters: Passivation in carbon dot synthesis—a critical review. *Journal of Materials Science*, *60*(11), 1–13. DOI: 10.1007/s10853-025-10764-2

Lee, H. R., Kim, Y. S., Lee, S. Y., Son, U. H., Lee, S., & Joh, H. (2024). Bifunctional effects of nitrogen-doped carbon quantum dots on cos2/mesoporous carbon composites for high-performance lithium-ion batteries. *Applied Surface Science*, *664*, 160228. DOI: 10.1016/j.apsusc.2024.160228

Lesch, R., Visser, E. D., Seroka, N. S., & Khotseng, L. (n.d.). *Biomass-Derived Carbon-Based Nanomaterials: Current Research*. Trends, and Challenges.

Leung, H. M., Chu, H. C., Mao, Z. W., & Lo, P. K. (2023). Versatile nanodiamond-based tools for therapeutics and bioimaging. *Chemical Communications*, *59*(15), 2039–2055. DOI: 10.1039/D2CC06495B PMID: 36723092

Levi, M. D., Levy, N., Sigalov, S., Salitra, G., Aurbach, D., & Maier, J. (2010). Electrochemical quartz crystal microbalance (eqcm) studies of ions and solvents insertion into highly porous activated carbons. *Journal of the American Chemical Society*, *132*(38), 13220–13222. DOI: 10.1021/ja104391g PMID: 20828121

Lewis, I. R., & Edwards, H. (2001). Handbook of Raman Spectroscopy : From the Research Laboratory to the Process Line. *Handbook of Raman Spectroscopy*. DOI: 10.1201/9781420029253

Li, S., Su, W., & Wu, H. (n.d.). *Targeted tumour theranostics in mice via carbon quantum dots structurally mimicking large amino acids*. DOI: 10.1038/s41551-020-0540-y

Li., ... (2023). Physiological Functions of Carbon Dots and Their Applications in Agriculture: A Review. In *Nanomaterials* (Vol. 13, Issue 19). Multidisciplinary Digital Publishing Institute (MDPI). DOI: 10.3390/nano13192684

Liao, B., Long, P., He, B., Yi, S., Ou, B., Shen, S., & Chen, J. (2013). Reversible fluorescence modulation of spiropyran-functionalized carbon nanoparticles. *Journal of Materials Chemistry. C, Materials for Optical and Electronic Devices*, *1*(23), 3716. DOI: 10.1039/c3tc00906h

Liao, H., & Zheng, H. (2016). Liquid cell transmission electron microscopy. *Annual Review of Physical Chemistry*, *67*(1), 719–747. DOI: 10.1146/annurev-physchem-040215-112501 PMID: 27215823

Liao, X., Wu, B., Li, H., Zhang, M., Cai, M., Lang, B., & Zhou, P. (2023). Fluorescent/colorimetric dual-mode discriminating Gln and Val enantiomers based on carbon dots. *Analytical Chemistry*, *95*(39), 14573–14581. DOI: 10.1021/acs.analchem.3c01854 PMID: 37729469

Libber, M., Gariya, N., & Kumar, M. (2024). A comprehensive analysis of supercapacitors with current limitations and emerging trends in research. *Journal of Solid State Electrochemistry*, *29*(2), 513–527. DOI: 10.1007/s10008-024-06107-x

Li, F., Li, T., Sun, C., Xia, J., Jiao, Y., & Xu, H. (2017). Facile synthesis of carbon dots with superior sensing ability. *Applied Nanoscience*, *8*(5), 1189–1196. DOI: 10.1007/s13204-018-0755-3

Li, H. (2024). A review of edlc and pseudocapacitance with synergistic integration of carbon-based and metal oxide materials for enhanced electrochemical energy storage. *Applied and Computational Engineering*, *61*(1), 231–244. DOI: 10.54254/2755-2721/61/20240968

Li, H., Liu, R., Liu, Y., Huang, H., Yu, H., Ming, H., Lian, S., Lee, S.-T., & Kang, Z. (2012). Carbon quantum dots/Cu_2O composites with protruding nanostructures and their highly efficient (near) infrared photocatalytic behavior. *Journal of Materials Chemistry*, *22*(34), 17470. DOI: 10.1039/c2jm32827e

Li, H., Zhang, Y., Wang, L., Tian, J., & Sun, X. (2011). Nucleic acid detection using carbon nanoparticles as a fluorescent sensing platform. *Chemical Communications (Cambridge)*, *47*(3), 961–963. DOI: 10.1039/C0CC04326E PMID: 21079843

Li, , Han, F., Cao, W., Zhang, G., Li, J., Zhou, J., Gong, X., Turnbull, G., Shu, W., Xia, L., Fang, B., Xing, X., & Li, B. (2020). Carbon quantum dots derived from lysine and arginine simultaneously scavenge bacteria and promote tissue repair. *Applied Materials Today*, *19*, 100601. Advance online publication. DOI: 10.1016/j.apmt.2020.100601

Li, , Kang, Z., Liu, Y., & Lee, S.-T. (2012a). Carbon nanodots: Synthesis, properties and applications. *Journal of Materials Chemistry*, *22*(46), 24230–24253. DOI: 10.1039/c2jm34690g

Li, L., & Dong, T. (2018). Photoluminescence tuning in carbon dots: Surface passivation or/and functionalization, heteroatom doping. *Journal of Materials Chemistry. C, Materials for Optical and Electronic Devices*, *6*(30), 7944–7970. DOI: 10.1039/C7TC05878K

Li, L., Liu, D., Shi, A., & You, T. (2018). Simultaneous stripping determination of cadmium and lead ions based on the N-doped carbon quantum dots-graphene oxide hybrid. *Sensors and Actuators. B, Chemical, 255*, 1762–1770. DOI: 10.1016/j.snb.2017.08.190

Lim., ... (2015). Carbon quantum dots and their applications. In *Chemical Society Reviews* (Vol. 44, Issue 1, pp. 362–381). Royal Society of Chemistry. DOI: 10.1039/C4CS00269E

Li, M., Ma, C., Wang, G., Zhang, X., Dong, X., & Ma, H. (2021). Controlling the up-conversion photoluminescence property of carbon quantum dots (CQDs) by modifying its surface functional groups for enhanced photocatalytic performance of CQDs/$BiVO_4$ under a broad-spectrum irradiation. *Research on Chemical Intermediates, 47*(8), 3469–3485. DOI: 10.1007/s11164-021-04459-x

Lim, C. S., Hola, K., Ambrosi, A., Zboril, R., & Pumera, M. (2015). Graphene and carbon quantum dots electrochemistry. *Electrochemistry Communications, 52*, 75–79. https://doi.org/https://doi.org/10.1016/j.elecom.2015.01.023. DOI: 10.1016/j.elecom.2015.01.023

Li, Q., & Jin, Q. (2025). Holistic comparison of hydrothermal carbonization versus slow pyrolysis for biochar production: Techno-economic analysis and life cycle assessment. *ACS Sustainable Chemistry & Engineering, 13*(31), 12382–12395. DOI: 10.1021/acssuschemeng.5c02249

Li, T., Murphy, S., Kiselev, B., Bakshi, K. S., Zhang, J., Eltahir, A., Zhang, Y., Chen, Y., Zhu, J., Davis, R. M., Madsen, L. A., Morris, J. R., Karolyi, D. R., LaConte, S. M., Sheng, Z., & Dorn, H. C. (2015). A New Interleukin-13 Amino-Coated Gadolinium Metallofullerene Nanoparticle for Targeted MRI Detection of Glioblastoma Tumor Cells. *Journal of the American Chemical Society, 137*(24), 7881–7888. DOI: 10.1021/jacs.5b03991 PMID: 26022213

Liu, G., Li, S., Cheng, M., Zhao, L., Zhang, B., Gao, Y., Yan, X., Liu, F., & Lu, G. (2018). Facile synthesis of nitrogen and sulfur co-doped carbon dots for multiple sensing capacities: alkaline fluorescence enhancement effect, temperature sensing, and selective detection of Fe^{3+} ions. *New Journal of Chemistry, 42*, 13147–13156. DOIhttps://DOI: 10.1039/C8NJ02086H

Liu, C., Zhang, P., Tian, F., Li, W., Li, F., & Liu, W. (2011). One-step synthesis of surface passivated carbon nanodots by microwave assisted pyrolysis for enhanced multicolor photoluminescence and bioimaging. *Journal of Materials Chemistry, 21*(35), 13163–13167. DOI: 10.1039/c1jm12744f

Liu, D., Pan, X., Zhao, X., & Zhang, H. (2022). Preparation and In Vitro and In Vivo Evaluation of Lung Cancer-Targeted Carbon Nanotube Ultrasound Contrast Agents. *Journal of Biomedical Nanotechnology*, *18*(6), 1684–1691. DOI: 10.1166/jbn.2022.3366

Liu, H., Jia, M., Zhu, Q., Cao, B., Chen, R., Wang, Y., Wu, F., & Xu, B. (2016). 3d-0d graphene-fe3o4 quantum dot hybrids as high-performance anode materials for sodium-ion batteries. *ACS Applied Materials &Amp. Interfaces*, *8*(40), 26878–26885. DOI: 10.1021/acsami.6b09496 PMID: 27642811

Liu, H., Liu, Z., Zhang, J., Zhi, L., & Wu, M. (2021). Boron and nitrogen co-doped carbon dots for boosting electrocatalytic oxygen reduction. *New Carbon Materials*, *36*(3), 585–593. DOI: 10.1016/S1872-5805(21)60043-4

Liu, H., Ye, T., & Mao, C. (2007). Fluorescent carbon nanoparticles derived from candle soot. *Angewandte Chemie International Edition*, *46*(34), 6473–6475. DOI: 10.1002/anie.200701271 PMID: 17645271

Liu, J., Bi, H., Cesar Morais, P., Zhang, X., Zhang, F., & Hu, L. (2017). Room-temperature Magnetism in Carbon Dots and Enhanced Ferromagnetism in Carbon Dots-Polyaniline Nanocomposite. *Scientific Reports*, *7*(1), 2165. DOI: 10.1038/s41598-017-01350-x PMID: 28526848

Liu, J., Li, J., Xu, L., Qiao, Y., & Chen, J. (2017). Facile Synthesis of N, B-Doped Carbon Dots and Their Application for Multisensor and Cellular Imaging. *Industrial & Engineering Chemistry Research*, *56*(14), 3905–3912. DOI: 10.1021/acs.iecr.6b04752

Liu, J., Li, R., & Yang, B. (2020). Carbon Dots: A New Type of Carbon-Based Nanomaterial with Wide Applications. *ACS Central Science*, *6*(12), 2179–2195. DOI: 10.1021/acscentsci.0c01306 PMID: 33376780

Liu, L., Anwar, S., Ding, H., Xu, M., Yin, Q., Xiao, Y., Yang, X., Yan, M., & Bi, H. (2019). Electrochemical sensor based on F, N-doped carbon dots decorated laccase for detection of catechol. *Journal of Electroanalytical Chemistry (Lausanne, Switzerland)*, *840*, 84–92. DOI: 10.1016/j.jelechem.2019.03.071

Liu, L., Zhang, X., Liu, Y., & Gong, X. (2025). Electrochemical energy storage devices—Batteries, supercapacitors, and battery–supercapacitor hybrid devices. *ACS Applied Electronic Materials*, *7*(6), 2233–2270. DOI: 10.1021/acsaelm.5c00069

Liu, M., Xu, Y., Niu, F., Gooding, J. J., & Liu, J. (2016). Carbon quantum dots directly generated from electrochemical oxidation of graphite electrodes in alkaline alcohols and the applications for specific ferric ion detection and cell imaging. *Analyst*, *141*(9), 2657–2664. DOI: 10.1039/C5AN02231B PMID: 26878217

Liu, Q., Chen, H., Mi, R., Min, X., Fang, M., Wu, X., Huang, Z., & Liu, Y. (2025). Biomass-Derived Carbon Dots: Preparation, Properties, and Applications. *Nanomaterials (Basel, Switzerland)*, *15*(16), 1279. DOI: 10.3390/nano15161279 PMID: 40863859

Liu, Q., Ge, K., Wu, X., Zhu, Z., Zhu, Y., & Bi, H. (2023). Preparation of Carbon Dots@r-GO Nanocomposite with an Enhanced Pseudo-Capacitance. *Molecules (Basel, Switzerland)*, *28*(2), 541. DOI: 10.3390/molecules28020541 PMID: 36677598

Liu, R., Wu, D., Liu, S., Koynov, K., Knoll, W., & Li, Q. (2009). An aqueous route to multicolor photoluminescent carbon dots using silica spheres as carriers. *Angewandte Chemie International Edition*, *48*(25), 4598–4601. DOI: 10.1002/anie.200900652 PMID: 19388019

Liu, S., Ao, Z., Zou, Y., Li, H., Chen, N., Wan, Y., & Lao, Y. (2023). Enhanced ionic conductivity of PEO-based solid electrolyte through synergistic dissociation effect. *Materials Letters*, *351*, 134981. DOI: 10.1016/j.matlet.2023.134981

Liu, S., Cao, X., Zhang, Y., Wang, K., Su, Q., Chen, J., He, Q., Liang, S., Cao, G., & Pan, A. (2020). Carbon quantum dot modified $Na_3V_2(PO_4)_2F_3$ as a high-performance cathode material for sodium-ion batteries. J Mater. *Journal of Materials Chemistry. A, Materials for Energy and Sustainability*, *8*(36), 18872–18879. DOI: 10.1039/D0TA04307A

Liu, S., Gao, D., Li, J., Hui, K. S., Yin, Y., Hui, K. N., & Jun, S. C. (2019). Phosphorus dual-site driven cos2@s, n co-doped porous carbon nanosheets for flexible quasi-solid-state supercapacitors. *Journal of Materials Chemistry. A, Materials for Energy and Sustainability*, *7*(46), 26618–26630. DOI: 10.1039/C9TA09646A

Liu, T., Cui, Z., Zhou, J., Wang, Y., & Zou, Z. (2017). Synthesis of Pyridinic-Rich N, S Co-doped Carbon Quantum Dots as Effective Enzyme Mimics. *Nanoscale Research Letters*, *12*(1), 375. DOI: 10.1186/s11671-017-2149-y PMID: 28549380

Liu, Y., Cai, D., Zheng, F., Qin, Z., Li, Y., Li, W., Li, A., Zhao, Y., & Zhang, J. (2024). A carbon quantum dot-decorated g-C_3N_4 composite as a sulfur hosting material for lithium–sulfur batteries. *Dalton Transactions (Cambridge, England)*, *53*(16), 7035–7043. DOI: 10.1039/D4DT00511B PMID: 38563460

Liu, Y., Roy, S., Sarkar, S., Xu, J., Zhao, Y., & Zhang, J. (2021). A review of carbon dots and their composite materials for electrochemical energy technologies. *Carbon Energy, 3*(5), 795–826. DOI: 10.1002/cey2.134

Liu, Y., Zhao, C., Sabirsh, A., Ye, L., Wu, X., Lu, H., & Liu, J. (2021). A novel graphene quantum dot-based mRNA delivery platform. *ChemistryOpen, 10*(7), 666–671. DOI: 10.1002/open.202000200 PMID: 33829677

Liu, Z., & Liang, X. J. (2012). Nano-Carbons as Theranostics. *Theranostics, 2*(3), 235–237. DOI: 10.7150/thno.4156 PMID: 22448193

Li, X., Liu, G., Zheng, H., Sun, K., Wan, L., Cao, J., Asif, S., Cao, Y., Si, W., Wang, F., & Bokhari, A. (2022). Recent advances on heteroatom-doped porous carbon—Based electrocatalysts for oxygen reduction reaction. *Energies, 16*(1), 128. DOI: 10.3390/en16010128

Li, Y., Lin, H., Luo, C., Wang, Y., Jiang, C., Qi, R., Huang, R., Travass-Sejdic, J., & Peng, H. (2017). Aggregation induced red shift emission of phosphorus doped carbon dots. *RSC Advances, 7*(51), 32225–32228. DOI: 10.1039/C7RA04781A

Li, Y., Li, S., Wang, Y., Wang, J., Liu, H., Liu, X., Wang, L., Liu, X., Xue, W., & Ma, N. (2017). Electrochemical synthesis of phosphorus-doped graphene quantum dots for free radical scavenging. *Physical Chemistry Chemical Physics, 19*(22), 11631–11638. DOI: 10.1039/C6CP06377B PMID: 28430285

Li, Y., Wang, S., Lv, Z., Wang, Z., Zhao, Y., Xie, Y., Xu, Y., Qian, L., Yang, Y., Zhao, Z., & Zhang, J. (2025). Transforming the synthesis of carbon nanotubes with machine learning models and automation. *Matter, 8*(1), 101913. DOI: 10.1016/j.matt.2024.11.007

Li, Y., Xu, Z., Qi, Z., Huang, X., Li, M., Liu, S., Yan, Y., & Gao, M. (2024). Application of Carbon Nanomaterials to Enhancing Tumor Immunotherapy: Current Advances and Prospects. *International Journal of Nanomedicine, 19*, 10899–10915. DOI: 10.2147/IJN.S480799 PMID: 39479174

Li, Z., Bommier, C., Chong, Z. S., Jian, Z., Surta, T. W., Wang, X., Xing, Z., Neuefeind, J. C., Stickle, W. F., Dolgos, M., Greaney, P. A., & Ji, X. (2017). Mechanism of Na-Ion Storage in Hard Carbon Anodes Revealed by Heteroatom Doping. *Advanced Energy Materials, 7*(18), 1602894. DOI: 10.1002/aenm.201602894

Li, Z., Ji, X., Xie, H., & Tang, B. Z. (2021). Aggregation-induced emission-active gels: Fabrications, functions, and applications. *Advanced Materials, 33*(33), 2100021. DOI: 10.1002/adma.202100021 PMID: 34216407

Li, Z., Liu, F., Chen, S., Zhai, F., Li, Y., Feng, Y., & Feng, W. (2021). Single Li ion conducting solid-state polymer electrolytes based on carbon quantum dots for Li-metal batteries. *Nano Energy*, *82*, 105698. DOI: 10.1016/j.nanoen.2020.105698

Lopez-Astacio, H. J. H., Vargas-Pérez, B. L., Valle-Perez, A. D., Pollock, C. J., & Cunci, L. (2024). Open-source electrochemical cell for in situ x-ray absorption spectroscopy in transmission and fluorescence modes. *Journal of Synchrotron Radiation*, *31*(2), 322–327. DOI: 10.1107/S1600577524000122 PMID: 38306299

Lou, Y., Ji, J., Qin, A., Liao, L., Li, Z., Chen, S., Zhang, K., & Ou, J. (2020). Cane molasses graphene quantum dots passivated by PEG functionalization for detection of metal ions. *ACS Omega*, *5*(12), 6763–6772. DOI: 10.1021/acsomega.0c00098 PMID: 32258911

Lu, J., Yang, J., Wang, J., Lim, A., Wang, S., & Loh, K. P. (2009). One-Pot Synthesis of Fluorescent Carbon Nanoribbons, Nanoparticles, and Graphene by the Exfoliation of Graphite in Ionic Liquids. *ACS Nano*, *3*(8), 2367–2375. DOI: 10.1021/nn900546b PMID: 19702326

Luo, B., Yang, H., Zhou, B., Ahmed, S. M., Zhang, Y., Liu, H., Liu, X., He, Y., & Xia, S. (2020). Facile synthesis of Luffa sponge activated carbon fiber-based carbon quantum dots with green fluorescence and their application in Cr (VI) determination. *ACS Omega*, *5*(10), 5540–5547. DOI: 10.1021/acsomega.0c00195 PMID: 32201847

Luo, P. G., Sahu, S., Yang, S.-T., Sonkar, S. K., Wang, J., Wang, H., LeCroy, G. E., Cao, L., & Sun, Y.-P. (2013). Carbon "quantum" dots for optical bioimaging. *Journal of Materials Chemistry. B*, *1*(16), 2116. DOI: 10.1039/c3tb00018d PMID: 32260843

Lu, S., Cheng, C., Shi, Y., Wu, Y., Zhang, Z., & Zhang, B. (2023). Unveiling the structural transformation and activity origin of heteroatom-doped carbons for hydrogen evolution. *Proceedings of the National Academy of Sciences of the United States of America*, *120*(20), e2300549120. Advance online publication. DOI: 10.1073/pnas.2300549120 PMID: 37155878

Lu, S., Xiao, Q., Yang, W., Wang, X., Guo, T., Xie, Q., & Ruan, Y. (2024). Multi-heteroatom-doped porous carbon with high surface adsorption energy of potassium derived from biomass waste for high-performance supercapacitors. *International Journal of Biological Macromolecules*, *258*, 128794. DOI: 10.1016/j.ijbiomac.2023.128794 PMID: 38110166

Lu, W., Gong, X., Nan, M., Liu, Y., Shuang, S., & Dong, C. (2015). Low temperature synthesis of phosphorous and nitrogen co-doped yellow fluorescent carbon dots for sensing and bioimaging. *Analytica Chimica Acta*, *898*, 116–127. DOI: 10.1016/j.aca.2015.09.050 PMID: 26526917

Lu, W., Hao, L., & Wang, Y. (2022). Highly active N, S Co-Doped UltramicroCo-dopedarbon for High-Performance Supercapacitor Electrodes. *Micromachines*, *13*(6), 905. DOI: 10.3390/mi13060905 PMID: 35744519

Lu, X., Li, S., Li, Y., Wu, F., Wu, C., & Bai, Y. (2024). From Lab to Application: Challenges and Opportunities in Achieving Fast Charging with Polyanionic Cathodes for Sodium-Ion Batteries. *Advanced Materials*, *36*(36), 2407359. Advance online publication. DOI: 10.1002/adma.202407359 PMID: 38936413

Lu, Y.-F., Lo, S.-T., Lin, J.-C., Zhang, W., Lu, J.-Y., Liu, F.-H., Tseng, C.-M., Lee, Y.-H., Liang, C.-T., & Li, L.-J. (2013). Nitrogen-Doped Graphene Sheets Grown by Chemical Vapor Deposition: Synthesis and Influence of Nitrogen Impurities on Carrier Transport. *ACS Nano*, *7*(8), 6522–6532. DOI: 10.1021/nn402102y PMID: 23879622

Lu, , & Zhou, L. (2019). One-step sonochemical synthesis of versatile nitrogen-doped carbon quantum dots for sensitive detection of Fe^{2+} ions and temperature in vitro. *Materials Science and Engineering C*, *101*, 352–359. DOI: 10.1016/j.msec.2019.03.109 PMID: 31029328

Lv, C., Zhou, X., Zhong, L., Yan, C., Srinivasan, M., Seh, Z. W., Liu, C., Pan, H., Li, S., Wen, Y., & Yan, Q. (2021). Machine learning: An advanced platform for materials development and state prediction in lithium-ion batteries. *Advanced Materials*, *34*(25), 2101474. Advance online publication. DOI: 10.1002/adma.202101474 PMID: 34490683

Lv, , Luo, Z., Yang, H., Liu, B., Weng, W., & Liu, J. (2010). Effect of processing conditions on sonochemical synthesis of nanosized copper aluminate powders. *Ultrasonics Sonochemistry*, *17*(2), 344–351. DOI: 10.1016/j.ultsonch.2009.06.006 PMID: 19570706

MacLeod, B. P., Parlane, F. G. L., Morrissey, T. D., Häse, F., Roch, L. M., Dettelbach, K. E., Moreira, R., Yunker, L. P. E., Rooney, M. B., Deeth, J. R., Lai, V., Ng, G. J., Situ, H., Zhang, R., Elliott, M. S., Haley, T. H., Dvořák, D., Aspuru-Guzik, A., Hein, J. E., & Berlinguette, C. P. (2020). Self-driving laboratory for accelerated discovery of thin-film materials. *Science Advances*, *6*(20), eaaz8867. Advance online publication. DOI: 10.1126/sciadv.aaz8867 PMID: 32426501

Magdy, , Belal, F., & Elmansi, H. (2023). Rapid microwave-assisted synthesis of nitrogen-doped carbon quantum dots as fluorescent nanosensors for the spectrofluorimetric determination of palbociclib: Application for cellular imaging and selective probing in living cancer cells. *RSC Advances*, *13*(7), 4156–4167. DOI: 10.1039/D2RA05759J PMID: 36744285

Magesh, . (2022). Recent Advances on Synthesis and Potential Applications of Carbon Quantum Dots. In *Frontiers in Materials* (Vol. 9). Frontiers Media S.A., DOI: 10.3389/fmats.2022.906838

Mahmood, A., Bai, Z., Wang, T., Lei, Y., Wang, S., Sun, B., Khan, H., Khan, K., Sun, K., & Wang, G. (2025). Enabling high-performance multivalent metal-ion batteries: Current advances and future prospects. *Chemical Society Reviews*, *54*(5), 2369–2435. DOI: 10.1039/D4CS00929K PMID: 39887968

Maity, P. P., Kapat, K., Poddar, P., Bora, H., Das, C. K., Das, P., & Dhara, S. (2023). Capra cartilage-derived peptide delivery via carbon nano-dots for cartilage regeneration. *Frontiers in Bioengineering and Biotechnology*, *11*, 1213932. DOI: 10.3389/fbioe.2023.1213932 PMID: 37701494

Maleki, R., Asadnia, M., & Razmjou, A. (2022). Artificial Intelligence-Based Material Discovery for Clean Energy Future. *Advanced Intelligent Systems*, *4*(10), 2200073. DOI: 10.1002/aisy.202200073

Malini et al., (2023). Activated carbon from biomass: Preparation, factors improving basicity and surface properties for enhanced CO_2 capture capacity - A review. *Journal of CO2 Utilization, 67*. DOI: 10.1016/j.jcou.2022.102318

Malitha, M. D., Molla, M. T. H., Bashar, M. A., Chandra, D., & Ahsan, M. S. (2024). Fabrication of a reusable carbon quantum dots (CQDs) modified nanocomposite with enhanced visible light photocatalytic activity. *Scientific Reports*, *14*(1), 1–18. DOI: 10.1038/s41598-024-66046-5 PMID: 39095409

Mangini, M., Verde, A., Boraschi, D., Puntes, V. F., Italiani, P., & De Luca, A. C. (2021). Interaction of nanoparticles with endotoxin: Importance in nanosafety testing and exploitation for endotoxin binding. *Nanotoxicology*, *15*(4), 558–576. DOI: 10.1080/17435390.2021.1898690 PMID: 33784953

Mankoti, M., Meena, S. S., & Mohanty, A. (2024). Exploring the potential of eco-friendly carbon dots in monitoring and remediation of environmental pollutants. *Environmental Science and Pollution Research International*, *31*(31), 43492–43523. DOI: 10.1007/s11356-024-33448-x PMID: 38713351

Mansuriya, B. D., & Altintas, Z. (2021). Carbon dots: Classification, properties, synthesis, characterization, and applications in health care—an updated review (2018–2021). *Nanomaterials (Basel, Switzerland)*, *11*(10), 2525. DOI: 10.3390/nano11102525 PMID: 34684966

Manzoor, S., Dar, A. H., Dash, K. K., Pandey, V. K., Srivastava, S., Bashir, I., & Khan, S. A. (2023). Carbon dots applications for development of sustainable technologies for food safety: A comprehensive review. *Applied Food Research, 3*(1), 100263. https://doi.org/https://doi.org/10.1016/j.afres.2023.100263. DOI: 10.1016/j.afres.2023.100263

Marcu, L., French, P. M. W., & Elson, D. S. (2015). *Fluorescence lifetime spectroscopy and imaging : principles and applications in biomedical diagnostics*. 554. https://books.google.com/books/about/Fluorescence_Lifetime_Spectroscopy_and_I.html?id=6xvcBQAAQBAJ

Maruthapandi, M., Sharma, K., Luong, J. H., & Gedanken, A. (2020). Antibacterial activities of microwave-assisted synthesized polypyrrole/chitosan and poly (pyrrole-N-(1-naphthyl) ethylenediamine) stimulated by C-dots. *Carbohydrate Polymers, 243*, 116474. DOI: 10.1016/j.carbpol.2020.116474 PMID: 32532398

Matek, B., Dorn, J., & Divita, F. (2019). *Life cycle assessment: C4V lithium-ion battery cells for electric vehicles. Final report* (Nos. 19–44). New York State Energy Research and Development Authority (NYSERDA). https://www.nyserda.ny.gov/-/media/Project/Nyserda/Files/Publications/Research/Transportation/19-44-Lifecycle-Assessment-of-CV4-Li-ion-Battery-Cells-for-Electric-Vehicles.pdf

Mauro, N., Utzeri, M. A., Varvarà, P., & Cavallaro, G. (2021). Functionalization of Metal and Carbon Nanoparticles with Potential in Cancer Theranostics. *Molecules 2021, Vol. 26, Page 3085, 26*(11), 3085. DOI: 10.3390/molecules26113085

Ma, Y., Chen, D., Fang, Z., Zheng, Y., Li, W., Xu, S., Lu, X., Shao, G., Liu, Q., & Yang, W. (2021). High energy density and extremely stable supercapacitors based on carbon aerogels with 100% capacitance retention up to 65,000 cycles. *Proceedings of the National Academy of Sciences of the United States of America, 118*(21), e2105610118. DOI: 10.1073/pnas.2105610118 PMID: 34011610

Ma, Y., Sung, K.-W., & Ahn, H.-J. (2023). N- and F-Co-Doped Carbon QuaCo-dopeds Coated on a Ni Foam Substrate as Current Collector for Highly Stable Li-Air Batteries. *International Journal of Energy Research, 2023*, 1–11. DOI: 10.1155/2023/5310171

Mehta, J., Dilbaghi, N., Singhal, N. K., Marrazza, G., Kaushik, A., & Kumar, S. (2023). Electrochemiluminescent quantum dots as emerging next generation sensing platforms. *Chemical Engineering Journal, 477*, 146958. DOI: 10.1016/j.cej.2023.146958

Meng, L., Xu, Q., Zhang, J., & Wang, X. (2024). Colloidal quantum dot materials for next-generation near-infrared optoelectronics. *Chemical Communications*, *60*(9), 1072–1087. DOI: 10.1039/D3CC04315K PMID: 38174780

Miao, S., Liang, K., Zhu, J., Yang, B., Zhao, D., & Kong, B. (2020). Hetero-atom-doped carbon dots: Doping strategies, properties and applications. *Nano Today*, *33*, 100879. DOI: 10.1016/j.nantod.2020.100879

Mimona, M. A., Rimon, M. I. H., Zohura, F. T., Sony, J. M., Rim, S. I., Arup, M. M. R., & Mobarak, M. H. (2025). Quantum dot nanomaterials: Empowering advances in optoelectronic devices. *Chemical Engineering Journal Advances*, *21*, 100704. DOI: 10.1016/j.ceja.2025.100704

Mindivan, F., & Göktaş, M. (2023). The green synthesis of carbon quantum dots (CQDs) and characterization of polycaprolactone (PCL/CQDs) films. *Colloids and Surfaces. A, Physicochemical and Engineering Aspects*, *677*, 132446. Advance online publication. DOI: 10.1016/j.colsurfa.2023.132446

Ming, F., Hou, J., Hou, C., Yang, M., Wang, X., Li, J., Huo, D., & He, Q. (2019). One-step synthesized fluorescent nitrogen doped carbon dots from thymidine for Cr (VI) detection in water. *Spectrochimica Acta. Part A: Molecular and Biomolecular Spectroscopy*, *222*, 117165. DOI: 10.1016/j.saa.2019.117165 PMID: 31185440

Mintz, K. J., Bartoli, M., Rovere, M., Zhou, Y., Hettiarachchi, S. D., Paudyal, S., Chen, J., Domena, J. B., Liyanage, P. Y., Sampson, R., Khadka, D., Pandey, R. R., Huang, S., Chusuei, C. C., Tagliaferro, A., & Leblanc, R. M. (2021). A deep investigation into the structure of carbon dots. *Carbon*, *173*, 433–447. DOI: 10.1016/j.carbon.2020.11.017

Mintz, K. J., Guerrero, B., & Leblanc, R. M. (2018). Photoinduced Electron Transfer in Carbon Dots with Long-Wavelength Photoluminescence. *The Journal of Physical Chemistry. C, Nanomaterials and Interfaces*, *122*(51), 29507–29515. DOI: 10.1021/acs.jpcc.8b06868

Min, Y., Qin, B., Si, C., Sun, X., & Li, B. (2024). Electrochemical reactions catalyzed by carbon dots from computational investigations: Functional groups, dopants, and defects. *Journal of Materials Chemistry. A, Materials for Energy and Sustainability*, *12*(5), 2520–2560. DOI: 10.1039/D3TA06361E

Mirseyed, P. S., Arjmand, S., Rahmandoust, M., Kheirabadi, S., & Anbarteh, R. (2024). Green synthesis of yeast cell wall-derived carbon quantum dots with multiple biological activities. *Heliyon*, *10*(9), e24827. DOI: 10.1016/j.heliyon.2024.e29440 PMID: 38699041

Mirtchev, P., Henderson, E. J., Soheilnia, N., Yip, C. M., & Ozin, G. A. (2012). Solution phase synthesis of carbon quantum dots as sensitizers for nanocrystalline TiO 2 solar cells. *Journal of Materials Chemistry, 22*(4), 1265–1269. DOI: 10.1039/C1JM14112K

Miruschenko, M. D., Kosolapova, K. D., Aleinik, I. A., Borodina, L. N., Vedernikova, A. A., Sokolova, A. V., Sandzhieva, M. A., Mitroshin, A. M., Yakimansky, A. V., Koroleva, A. V., Zhizhin, E. V., Cherevkov, S. A., Langer, M., Otyepka, M., Ushakova, E. V., & Rogach, A. L. (2025). Functionalization of Hydrophilic and Amphiphilic Carbon Dots with Polyethylene Glycol for Electroluminescent Devices. *Small Structures, 6*(3), 2400528. DOI: 10.1002/sstr.202400528

Mmelesi, O. K., Mguni, L. L., Li, F., Nkosi, B., & Liu, X. (2024). Recent development in fluorescent carbon quantum dots-based photocatalysts for water and energy applications. *Materials Science in Semiconductor Processing, 181*, 108661. DOI: 10.1016/j.mssp.2024.108661

Modi, P. D., Mehta, V. N., Prajapati, V. S., Patel, S., & Rohit, J. V. (2023). Bottom-up approaches for the preparation of carbon dots. In Kailasa, S. K., & Hussain, C. M. (Eds.), *Carbon dots in analytical chemistry* (pp. 15–29). Elsevier., DOI: 10.1016/B978-0-323-98350-1.00022-0

Moghimi, M., Akpa, P. A., Peter, I. E., Onwuka, A. M., Obi, B. C., Akunne, M. O., Nworu, C. S., Ejikeme, P. M., Akunne, T. C., Attama, A. A., Akah, P. A., & Ng, A. A. (2023). Nanotheranostics: Platforms, Current Applications, and Mechanisms of Targeting in Breast and Prostate Cancers. *Journal of Nanotheranostics 2023, Vol. 4, Pages 346-383, 4*(3), 346–383. DOI: 10.3390/jnt4030016

Mohammadpour, , Kazemi, A., Baghapour, M. A., Samaei, M. R., Kannan, K., & Mousavi Khaneghah, A. (2024). Bioengineered FeZn/GA@ Cu nanocomposite utilizing spent coffee ground extract and gum arabic: Enhanced nitrate removal via (RSM) and machine learning optimization. *International Journal of Biological Macromolecules, 277*, 134060. DOI: 10.1016/j.ijbiomac.2024.134060 PMID: 39097464

Molaei, M. J. (2020). Principles, mechanisms, and application of carbon quantum dots in sensors: A review. *Analytical Methods, 12*(10), 1266–1287. DOI: 10.1039/C9AY02696G

Mondal, M., & Pramanik, S. (2023). A mechanism for excitation-dependent emission from carbon nanodots. *Materials Letters: X, 18,* 100195. DOI: 10.1016/j.mlblux.2023.100195

Mondal, S., Rana, U., & Malik, S. (2015). Graphene quantum dot-doped polyaniline nanofiber as high performance supercapacitor electrode materials. *Chemical Communications, 51*(62), 12365–12368. DOI: 10.1039/C5CC03981A PMID: 26095974

Muazim, K., & Hussain, Z. (2017). Graphene oxide — A platform towards theranostics. *Materials Science and Engineering C, 76*, 1274–1288. DOI: 10.1016/j.msec.2017.02.121 PMID: 28482495

Mumtaz, M., Hussain, N., Salam, S., & Bilal, M. (2022). Multifunctional nanodiamonds as emerging platforms for cancer treatment, and targeted delivery of genetic factors and protein medications—a review. *Journal of Materials Science 2022 57:17, 57*(17), 8064–8099. DOI: 10.1007/s10853-022-07168-x

Murakami, Y., Chiashi, S., Miyauchi, Y., Hu, M., Ogura, M., Okubo, T., & Maruyama, S. (2004). Growth of vertically aligned single-walled carbon nanotube films on quartz substrates and their optical anisotropy. *Chemical Physics Letters, 385*(3-4), 298–303. DOI: 10.1016/j.cplett.2003.12.095

Murali, A. S., Sajeevan, B., Kumar, S. M. S., Govindasamy, M., & Saraswathyamma, B. (2024). Facile formation of surface functionalised α-Fe2O3 nanoparticles and their role in developing dual sensors towards cysteamine quantification. *Materials Research Bulletin, 174*, 112734. DOI: 10.1016/j.materresbull.2024.112734

Muthukutty, B., Sathish Kumar, P., Lee, D., & Lee, S. (2024). Multichannel Carbon Nanofibers: Pioneering the Future of Energy Storage. *ACS Nano, 18*(40), 27287–27316. DOI: 10.1021/acsnano.4c11146 PMID: 39324479

Muthusankar, G., Devi, R. K., & Gopu, G. (2020). Nitrogen-doped carbon quantum dots embedded Co3O4 with multiwall carbon nanotubes: An efficient probe for the simultaneous determination of anticancer and antibiotic drugs. *Biosensors & Bioelectronics, 150*, 111947. DOI: 10.1016/j.bios.2019.111947 PMID: 31818763

Muthusankar, G., Rajkumar, C., Chen, S.-M., Karkuzhali, R., Gopu, G., Sangili, A., Sengottuvelan, N., & Sankar, R. (2019). Sonochemical driven simple preparation of nitrogen-doped carbon quantum dots/SnO2 nanocomposite: A novel electrocatalyst for sensitive voltammetric determination of riboflavin. *Sensors and Actuators. B, Chemical, 281*, 602–612. DOI: 10.1016/j.snb.2018.10.145

Muthusankar, G., Sangili, A., Chen, S.-M., Karkuzhali, R., Sethupathi, M., Gopu, G., Karthick, S., Devi, R. K., & Sengottuvelan, N. (2018). In situ assembly of sulfur-doped carbon quantum dots surrounded iron (III) oxide nanocomposite; a novel electrocatalyst for highly sensitive detection of antipsychotic drug olanzapine. *Journal of Molecular Liquids, 268*, 471–480. DOI: 10.1016/j.molliq.2018.07.059

Muthusankar, G., Sasikumar, R., Chen, S.-M., Gopu, G., Sengottuvelan, N., & Rwei, S.-P. (2018). Electrochemical synthesis of nitrogen-doped carbon quantum dots decorated copper oxide for the sensitive and selective detection of non-steroidal anti-inflammatory drug in berries. *Journal of Colloid and Interface Science*, *523*, 191–200. DOI: 10.1016/j.jcis.2018.03.095 PMID: 29621646

Muthusankar, G., Sethupathi, M., Chen, S.-M., Devi, R. K., Vinoth, R., Gopu, G., Anandhan, N., & Sengottuvelan, N. (2019). N-doped carbon quantum dots@ hexagonal porous copper oxide decorated multiwall carbon nanotubes: A hybrid composite material for an efficient ultra-sensitive determination of caffeic acid. *Composites. Part B, Engineering*, *174*, 106973. DOI: 10.1016/j.compositesb.2019.106973

Nagaiah, T. C., Bordoloi, A., Sánchez, M. D., Muhler, M., & Schuhmann, W. (2012). Mesoporous nitrogen-rich carbon materials as catalysts for the oxygen reduction reaction in alkaline solution. *ChemSusChem*, *5*(4), 637–641. DOI: 10.1002/cssc.201100284 PMID: 22431491

Nahirniak, S., Ray, A., & Saruhan, B. (2023). Challenges and Future Prospects of the MXene-Based Materials for Energy Storage Applications. *Batteries*, *9*(2), 126. DOI: 10.3390/batteries9020126

Nandy, A., Duan, C., Taylor, M. G., Liu, F., Steeves, A. H., & Kulik, H. J. (2021). Computational discovery of transition-metal complexes: From high-throughput screening to machine learning. *Chemical Reviews*, *121*(16), 9927–10000. DOI: 10.1021/acs.chemrev.1c00347 PMID: 34260198

Nasrin, A., Hassan, M., Mann, G., & Gomes, V. G. (2020). Conjugated ternary doped carbon dots from vitamin B derivative: Multispectral nanoprobes for targeted melanoma bioimaging and photosensitization. *Journal of Luminescence*, *217*, 116811. DOI: 10.1016/j.jlumin.2019.116811

Naushad, M., Ahamad, T., Ubaidullah, M., Ahmed, J., Ghafar, A. A., Al-Sheetan, K. M., & Arunachalam, P. (2021). Nitrogen-doped carbon quantum dots (N-CQDs)/Co3O4 nanocomposite for high performance supercapacitor. *Journal of King Saud University. Science*, *33*(1), 101252. DOI: 10.1016/j.jksus.2020.101252

Nazar, , Hasan, M., Wirjosentono, B., Gani, B. A., & Nada, C. E. (2024). Microwave Synthesis of Carbon Quantum Dots from Arabica Coffee Ground for Fluorescence Detection of Fe^{3+}, Pb^{2+}, and Cr^{3+}. *ACS Omega*, *9*(18), 20571–20581. DOI: 10.1021/acsomega.4c02254 PMID: 38737012

Ngafwan, N., Rasyid, H., Salaam Abood, E., Kamal Abdelbasset, W., Al-Shawi, S. G., Bokov, D., & Jalil, A. T. (2022). Study on novel fluorescent carbon nanomaterials in food analysis. *Food Science and Technology (Brazil), 42*. https://doi.org/ DOI: 10.1590/FST.37821

Nguyen, D. N., Roh, S. H., Kim, D. H., Lee, J. Y., Wang, D. H., & Kim, J. K. (2021). Molecular manipulation of PEDOT: PSS for efficient hole transport by incorporation of N-doped carbon quantum dots. *Dyes and Pigments*, *194*, 109610. DOI: 10.1016/j.dyepig.2021.109610

Nguyen, K. G., Huš, M., Baragau, I., Bowen, J., Heil, T., Nicolaev, A., Abramiuc, L. E., Sapelkin, A., Sajjad, M. T., & Kellici, S. (2024). Engineering Nitrogen-Doped Carbon Quantum Dots: Tailoring Optical and Chemical Properties through Selection of Nitrogen Precursors. *Small*, *20*(24), 2310587. DOI: 10.1002/smll.202310587 PMID: 38546418

Niu, L., Yang, L., Yang, J., Chen, M., Zeng, L., Duan, P., Wu, T., Pameté, E., Presser, V., & Feng, G. (2023). Understanding the charging of supercapacitors by electrochemical quartz crystal microbalance. *Industrial Chemistry & Materials*, *1*(2), 175–187. DOI: 10.1039/D2IM00038E

Niveria, K., Singh, P., Yadav, M., & Verma, A. K. (2023). Quantum Dot (QD)-Induced Toxicity and Biocompatibility. In Korotcenkov, G. (Ed.), *Handbook of II-VI Semiconductor-Based Sensors and Radiation Detectors*. Springer., DOI: 10.1007/978-3-031-19531-0_8

Orecchioni, M., Cabizza, R., Bianco, A., & Delogu, L. G. (2015). Graphene as Cancer Theranostic Tool: Progress and Future Challenges. *Theranostics*, *5*(7), 710–723. DOI: 10.7150/thno.11387 PMID: 25897336

Ozyurt, D., Al Kobaisi, M., Hocking, R. K., & Fox, B. (2023). Properties, synthesis, and applications of carbon dots: A review. *Carbon Trends*, *12*, 100276. DOI: 10.1016/j.cartre.2023.100276

Padmapriya, A., Thiyagarajan, P., Devendiran, M., Kalaivani, R. A., & Shanmugharaj, A. M. (2023). Electrochemical sensor based on N,P–doped carbon quantum dots derived from the banana flower bract (Musa acuminata) biomass extract for selective and picomolar detection of dopamine. *Journal of Electroanalytical Chemistry (Lausanne, Switzerland)*, *943*, 117609. https://doi.org/https://doi.org/10.1016/j .jelechem.2023.117609. DOI: 10.1016/j.jelechem.2023.117609

Pajewska-Szmyt, , Buszewski, B., & Gadzała-Kopciuch, R. (2020). Sulphur and nitrogen doped carbon dots synthesis by microwave assisted method as quantitative analytical nano-tool for mercury ion sensing. *Materials Chemistry and Physics*, *242*, 122484. Advance online publication. DOI: 10.1016/j.matchemphys.2019.122484

Pakapongpan, S., Poo-arporn, Y., Ninket, S., & Poo-arporn, R. P. (2024). A disposable electrochemical sensor for amyloid-β42 protein based on molecular imprinted polymers with nitrogen doped carbon dots-graphene nanohybrid. *Microchemical Journal*, *206*, 111559. DOI: 10.1016/j.microc.2024.111559

Pal, A., Ahmad, K., Dutta, D., & Chattopadhyay, A. (2019). Boron doped carbon dots with unusually high photoluminescence quantum yield for ratiometric intracellular pH sensing. *ChemPhysChem*, *20*(8), 1018–1027. DOI: 10.1002/cphc.201900140 PMID: 30891892

Pallavolu, M. R., Prabhu, S., Nallapureddy, R. R., Kumar, A. S., Banerjee, A. N., & Joo, S. W. (2023). Bio-derived graphitic carbon quantum dot encapsulated S- and N-doped graphene sheets with unusual battery-type behavior for high-performance supercapacitor. *Carbon*, *202*, 93–102. DOI: 10.1016/j.carbon.2022.10.077

Pal, M., Pal, P., & Nandi, M. (2025). B/n-codoped porous carbco-dopedmetal-free and binder-free electrode material for supercapacitors. *Chemistry (Weinheim an der Bergstrasse, Germany)*, *31*(34), e202500800. Advance online publication. DOI: 10.1002/chem.202500800 PMID: 40351145

Pan, D., Zhang, J., Li, Z., Wu, C., Yan, X., & Wu, M. (2010). Observation of pH, solvent, spin and excitation dependent blue photoluminescence from carbon nanoparticles. *Chemical Communications*, *46*(21), 3681–3683. DOI: 10.1039/c000114g PMID: 20396809

Pandit, S., Behera, P., Sahoo, J., & De, M. (2019). In situ synthesis of amino acid functionalized carbon dots with tunable properties and their biological applications. *ACS Applied Bio Materials*, *2*(8), 3393–3403. DOI: 10.1021/acsabm.9b00374 PMID: 35030781

Pandiyan, S., Arumugam, L., Srirengan, S. P., Pitchan, R., Sevugan, P., Kannan, K., Pitchan, G., Hegde, T. A., & Gandhirajan, V. (2020). Biocompatible carbon quantum dots derived from sugarcane industrial wastes for effective nonlinear optical behavior and antimicrobial activity applications. *ACS Omega*, *5*(47), 30363–30372. DOI: 10.1021/acsomega.0c03290 PMID: 33283084

Panigrahi, S. (2025). *Carbon Quantum Dots Prepared by Hydrothermal Method for Drug Delivery, Photocatalysis, and Bioimaging Applications*. Chemistry. DOI: 10.26434/chemrxiv-2025-w4117

Pankaj, A., Tewari, K., Singh, S., & Singh, S. P. (2018). Waste candle soot derived nitrogen doped carbon dots based fluorescent sensor probe: An efficient and inexpensive route to determine Hg (II) and Fe (III) from water. *Journal of Environmental Chemical Engineering*, *6*(4), 5561–5569. DOI: 10.1016/j.jece.2018.08.059

Pant, M., Singh, R., Negi, P., Tiwari, K., & Singh, Y. (2021). A comprehensive review on carbon nano-tube synthesis using chemical vapor deposition. *Materials Today: Proceedings*, *46*, 11250–11253. DOI: 10.1016/j.matpr.2021.02.646

Pan, Z., Brett, D. J. L., He, G., & Parkin, I. P. (2022). Progress and Perspectives of Organosulfur for Lithium–Sulfur Batteries. *Advanced Energy Materials*, *12*(8), 2103483. Advance online publication. DOI: 10.1002/aenm.202103483

Pan, Z., Yang, J., Kong, J., Loh, X. J., Wang, J., & Liu, Z. (2022). "Porous and Yet Dense" Electrodes for High-Volumetric-Performance Electrochemical Capacitors: Principles, Advances, and Challenges. *Advancement of Science*, *9*(4), 2103953. DOI: 10.1002/advs.202103953 PMID: 34796698

Papi, R. M., Tasioulis, K. S., Kechagioglou, P. V., Papaioannou, M. A., Andriotis, E. G., & Kyriakidis, D. A. (2024). Carbon Nanotube-Mediated Delivery of PTEN Variants: In Vitro Antitumor Activity in Breast Cancer Cells. *Molecules (Basel, Switzerland)*, *29*(12), 2785. DOI: 10.3390/molecules29122785 PMID: 38930850

Parashar, A. K., Saraogi, G. K., Jain, P. K., Kurmi, B., Shrivastava, V., & Arora, V. (2024). Polymer-drug conjugates: Revolutionizing nanotheranostic agents for diagnosis and therapy. *Discover. Oncology*, *15*(1), 641. DOI: 10.1007/s12672-024-01509-9 PMID: 39527173

Park, S. K., Lee, H., Choi, M. S., Suh, D. H., Nakhanivej, P., & Park, H. S. (2018). Straightforward and controllable synthesis of heteroatom-doped carbon dots and nanoporous carbons for surface-confined energy and chemical storage. *Energy Storage Materials*, *12*, 331–340. DOI: 10.1016/j.ensm.2017.10.008

Parra, , Silva, K., Valezin, P., Martins, R., Gomes, R., Pereira, R., de Melo, F., Morandim-Giannetti, A., dos Santos, R., Panzarini, L. C., & Toyama, M. (2020). Preparation of fluorescent carbon-based dots from waste tire pyrolysis. *Journal of the Brazilian Chemical Society*, *31*(11), 2224–2231. DOI: 10.21577/0103-5053.20200072

Patil, S., Mishra, V. S., Yadav, N., Reddy, P. C., & Lochab, B. (2022). Dendrimer-Functionalized Nanodiamonds as Safe and Efficient Drug Carriers for Cancer Therapy: Nucleus Penetrating Nanoparticles. *ACS Applied Bio Materials*, *5*(7), 3438–3451. DOI: 10.1021/acsabm.2c00373 PMID: 35754387

Patra, R., Panda, P. K., Hung, C.-Y., Dash, P., Yang, P.-C., & Hsieh, C.-T. (2024). Facile synthesis of yellow emissive heteroatom doped carbon quantum dots fluorescent probe with superior quantum yield for fast-track ionic detection in aqueous media. *Materials Research Bulletin*, *180*, 113040. DOI: 10.1016/j.materresbull.2024.113040

Pavia, D. L., Lampman, G. M., Kriz, G. S., & Vyvyan, J. R. (2015). Introduction to Spectrometry. *Cengage Learning*, 1–690. https://www.researchgate.net/publication/44372742_Introduction_to_spectroscopy_Donald_L_Pavia_Gary_M_Lampman_George_S_Kriz_and_James_R_Vyvyan

Peng, H., & Sejdic, J. T. (2009). Simple aqueous solution route to luminescent carbogenic dots from carbohydrates. *Chemistry of Materials*, *21*(23), 5563–5565. DOI: 10.1021/cm901593y

Peng, J., Gao, W., Gupta, B. K., Liu, Z., Romero-Aburto, R., Ge, L., Song, L., Alemany, L. B., Zhan, X., Gao, G., Vithayathil, S. A., Kaipparettu, B. A., Marti, A. A., Hayashi, T., Zhu, J.-J., & Ajayan, P. M. (2012). Graphene Quantum Dots Derived from Carbon Fibers. *Nano Letters*, *12*(2), 844–849. DOI: 10.1021/nl2038979 PMID: 22216895

Peng, Z., Ji, C., Zhou, Y., Zhao, T., & Leblanc, R. M. (2020). Polyethylene glycol (PEG) derived carbon dots: Preparation and applications. *Applied Materials Today*, *20*, 100677. DOI: 10.1016/j.apmt.2020.100677

Permatasari, F. A., Irham, M. A., Bisri, S. Z., & Iskandar, F. (2021). Carbon-Based Quantum Dots for Supercapacitors: Recent Advances and Future Challenges. *Nanomaterials (Basel, Switzerland)*, *11*(1), 91. DOI: 10.3390/nano11010091 PMID: 33401630

Phang, S. J., Lee, J., Wong, V.-L., Tan, L.-L., & Chai, S.-P. (2022). Synergistic effects of the hybridization between boron-doped carbon quantum dots and n/n-type g-C3N4 homojunction for boosted visible-light photocatalytic activity. *Environmental Science and Pollution Research International*, *29*(27), 41272–41292. DOI: 10.1007/s11356-021-18253-0 PMID: 35088270

Phang, S. J., & Tan, L. L. (2019). Recent advances in carbon quantum dot (CQD)-based two dimensional materials for photocatalytic applications. *Catalysis Science & Technology*, *9*(21), 5882–5905. DOI: 10.1039/C9CY01452G

Phogat, P., Dey, S., & Wan, M. (2025). Powering the sustainable future: A review of emerging battery technologies and their environmental impact. *RSC Sustainability*, *3*(8), 3266–3306. DOI: 10.1039/D5SU00127G

Pholauyphon, W., Bulakhe, R. N., Manyam, J., In, I., & Paoprasert, P. (2022). High-performance supercapacitors using carbon dots/titanium dioxide composite electrodes and carbon dot-added sulfuric acid electrolyte. *Journal of Electroanalytical Chemistry (Lausanne, Switzerland)*, *910*, 116177. DOI: 10.1016/j.jelechem.2022.116177

Pichot, V., Risse, B., Schnell, F., Mory, J., & Spitzer, D. (2013). Understanding ultrafine nanodiamond formation using nanostructured explosives. *Scientific Reports*, *3*(1), 1–6. DOI: 10.1038/srep02159 PMID: 23831716

Pillar-Little, T. J.Jr. (2018). Carbon Quantum Dots: Bridging the Gap Between Chemical Structure and Material Properties [University of Kentucky Libraries; PDF]. DOI: 10.13023/ETD.2018.108

Pillar-Little, T. J., Wanninayake, N., Nease, L., Heidary, D. K., Glazer, E. C., & Kim, D. Y. (2018). Superior photodynamic effect of carbon quantum dots through both type I and type II pathways: Detailed comparison study of top-down-synthesized and bottom-up-synthesized carbon quantum dots. *Carbon*, *140*, 616–623. https://doi.org/https://doi.org/10.1016/j.carbon.2018.09.004. DOI: 10.1016/j.carbon.2018.09.004

Poh, H. L., & Pumera, M. (2015). p-Element-Doped Graphene: Heteroatoms for Electrochemical Enhancement. *ChemElectroChem*, *2*(2), 190–199. https://doi.org/https://doi.org/10.1002/celc.201402307. DOI: 10.1002/celc.201402307

Pollice, R., Gomes, G. P., Aldeghi, M., Hickman, R. J., Krenn, M., Lavigne, C., Lindner-D'Addario, M., Nigam, A., Ser, C. T., Yao, Z., & Aspuru-Guzik, A. (2021). Data-driven strategies for accelerated materials design. *Accounts of Chemical Research*, *54*(4), 849–860. DOI: 10.1021/acs.accounts.0c00785 PMID: 33528245

Poolakkandy, R. R., & Menamparambath, M. M. (2020). Soft-template-assisted synthesis: A promising approach for the fabrication of transition metal oxides. *Nanoscale Advances*, *2*(11), 5015–5045. DOI: 10.1039/D0NA00599A PMID: 36132034

Porada, S., Schipper, F., Aslan, M., Antonietti, M., Presser, V., & Fellinger, T. (2015). Capacitive deionization using biomass-based microporous salt-templated heteroatom-doped carbons. *ChemSusChem*, *8*(11), 1867–1874. DOI: 10.1002/cssc.201500166 PMID: 25970654

Prabakaran, P., Arumugam, G., Ramu, P., Selvaraj, M., Assiri, M. A., Rokhum, S. L., Periyasamy, S., Arjunan, S., & Rajendran, R. (2023). Preparation and characterization of carbon quantum dots grafted Co_3O_4 nanocomposite for supercapacitors application. *Surfaces and Interfaces*, *40*, 103153. DOI: 10.1016/j.surfin.2023.103153

Pradhan, A. K., Halder, S., & Chakraborty, C. (2025). Pt-nanoparticles on ZnO/carbon quantum dots: A trifunctional nanocomposite with superior electrocatalytic activity boosting direct methanol fuel cells and zinc–air batteries. J Mater. *Journal of Materials Chemistry. A, Materials for Energy and Sustainability*, *13*(1), 243–256. DOI: 10.1039/D4TA05630B

Prasath, A., Athika, M., Duraisamy, E., Sharma, A. S., & Elumalai, P. (2018). Carbon-Quantum-Dot-Derived Nanostructured MnO_2 and Its Symmetrical Supercapacitor Performances. *ChemistrySelect*, *3*(30), 8713–8723. DOI: 10.1002/slct.201801950

Preethi, M., Viswanathan, C., & Ponpandian, N. (2022). A metal-free, dual catalyst for the removal of Rhodamine B using novel carbon quantum dots from muskmelon peel under sunlight and ultrasonication: A green way to clean the environment. *Journal of Photochemistry and Photobiology A Chemistry*, *426*, 113765. DOI: 10.1016/j.jphotochem.2021.113765

Prekodravac, , Vasiljević, B., Marković, Z., Jovanović, D., Kleut, D., Špitalský, Z., Mičušik, M., Danko, M., Bajuk-Bogdanović, D., & Todorović-Marković, B. (2019). Green and facile microwave assisted synthesis of (metal-free) N-doped carbon quantum dots for catalytic applications. *Ceramics International*, *45*(14), 17006–17013. DOI: 10.1016/j.ceramint.2019.05.250

Priya, A. S., Premanand, R., Ragupathi, I., Bhaviripudi, V. R., Aepuru, R., Kannan, K., & Shanmugaraj, K. (2024). Comprehensive review of hydrogel synthesis, characterization, and emerging applications. *Journal of Composites Science*, *8*(11), 457. DOI: 10.3390/jcs8110457

Priyadarshni, N., Singh, R., & Mishra, M. K. (2024). Nanodiamonds: Next generation nano-theranostics for cancer therapy. *Cancer Letters*, *587*, 216710. DOI: 10.1016/j.canlet.2024.216710 PMID: 38369006

Puchaicela, M., Lara, D., Cevallos, V. J., Garzón, A., Gardener, J., Solorzano, G., Spencer, L., & Chimborazo, J. (2025). Synthesis and physical characterization of carbon quantum dots from watermelon seed towards a biological application. *Carbon Trends*, *20*, 100493. DOI: 10.1016/j.cartre.2025.100493

Pundi, A., & Chang, C.-J. (2022). Recent advances in synthesis, modification, characterization, and applications of carbon dots. *Polymers*, *14*(11), 2153. DOI: 10.3390/polym14112153 PMID: 35683827

Qian, Z., Ma, J., Shan, X., Shao, L., Zhou, J., Chen, J., & Feng, H. (2013). Surface functionalization of graphene quantum dots with small organic molecules from photoluminescence modulation to bioimaging applications: An experimental and theoretical investigation. *RSC Advances*, *3*(34), 14571–14579. DOI: 10.1039/c3ra42066c

Qian, Z., Shan, X., Chai, L., Ma, J., Chen, J., & Feng, H. (2014). Si-Doped Carbon Quantum Dots: A Facile and General Preparation Strategy, Bioimaging Application, and Multifunctional Sensor. *ACS Applied Materials & Interfaces*, *6*(8), 6797–6805. DOI: 10.1021/am500403n PMID: 24707855

Qi, L.-Y., Zhang, Y.-W., Zuo, Z.-C., Xin, Y.-L., Yang, C.-K., Wu, B., Zhang, X.-X., & Zhou, H.-H. (2016). *In situ* quantization of ferroferric oxide embedded in 3D microcarbon for ultrahigh performance sodium-ion batteries. J Mater. *Journal of Materials Chemistry. A, Materials for Energy and Sustainability*, *4*(22), 8822–8829. DOI: 10.1039/C6TA01836J

Qin, N., Tong, B., Ling, X., Shi, J., Wei, W., & Mi, L. (2023). Fabrication of nitrogen-doped carbon-coated nis1.97 quantum dots for advanced magnesium-ion batteries. *Langmuir*, *39*(45), 16111–16117. DOI: 10.1021/acs.langmuir.3c02250 PMID: 37924327

Quan, Y., Wang, G., Lu, L., Wang, Z., Xu, H., Liu, S., & Wang, D. (2020). High-performance pseudo-capacitor energy storage device based on a hollow-structured copper sulfide nanoflower and carbon quantum dot nanocomposite. *Electrochimica Acta*, *353*, 136606. DOI: 10.1016/j.electacta.2020.136606

Quarti, M., Bayer, A., & Bessler, W. G. (2023). Trade-off between energy density and fast-charge capability of lithium-ion batteries: A model-based design study of cells with thick electrodes. *Electrochemical Science Advances*, *3*(1), e2100161. DOI: 10.1002/elsa.202100161

Quesada-González, D., & Merkoçi, A. (2025). Quantum dots for biosensing: Classification and applications. *Biosensors & Bioelectronics*, *273*, 117180. DOI: 10.1016/j.bios.2025.117180 PMID: 39864309

Rafiq, K., Sadia, I., Abid, M. Z., Waleed, M. Z., Rauf, A., & Hussain, E. (2024). Scientific Insights into the Quantum Dots (QDs)-Based Electrochemical Sensors for State-of-the-Art Applications. *ACS Biomaterials Science & Engineering*, *10*(12), 7268–7313. DOI: 10.1021/acsbiomaterials.4c01256 PMID: 39499739

Rajakumar, G., Zhang, X. H., Gomathi, T., Wang, S. F., Ansari, M. A., Mydhili, G., Nirmala, G., Alzohairy, M. A., & Chung, I. M. (2020). Current Use of Carbon-Based Materials for Biomedical Applications—A Prospective and Review. *Processes 2020, Vol. 8, Page 355*, *8*(3), 355. DOI: 10.3390/pr8030355

Ramezani, , Qorbanpour, M., & Rahbar, N. (2018). Green synthesis of carbon quantum dots using quince fruit (Cydonia oblonga) powder as carbon precursor: Application in cell imaging and As^{3+} determination. *Colloids and Surfaces. A, Physicochemical and Engineering Aspects*, *549*, 58–66. DOI: 10.1016/j.colsurfa.2018.04.006

Ramírez Sanabria, A. E., Narváez, S. U., Gutiérrez Portilla, J. V., & Torres Rodríguez, G. A. (2025). Sonochemical synthesis of CQDs from coffee husk: Insights in aggregation mechanism, optimization and sustainability analysis. *Heliyon, 11*(1), e41000. DOI: 10.1016/j.heliyon.2024.e41000 PMID: 39807508

Ran, X., Qu, Q., Qian, X., Xie, W., Li, S., Li, L., & Yang, L. (2018). Water-soluble pillar [6] arene functionalized nitrogen-doped carbon quantum dots with excellent supramolecular recognition capability and superior electrochemical sensing performance towards TNT. *Sensors and Actuators. B, Chemical, 257*, 362–371. DOI: 10.1016/j.snb.2017.10.185

Rao, K. K., & Prakash, B. (2025). Analyzing Functionalized Carbon Quantum Dots in Electrode Engineering for the Precise Electrochemical Detection of a Neurotransmitter: Achieving Laboratory Accuracy with Portable Sensors. *ACS Sensors, 10*(5), 3423–3432. DOI: 10.1021/acssensors.4c03298 PMID: 40261822

Rao, , Yuan, M., Jiang, H., Li, L., & Liu, Z. (2019). A universal strategy to obtain chiroptical carbon quantum dots through the optically active surface passivation procedure. *New Journal of Chemistry, 43*(35), 13735–13740. DOI: 10.1039/C9NJ03434J

Rasal, A. S., Yadav, S., Yadav, A., Kashale, A. A., Manjunatha, S. T., Altaee, A., & Chang, J. Y. (2021). Carbon quantum dots for energy applications: A review. *ACS Applied Nano Materials, 4*(7), 6515–6541. DOI: 10.1021/acsanm.1c01372

RasulKhan, B. B., Ponnaiah, S. K., Periakaruppan, P., Venkatachalam, G., & Balasubramanian, J.RasulKhan. (2020). A new CQDs/f-MWCNTs/GO nanocomposite electrode for arsenic (10^{-12} M) quantification in bore-well water and industrial effluents. *New Journal of Chemistry, 44*(42), 18149–18156. DOI: 10.1039/D0NJ04252H

Raza, W., Ali, F., Raza, N., Luo, Y., Kim, K.-H., Yang, J., Kumar, S., Mehmood, A., & Kwon, E. E. (2018). Recent advancements in supercapacitor technology. *Nano Energy, 52*, 441–473. DOI: 10.1016/j.nanoen.2018.08.013

Redondo-Fernandez, G., Cigales Canga, J., Soldado, A., Ruiz Encinar, J., & Costa-Fernandez, J. M. (2023). Functionalized heteroatom-doped carbon dots for biomedical applications: A review. *Analytica Chimica Acta, 1284*, 341874. DOI: 10.1016/j.aca.2023.341874 PMID: 37996151

Reifenberger, R. (2016). *Fundamentals of atomic force microscopy: part I: foundations*. https://doi.org/DOI: 10.1557/MRS

Ren, H., Yuan, Y., Labidi, A., Dong, Q., Zhang, K., Lichtfouse, E., & Wang, C. (2024). Biological imaging in vitro and in vivo. *Chinese Chemical Letters, 34*(6), 107998. DOI: 10.1016/j.cclet.2022.107998

Ren, J., Malfatti, L., & Innocenzi, P. (2021). Citric acid derived carbon dots, the challenge of understanding the synthesis-structure relationship. *Journal of Carbon Research*, *7*(2), 2. Advance online publication. DOI: 10.3390/c7010002

Ren, , Zhang, F., Guo, B., Gao, N., & Zhang, X. (2019). Synthesis of N-doped Micropore carbon quantum dots with high quantum yield and dual-wavelength photoluminescence emission from biomass for cellular imaging. *Nanomaterials (Basel, Switzerland)*, *9*(4), 495. Advance online publication. DOI: 10.3390/nano9040495 PMID: 30939724

Ren, , Zhang, Z., Zhao, P., Shi, J., Han, K., Yang, Z., Gao, D., & Bi, F. (2019). Facile and one-step preparation carbon quantum dots from biomass residue and their applications as efficient surfactants. *Journal of Dispersion Science and Technology*, *40*(5), 627–633. DOI: 10.1080/01932691.2018.1475239

Rhili, K., Chergui, S., Abergo-Martinez, J. C., El Douhaibi, A. S., & Siaj, M. (2023). Phosphorus and nitrogen co-doped reduced graphene oxide as superior electrode nanomaterials for supercapacitors. *Materials Advances*, *4*(21), 5263–5272. DOI: 10.1039/D3MA00553D

Rimal, V., Shishodia, S., & Srivastava, P. K. (2020). Novel synthesis of high-thermal stability carbon dots and nanocomposites from oleic acid as an organic substrate. *Applied Nanoscience*, *10*(2), 455–464. DOI: 10.1007/s13204-019-01178-z

Rocco, D., Moldoveanu, V. G., Feroci, M., Bortolami, M., & Vetica, F. (2023). Electrochemical Synthesis of Carbon Quantum Dots. *ChemElectroChem*, *10*(3), e202201104. DOI: 10.1002/celc.202201104 PMID: 37502311

Rohith, R., Thejas Prasannakumar, A., V., M., Thomas, M., R.Mohan, R., J.Varma, S., 2024. An insight into the electrochemical performance of nanostructured V_2O_5 in aqueous neutral electrolytes and fabrication of 2V, high energy density, symmetric supercapacitor. Electrochim Acta 503, 144911. DOI: 10.1016/j.electacta.2024.144911

Romero, M. P., Buzza, H. H., Stringasci, M. D., Estevão, B. M., Silva, C. C., Pereira-Da-Silva, M. A., Inada, N. M., & Bagnato, V. S. (2021). *Graphene Oxide Theranostic Effect: Conjugation of Photothermal and Photodynamic Therapies Based on an in vivo Demonstration*. DOI: 10.2147/IJN.S287415

Rong, C., Huang, Y., & Zheng, X. (2025). Sensitive determination of hydroquinone and catechol using an electrochemical sensor based on nitrogen-doped malic acid carbon quantum dots. *Materials Chemistry and Physics*, *329*, 130077. DOI: 10.1016/j.matchemphys.2024.130077

Rong, M. C., Zhang, K. X., Wang, Y. R., & Chen, X. (2017). The synthesis of B, N-carbon dots by a combustion method and the application of fluorescence detection for Cu^{2+}. *Chinese Chemical Letters*, *28*(7), 1119–1124. DOI: 10.1016/j.cclet.2016.12.009

Roshni, V., & Divya, O. (2017). One-step microwave-assisted green synthesis of luminescent N-doped carbon dots from sesame seeds for selective sensing of Fe (III). *Current Science*, *112*(2), 385–390. DOI: 10.18520/cs/v112/i02/385-390

Roszkowski, S., Durczyńska, Z., & Szablewska, S. (2025). Targeted nanodelivery systems for personalized cancer therapy. *Reports of Practical Oncology and Radiotherapy : Journal of Greatpoland Cancer Center in Poznan and Polish Society of Radiation Oncology*, *29*(6), 776–788. DOI: 10.5603/rpor.103524 PMID: 40104662

Rüther, T., Bhatt, A. I., Best, A. S., Harris, K. R., & Hollenkamp, A. F. (2020). Electrolytes for Lithium (Sodium) Batteries Based on Ionic Liquids: Highlighting the Key Role Played by the Anion. *Batteries & Supercaps*, *3*(9), 793–827. DOI: 10.1002/batt.202000022

Ru, Y., Ai, L., Jia, T., Liu, X., Lu, S., Tang, Z., & Yang, B. (2020). Recent advances in chiral carbonized polymer dots: From synthesis and properties to applications. *Nano Today*, *34*, 100953. DOI: 10.1016/j.nantod.2020.100953

Ru, Y., Waterhouse, G. I. N., & Lu, S. (2022). Aggregation in carbon dots. *Aggregate*, *3*(6), e296. DOI: 10.1002/agt2.296

Saari, J., Martinez, C. M., Kuparinen, K., Lipiäinen, S., Hamaguchi, M., Cardoso, M., Vakkilainen, E., & Haider, M. (2023). Techno-economic analysis of hydrothermal carbonization of pulp mill biosludge. *Tappi Journal*, *22*(3), 206–216. DOI: 10.32964/TJ22.3.206

Sabet, M., & Mahdavi, K. (2019). Green synthesis of high photoluminescence nitrogen-doped carbon quantum dots from grass via a simple hydrothermal method for removing organic and inorganic water pollutions. *Applied Surface Science*, *463*, 283–291. DOI: 10.1016/j.apsusc.2018.08.223

Safari, M., Moghaddam, A., Salehi Moghaddam, A., Absalan, M., Kruppke, B., Ruckdäschel, H., & Khonakdar, H. A. (2023). Carbon-based biosensors from graphene family to carbon dots: A viewpoint in cancer detection. *Talanta*, *258*, 124399. Advance online publication. DOI: 10.1016/j.talanta.2023.124399 PMID: 36870153

Sahu, S., Behera, B., Maiti, T. K., & Mohapatra, S. (2012). Simple one-step synthesis of highly luminescent carbon dots from orange juice: Application as excellent bio-imaging agents. *Chemical Communications*, *48*(70), 8835–8837. DOI: 10.1039/c2cc33796g PMID: 22836910

Sainio, S., Wester, N., Aarva, A., Titus, C. J., Nordlund, D., Kauppinen, E. I., Leppänen, E., Palomäki, T., Koehne, J. E., Pitkänen, O., Kordas, K., Kim, M., Lipsanen, H., Mozetič, M., Caro, M. A., Meyyappan, M., Koskinen, J., & Laurila, T. (2021). Trends in Carbon, Oxygen, and Nitrogen Core in the X-ray Absorption Spectroscopy of Carbon Nanomaterials: A Guide for the Perplexed. *The Journal of Physical Chemistry. C, Nanomaterials and Interfaces*, *125*(1), 973–988. DOI: 10.1021/acs.jpcc.0c08597

Saleh Mohammadnia, M., Roghani-Mamaqani, H., Ghalkhani, M., & Hemmati, S. (2023a). A modified electrochemical sensor based on N, S-doped carbon dots/carbon nanotube-poly (amidoamine) dendrimer hybrids for imatinib mesylate determination. *Biosensors (Basel)*, *13*(5), 547. DOI: 10.3390/bios13050547 PMID: 37232908

Salman (2023). A Novel Design Eco-friendly Microwave-assisted Cu–N@CQDs Sensor for the Quantification of Eravacycline via Spectrofluorimetric Method; Application to Greenness Assessments, Dosage Form and Biological Samples. *Journal of Fluorescence, 33*(5), 1887–1896. DOI: 10.1007/s10895-023-03190-7

Salman, , Batakoushy, H. A., Saraya, R. E., Hassan, A. I., Al-Harrasi, A., & Ibrahim, A. E. (2025). Comprehensive investigation of Prunus armeniaca for natural green synthesis of carbon quantum dots; Applications as fluorescent nano-probes for ramipril. *Talanta*, *292*, 128014. Advance online publication. DOI: 10.1016/j.talanta.2025.128014 PMID: 40154049

Salman, , Hassan, A. I., Hassan, Y. F., Saraya, R. E., & Batakoushy, H. A. (2023). Rapid One-Pot Microwave Assisted Green Synthesis Nitrogen Doped Carbon Quantum Dots as Fluorescent Precursor for Estimation of Modafinil as Post-Covid Neurological Drug in Human Plasma with Greenness Assessments. *Journal of Fluorescence*, *33*(3), 1101–1110. DOI: 10.1007/s10895-022-03128-5 PMID: 36576682

Samantara, A. K., Sahu, S. C., Ghosh, A., & Jena, B. K. (2015). Sandwiched graphene with nitrogen, sulphur co-doped cqds: An efficient metal-free material for energy storage and conversion applications. *Journal of Materials Chemistry. A, Materials for Energy and Sustainability*, *3*(33), 16961–16970. DOI: 10.1039/C5TA03376D

Sarath Kumar, C. B., Reji, R. P., Sivalingam, Y., Kawazoe, Y., & Surya, V. J. (2024). Carbon and boron nitride quantum dots as optical sensor probes for selective detection of toxic metals in drinking water: A quantum chemical prediction through structure- and morphology-dependent electronic and optical properties. *RSC Advances*, *14*(38), 28182–28200. DOI: 10.1039/D4RA04843A PMID: 39234523

Sarkar, S., Srivastava, T. P., Sahoo, O. S., Shankar, A., Rai, A., Pethusamy, K., & Karmakar, S. (2024). Applications of quantum dots in preventive oncology. *Asian Pacific Journal of Cancer Prevention : APJCP*, *25*(3), 747–756. DOI: 10.31557/APJCP.2024.25.3.747 PMID: 38546057

Sarkar, S., Sudolská, M., Dubecký, M., Reckmeier, C. J., Rogach, A. L., Zbořil, R., & Otyepka, M. (2016). Graphitic nitrogen doping in carbon dots causes red-shifted absorption. *The Journal of Physical Chemistry. C, Nanomaterials and Interfaces*, *120*(2), 1303–1308. DOI: 10.1021/acs.jpcc.5b10186

Saroja, A. P. V. K., & Garapati, M. S. (2020). Facile synthesis of heteroatom doped and undoped graphene quantum dots as active materials for reversible lithium and sodium ions storage. *Applied Surface Science*, *504*, 144430. DOI: 10.1016/j.apsusc.2019.144430

Sciortino, E., Marino, B., Dam, P., Schall, M., Cannas, F., & Messina, F. (2016). Solvatochromism unravels the emission mechanism of carbon nanodots. *The Journal of Physical Chemistry Letters*, *7*(17), 3419–3423. DOI: 10.1021/acs.jpclett.6b01590 PMID: 27525451

Sead, F. F., Jadeja, Y., & Kumar, A., MM, R., Kundlas, M., Saini, S., Joshi K. K. & Noorizadeh, H. (. (2025). Carbon quantum dots for sustainable energy: Enhancing electrocatalytic reactions through structural innovation. *Nanoscale Advances*, *7*(13), 3961–3998. DOI: 10.1039/D5NA00205B PMID: 40444140

Sead, F. F., Makasana, J., Saraswat, S. K., Rekha, M. M., Kundlas, M., Saini, S., Joshi, K. K., & Noorizadeh, H. (2025). Electrochemical behavior of carbon quantum dots as electrolyte additives for enhanced battery and supercapacitor performance. *Materials Technology*, *40*(1), 2500524. Advance online publication. DOI: 10.1080/10667857.2025.2500524

Selvam, R., Pearl, W. G., Perevedentseva, E., Karmenyan, A., & Cheng, C. L. (2024). Fe-doped nanodiamond-based photo-Fenton catalyst for dual-modal fluorescence imaging and improved chemotherapeutic efficacy against tumor hypoxia. *RSC Advances*, *14*(6), 4285–4300. DOI: 10.1039/D3RA08465E PMID: 38298935

Semaltianos (2010). Nanoparticles by laser ablation. *Critical Reviews in Solid State and Materials Sciences, 35*(2), 105–124. DOI: 10.1080/10408431003788233

Semenov, K. N., Shemchuk, O. S., Ageev, S. V., Andoskin, P. A., Iurev, G. O., Murin, I. V., Kozhukhov, P. K., Maystrenko, D. N., Molchanov, O. E., Kholmurodova, D. K., Rizaev, J. A., & Sharoyko, V. V. (2024). Development of Graphene-Based Materials with the Targeted Action for Cancer Theranostics. *Biochemistry (Moscow) 2024 89:8, 89*(8), 1362–1391. DOI: 10.1134/S0006297924080029

Sendão, R., Yuso, M. V. M., Algarra, M., Esteves da Silva, J. C. G., & Pinto da Silva, L. (2020). Comparative life cycle assessment of bottom-up synthesis routes for carbon dots derived from citric acid and urea. *Journal of Cleaner Production, 254*, 120080. DOI: 10.1016/j.jclepro.2020.120080

Sen, S., & Bose, A. (2025). Carbon dots: A review of innovations, applications, challenges, and future prospects. *Inorganic Chemistry Communications, 173*, 113852. DOI: 10.1016/j.inoche.2024.113852

Sethulekshmi, A. S., Aparna, A., Parvathi, P., Pathak, R., Punetha, V. D., Selvaraj, M., & Saritha, A. (2025). Advances in doped carbon quantum dots: Synthesis, mechanisms, and applications in sensing technologies. *Chemical Engineering Journal, 514*, 163262. https://doi.org/https://doi.org/10.1016/j.cej.2025.163262. DOI: 10.1016/j.cej.2025.163262

Shafiee, A., Iravani, S., & Varma, R. S. (2022). Graphene and graphene oxide with anticancer applications: Challenges and future perspectives. *MedComm, 3*(1), e118. DOI: 10.1002/mco2.118 PMID: 35281783

Shahba et al., (2022). *Two-Step and Green Synthesis of Highly Fluorescent Carbon Quantum Dots and Carbon Nanofibers from Pine Fruit.* https://doi.org/DOI: 10.1007/s10895-020-02562-7/Published

Shahbaz, M., Salma, U., Alam, M. Z., Mujeeb, M. A., Ansari, R., Basha, H. A., Alimuddin, , & Khan, S. A. (2025, February 11). ul H., Basha, H. A., Alimuddin, & Khan, S. A. (2025). Fluorescent/ Photoluminescent Carbon Dots as a Sensor for the Selective and Sensitive Detection of Fe3+/Fe2+ Metal Ions. A Review of the Last Decade. *Journal of Fluorescence, 35*(9), 7451–7474. Advance online publication. DOI: 10.1007/s10895-025-04166-5 PMID: 39932667

Shah, S. S. (2024). Biomass-derived carbon materials for advanced metal-ion hybrid supercapacitors: A step towards more sustainable energy. *Batteries, 10*(5), 168. DOI: 10.3390/batteries10050168

Shaikh, A. F., Tamboli, M. S., Patil, R. H., Bhan, A., Ambekar, J. D., & Kale, B. B. (2019). Bioinspired Carbon Quantum Dots: An Antibiofilm Agents. *Journal of Nanoscience and Nanotechnology, 19*(4), 2339–2345. DOI: 10.1166/jnn.2019.16537 PMID: 30486995

Shaker, M., Ng, S., Sadeghi Ghazvini, A. A., Javanmardi, S., Gaho, M. A., Jin, Z., & Ge, Q. (2024). Carbon/graphene quantum dots as electrolyte additives for batteries and supercapacitors: A review. *Journal of Energy Storage*, *85*, 111040. DOI: 10.1016/j.est.2024.111040

Shang, , Cai, T., Zhang, Y., Liu, D., & Liu, S. (2018). Facile one pot pyrolysis synthesis of carbon quantum dots and graphene oxide nanomaterials: All carbon hybrids as eco-environmental lubricants for low friction and remarkable wear-resistance. *Tribology International*, *118*, 373–380. DOI: 10.1016/j.triboint.2017.09.029

Shanmugasundaram, E., Vellaisamy, K., Ganesan, V., Narayanan, V., Saleh, N., & Thambusamy, S. (2024). Dual Applications of Cobalt-Oxide-Grafted Carbon Quantum Dot Nanocomposite for Two Electrode Asymmetric Supercapacitors and Photocatalytic Behavior. *ACS Omega*, *9*(12), 14101–14117. DOI: 10.1021/acsomega.3c09594 PMID: 38559980

Shariq, M., AlGhamdi, G. S., Alotaibi, R., Almuqati, N. S., Toraba, M., Albarqan, M., Abdullah, S. B., Alshamrani, K. M., & Ali, S. K. (2025). Design, synthesis, and structural engineering of carbon quantum dots and multifunctional N, P, B, S-doped CQDs for photocatalytic water purification. *Journal of Water Process Engineering*, *77*, 108450. DOI: 10.1016/j.jwpe.2025.108450

Sharma, D. K. (2025). Recent advancements in nanoparticles for cancer treatment. *Medical Oncology 2025 42:3*, *42*(3), 1–17. DOI: 10.1007/s12032-025-02609-4

Sharma, A. K., Kuamri, N., Chauhan, P., Thakur, S., Kumar, S., & Shandilya, M. (2025). Comprehensive Insights into Carbon Quantum Dots: Synthesis Strategies and Multidomain Applications. *Journal of Fluorescence*. Advance online publication. DOI: 10.1007/s10895-025-04432-6 PMID: 40616654

Sharma, S. K., Sharma, G., Gaur, A., Arya, A., Mirsafi, F. S., Abolhassani, R., Rubahn, H.-G., Yu, J.-S., & Mishra, Y. K. (2022). Progress in electrode and electrolyte materials: Path to all-solid-state Li-ion batteries. *Energy Advances*, *1*(8), 457–510. DOI: 10.1039/D2YA00043A

Sheikhpour, M., Naghinejad, M., Kasaeian, A., Lohrasbi, A., Shahraeini, S. S., & Zomorodbakhsh, S. (2020). The applications of carbon nanotubes in the diagnosis and treatment of lung cancer: A critical review. *International Journal of Nanomedicine*, *15*, 7063–7078. DOI: 10.2147/IJN.S263238 PMID: 33061368

Shen, C., Wang, J., Cao, Y., & Lu, Y. (2015). Facile access to B-doped solid-state fluorescent carbon dots toward light emitting devices and cell imaging agents. *Journal of Materials Chemistry. C, Materials for Optical and Electronic Devices*, *3*(26), 6668–6675. DOI: 10.1039/C5TC01156F

Shen, , Jia, P.-Y., Chen, D.-S., & Wang, L.-N. (2021). Hydrothermal synthesis of N-doped carbon quantum dots and their application in ion-detection and cell-imaging. *Spectrochimica Acta. Part A: Molecular and Biomolecular Spectroscopy, 248*, 119282. Advance online publication. DOI: 10.1016/j.saa.2020.119282 PMID: 33316652

Shen, Z., Chen, L., Cheng, H., & Qi, Z. (2023). Selective synthesis of glyceryl monolaurate intensified by boric acid based deep eutectic solvent. *Catalysis Today, 407*, 291–300. DOI: 10.1016/j.cattod.2021.05.018

Shi, F., Xing, B., Zeng, H., Guo, H., Qu, X., Huang, G., Cao, Y., Li, P., Feng, L., & Zhang, C. (2024). Facile synthesis of ultrathin carbon nanosheets through NaCl-KCl templates coupled with ice-induced assembly strategy from carbon quantum dots as lithium-ion batteries anodes. *Journal of Alloys and Compounds, 976*, 173325. DOI: 10.1016/j.jallcom.2023.173325

Shi, F., Xing, B., Zeng, H., Guo, H., Qu, X., Huang, G., Cao, Y., Li, P., & Zhang, C. (2023). Ice template induced assembly strategy for preparation of 3D porous carbon frameworks from low-cost carbon quantum dots for high-performance lithium-ion batteries. *Journal of Energy Storage, 70*, 107982. DOI: 10.1016/j.est.2023.107982

Shin, D.-Y., Sung, K.-W., & Ahn, H.-J. (2021). Fluorine-doped carbon quantum dot interfacial layer on stockade-like etched copper foil for boosting Li-ion storage. *Chemical Engineering Journal, 413*, 127563. DOI: 10.1016/j.cej.2020.127563

Shi, R., Li, Z., Yu, H., Shang, L., Zhou, C., Waterhouse, G. I. N., Wu, L., & Zhang, T. (2017). Effect of nitrogen doping level on the performance of n-doped carbon quantum dot/tio2 composites for photocatalytic hydrogen evolution. *ChemSusChem, 10*(22), 4650–4656. DOI: 10.1002/cssc.201700943 PMID: 28671326

Sindhwani, S., Syed, A. M., Ngai, J., Kingston, B. R., Maiorino, L., Rothschild, J., MacMillan, P., Zhang, Y., Rajesh, N. U., Hoang, T., Wu, J. L. Y., Wilhelm, S., Zilman, A., Gadde, S., Sulaiman, A., Ouyang, B., Lin, Z., Wang, L., Egeblad, M., & Chan, W. C. W. (2020). The entry of nanoparticles into solid tumours. *Nature Materials 2020 19:5, 19*(5), 566–575. DOI: 10.1038/s41563-019-0566-2

Singh, R., & Kumar, S. (2022). Cancer Targeting and Diagnosis: Recent Trends with Carbon Nanotubes. *Nanomaterials 2022, Vol. 12, Page 2283, 12*(13), 2283. DOI: 10.3390/nano12132283

Singh, H., Bamrah, A., Khatri, M., & Bhardwaj, N. (2020a). One-pot hydrothermal synthesis and characterization of carbon quantum dots (CQDs). *Materials Today: Proceedings, 28*, 1891–1894. https://doi.org/https://doi.org/10.1016/j.matpr.2020.05.297. DOI: 10.1016/j.matpr.2020.05.297

Singh, I., Arora, R., Dhiman, H., & Pahwa, R. (2018). Carbon quantum dots: Synthesis, characterization and biomedical applications. *Turkish Journal of Pharmaceutical Sciences*, *15*(2), 219–230. DOI: 10.4274/tjps.63497 PMID: 32454664

Singh, V., Kashyap, S., Yadav, U., Srivastava, A., Singh, A. V., Singh, R. K., Singh, S. K., & Saxena, P. S. (2019). Nitrogen doped carbon quantum dots demonstrate no toxicity under in vitro conditions in a cervical cell line and in vivo in Swiss albino mice. *Toxicology Research*, *8*(3), 395–406. DOI: 10.1039/C8TX00260F PMID: 31160973

Siwal, S. S., & Bishnoi, P. (2025). *Nanoscopic Wonders: Carbon Quantum Dots as Catalysts and Charge Carriers in Advanced Energy Storage Systems*. CleanMat., DOI: 10.1002/clem.70003

Slewa (2024). Antifungal films for strawberry packaging using carbon quantum dots derived from lemon and onion juice via green hydrothermal method. *Food Bioscience, 61*. DOI: 10.1016/j.fbio.2024.104653

Smagulova, S., Egorova, M., & Tomskaya, A. (2019). Investigation of the properties of carbon quantum dots synthesized by the hydrothermal method. *IOP Conference Series. Materials Science and Engineering*, *693*(1), 012031. Advance online publication. DOI: 10.1088/1757-899X/693/1/012031

Sohal, N., Maity, B., & Basu, S. (2021). Recent advances in heteroatom-doped graphene quantum dots for sensing applications. *RSC Advances*, *11*(41), 25586–25615. DOI: 10.1039/D1RA04248C PMID: 35478909

Somaraj, G., Mathew, S., Abraham, T., Ambady, K. G., Mohan, C., & Mathew, B. (2022). Nitrogen and sulfur co-doped carbon quantum dots for sensing applications: A review. *ChemistrySelect*, *7*(19), e202200473. Advance online publication. DOI: 10.1002/slct.202200473

Sonawane et al., (2018). Nanocomposites and Its Applications. *Applications of Nanomaterials: Advances and Key Technologies*, 1–22. DOI: 10.1016/B978-0-08-101971-9.00001-6

Song, S., Wu, K., Wu, H., Guo, J., & Zhang, L. (2019). Multi-shelled zno decorated with nitrogen and phosphorus co-doped carbon quantum dots: Synthesis and enhanced photodegradation activity of methylene blue in aqueous solutions. *RSC Advances*, *9*(13), 7362–7374. DOI: 10.1039/C9RA00168A PMID: 35519954

Sousa, H. B. A., Martins, C. S. M., & Prior, J. A. V. (2021). You Don't Learn That in School: An Updated Practical Guide to Carbon Quantum Dots. *Nanomaterials (Basel, Switzerland)*, *11*(3), 611. DOI: 10.3390/nano11030611 PMID: 33804394

Strickland, S., Fourroux, L., & Pappas, D. (2025). Effect of precursors on carbon dot functionalization and applications: A review. *Analyst, 150*(8), 1448–1469. DOI: 10.1039/D4AN01554A PMID: 40105280

Suman, R., Rani, G., Bhawna, , Kumar, H., Devi, J., & Siddharth, . (2025). B, S doped zero-dimensional carbon quantum dots from bio-waste for high density supercapacitors. *Carbon Letters*. Advance online publication. DOI: 10.1007/s42823-025-00945-z

Sung, K.-W., Ko, K.-Y., & Ahn, H.-J. (2023). Tailored sulfur and nitrogen co-doped carbon quantum dot interfacial layer on copper foil for highly stable and ultrafast lithium-ion capacitors. *Journal of Energy Storage*, *72*, 108797. DOI: 10.1016/j.est.2023.108797

Sun, , Guo, S., Qin, Q., Liao, Y., Li, M., & Du, F. (2022). Box–Behnken Design Optimizing Sugarcane Bagasse-Based Nitrogen-Doped Carbon Quantum Dots Preparation and Application in Ferric Ion Detection. *Chemosensors (Basel, Switzerland)*, *10*(11), 453. Advance online publication. DOI: 10.3390/chemosensors10110453

Sun, H., Ji, H., Ju, E., Guan, Y., Ren, J., & Qu, X. (2015). Synthesis of fluorinated and nonfluorinated graphene quantum dots through a new top-down strategy for long-time cellular imaging. *Chemistry (Weinheim an der Bergstrasse, Germany)*, *21*(9), 3791–3797. DOI: 10.1002/chem.201406345 PMID: 25614445

Sun, Y. P., Zhou, B., Lin, Y., Wang, W., Fernando, K. A. S., Pathak, P., Meziani, M. J., Harruff, B. A., Wang, X., Wang, H., Lu, P. G., Yang, H., Kose, M. E., Chen, B., Veca, L. M., & Xie, S. Y. (2006). Quantum-sized carbon dots for bright and colorful photoluminescence. *Journal of the American Chemical Society*, *128*(24), 7756–7757. DOI: 10.1021/ja062677d PMID: 16771487

Surendran, P., Lakshmanan, A., Priya, S. S., Balakrishnan, K., Rameshkumar, P., Kannan, K., Mahalakshmi, K., Gayathri, V., & Vinitha, G. (2024). Synthesis of fluorescent carbon quantum dots from Manihot esculenta waste peels for nonlinear optical and biological applications. *Chemical Physics Impact*, *8*, 100515. DOI: 10.1016/j.chphi.2024.100515

Surendran, P., Lakshmanan, A., Sakthy Priya, S., Geetha, P., Rameshkumar, P., Kannan, K., Hegde, T. A., & Vinitha, G. (2021). Fluorescent carbon quantum dots from Ananas comosus waste peels: A promising material for NLO behaviour, antibacterial, and antioxidant activities. *Inorganic Chemistry Communications*, *124*, 108397. DOI: 10.1016/j.inoche.2020.108397

Surendran, P., Lakshmanan, A., Vinitha, G., Ramalingam, G., & Rameshkumar, P. (2020). Bioinspired fluorescence carbon quantum dots extracted from natural honey: Efficient material for photonic and antibacterial applications. *Nano-Structures & Nano-Objects*, *24*, 100589. DOI: 10.1016/j.nanoso.2020.100589

Surendran, P., Lakshmanan, A., Vinitha, G., Ramalingam, G., & Rameshkumar, P. (2020). Facile preparation of high fluorescent carbon quantum dots from orange waste peels for nonlinear optical applications. *Luminescence*, *35*(2), 196–202. DOI: 10.1002/bio.3713 PMID: 31591819

Sushko, E. S., Vnukova, N. G., Churilov, G. N., & Kudryasheva, N. S. (2022). Endohedral Gd-Containing Fullerenol: Toxicity, Antioxidant Activity and Regulation of Reactive Oxygen Species in Cellular and Enzymatic Systems. *International Journal of Molecular Sciences*, *23*(9), 5152. DOI: 10.3390/ijms23095152 PMID: 35563539

Susi, T., Pichler, T., & Ayala, P. (2015). X-ray photoelectron spectroscopy of graphitic carbon nanomaterials doped with heteroatoms. *Beilstein Journal of Nanotechnology*, *6*, 177–192. DOI: 10.3762/bjnano.6.17 PMID: 25671162

Suzuki, K., Miyamura, H., & Balachandran, J. (2021). One-pot hydrothermal synthesis of carbon dots-immobilized hydrozincite for ZnO-based nanocomposite lighting applications. *Journal of Asian Ceramic Societies*, *9*(4), 1473–1480. DOI: 10.1080/21870764.2021.1995939

Tade, R. S., & Patil, P. O. (2020). Theranostic Prospects of Graphene Quantum Dots in Breast Cancer. *ACS Biomaterials Science & Engineering*, *6*(11), 5987–6008. DOI: 10.1021/acsbiomaterials.0c01045 PMID: 33449670

Talat, M., Awasthi, K., Singh, V. K., & Srivastava, O. N. (2020). Investigations on Exotic Forms of Carbon: Nanotubes, Graphene, Fullerene, and Quantum Dots. In *Functionalized Nanomaterials I* (pp. 125–134). CRC Press., DOI: 10.1201/9781351021623-8

Tang, B. Z., & Qin, A. (2013). *Aggregation-induced emission: Applications*. John Wiley & Sons.

Tang, L., Ji, R., Cao, X., Lin, J., Jiang, H., Li, X., Teng, K. S., Luk, C. M., Zeng, S., Hao, J., & Lau, S. P. (2012). Deep ultraviolet photoluminescence of water-soluble self-passivated graphene quantum dots. *ACS Nano*, *6*(6), 5102–5110. DOI: 10.1021/nn300760g PMID: 22559247

Tang, L., Xiao, Q., Mei, Y., He, S., Zhang, Z., Wang, R., & Wang, W. (2021). Insights on functionalized carbon nanotubes for cancer theranostics. *Journal of Nanobiotechnology*, *19*(1), 423. Advance online publication. DOI: 10.1186/s12951-021-01174-y PMID: 34915901

Tan, , Zuo, S., Zhao, Y., & Shen, B. (2019). Preparation of multicolored carbon quantum dots using HNO3/HClO4 oxidation of graphitized carbon. *Journal of Materials Research*, *34*(20), 3428–3438. DOI: 10.1557/jmr.2019.261

Tao, C., & Rouhi, J. (2023). A biosensor based on graphene oxide nanocomposite for determination of carcinoembryonic antigen in colorectal cancer biomarker. *Environmental Research*, *238*, 117113. DOI: 10.1016/j.envres.2023.117113 PMID: 37696325

Tavan, M., Yousefian, Z., Mirjalili, M. H., Bakhtiar, Z., Rahmandoust, M., & Mirjalili, M. H. (2025). Carbon quantum dots: Multifunctional fluorescent nanomaterials for sustainable advances in biomedicine and agriculture. *Industrial Crops and Products*, *231*, 121207. DOI: 10.1016/j.indcrop.2025.121207

Tegafaw, T., Mulugeta, E., Zhao, D., Liu, Y., Chen, X., Baek, A., & Lee, G. H. (2025). Surface modification, toxicity, and applications of carbon dots to cancer theranosis: A review. *Nanomaterials (Basel, Switzerland)*, *15*(11), 781. DOI: 10.3390/nano15110781 PMID: 40497830

Tepale-Cortés, A., Moreno-Saavedra, H., Hernandez-Tenorio, C., Rojas-Ramírez, T., Illescas, J., Tepale-Cortés, A., Moreno-Saavedra, H., Hernandez-Tenorio, C., Rojas-Ramírez, T., & Illescas, J. (2021). Multi-walled Carbon Nanotubes Synthesis by Arc Discharge Method in a Glass Chamber. *Journal of the Mexican Chemical Society*, *65*(4), 480–490. DOI: 10.29356/jmcs.v65i4.1486

Tian, . (2021). Carbon quantum dots for advanced electrocatalysis. In *Journal of Energy Chemistry* (Vol. 55, pp. 279–294). Elsevier B.V., DOI: 10.1016/j.jechem.2020.06.057

Tian, J., An, M., Zhao, X., Wang, Y., & Hasan, M. (2023). Advances in Fluorescent Sensing Carbon Dots: An Account of Food Analysis. *ACS Omega*, *8*(10), 9031–9039. DOI: 10.1021/acsomega.2c07986 PMID: 36936334

Tian, , Wang, Y., & Zhang, Y. (2018). Synthesis of Fluorescent Nitrogen-Doped Carbon Quantum Dots for Selective Detection of Picric Acid in Water Samples. *Journal of Nanoscience and Nanotechnology*, *18*(12), 8111–8117. DOI: 10.1166/jnn.2018.16342 PMID: 30189927

Tian, X. T., & Yin, X. B. (2019). Carbon dots, unconventional preparation strategies, and applications beyond photoluminescence. *Small, 15*(48), 1901803. DOI: 10.1002/smll.201901803 PMID: 31240816

Timoshenko, J., & Cuenya, B. R. (2020). In situ/operando electrocatalyst characterization by x-ray absorption spectroscopy. *Chemical Reviews, 121*(2), 882–961. DOI: 10.1021/acs.chemrev.0c00396 PMID: 32986414

Tu, H., Zhang, Y., Wu, J., Li, Y., Liu, H., Deng, W., Zou, G., Hou, H., & Ji, X. (2024). Revealing electrochemical process of functional carbon dots stabilized sodium metal anode: Co-deposition and strengthened sei films. *Advanced Functional Materials, 35*(3), 2413488. Advance online publication. DOI: 10.1002/adfm.202413488

Tu, Z., Yang, G., Song, H., & Wang, C. (2016). Amorphous zno quantum dot/mesoporous carbon bubble composites for a high-performance lithium-ion battery anode. *ACS Applied Materials &Amp. Interfaces, 9*(1), 439–446. DOI: 10.1021/acsami.6b13113 PMID: 27966898

U.S. Department of Energy. Industrial Efficiency & Decarbonization Office. (2023). *Techno-economic, energy, & carbon heuristic tool for early-stage technologies (TECHTEST) tool.* https://www.energy.gov/eere/iedo/techno-economic-energy-carbon-heuristic-tool-early-stage-technologies-techtest-tool

Uppugalla, S., Pothu, R., Boddula, R., Desai, M. A., & Al-Qahtani, N. (2023). Nitrogen and sulfur co-doped activated carbon nanosheets for high-performance coin cell supercapacitor device with outstanding cycle stability. *Emergent Materials, 6*(4), 1167–1176. DOI: 10.1007/s42247-023-00503-1

Upreti, G., Dhingra, R., Naidu, S., Atuahene, I., & Sawhney, R. (2015). Life cycle assessment of nanomaterials. *Green Processes for Nanotechnology*, 393–408. DOI: 10.1007/978-3-319-15461-9_14

Uriarte, D., Domini, C., & Garrido, M. (2019). New carbon dots based on glycerol and urea and its application in the determination of tetracycline in urine samples. *Talanta, 201*, 143–148. DOI: 10.1016/j.talanta.2019.04.001 PMID: 31122404

Usman, M., & Cheng, S. (2024). Recent trends and advancements in green synthesis of biomass-derived carbon dots. *Eng, 5*(3), 2223–2263. DOI: 10.3390/eng5030116

Valeur, B., & Berberan-Santo, M. N. (2012). *Molecular Fluorescence - Principal and Applications*. 331.

Van Tam, T., Kang, S. G., Babu, K. F., Oh, E. S., Lee, S. G., & Choi, W. M. (2017). Synthesis of B-doped graphene quantum dots as a metal-free electrocatalyst for the oxygen reduction reaction. *Journal of Materials Chemistry. A, Materials for Energy and Sustainability*, *5*(21), 10537–10543. DOI: 10.1039/C7TA01485F

Varadarajan, S., Kavitha, A., Selvaraju, P., Muthu, S. E., Gurushankar, K., Shanmugan, S., & Kannan, K. (2024). Enhanced Photocatalytic Hydrogen Evolution by TiO2: A Synergistic Approach with Defect-Rich SnS2 and Ti3C2 MXene Cocatalysts. *Hydrogen*, *5*(4), 940–957. DOI: 10.3390/hydrogen5040050

Veerakumari, P., Rao, A. V., Devi, K. A., Rao, P. V., Raziya, S., Shanmugan, S., & Kannan, K. (2025). A Review on Tailoring Metal–Organic Frameworks for Electrochemical Applications: Advances in Synthesis and Structure for Energy Storage. *iScience*.

Veerasubramani, G. K., Park, M., Choi, J., & Kim, D. (2020). Ultrasmall sns quantum dots anchored onto nitrogen-enriched carbon nanospheres as an advanced anode material for sodium-ion batteries. *ACS Applied Materials &Amp. Interfaces*, *12*(6), 7114–7124. DOI: 10.1021/acsami.9b18997 PMID: 31944653

Velmurugan, S., Traiwatcharanon, P., Jiwanti, P. K., Cheng, S.-H., & Wongchoosuk, C. (2024). Highly selective carbendazim fungicide sensing performance of nitrogen-doped carbon quantum dots encapsulated aminated-UiO-66 zirconium metal-organic framework electrocatalyst. *Electrochimica Acta*, *479*, 143911. DOI: 10.1016/j.electacta.2024.143911

Vercelli, , Donnini, R., Ghezzi, F., Sansonetti, A., Giovanella, U., & La Ferla, B. (2021). Nitrogen-doped carbon quantum dots obtained hydrothermally from citric acid and urea: The role of the specific nitrogen centers in their electrochemical and optical responses. *Electrochimica Acta*, *387*, 138557. Advance online publication. DOI: 10.1016/j.electacta.2021.138557

Verma, C., Dubey, S., Alfantazi, A., & Rhee, K. Y. (2024). Heteroatoms-doped carbon dots: Fundamental, properties, coordination bonding and corrosion protection. *Journal of Industrial and Engineering Chemistry*, *133*, 90–111. DOI: 10.1016/j.jiec.2023.12.018

Vibhute, , Nille, O., Kolekar, G., Rohiwal, S., Patil, S., Lee, S., & Tiwari, A. P. (2022). Fluorescent Carbon Quantum Dots Functionalized by Poly L-Lysine: Efficient Material for Antibacterial, Bioimaging and Antiangiogenesis Applications. *Journal of Fluorescence*, *32*(5), 1789–1800. DOI: 10.1007/s10895-022-02977-4 PMID: 35689742

Wang, Z., Changotra, R., Dasog, M., Singh Selopal, G., Yang, J., & He, Q. S. (2025). Carbon quantum dots: Synthesis via hydrothermal processing, doping strategies, integration with photocatalysts, and their application in photocatalytic hydrogen production. *Sustainable Materials and Technologies, 44*, e01386. https://doi.org/https://doi.org/10.1016/j.susmat.2025.e01386

Wang, B., Wei, Z., Sui, L., Yu, J., Zhang, B., Wang, X., & Tian, Y. (2022). Electron–phonon coupling-assisted universal red luminescence of o-phenylenediamine-based carbon dots. *Light, Science & Applications, 11*(1), 172. DOI: 10.1038/s41377-022-00865-x PMID: 35668065

Wang, C., Sun, D., Zhuo, K., Zhang, H., & Wang, J. (2014). Simple and green synthesis of nitrogen-, sulfur-, and phosphorus-co-doped carbon dots with tunable luminescence properties and sensing application. *RSC Advances, 4*(96), 54060–54065. DOI: 10.1039/C4RA10885J

Wang, C., Wu, X., Li, X., Wang, W., Wang, L., Gu, M., & Li, Q. (2012). Upconversion fluorescent carbon nanodots enriched with nitrogen for light harvesting. *Journal of Materials Chemistry, 22*(31), 15522–15525. DOI: 10.1039/c2jm30935a

Wang, C., Xu, Z., & Zhang, C. (2015). Polyethyleneimine-Functionalized Fluorescent Carbon Dots: Water Stability, pH Sensing, and Cellular Imaging. *ChemNanoMat : Chemistry of Nanomaterials for Energy, Biology and More, 1*(2), 122–127. DOI: 10.1002/cnma.201500009

Wang, D., Wang, Q., Tan, M., Wang, S., Luo, S., Hou, P., Zhang, Y., Yan, S., & Liu, X. (2022). Biomass CQDs derivate carbon as high-performance anode for K-ion battery. *Journal of Alloys and Compounds, 922*, 166260. DOI: 10.1016/j.jallcom.2022.166260

Wang, E., Wan, J., Guo, Y., Zhang, Q., He, W., Zhang, C., Chen, W., Yan, H., Xue, D., Fang, T., Wang, F., Wen, R., Xin, S., Yin, Y., & Guo, Y. (2022). Mitigating electron leakage of solid electrolyte interface for stable sodium-ion batteries. *Angewandte Chemie International Edition, 62*(4), e202216354. Advance online publication. DOI: 10.1002/anie.202216354 PMID: 36440597

Wang, , Guo, J., Zhou, Y., Zhang, X., Gao, C., Xiong, R., Song, H., Chen, X., & Luo, J. (2025). Enhancing oil shale pyrolysis using metal-doped carbon quantum dot catalysts: A comprehensive behavioral, kinetic, and mechanistic analysis. *Fuel, 381*, 133464. Advance online publication. DOI: 10.1016/j.fuel.2024.133464

Wang, H., Li, Q., Alam, P., Bai, H., Bhalla, V., Bryce, M. R., & Chen, X. (2023). Aggregation-induced emission (AIE), life and health. *ACS Nano, 17*(15), 14347–14405. DOI: 10.1021/acsnano.3c03925 PMID: 37486125

Wang, , & Hu, A. (2014). Carbon quantum dots: Synthesis, properties and applications. *Journal of Materials Chemistry. C, Materials for Optical and Electronic Devices*, *2*(34), 6921–6939. DOI: 10.1039/C4TC00988F

Wang, L., Wang, M., Zhou, B., Zhou, F., Murray, C., Towner, R. A., Smith, N., Saunders, D., Xie, G., & Chen, W. R. (2019). PEGylated reduced-graphene oxide hybridized with Fe3O4 nanoparticles for cancer photothermal-immunotherapy. *Journal of Materials Chemistry. B*, *7*(46), 7406–7414. DOI: 10.1039/C9TB00630C PMID: 31710067

Wang, M., Fang, J., Hu, L., Lai, Y., & Liu, Z. (2017). Defects-rich graphene/carbon quantum dot composites as highly efficient electrocatalysts for aqueous zinc/air batteries. *International Journal of Hydrogen Energy*, *42*(33), 21305–21310. DOI: 10.1016/j.ijhydene.2017.07.045

Wang, Q., Wang, H., Liu, D., Du, P., & Liu, P. (2017). Synthesis of flake-shaped nitrogen-doped carbon quantum dot/polyaniline (N-CQD/PANI) nanocomposites via rapid-mixing polymerization and their application as electrode materials in supercapacitors. *Synthetic Metals*, *231*, 120–126. DOI: 10.1016/j.synthmet.2017.06.018

Wang, S., McCoy, C. P., Li, P., Li, Y., Zhao, Y., Andrews, G. P., & Ge, Y. (2024). Carbon dots in photodynamic/photothermal antimicrobial therapy. *Nanomaterials (Basel, Switzerland)*, *14*(15), 1250. DOI: 10.3390/nano14151250 PMID: 39120355

Wang, S., Wang, D., Wang, G., Zhang, M., Sun, Y., & Ding, J. (2025). Antibacterial carbon dots. *Materials Today. Bio*, *30*, 101383. DOI: 10.1016/j.mtbio.2024.101383 PMID: 39811607

Wang, , Shen, J., Ma, Y., Qin, X., Qin, X., Yang, F., Ostrikov, K. K., Zhang, Q., He, J., & Zhong, X. (2024). Cancer-targeting carbon quantum dots synthesized by plasma electrochemical method for red-light-activated photodynamic therapy. *Plasma Processes and Polymers*, *21*(4), 2300174. Advance online publication. DOI: 10.1002/ppap.202300174

Wang, T., Liu, P., Chen, X., Guo, Y., He, C., Guo, J., Liu, W., Gao, S., Lv, Y., & Liu, K. (2025). Enhancing aqueous zinc-ion energy storage performance with ion-mediating carbon quantum dots-modified separators regulating zinc deposition behavior. *Journal of Colloid and Interface Science*, *678*, 301–310. DOI: 10.1016/j.jcis.2024.08.175 PMID: 39208758

Wang, X. W., Zheng, D. L., Yang, P. Z., Wang, X. E., Zhu, Q. Q., Ma, P. F., & Sun, L. Y. (2017). Preparation and electrochemical properties of NiO-Co_3O_4 composite as electrode materials for supercapacitors. *Chemical Physics Letters*, *667*, 260–266. DOI: 10.1016/j.cplett.2016.10.079

Wang, X., Feng, Y., Dong, P., & Huang, J. (2019). A Mini Review on Carbon Quantum Dots: Preparation, Properties, and Electrocatalytic Application. *Frontiers in Chemistry*, *7*, 671. Advance online publication. DOI: 10.3389/fchem.2019.00671 PMID: 31637234

Wang, Y., Jing, Y., Wang, L., Kong, W., Wang, S., Wang, Z., Li, Y., & Lu, Q. (2019). Photoluminescence of graphene quantum dots doped with different elements. *Kexue Tongbao*, *64*(4), 411–418. DOI: 10.1360/N972018-00986

Wang, Z., Changotra, R., Dasog, M., Singh Selopal, G., Yang, J., & He, Q. S. (2025). Carbon quantum dots: Synthesis via hydrothermal processing, doping strategies, integration with photocatalysts, and their application in photocatalytic hydrogen production. *Sustainable Materials and Technologies*, *44*, e01386. DOI: 10.1016/j.susmat.2025.e01386

Wang, Z., Che, H., Lu, W., Chao, Y., Wang, L., Liang, B., Liu, J., Xu, Q., & Cui, X. (2023). Application of Inorganic Quantum Dots in Advanced Lithium–Sulfur Batteries. *Advancement of Science*, *10*(19), 2301355. Advance online publication. DOI: 10.1002/advs.202301355 PMID: 37088862

Wang, Z., Xie, Y., Lei, Z., Lu, Y., Wei, G., Liu, S., Xu, C., Zhang, Z., Wang, X., Rao, L., & Chen, J. (2019). Quantitative analysis of surface sites on carbon dots and their interaction with metal ions by a potentiometric titration method. *Analytical Chemistry*, *91*(15), 9690–9697. DOI: 10.1021/acs.analchem.9b01225 PMID: 31245999

Wareing, T., Gentile, P., & Phan, A. N. (2021). Biomass-based carbon dots: Current development and future perspectives. *ACS Nano*, *15*(10), 15471–15501. DOI: 10.1021/acsnano.1c03886 PMID: 34559522

Watts, J. F., & Wolstenholme, J. (2003). An Introduction to Surface Analysis by XPS and AES. *An Introduction to Surface Analysis by XPS and AES*. DOI: 10.1002/0470867930

Weaving, J. S., Lim, A., Millichamp, J., Neville, T. P., Ledwoch, D., Kendrick, E., McMillan, P. F., Shearing, P. R., Howard, C. A., & Brett, D. J. L. (2020). Elucidating the sodiation mechanism in hard carbon by operando raman spectroscopy. *ACS Applied Energy Materials*, *3*(8), 7474–7484. DOI: 10.1021/acsaem.0c00867

Weerasinghe, , Scott, J., Deshan, A. D. K., Chen, D., Singh, A., Sen, S., Sonar, P., Vasilev, K., Li, Q., & Ostrikov, K. K. (2022). Monochromatic Blue and Switchable Blue-Green Carbon Quantum Dots by Room-Temperature Air Plasma Processing. *Advanced Materials Technologies*, *7*(4), 2100586. Advance online publication. DOI: 10.1002/admt.202100586

Wei, F., Zeng, Y., Guo, Y., Li, J., Zhu, S., Gao, S., Zhang, H., & He, X. (2023). Recent progress on the heteroatom-doped carbon cathode for zinc ion hybrid capacitors. *Chemical Engineering Journal*, *468*, 143576. DOI: 10.1016/j.cej.2023.143576

Wen, J., Li, M., Xiao, J., Liu, C., Li, Z., Xie, Y., Ning, P., Cao, H., & Zhang, Y. (2016). Novel oxidative cutting graphene oxide to graphene quantum dots for electrochemical sensing application. *Materials Today. Communications*, *8*, 127–133. DOI: 10.1016/j.mtcomm.2016.07.006

Wiggins-Camacho, J. D., & Stevenson, K. J. (2009). Effect of nitrogen concentration on capacitance, density of states, electronic conductivity, and morphology of n-doped carbon nanotube electrodes. *The Journal of Physical Chemistry. C, Nanomaterials and Interfaces*, *113*(44), 19082–19090. DOI: 10.1021/jp907160v

Wikimedia Commons contributors. (2020). *Infographic – Lifecycle analysis of engineered nanomaterials (ENM)* [Graphic]. https://commons.wikimedia.org/wiki/File:Infographic_%E2%80%93_Lifecycle_analysis_of_engineered_nanomaterials_(ENM).jpg

Williams, D. B., & Carter, C. B. (2009). The Transmission Electron Microscope. *Transmission Electron Microscopy*, 3–22. DOI: 10.1007/978-0-387-76501-3_1

Won, J. E., Park, M., Hong, S. H., Kim, Y. S., & Song, H. (2025). Quantum dots as biocompatible small RNA nanocarriers modulating macrophage polarization to treat Asherman's syndrome. *NPJ Regenerative Medicine*, *10*(1), 15. DOI: 10.1038/s41536-025-00403-4 PMID: 40133362

Wu, Y. F., Wu, H. C., Kuan, C. H., Lin, C. J., Wang, L. W., Chang, C. W., & Wang, T. W. (2016). Multi-functionalized carbon dots as theranostic nanoagent for gene delivery in lung cancer therapy. *Scientific Reports 2016 6:1*, *6*(1), 1–12. DOI: 10.1038/srep21170

Wu, C., Yin, Z., Wang, W., Liu, Y., Yu, J., & Liu, Y. (2022). Phase behavior and aggregate transition based on co-assembly of negatively charged carbon dots and a ph-responsive tertiary amine cationic surfactant. *Langmuir*, *38*(45), 13771–13781. DOI: 10.1021/acs.langmuir.2c01895 PMID: 36318637

Wu, M., Li, J., Wu, Y., Gong, X., & Wu, M. (2023). Design of a synthetic strategy to achieve enhanced fluorescent carbon dots with sulfur and nitrogen codoping and its muco-dopingional applications. *Small*, *19*(42), 2302764. Advance online publication. DOI: 10.1002/smll.202302764 PMID: 37330653

Wu, , Mu, L., Luo, X., Zhang, Y., Huang, J., Peng, H., & Tao, J. (2025). Effect of temperature on synthesis of carbon quantum dots and biochar through one-step hydrothermal treatment of distillers' grains. *Industrial Crops and Products, 227*, 120832. Advance online publication. DOI: 10.1016/j.indcrop.2025.120832

Wu, S., Wang, X., Zhu, Y., He, P., Yu, X., Qin, F., Yao, Y., & Ren, L. (2023). Different types of nitrogen-doped cqds loaded by cop used as oer electrocatalysts. *Inorganic Chemistry Communications, 153*, 110872. DOI: 10.1016/j.inoche.2023.110872

Wu, S., & Zhu, Y. (2016). Highly densified carbon electrode materials towards practical supercapacitor devices. *Science China Materials, 60*(1), 25–38. DOI: 10.1007/s40843-016-5109-4

Xia, C., Zhu, S., Feng, T., Yang, M., & Yang, B. (2019). Evolution and Synthesis of Carbon Dots: From Carbon Dots to Carbonized Polymer Dots. *Advancement of Science, 6*(23), 1901316. DOI: 10.1002/advs.201901316 PMID: 31832313

Xie, W., Zhang, Q., Song, S., Cheng, X., Yang, Y., Wang, L., Ouyang, X., Xie, S., & Huang, J. (2023). High S-doped amorphous carbon/carbon quantum dots coupled micro-frame for highly efficient potassium storage. *Journal of Colloid and Interface Science, 652*, 1522–1532. DOI: 10.1016/j.jcis.2023.08.123 PMID: 37660609

Xue, L., Li, Y., Hu, A., Zhou, M., Chen, W., Lei, T., Yan, Y., Huang, J., Yang, C., Wang, X., Hu, Y., & Xiong, J. (2022). In situ/operando raman techniques in lithium–sulfur batteries. *Small Structures, 3*(3), 2100170. Advance online publication. DOI: 10.1002/sstr.202100170

Xu, J., Hou, K., Ju, Z., Ma, C., Wang, W., Wang, C., Cao, J., & Chen, Z. (2017). Synthesis and electrochemical properties of carbon dots/manganese dioxide (CQDs/MnO_2) nanoflowers for supercapacitor applications. *Journal of the Electrochemical Society, 164*(2), A430–A437. DOI: 10.1149/2.1241702jes

Xu, J., Xue, Y., Cao, J., Wang, G., Li, Y., Wang, W., & Chen, Z. (2016). Carbon quantum dots/nickel oxide (CQDs/NiO) nanorods with high capacitance for supercapacitors. *RSC Advances, 6*(7), 5541–5546. DOI: 10.1039/C5RA24192H

Xu, J., Zhou, P., Yuan, L., Liu, X., Ma, J., & Zhang, C. (2021). Dual lignin valorization enabled by carbon quantum dots and lithium-sulfur cathode. *Industrial Crops and Products, 170*, 113801. DOI: 10.1016/j.indcrop.2021.113801

Xu, L., Tu, H., Zhu, F., Xiang, Y., Luo, Z., Fang, S., Deng, W., Zou, G., Hou, H., & Ji, X. (2022). Carbon dots for ultrastable solid-state batteries. *SmartMat, 3*(2), 286–297. DOI: 10.1002/smm2.1097

Xu, Q., Liu, Y., Gao, C., Wei, J., Zhou, H., Chen, Y., Dong, C., Sreenivasan, S. T., Li, N., & Xia, Z. (2015). Synthesis, mechanistic investigation, and application of photoluminescent sulfur and nitrogen co-doped carbon dots. *Journal of Materials Chemistry. C, Materials for Optical and Electronic Devices*, *3*(38), 9885–9893. DOI: 10.1039/C5TC01912E

Xu, Q., Niu, Y., Li, J., Yang, Z., Gao, J., Ding, L., Ni, H., Zhu, P., Liu, Y., Tang, Y., Lv, Z. P., Peng, B., Hu, T. S., Zhou, H., & Xu, C. (2022). Recent progress of quantum dots for energy storage applications. *Carbon Neutrality*, *1*(1), 13. DOI: 10.1007/s43979-022-00002-y

Xu, X., Ray, R., Gu, Y., Ploehn, H. J., Gearheart, L., Raker, K., & Scrivens, W. A. (2004). Electrophoretic analysis and purification of fluorescent single–walled carbon nanotube fragments. *Journal of the American Chemical Society*, *126*(40), 12736–12737. DOI: 10.1021/ja040082h PMID: 15469243

Xu, Y., Wang, S., Hou, X., Sun, Z., Jiang, Y., Dong, Z., Tao, Q., Man, J., & Cao, Y. (2018). Coal-derived nitrogen, phosphorus and sulfur co-doped graphene quantum dots: A promising ion fluorescent probe. *Applied Surface Science*, *445*, 519–526. DOI: 10.1016/j.apsusc.2018.03.156

Yadav et al., (2024). Emerging ecofriendly approaches for biomass conversion and waste management. *Sustainable Management of Agro-Food Waste: Fundamental Aspects and Practical Applications*, 39–55. DOI: 10.1016/B978-0-443-23679-2.00004-5

Yadav, P. K., Chandra, S., Kumar, V., Kumar, D., & Hasan, S. H. (2023). Carbon quantum dots: Synthesis, structure, properties, and catalytic applications for organic synthesis. *Catalysts*, *13*(2), 422. DOI: 10.3390/catal13020422

Yadav, R., & Lahariya, V. (2025a). Investigation of Fluorescence Sensing Capabilities in Boron-Nitrogen Co-Doped Carbon Quantum Dots Towards Fe (III) and Hg(II) Ions. *Journal of Fluorescence*. Advance online publication. DOI: 10.1007/s10895-025-04487-5 PMID: 40824367

Yan et al., (2012). Functionalized Carbon Nanotubes and Their Enhanced Polymers. *Polymer Science: A Comprehensive Reference: Volume 1-10, 1–10*, 439–478. DOI: 10.1016/B978-0-444-53349-4.00217-X

Yan, F., Jiang, Y., Sun, X., Bai, Z., Zhang, Y., & Zhou, X. (2018). Surface modification and chemical functionalization of carbon dots: A review. *Microchimica Acta*, *185*(9), 424. DOI: 10.1007/s00604-018-2953-9 PMID: 30128831

Yang, D., Cui, H., Soomro, R. A., Zhang, P., & Xu, B. (2025). Flexible, highly conductive, and microscopically disordered MXene film electrode constructed via in-situ carbon dots intercalation toward high-performatowardspercapacitors. *Energy Materials*, *5*(1). Advance online publication. DOI: 10.20517/energymater.2024.45

Yang, , Duan, J., Yang, W., Li, X., Mo, J., Yang, P., & Tang, Q. (2018). Nitrogen-doped carbon quantum dots from biomass via simple one-pot method and exploration of their application. *Applied Surface Science*, *434*, 1079–1085. DOI: 10.1016/j.apsusc.2017.11.040

Yang, H. L., Bai, L. F., Geng, Z. R., Chen, H., Xu, L. T., Xie, Y. C., Wang, D. J., Gu, H. W., & Wang, X. M. (2023). Carbon quantum dots: Preparation, optical properties, and biomedical applications. *Materials Today. Advances*, *18*, 100376. DOI: 10.1016/j.mtadv.2023.100376

Yang, , Lin, W., Lai, H., Tong, J., Lei, J., Yuan, M., Zhang, Y., & Cui, C. (2021). Understanding the relationship between particle size and ultrasonic treatment during the synthesis of metal nanoparticles. *Ultrasonics Sonochemistry*, *73*, 105497. Advance online publication. DOI: 10.1016/j.ultsonch.2021.105497 PMID: 33677187

Yang, R., Zhang, Y., Chen, X., Song, L., Hu, Y., & Song, Y. (2024). Locally boosted Li_2S nucleation on VO_2 by loading carbon quantum dots for soft-packaged lithium–sulfur pouch cells. *Chemical Communications*, *60*(68), 9078–9081. DOI: 10.1039/D4CC02263G PMID: 39105356

Yang, S., Xu, Z., Wang, S., Sun, J., Zhao, D., Cao, B., & Wang, X. (2024). Hydrophilic and nanocrystalline carbon quantum dots enable highly reversible zinc-ion batteries. *Green Chemistry*, *26*(12), 7293–7301. DOI: 10.1039/D4GC01434K

Yang, S., Ye, X., Shen, N., Sun, T., Yu, F., Liu, H., Guo, C., & Li, J. (2025). Carbon Dots Enabling High-Performances Sodium-Ion Batteries. *Chemistry (Weinheim an der Bergstrasse, Germany)*, *31*(2), e202402794. Advance online publication. DOI: 10.1002/chem.202402794 PMID: 39467772

Yang, Y., Liu, X., Dai, Z., Yuan, F., Bando, Y., Golberg, D., & Wang, X. (2017). In situ electrochemistry of rechargeable battery materials: Status report and perspectives. *Advanced Materials*, *29*(31), 1606922. Advance online publication. DOI: 10.1002/adma.201606922 PMID: 28627135

Yan, , Yang, Y., Ma, C.-Q., Liu, X., Wang, H., & Xu, B. (2016). Synthesis of carbon quantum dots by chemical vapor deposition approach for use in polymer solar cell as the electrode buffer layer. *Carbon*, *109*, 598–607. DOI: 10.1016/j.carbon.2016.08.058

Yao, J.-J., Liu, C., Li, J.-Y., Hu, Z.-L., Zhou, R.-Y., Guo, C.-C., Liu, X.-R., Yang, F.-F., & Zhu, Y.-R. (2023). 3D nitrogen and boron dual-doped carbon quantum dots/reduced graphene oxide aerogel for advanced aqueous and flexible quasi-solid-state zinc-ion hybrid capacitors. *Rare Metals*, *42*(7), 2307–2323. DOI: 10.1007/s12598-023-02265-5

Yao, N., & Wang, Z. L. (2005). *Handbook of microscopy for nanotechnology*. 731. DOI: 10.1007/1-4020-8006-9

Yashwanth, H., Rondiya, S. R., Dzade, N. Y., Dhole, S., Phase, D. M., & Hareesh, K. (2020). Enhanced photocatalytic activity of n, p, co-doped carbon quantum dots: An insight from experimental and computational approach. *Vacuum*, *180*, 109589. DOI: 10.1016/j.vacuum.2020.109589

Yilmaz, M., Raina, S., Hsu, S. H., & Kang, W. P. (2017). Growing micropatterned CNT arrays on aluminum substrates using hot-filament CVD process. *Materials Letters*, *209*, 376–378. DOI: 10.1016/j.matlet.2017.08.061

Yin, T., Guo, Y., Huang, X., Yang, X., Qin, L., Ning, T., Tan, L., Li, L., & Zou, K. (2025). Heteroatom Doping Strategy of Advanced Carbon for Alkali Metal-Ion Capacitors. *Batteries*, *11*(2), 69. DOI: 10.3390/batteries11020069

Yola, M. L., & Atar, N. (2019). Development of molecular imprinted sensor including graphitic carbon nitride/N-doped carbon dots composite for novel recognition of epinephrine. *Composites. Part B, Engineering*, *175*, 107113. DOI: 10.1016/j.compositesb.2019.107113

Yuan, T., Yuan, F., Sui, L., Zhang, Y., Li, Y., Li, X., Tan, Z., & Fan, L. (2023). Carbon quantum dots with near-unity quantum yield bandgap emission for electroluminescent light-emitting diodes. *Angewandte Chemie*, *135*(20), e202218568. Advance online publication. DOI: 10.1002/ange.202218568 PMID: 36924197

Yuan, Y., Pu, S. D., Gao, X., & Robertson, A. W. (2023). The application of in situ liquid cell tem in advanced battery research. *Energy Materials*. Advance online publication. DOI: 10.20517/energymater.2023.14

Yu, J., Yong, X., Tang, Z., Yang, B., & Lu, S. (2021). Theoretical understanding of structure–property relationships in luminescence of carbon dots. *The Journal of Physical Chemistry Letters*, *12*(32), 7671–7687. DOI: 10.1021/acs.jpclett.1c01856 PMID: 34351771

Yu, K., & Schanze, K. S. (2023). Commemorating the Nobel Prize in Chemistry 2023 for the discovery and synthesis of quantum dots. *ACS Central Science*, *9*(11), 1989–1992. DOI: 10.1021/acscentsci.3c01296 PMID: 38033796

Yu, L., Hsieh, C.-T., Keffer, D. J., Chen, H., Goenaga, G. A., Dai, S., Zawodzinski, T. A., & Harper, D. P. (2021). Hierarchical Lignin-Based Carbon Matrix and Carbon Dot Composite Electrodes for High-Performance Supercapacitors. *ACS Omega*, *6*(11), 7851–7861. DOI: 10.1021/acsomega.1c00448 PMID: 33778297

Yuncu, H., Nadaroglu, H., & Bozkurt, E. (2024). Eco-friendly synthesis of Carbon Quantum Dots (CQDs) from hazelnut husk for sensitive Aflatoxin B1 (AFB1) detection. *Toxicology Reports*, *13*, 101824. DOI: 10.1016/j.toxrep.2024.101824 PMID: 39649378

Yun, X., Li, J., Chen, X., Chen, H., Li, X., Xiang, K., Chen, W., Liao, H., & Zhu, Y. (2019). Porous fe2o3 modified by nitrogen-doped carbon quantum dots/reduced graphene oxide composite aerogel as a high-capacity and high-rate anode material for alkaline aqueous batteries. *ACS Applied Materials &Amp. Interfaces*, *11*(40), 36970–36984. DOI: 10.1021/acsami.9b12827 PMID: 31487152

Yun, Y. S., Kim, D. H., Hong, S. J., Park, M., Park, Y., Kim, B. H., Jin, H., & Kang, K. (2015). Microporous carbon nanosheets with redox-active heteroatoms for pseudocapacitive charge storage. *Nanoscale*, *7*(37), 15051–15058. DOI: 10.1039/C5NR04231C PMID: 26315977

Yu, S., Lu, S., Tan, D., & Zhu, Y. (2022a). Nitrogen and phosphorus co-doped carbon dots for developing highly flame retardant poly (vinyl alcohol) composite films. *European Polymer Journal*, *164*, 110970. https://doi.org/https://doi.org/10.1016/j.eurpolymj.2021.110970. DOI: 10.1016/j.eurpolymj.2021.110970

Yu, W.-J., He, W., Wang, C., Liu, F., Zhu, L., Tian, Q., Tong, H., & Guo, X. (2022). Confinement of TiO_2 quantum dots in graphene nanoribbons for high-performance lithium and sodium ion batteries. *Journal of Alloys and Compounds*, *898*, 162856. DOI: 10.1016/j.jallcom.2021.162856

Yu, Y., Chen, Z., Yu, H., Zhang, X., Liu, T., Xia, M., Zheng, R., Shui, M., & Shu, J. (2020). Heteroatom-doped carbon-based materials for lithium and sodium ion batteries. *Energy Storage Materials*, *32*, 65–90. DOI: 10.1016/j.ensm.2020.07.027

Zaman, K.. (2021). Properties and molecular structure of carbon quantum dots derived from empty fruit bunch biochar using a facile microwave-assisted method for the detection of Cu2+ ions. *Optical Materials*, *112*, 110801. Advance online publication. DOI: 10.1016/j.optmat.2021.110801

Zhai, X., Zhang, P., Liu, C., Bai, T., Li, W., Dai, L., & Liu, W. (2012). Highly luminescent carbon nanodots by microwave-assisted pyrolysis. *Chemical Communications*, *48*(64), 7955. DOI: 10.1039/c2cc33869f PMID: 22763501

Zhai, Y., Zhang, B., Shi, R., Zhang, S., Liu, Y., Wang, B., Zhang, K., Waterhouse, G. I. N., Zhang, T., & Lu, S. (2021). Carbon dots as new building blocks for electrochemical energy storage and electrocatalysis. *Advanced Energy Materials*, *12*(6), 2103426. Advance online publication. DOI: 10.1002/aenm.202103426

Zhang, H. (2008). Gd@C82 Fullerenol. *Molecular Imaging and Contrast Agent Database (MICAD)*. https://www.ncbi.nlm.nih.gov/books/NBK23539/

Zhang, B., Liu, C. Y., & Liu, Y. (2010). A novel one-step approach to synthesize fluorescent carbon nanoparticles. *European Journal of Inorganic Chemistry*, *4411–4414*(28), 4411–4414. Advance online publication. DOI: 10.1002/ejic.201000622

Zhang, F., Zhang, T., Yang, X., Zhang, L., Leng, K., Huang, Y., & Chen, Y. (2013). A high-performance supercapacitor-battery hybrid energy storage device based on graphene-enhanced electrode materials with ultrahigh energy density. *Energy & Environmental Science*, *6*(5), 1623. DOI: 10.1039/c3ee40509e

Zhang, J., Geng, W., Shi, L., Yang, C. C., Zhang, X., Geng, Y., Hirani, R. A. K., Xu, X., Wei, J., Jing, Y., Zhang, S., Zhang, H., Wang, S., & Sun, H. (2022). One–pot synthesis of boron and nitrogen co–doped nanocarboco-dopedfficient catalytic reduction of nitrophenols. *Chemical Engineering Journal*, *439*, 135733. DOI: 10.1016/j.cej.2022.135733

Zhang, J., Wei, S., Wang, H., Liu, H., Zhang, Y., Liu, S., Wang, Z., & Lu, X. (2021). Carbon Quantum Dots Promote Coupled Valence Engineering of V_2O_5 Nanobelts for High-Performance Aqueous Zinc-Ion Batteries. *ChemSusChem*, *14*(9), 2076–2083. DOI: 10.1002/cssc.202100223 PMID: 33751841

Zhang, P., & Han, H. (2012). Compact PEGylated polymer-caged quantum dots with improved stability. *Colloids and Surfaces. A, Physicochemical and Engineering Aspects*, *402*, 72–79. DOI: 10.1016/j.colsurfa.2012.03.022

Zhang, R., Wang, Y., Jia, M., Xu, J., & Pan, E. (2018). One-pot hydrothermal synthesis of ZnS quantum dots/graphene hybrids as a dual anode for sodium ion and lithium ion batteries. *Applied Surface Science*, *437*, 375–383. DOI: 10.1016/j.apsusc.2017.12.110

Zhang, X. Q., Lam, R., Xu, X., Chow, E. K., Kim, H. J., & Ho, D. (2011). Multimodal Nanodiamond Drug Delivery Carriers for Selective Targeting, Imaging, and Enhanced Chemotherapeutic Efficacy. *Advanced Materials*, *23*(41), 4770–4775. DOI: 10.1002/adma.201102263 PMID: 21932280

Zhao, C., Song, X., Liu, Y., Fu, Y., Ye, L., Wang, N., & Liu, J. (2020). Synthesis of graphene quantum dots and their applications in drug delivery. *Journal of Nanobiotechnology*, *18*(1), 142. DOI: 10.1186/s12951-020-00698-z PMID: 33008457

Zhao, , Chen, Z., Zhao, C., Gao, N., Ren, J., & Qu, X. (2015). Recent advances in bioapplications of C-dots. *Carbon*, *85*, 309–327. DOI: 10.1016/j.carbon.2014.12.045

Zhao, D. L., & Chung, T.-S. (2018). Applications of carbon quantum dots (CQDs) in membrane technologies: A review. *Water Research*, *147*, 43–49. https://doi.org/https://doi.org/10.1016/j.watres.2018.09.040. DOI: 10.1016/j.watres.2018.09.040 PMID: 30296608

Zhao, F., Li, X., Zuo, M., Liang, Y., Qin, P., Wang, H., Wu, Z., Luo, L., Liu, C., & Leng, L. (2023). Preparation of photocatalysts decorated by carbon quantum dots (CQDs) and their applications: A review. *Journal of Environmental Chemical Engineering*, *11*(2), 109487. https://doi.org/https://doi.org/10.1016/j.jece.2023.109487. DOI: 10.1016/j.jece.2023.109487

Zhao, P., Song, X., Dong, M., Sun, H., Wu, W., Zhang, R., Sun, M., & Zhao, X. (2022a). Preparation and characterization of CQDs/SBS composites and its application performance as asphalt modifier. *Construction & Building Materials*, *320*, 126312. https://doi.org/https://doi.org/10.1016/j.conbuildmat.2022.126312. DOI: 10.1016/j.conbuildmat.2022.126312

Zhao, P., Zhang, Q., Cao, J., Qian, C., Ye, J., Xu, S., & Li, Y. (2022). Facile and green synthesis of highly fluorescent carbon quantum dots from water hyacinth for the detection of ferric iron and cellular imaging. *Nanomaterials (Basel, Switzerland)*, *12*(9), 1528. DOI: 10.3390/nano12091528 PMID: 35564237

Zhao, W., Dong, H., Xing, Z., Zhou, L., Chen, S., Liu, H., Dou, S., Zhao, Z., Xia, H., Chou, S., & Chen, M. (2024). Advances and Challenges of Porous Structure on Solid–Liquid Interfaces in Polyanionic Sodium-Ion Batteries. *Advanced Energy Materials*, *14*(42), 2402720. Advance online publication. DOI: 10.1002/aenm.202402720

Zhao, X., Lei, Z., Hou, H., Lin, G., Sun, Y., Li, H., Yang, X., Su, Y., Ali, K. M. Y., Hu, E., Wang, H., Wang, Q., & Hu, E. (2025). Purification of high–toxicity beryllium–containing water resources using carbon dot–enhanced phosphoric acid/chitosan composite gels. *Separation and Purification Technology*, *358*, 130299. DOI: 10.1016/j.seppur.2024.130299

Zhao, Y., Zou, S., Huo, D., Hou, C., Yang, M., Li, J., & Bian, M. (2019). Simple and sensitive fluorescence sensor for methotrexate detection based on the inner filter effect of N, S co-doped carbon quantum dots. *Analytica Chimica Acta*, *1047*, 179–187. DOI: 10.1016/j.aca.2018.10.005 PMID: 30567648

Zhao, Z., & Xie, Y. (2017). Enhanced electrochemical performance of carbon quantum dots-polyaniline hybrid. *Journal of Power Sources*, *337*, 54–64. DOI: 10.1016/j.jpowsour.2016.10.110

Zheng, L., Chi, Y., Dong, Y., Lin, J., & Wang, B. (2009). Electrochemiluminescence of water-soluble carbon nanocrystals released electrochemically from graphite. *Journal of the American Chemical Society*, *131*(13), 4564–4565. DOI: 10.1021/ja809073f PMID: 19296587

Zheng, M., Li, Y., Zhang, Y., & Xie, Z. (2016). Solvatochromic fluorescent carbon dots as optic noses for sensing volatile organic compounds. *RSC Advances*, *6*(86), 83501–83504. DOI: 10.1039/C6RA16055G

Zheng, M., Xie, Z., Qu, D., Li, D., Du, P., Jing, X., & Sun, Z. (2013). On–Off–On Fluorescent Carbon Dot Nanosensor for Recognition of Chromium(VI) and Ascorbic Acid Based on the Inner Filter Effect. *ACS Applied Materials & Interfaces*, *5*(24), 13242–13247. DOI: 10.1021/am4042355 PMID: 24274643

Zhen, M., Xu, Y., Wang, C., & Bai, C. (2024). Fullerene-Based Immunoregulatory Nanomaterials for Immunotherapy of Tumor and Immune-Related Inflammatory Diseases. *Advanced Functional Materials*, *34*(49), 2409319. DOI: 10.1002/adfm.202409319

Zhong, L., Sun, C., Zhao, X., & Zhao, Q. (2025). Dual Roles of Carbon Quantum Dots from Green Carbon Sources: A Fluorescence Sensor for Fe3+ Ions, UV and High-Energy Blue Light Screening. *Nanomaterials (Basel, Switzerland)*, *15*(6), 436. DOI: 10.3390/nano15060436 PMID: 40137609

Zhou, J., Booker, C., Li, R., Zhou, X., Sham, T. K., Sun, X., & Ding, Z. (2007). An electrochemical avenue to blue luminescent nanocrystals from multiwalled carbon nanotubes (MWCNTs). *Journal of the American Chemical Society*, *129*(4), 744–745. DOI: 10.1021/ja0669070 PMID: 17243794

Zhou, J., Shan, X., Ma, J., Gu, Y., Qian, Z., Chen, J., & Feng, H. (2014). Facile synthesis of P-doped carbon quantum dots with highly efficient photoluminescence. *RSC Advances*, *4*(11), 5465–5468. DOI: 10.1039/c3ra45294h

Zhou, J., Zhou, H., Tang, J., Deng, S., Yan, F., Li, W., & Qu, M. (2017). Carbon dots doped with heteroatoms for fluorescent bioimaging: A review. *Microchimica Acta*, *184*(2), 343–368. DOI: 10.1007/s00604-016-2043-9

Zhou, , Qiao, M., Zhang, L., Sun, L., Zhang, Y., & Liu, W. (2019). Green and efficient synthesis of carbon quantum dots and their luminescent properties. *Journal of Luminescence*, *206*, 158–163. DOI: 10.1016/j.jlumin.2018.10.057

Zhou, W., Zhuang, J., Li, W., Hu, C., Lei, B., & Liu, Y. (2017). Towards efficient dual-emissive carbon dots through sulfur and nitrogen co-doping. *Journal of Materials Chemistry. C, Materials for Optical and Electronic Devices*, *5*(32), 8014–8021. DOI: 10.1039/C7TC01819C

Zhu et al., (2000). *Graphic abstract Mild acidolysis-assisted hydrothermal carbonization of lignin for simultaneous preparation of green and blue fluorescent carbon quantum dots.*

Zhu, H., Liu, A., Xu, Y., Shan, F., Li, A., Wang, J., Yang, W., Barrow, C., & Liu, J. (2015). Graphene quantum dots directly generated from graphite via magnetron sputtering and the application in thin-film transistors. *Carbon*, *88*, 225–232. DOI: 10.1016/j.carbon.2015.03.007

Zhu, H., Wang, X., Li, Y., Wang, Z., Yang, F., & Yang, X. (2009). Microwave synthesis of fluorescent carbon nanoparticles with electrochemiluminescence properties. *Chemical Communications*, *5118–5120*(34), 5118. Advance online publication. DOI: 10.1039/b907612c PMID: 20448965

Zhu, L., Shen, D., Wu, C., & Gu, S. (2020). State-of-the-art on the preparation, modification, and application of biomass-derived carbon quantum dots. *Industrial &Amp. Industrial & Engineering Chemistry Research*, *59*(51), 22017–22039. DOI: 10.1021/acs.iecr.0c04760

Zhu, Y., Li, J., Yun, X., Chen, X., Yao, J., & Lu, Z. (2021). Design and Fabrication of Advanced Cathode and Anode Materials for Hybrid Supercapacitors Based on Graphitic Carbon Quantum Dot-Decorated Reduced Graphene Oxide Composite Aerogels. *ACS Applied Energy Materials*, *4*(1), 714–729. DOI: 10.1021/acsaem.0c02593

Zou, S., Hou, C., Fa, H., Zhang, L., Ma, Y., Dong, L., Li, D., Hou, D., & Yang, M. (2017). An efficient fluorescent probe for fluazinam using N, S co-doped carbon dots from l-cysteine. *Sensors and Actuators. B, Chemical*, *239*, 1033–1041. DOI: 10.1016/j.snb.2016.07.169

Zuo, K., Liu, W., Liu, X., & Liu, X. (2023). Phosphorescence of carbon dot: The intrinsic mechanism and recent progress. *Carbon Trends*, *12*, 100278. DOI: 10.1016/j.cartre.2023.100278

Zuo, W., Li, R., Zhou, C., Li, Y., Xia, J., & Liu, J. (2017). Battery-Supercapacitor Hybrid Devices: Recent Progress and Future Prospects. *Advancement of Science*, *4*(7), 1600539. Advance online publication. DOI: 10.1002/advs.201600539 PMID: 28725528

Zuo, W., Tang, L., Xiang, J., Ji, R., Luo, L., Rogee, L., & Lau, S. P. (2017). Nitrogen-doped hollow carbon nanoparticles as efficient counter electrodes in quantum dot sensitized solar cells. *Applied Physics Letters, 110*(21), 211901. DOI: 10.1016/j.solener.2024.112699

About the Contributors

P. Surendran is an accomplished physicist with a strong academic and research background. Born in 1992, he completed his Master of Science in Physics at Periyar E.V.R. College (Autonomous), Tiruchirappalli, India. He obtained his Ph.D. from Bharathidasan University in 2022. He is presently serving as a Guest Lecturer in the PG and Research Department of Physics, Arignar Anna Government Arts College, Namakkal, Tamil Nadu, India. His academic journey has been marked by exceptional achievements. He was awarded with a National Fellowship for Higher Education by the UGC, New Delhi, India, and received multiple accolades, including the Best Research Scholar Award and the Best Oral Presentation Award. He is an active reviewer for several international peer-reviewed journals. With a passion for research, he has published 22 peer-reviewed research articles, contributing significantly to the scientific community. His work has garnered 859 citations, reflecting the impact and relevance of his research. His primary research interests lie in converting bio-waste materials into Carbon Quantum Dots (CQDs) and exploring their optoelectronic applications. His groundbreaking research continues to open new avenues in both environmental sustainability and advanced material science.

Karthik Kannan is currently serving as a Research Faculty in the Department of Mechanical Engineering at National Chung Cheng University, Taiwan, and as Adjunct Professor at Saveetha Institute of Technical and Medical Sciences, Chennai, India and Karpagam Academy of Higher Education, Coimbatore, India. Dr. Kannan is the recipient of the prestigious TEGC (Science and Technology specialization), approved by the NSTC. He was also honored with the highly competitive Brain Pool Fellowship from the NRF, which he held at Kumoh National Institute of Technology, South Korea. Dr. Karthik Kannan has an extensive publication record with over 170 peer-reviewed research articles with 8,076 citations (Google Scholar, November 2025), with an h-index of 57 and an i10-index of 125. Notably, Stanford University has listed him among the World's Top 2% Scientists for six consecutive years (2020–2025), with distinction in both single-year and career-long impact rankings (1960–2023).

Santhy A currently serves as Assistant Professor (Level 11) at the Department of Chemistry, School of Physical Sciences, Amrita Vishwa Vidyapeetham, Amritapuri Campus. She joined the department of Chemistry, Amrita in the year 2014. She completed her M.Sc. from the University of Kerala in 2011 and has qualified CSIR-NET in the year 2013. Dr. Santhy completed her Ph.D. in the area of 'Electrochemical Sensors' from Amrita Vishwa Vidyapeetham in the year 2023. To date, she has published 15 papers in international peer-reviewed journals,10 conference papers and 6 book chapters. Her present research interest is focused on the development of electrochemical sensors for biomolecules, environmental pollutants, pharmaceuticals etc., and synthesis of various composite materials for electrocatalytic and energy storage applications.

Saritha A. currently serves as Professor in the department of Chemistry, Amrita School of Physical Sciences, Amritapuri. Her research interests include Polymer micro and nano composites, rubber nanocomposites, 2D materials, green composites, green synthesis of nanomaterials, electrospun fiber composites, nanoclusters, chemical modification of nano clay, carbon nano tubes, graphene-based composites, epoxy nanocomposites, fabrication of nano composite membranes for water filtration and fabrication of VOC barrier membranes. She has 95 international publications, 61 book chapters, 8 conference papers and 5 edited books to her credit. Orcid id- https://orcid.org/0000-0002-2253-8050 Mail id- sarithatvla@gmail.com

A. Sakunthala is a materials scientist and electrochemist with expertise in solid-state electrolytes and nanomaterial electrodes for next-generation energy storage systems. She has over 110 research publications in leading international journals, including high-capacity lithium-ion, sodium-ion, and magnesium-ion batteries, as well as high-performance supercapacitors with a focus on Electric Double Layer Capacitors (EDLCs) for large-scale energy storage applications. Her work involves the design, synthesis, and characterization of materials for energy storage, with a strong emphasis on understanding and optimizing electrochemical performance. Dr. Sakunthala also oversees a fully-equipped laboratory for coin-cell fabrication and electrochemical characterization, including AC impedance spectroscopy studies. In addition, she has delved into the realm of thin-film materials, contributing to the development of next-generation thin-film batteries. Dr. Sakunthala's research advances the field of energy storage, bridging the gap between materials science, electrochemistry, and device engineering for sustainable energy solutions.

Aishwarya GV is a biomedical engineer specialising in medical device design and development, driven by a passion for mental health rehabilitation, exploring computational neuroscience, biomaterials and regulatory sciences. Currently working on a project supporting individuals with stress and anxiety disorders and promoting mental well-being.

Rekha R currently pursuing 4th year of study in Biomedical Engineering. I has a keen interest in the field of biomaterials and its applications in healthcare. I had actively participated in projects and academic research during my undergraduate studies. Future goal is to pursue research and higher studies in the biomaterial domain.

Harshavardhini R.K. is a Biomedical Engineering student with ongoing research in the field of biomaterials. Her academic and research interests lie in biomaterials, material science, regenerative medicine, and tissue engineering, with a strong passion for advancing healthcare through innovative materials and technologies.

Koushik Saikia is currently working as an Assistant Professor at the Department of Physics, The Assam Royal Global University, Guwahati, Assam, India. He obtained his M.Sc. in Nanoscience and Technology from the Department of Physics, Tezpur University, in 2011, followed by a Ph.D. in Physics from the same department in 2016. His research interests focus on the self-assembly of nanoparticles and quantum dots. Dr. Saikia has published 14 peer-reviewed articles in reputed journals and holds one granted patent. Prior to his current position, he worked as a Project Scientist at the Centre for Nanotechnology, Indian Institute of Technology Guwahati, and subsequently as a National Postdoctoral Fellow (NPDF) at the Department of Physics at the same institute during 2017–2019. He also served as an Assistant Professor at Mariyam Ajmal Women's College of Science and Technology, Hojai, Assam for one year, and as a Research Assistant at Tezpur University, Assam for four and half years.

Matbiangthew Shadap is a Research Scholar in the Division of Physical Sciences at Karunya Institute of Technology and Sciences, India. He holds a Master's degree in Physics from the National Institute of Technology Meghalaya. His research focuses on energy storage devices, polymer electrolytes, and electrode materials for electrochemical systems. He has published eight peer-reviewed papers, with four more under review, and is a recipient of the National Fellowship for Scheduled Tribe for his doctoral research.

Subin Stephen is a researcher in the Division of Physical Sciences, Karunya Institute of Technology and Sciences, Coimbatore. His work focuses on nanomaterials for energy storage devices, including manganese oxides, carbon-based composites, with applications in supercapacitors. His research interests also include the development of hybrid nanostructures and sustainable electrode materials for next-generation energy devices. He has co-authored one research article and one book chapter, and he has one additional manuscript currently under review. He continues to contribute to innovative, scalable, and environmentally friendly material development for practical energy and environmental applications.

Index

A

B

C

D

E

F

G

H

N

O

P

Q

www.ingramcontent.com/pod-product-compliance
Lightning Source LLC
LaVergne TN
LVHW080911300926
849162LV00003B/639